Automatic Control Systems

Third Edition

BENJAMIN C. KUO

Professor of Electrical Engineering
University of Illinois at Urbana-Champaign

PRENTICE-HALL, INC., *Englewood Cliffs, New Jersey*

Library of Congress Cataloging in Publication Data

KUO, BENJAMIN C
  Automatic control systems.

  Includes index.
  1. Automatic control. 2. Control theory.
I. Title.
TJ213.K8354 1975   629.8'3   74–26544
ISBN 0–13–054973–8

© 1975 by Prentice-Hall, Inc.
Englewood Cliffs, New Jersey

10 9 8 7 6 5 4 3 2

Printed in the United States of America

PRENTICE-HALL INTERNATIONAL, INC., *London*
PRENTICE-HALL OF AUSTRALIA, PTY. LTD., *Sydney*
PRENTICE-HALL OF CANADA, LTD., *Toronto*
PRENTICE-HALL OF INDIA PRIVATE LIMITED, *New Delhi*
PRENTICE-HALL OF JAPAN, INC., *Tokyo*

Contents

Preface

The first edition of this book, published in 1962, was characterized by having chapters on sampled-data and nonlinear control systems. The treatment of the analysis and design of control systems was all classical.

The two major changes in the second edition, published in 1967, were the inclusion of the state variable technique and the integration of the discrete-data systems with the continuous data system. The chapter on nonlinear systems was eliminated in the second edition to the disappointment of some users of that text. At the time of the revision the author felt that a comprehensive treatment on the subject of nonlinear systems could not be made effectively with the available space.

The third edition is still written as an introductory text for a senior course on control systems. Although a great deal has happened in the area of modern control theory in the past ten years, preparing suitable material for a modern course on introductory control systems remains a difficult task. The problem is a complicated one because it is difficult to teach the topics concerned with new developments in modern control theory at the undergraduate level. The unique situation in control systems has been that many of the practical problems are still being solved in the industry by the classical methods. While some of the techniques in modern control theory are much more powerful and can solve more complex problems, there are often more restrictions when it comes to practical applications of the solutions. However, it should be recognized that a modern control engineer should have an understanding of the classical as well as the modern control methods. The latter will enhance and broaden one's perspective in solving a practical problem. It is the author's opinion that one should strike a balance in the teaching of control systems theory at the beginning and intermediate levels. Therefore in this current edition, equal emphasis is placed on the classical methods and the modern control theory.

A number of introductory books with titles involving modern control theory have been published in recent years. Some authors have attempted to unify and integrate the classical control with the modern control, but according to the critics and reviews, most have failed. Although such a goal is highly desirable, if only from the standpoint of presentation, there does not seem to be a good solution. It is possible that the objective may not be achieved until new theories and new techniques are developed for this purpose. The fact remains that control systems, in some way, may be regarded as a science of learning how to solve one problem—control, in many different ways. These different ways of solution may be compared and weighed against each other, but it may not be possible to unify all the approaches. The approach used in this text is to present the classical method and the modern approach independently, and whenever possible, the two approaches are considered as alternatives, and the advantages and disadvantages of each are weighed. Many illustrative examples are carried out by both methods.

Many existing text books on control systems have been criticized for not including adequate practical problems. One reason for this is, perhaps, that many text book writers are theorists, who lack the practical background and experience necessary to provide real-life examples. Another reason is that the difficulty in the control systems area is compounded by the fact that most real-life problems are highly complex, and are rarely suitable as illustrative examples at the introductory level. Usually, much of the realism is lost by simplifying the problem to fit the nice theorems and design techniques developed in the text material. Nevertheless, the majority of the students taking a control system course at the senior level do not pursue a graduate career, and they must put their knowledge to immediate use in their new employment. It is extremely important for these students, as well as those who will continue, to gain an actual feel of what a real control system is like. Therefore, the author has introduced a number of practical examples in various fields in this text. The homework problems also reflect the attempt of this text to provide more real-life problems.

The following features of this new edition are emphasized by comparison with the first two editions:

1. Equal emphasis on classical and modern control theory.
2. Inclusion of sampled-data and nonlinear systems.
3. Practical system examples and homework problems.

The material assembled in this book is an outgrowth of a senior-level control system course taught by the author at the University of Illinois at Urbana-Champaign for many years. Moreover, this book is written in a style adaptable for self-study and reference.

Chapter 1 presents the basic concept of control systems. The definition of feedback and its effects are covered. Chapter 2 presents mathematical foundation and preliminaries. The subjects included are Laplace transform, z-transform, matrix algebra, and the applications of the transform methods. Transfer function and signal flow graphs are discussed in Chapter 3. Chapter 4 introduces the state variable approach to dynamical systems. The concepts and definitions of controllability and observability are introduced at the early stage. These subjects are later being used for the analysis and design of linear control systems. Chapter 5 discusses the mathematical modeling of physical systems. Here, the emphasis is on electromechanical systems. Typical transducers and control systems used in practice are illustrated. The treatment cannot be exhaustive as there are numerous types of devices and control systems. Chapter 6 gives the time response considerations of control systems. Both the classical and the modern approach are used. Some simple design considerations in the time domain are pointed out. Chapters 7, 8, and 9 deal with topics on stability, root locus, and frequency response of control systems.

In Chapter 10, the design of control systems is discussed, and the approach is basically classical. Chapter 11 contains some of the optimal control subjects which, in the author's opinion, can be taught at the undergraduate level if time permits. The text does contain more material than can be covered in one semester.

One of the difficulties in preparing this book was the weighing of what subjects to cover. To keep the book to a reasonable length, some subjects, which were in the original draft, had to be left out of the final manuscript. These included the treatment of signal flow graphs and time-domain analysis, of discrete-data systems, the second method of Liapunov's stability method, describing function analysis, state plane analysis, and a few selected topics on implementing optimal control. The author feels that the inclusion of these subjects would add materially to the spirit of the text, but at the cost of a higher price.

The author wishes to express his sincere appreciation to Dean W. L. Everitt (emeritus), Professors E. C. Jordan, O. L. Gaddy, and E. W. Ernst, of the University of Illinois, for their encouragement and interest in the project. The author is grateful to Dr. Andrew Sage of the University of Virginia and Dr. G. Singh of the University of Illinois for their valuable suggestions. Special thanks also goes to Mrs. Jane Carlton who typed a good portion of the manuscript and gave her invaluable assistance in proofreading.

BENJAMIN C. KUO

Urbana, Illinois

1

Introduction

1.1 Control Systems

In recent years, automatic control systems have assumed an increasingly important role in the development and advancement of modern civilization and technology. Domestically, automatic controls in heating and air conditioning systems regulate the temperature and the humidity of modern homes for comfortable living. Industrially, automatic control systems are found in numerous applications, such as quality control of manufactured products, automation, machine tool control, modern space technology and weapon systems, computer systems, transportation systems, and robotics. Even such problems as inventory control, social and economic systems control, and environmental and hydrological systems control may be approached from the theory of automatic control.

The basic control system concept may be described by the simple block diagram shown in Fig. 1-1. The objective of the system is to control the variable c in a prescribed manner by the actuating signal e through the elements of the control system.

In more common terms, the controlled variable is the *output* of the system, and the actuating signal is the *input*. As a simple example, in the steering control of an automobile, the direction of the two front wheels may be regarded as the controlled variable c, the output. The position of the steering wheel is the input, the actuating signal e. The controlled process or system in this case is composed of the steering mechanisms, including the dynamics of the entire automobile. However, if the objective is to control the speed of the automobile, then the amount of pressure exerted on the accelerator is the actuating signal, with the speed regarded as the controlled variable.

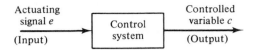

Fig. 1-1. Basic control system.

There are many situations where several variables are to be controlled simultaneously by a number of inputs. Such systems are referred to as *multivariable systems*.

Open-Loop Control Systems (Nonfeedback Systems)

The word *automatic* implies that there is a certain amount of sophistication in the control system. By automatic, it generally means that the system is usually capable of adapting to a variety of operating conditions and is able to respond to a class of inputs satisfactorily. However, not any type of control system has the automatic feature. Usually, the automatic feature is achieved by feeding the output variable back and comparing it with the command signal. When a system does not have the feedback structure, it is called an *open-loop system*, which is the simplest and most economical type of control system. Unfortunately, open-loop control systems lack accuracy and versatility and can be used in none but the simplest types of applications.

Consider, for example, control of the furnace for home heating. Let us assume that the furnace is equipped only with a timing device, which controls the on and off periods of the furnace. To regulate the temperature to the proper level, the human operator must estimate the amount of time required for the furnace to stay on and then set the timer accordingly. When the preset time is up, the furnace is turned off. However, it is quite likely that the house temperature is either above or below the desired value, owing to inaccuracy in the estimate. Without further deliberation, it is quite apparent that this type of control is inaccurate and unreliable. One reason for the inaccuracy lies in the fact that one may not know the exact characteristics of the furnace. The other factor is that one has no control over the outdoor temperature, which has a definite bearing on the indoor temperature. This also points to an important disadvantage of the performance of an open-loop control system, in that the system is not capable of adapting to variations in environmental conditions or to external disturbances. In the case of the furnace control, perhaps an experienced person can provide control for a certain desired temperature in the house; but if the doors or windows are opened or closed intermittently during the operating period, the final temperature inside the house will not be accurately regulated by the open-loop control.

An electric washing machine is another typical example of an open-loop system, because the amount of wash time is entirely determined by the judgment and estimation of the human operator. A true automatic electric washing machine should have the means of checking the cleanliness of the clothes continuously and turn itself off when the desired degree of cleanliness is reached.

Although open-loop control systems are of limited use, they form the basic elements of the closed-loop control systems. In general, the elements of an open-loop control system are represented by the block diagram of Fig. 1-2. An input signal or command *r* is applied to the controller, whose output acts as the actuating signal *e*; the actuating signal then actuates the controlled process and hopefully will drive the controlled variable *c* to the desired value.

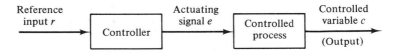

Fig. 1-2. Block diagram of an open-loop control system.

Closed-Loop Control Systems (Feedback Control Systems)

What is missing in the open-loop control system for more accurate and more adaptable control is a link or feedback from the output to the input of the system. In order to obtain more accurate control, the controlled signal $c(t)$ must be fed back and compared with the reference input, and an actuating signal proportional to the difference of the output and the input must be sent through the system to correct the error. A system with one or more feedback paths like that just described is called a *closed-loop system*. Human beings are probably the most complex and sophisticated feedback control system in existence. A human being may be considered to be a control system with many inputs and outputs, capable of carrying out highly complex operations.

To illustrate the human being as a feedback control system, let us consider that the objective is to reach for an object on a desk. As one is reaching for the object, the brain sends out a signal to the arm to perform the task. The eyes serve as a sensing device which feeds back continuously the position of the hand. The distance between the hand and the object is the error, which is eventually brought to zero as the hand reaches the object. This is a typical example of closed-loop control. However, if one is told to reach for the object and then is blindfolded, one can only reach toward the object by estimating its exact position. It is quite possible that the object may be missed by a wide margin. With the eyes blindfolded, the feedback path is broken, and the human is operating as an open-loop system. The example of the reaching of an object by a human being is described by the block diagram shown in Fig. 1-3.

As another illustrative example of a closed-loop control system, Fig. 1-4

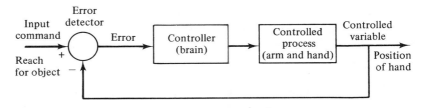

Fig. 1-3. Block diagram of a human being as a closed-loop control system.

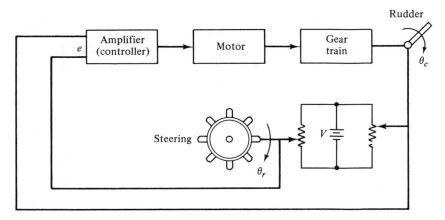

Fig. 1-4. Rudder control system.

shows the block diagram of the rudder control system of a ship. In this case the objective of control is the position of the rudder, and the reference input is applied through the steering wheel. The error between the relative positions of the steering wheel and the rudder is the signal, which actuates the controller and the motor. When the rudder is finally aligned with the desired reference direction, the output of the error sensor is zero. Let us assume that the steering wheel position is given a sudden rotation of R units, as shown by the time signal in Fig. 1-5(a). The position of the rudder as a function of time, depending upon the characteristics of the system, may typically be one of the responses shown in Fig. 1-5(b). Because all physical systems have electrical and mechanical inertia, the position of the rudder cannot respond instantaneously to a step input, but will, rather, move gradually toward the final desired position. Often, the response will oscillate about the final position before settling. It is apparent that for the rudder control it is desirable to have a nonoscillatory response.

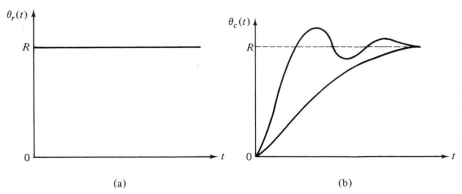

(a)   (b)

Fig. 1-5. (a) Step displacement input of rudder control system. (b) Typical output responses.

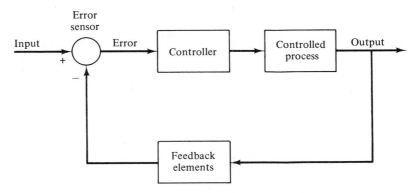

Fig. 1-6. Basic elements of a feedback control system.

The basic elements and the block diagram of a closed-loop control system are shown in Fig. 1-6. In general, the configuration of a feedback control system may not be constrained to that of Fig. 1-6. In complex systems there may be a multitude of feedback loops and element blocks.

Figure 1-7(a) illustrates the elements of a tension control system of a windup process. The unwind reel may contain a roll of material such as paper or cable which is to be sent into a processing unit, such as a cutter or a printer, and then collects it by winding it onto another roll. The control system in this case is to maintain the tension of the material or web at a certain prescribed tension to avoid such problems as tearing, stretching, or creasing.

To regulate the tension, the web is formed into a half-loop by passing it down and around a weighted roller. The roller is attached to a pivot arm, which allows free up-and-down motion of the roller. The combination of the roller and the pivot arm is called the *dancer*.

When the system is in operation, the web normally travels at a constant speed. The ideal position of the dancer is horizontal, producing a web tension equal to one-half of the total weight W of the dancer roll. The electric brake on the unwind reel is to generate a restraining torque to keep the dancer in the horizontal position at all times.

During actual operation, because of external disturbances, uncertainties and irregularities of the web material, and the decrease of the effective diameter of the unwind reel, the dancer arm will not remain horizontal unless some scheme is employed to properly sense the dancer-arm position and control the restraining braking torque.

To obtain the correction of the dancing-arm-position error, an angular sensor is used to measure the angular deviation, and a signal in proportion to the error is used to control the braking torque through a controller. Figure 1-7(b) shows a block diagram that illustrates the interconnections between the elements of the system.

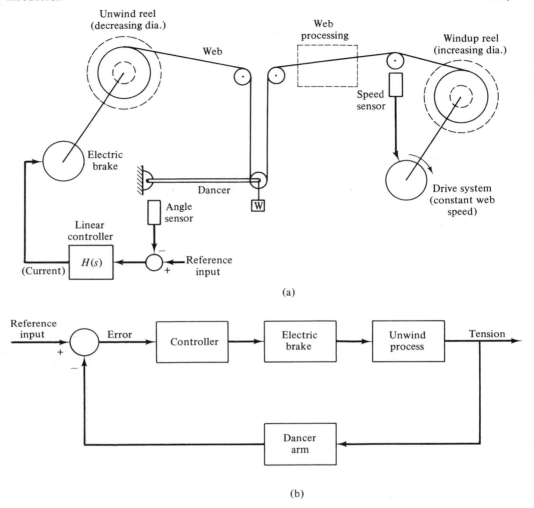

Fig. 1-7. (a) Tension control system. (b) Block diagram depicting the basic elements and interconnections of a tension control system.

1.2 What Is Feedback and What Are Its Effects?

The concept of feedback plays an important role in control systems. We demonstrated in Section 1.1 that feedback is a major requirement of a closed-loop control system. Without feedback, a control system would not be able to achieve the accuracy and reliability that are required in most practical applications. However, from a more rigorous standpoint, the definition and the significance of feedback are much deeper and more difficult to demonstrate than the few examples given in Section 1.1. In reality, the reasons for using feedback carry far more meaning than the simple one of comparing the input with the output in order to reduce the error. The reduction of system error is merely one of the many effects that feedback may bring upon a system. We shall now show that feedback also has effects on such system performance characteristics as stability, bandwidth, overall gain, impedance, and sensitivity.

To understand the effects of feedback on a control system, it is essential that we examine this phenomenon with a broad mind. When feedback is deliberately introduced for the purpose of control, its existence is easily identified. However, there are numerous situations wherein a physical system that we normally recognize as an inherently nonfeedback system may turn out to have feedback when it is observed in a certain manner. In general we can state that whenever a closed sequence of *cause-and-effect relation* exists among the variables of a system, feedback is said to exist. This viewpoint will inevitably admit feedback in a large number of systems that ordinarily would be identified as nonfeedback systems. However, with the availability of the feedback and control system theory, this general definition of feedback enables numerous systems, with or without physical feedback, to be studied in a systematic way once the existence of feedback in the above-mentioned sense is established.

We shall now investigate the effects of feedback on the various aspects of system performance. Without the necessary background and mathematical foundation of linear system theory, at this point we can only rely on simple static system notation for our discussion. Let us consider the simple feedback system configuration shown in Fig. 1-8, where r is the input signal, c the output signal, e the error, and b the feedback signal. The parameters G and H may be considered as constant gains. By simple algebraic manipulations it is simple to show that the input–output relation of the system is

$$M = \frac{c}{r} = \frac{G}{1 + GH} \qquad (1\text{-}1)$$

Using this basic relationship of the feedback system structure, we can uncover some of the significant effects of feedback.

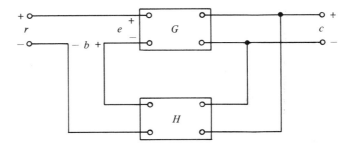

Fig. 1-8. Feedback system.

Effect of Feedback on Overall Gain

As seen from Eq. (1-1), feedback affects the gain G of a nonfeedback system by a factor of $1 + GH$. The reference of the feedback in the system of Fig. 1-8 is negative, since a minus sign is assigned to the feedback signal. The quantity GH may itself include a minus sign, so the general effect of feedback is that it may increase or decrease the gain. In a practical control system, G and H are functions of frequency, so the magnitude of $1 + GH$ may be greater than 1 in one frequency range but less than 1 in another. Therefore, feedback could increase the gain of the system in one frequency range but decrease it in another.

Effect of Feedback on Stability

Stability is a notion that describes whether the system will be able to follow the input command. In a nonrigorous manner, a system is said to be unstable if its output is out of control or increases without bound.

To investigate the effect of feedback on stability, we can again refer to the expression in Eq. (1-1). If $GH = -1$, the output of the system is infinite for any finite input. Therefore, we may state that feedback can cause a system that is originally stable to become unstable. Certainly, feedback is a two-edged sword; when it is improperly used, it can be harmful. It should be pointed out, however, that we are only dealing with the static case here, and, in general $GH = -1$ is not the only condition for instability.

It can be demonstrated that one of the advantages of incorporating feedback is that it can stabilize an unstable system. Let us assume that the feedback system in Fig. 1-8 is unstable because $GH = -1$. If we introduce another feedback loop through a negative feedback of F, as shown in Fig. 1-9, the input–output relation of the overall system is

$$\frac{c}{r} = \frac{G}{1 + GH + GF} \tag{1-2}$$

It is apparent that although the properties of G and H are such that the inner-loop feedback system is unstable, because $GH = -1$, the overall system can be stable by properly selecting the outer-loop feedback gain F.

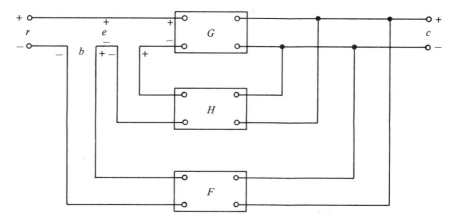

Fig. 1-9. Feedback system with two feedback loops.

Effect of Feedback on Sensitivity

Sensitivity considerations often play an important role in the design of control systems. Since all physical elements have properties that change with environment and age, we cannot always consider the parameters of a control system to be completely stationary over the entire operating life of the system. For instance, the winding resistance of an electric motor changes as the temperature of the motor rises during operation. In general, a good control system should be very insensitive to these parameter variations while still able to follow the command responsively. We shall investigate what effect feedback has on the sensitivity to parameter variations.

Referring to the system in Fig. 1-8, we consider G as a parameter that may vary. The sensitivity of the gain of the overall system M to the variation in G is defined as

$$S_G^M = \frac{\partial M/M}{\partial G/G} \tag{1-3}$$

where ∂M denotes the incremental change in M due to the incremental change in G; $\partial M/M$ and $\partial G/G$ denote the percentage change in M and G, respectively. The expression of the sensitivity function S_G^M can be derived by using Eq. (1-1). We have

$$S_G^M = \frac{\partial M}{\partial G} \frac{G}{M} = \frac{1}{1 + GH} \tag{1-4}$$

This relation shows that the sensitivity function can be made arbitrarily small by increasing GH, provided that the system remains stable. It is apparent that in an open-loop system the gain of the system will respond in a one-to-one fashion to the variation in G.

In general, the sensitivity of the system gain of a feedback system to parameter variations depends on where the parameter is located. The reader may derive the sensitivity of the system in Fig. 1-8 due to the variation of H.

Effect of Feedback on External Disturbance or Noise

All physical control systems are subject to some types of extraneous signals or noise during operation. Examples of these signals are thermal noise voltage in electronic amplifiers and brush or commutator noise in electric motors.

The effect of feedback on noise depends greatly on where the noise is introduced into the system; no general conclusions can be made. However, in many situations, feedback can reduce the effect of noise on system performance.

Let us refer to the system shown in Fig. 1-10, in which r denotes the command signal and n is the noise signal. In the absence of feedback, $H = 0$, the output c is

$$c = G_1 G_2 e + G_2 n \tag{1-5}$$

where $e = r$. The signal-to-noise ratio of the output is defined as

$$\frac{\text{output due to signal}}{\text{output due to noise}} = \frac{G_1 G_2 e}{G_2 n} = G_1 \frac{e}{n} \tag{1-6}$$

To increase the signal-to-noise ratio, evidently we should either increase the magnitude of G_1 or e relative to n. Varying the magnitude of G_2 would have no effect whatsoever on the ratio.

With the presence of feedback, the system output due to r and n acting

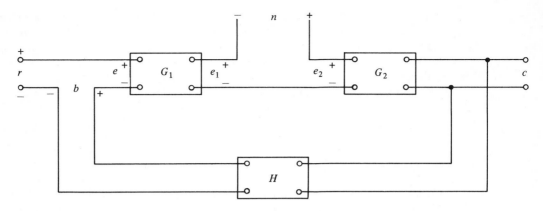

Fig. 1-10. Feedback system with a noise signal.

simultaneously is

$$c = \frac{G_1 G_2}{1 + G_1 G_2 H} r + \frac{G_2}{1 + G_1 G_2 H} n \tag{1-7}$$

Simply comparing Eq. (1-7) with Eq. (1-5) shows that the noise component in the output of Eq. (1-7) is reduced by the factor $1 + G_1 G_2 H$, but the signal component is also reduced by the same amount. The signal-to-noise ratio is

$$\frac{\text{output due to signal}}{\text{output due to noise}} = \frac{G_1 G_2 r/(1 + G_1 G_2 H)}{G_2 n/(1 + G_1 G_2 H)} = G_1 \frac{r}{n} \tag{1-8}$$

and is the same as that without feedback. In this case feedback is shown to have no direct effect on the output signal-to-noise ratio of the system in Fig. 1-10. However, the application of feedback suggests a possibility of improving the signal-to-noise ratio under certain conditions. Let us assume that in the system of Fig. 1-10, if the magnitude of G_1 is increased to G_1' and that of the input r to r', with all other parameters unchanged, the output due to the input signal acting alone is at the same level as that when feedback is absent. In other words, we let

$$c\,|_{n=0} = \frac{G_1' G_2 r'}{1 + G_1' G_2 H} = G_1 G_2 r \tag{1-9}$$

With the increased G_1, G_1', the output due to noise acting alone becomes

$$c\,|_{r=0} = \frac{G_2 n}{1 + G_1' G_2 H} \tag{1-10}$$

which is smaller than the output due to n when G_1 is not increased. The signal-to-noise ratio is now

$$\frac{G_1 G_2 r}{G_2 n/(1 + G_1' G_2 H)} = \frac{G_1 r}{n}(1 + G_1' G_2 H) \tag{1-11}$$

which is greater than that of the system without feedback by a factor of $(1 + G_1' G_2 H)$.

In general, feedback also has effects on such performance characteristics as bandwidth, impedance, transient response, and frequency response. These effects will become known as one progresses into the ensuing material of this text.

1.3 Types of Feedback Control Systems

Feedback control systems may be classified in a number of ways, depending upon the purpose of the classification. For instance, according to the method of analysis and design, feedback control systems are classified as linear and non-linear, time varying or time invariant. According to the types of signal found in the system, reference is often made to continuous-data and discrete-data systems, or modulated and unmodulated systems. Also, with reference to the type of system components, we often come across descriptions such as electromechanical control systems, hydraulic control systems, pneumatic systems, and biological control systems. Control systems are often classified according to the main purpose of the system. A positional control system and a velocity control system control the output variables according to the way the names imply. In general, there are many other ways of identifying control systems according to some special features of the system. It is important that some of these more common ways of classifying control systems are known so that proper perspective is gained before embarking on the analysis and design of these systems.

Linear Versus Nonlinear Control Systems

This classification is made according to the methods of analysis and design. Strictly speaking, linear systems do not exist in practice, since all physical systems are nonlinear to some extent. Linear feedback control systems are idealized models that are fabricated by the analyst purely for the simplicity of analysis and design. When the magnitudes of the signals in a control system are limited to a range in which system components exhibit linear characteristics (i.e., the principle of superposition applies), the system is essentially linear. But when the magnitudes of the signals are extended outside the range of the linear operation, depending upon the severity of the nonlinearity, the system should no longer be considered linear. For instance, amplifiers used in control systems often exhibit saturation effect when their input signals become large; the magnetic field of a motor usually has saturation properties. Other common nonlinear effects found in control systems are the backlash or dead play between coupled gear members, nonlinear characteristics in springs, nonlinear frictional force or torque between moving members, and so on. Quite often, nonlinear characteristics are intentionally introduced in a control system to improve its performance or provide more effective control. For instance, to achieve minimum-time control, an on–off (bang-bang or relay) type of controller is used. This type of control is found in many missile or spacecraft control systems. For instance, in the attitude control of missiles and spacecraft, jets are mounted on the sides of the vehicle to provide reaction torque for attitude control. These jets are often controlled in a full-on or full-off fashion, so a fixed amount of air is applied from a given jet for a certain time duration to control the attitude of the space vehicle.

For linear systems there exists a wealth of analytical and graphical techniques for design and analysis purposes. However, nonlinear systems are very difficult to treat mathematically, and there are no general methods that may be used to solve a wide class of nonlinear systems.

Time-Invariant Versus Time-Varying Systems

When the parameters of a control system are stationary with respect to time during the operation of the system, we have a time-invariant system. Most physical systems contain elements that drift or vary with time to some extent. If the variation of parameter is significant during the period of operation, the system is termed a time-varying system. For instance, the radius of the unwind reel of the tension control system in Fig. 1-7 decreases with time as the material is being transferred to the windup reel. Although a time-varying system without nonlinearity is still a linear system, its analysis is usually much more complex than that of the linear time-invariant systems.

Continuous-Data Control Systems

A continuous-data system is one in which the signals at various parts of the system are all functions of the continuous time variable t. Among all continuous-data control systems, the signals may be further classified as ac or dc. Unlike the general definitions of ac and dc signals used in electrical engineering, ac and dc control systems carry special significances. When one refers to an ac control system it usually means that the signals in the system are modulated by some kind of modulation scheme. On the other hand, when a dc control system is referred to, it does not mean that all the signals in the system are of the direct-current type; then there would be no control movement. A dc control system simply implies that the signals are unmodulated, but they are still ac by common definition. The schematic diagram of a closed-loop dc control system is shown in Fig. 1-11. Typical waveforms of the system in response to a step

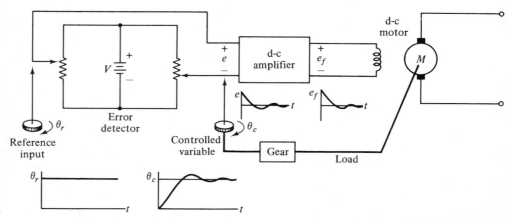

Fig. 1-11. Schematic diagram of a typical dc closed-loop control system.

function input are shown in the figure. Typical components of a dc control system are potentiometers, dc amplifiers, dc motors, and dc tachometers.

The schematic diagram of a typical ac control system is shown in Fig. 1-12. In this case the signals in the system are modulated; that is, the information is transmitted by an ac carrier signal. Notice that the output controlled variable still behaves similar to that of the dc system if the two systems have the same control objective. In this case the modulated signals are demodulated by the low-pass characteristics of the control motor. Typical components of an ac control system are synchros, ac amplifiers, ac motors, gyroscopes, and accelerometers.

In practice, not all control systems are strictly of the ac or the dc type. A system may incorporate a mixture of ac and dc components, using modulators and demodulators to match the signals at various points of the system.

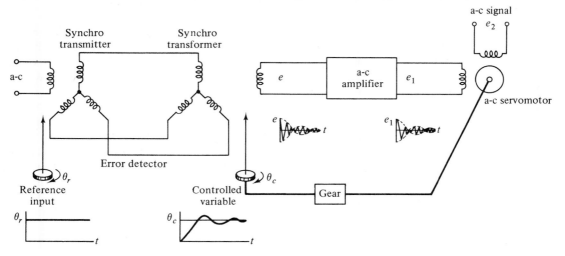

Fig. 1-12. Schematic diagram of a typical ac closed-loop control system.

Sampled-Data and Digital Control Systems

Sampled-data and digital control systems differ from the continuous-data systems in that the signals at one or more points of the system are in the form of either a pulse train or a digital code. Usually, sampled-data systems refer to a more general class of systems whose signals are in the form of pulse data, where a digital control system refers to the use of a digital computer or controller in the system. In this text the term "discrete-data control system" is used to describe both types of systems.

In general a sampled-data system receives data or information only intermittently at specific instants of time. For instance, the error signal in a control system may be supplied only intermittently in the form of pulses, in which case the control system receives no information about the error signal during the periods between two consecutive pulses. Figure 1-13 illustrates how a typical sampled-data system operates. A continuous input signal $r(t)$ is applied to the

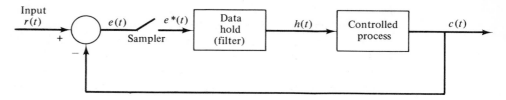

Fig. 1-13. Block diagram of a sampled-data control system.

system. The error signal $e(t)$ is sampled by a sampling device, the sampler, and the output of the sampler is a sequence of pulses. The sampling rate of the sampler may or may not be uniform. There are many advantages of incorporating sampling in a control system, one of the most easily understood of these being that sampling provides time sharing of an expensive equipment among several control channels.

Because digital computers provide many advantages in size and flexibility, computer control has become increasingly popular in recent years. Many airborne systems contain digital controllers that can pack several thousand discrete elements in a space no larger than the size of this book. Figure 1-14 shows the basic elements of a digital autopilot for a guided missile.

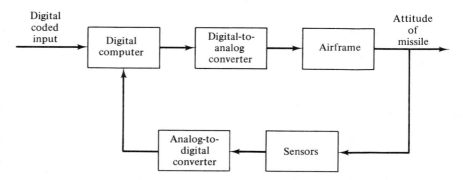

Fig. 1-14. Digital autopilot system for a guided missile.

2

Mathematical Foundation

2.1 Introduction

The study of control systems relies to a great extent on the use of applied mathematics. For the study of classical control theory, the prerequisites include such subjects as complex variable theory, differential equations, Laplace transform, and z-transform. Modern control theory, on the other hand, requires considerably more intensive mathematical background. In addition to the above-mentioned subjects, modern control theory is based on the foundation of matrix theory, set theory, linear algebra, variational calculus, various types of mathematical programming, and so on.

2.2 Complex-Variable Concept

Complex-variable theory plays an important role in the analysis and design of control systems. When studying linear continuous-data systems, it is essential that one understands the concept of complex variable and functions of a complex variable when the transfer function method is used.

Complex Variable

A complex variable s is considered to have two components: a real component σ, and an imaginary component ω. Graphically, the real component is represented by an axis in the horizontal direction, and the imaginary component is measured along a vertical axis, in the complex s-plane. In other words, a complex variable is always defined by a point in a complex plane that has a σ axis and a $j\omega$ axis. Figure 2-1 illustrates the complex s-plane, in which any

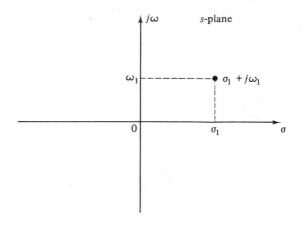

Fig. 2-1. Complex s-plane.

arbitrary point, $s = s_1$, is defined by the coordinates $\sigma = \sigma_1$ and $\omega = \omega_1$, or simply $s_1 = \sigma_1 + j\omega_1$.

Functions of a Complex Variable

The function $G(s)$ is said to be a function of the complex variable s if for every value of s there is a corresponding value (or there are corresponding values) of $G(s)$. Since s is defined to have real and imaginary parts, the function $G(s)$ is also represented by its real and imaginary parts; that is,

$$G(s) = \text{Re } G + j \text{ Im } G \tag{2-1}$$

where Re G denotes the real part of $G(s)$ and Im G represents the imaginary part of G. Thus, the function $G(s)$ can also be represented by the complex G-plane whose horizontal axis represents Re G and whose vertical axis measures the imaginary component of $G(s)$. If for every value of s (every point in the s-plane) there is only one corresponding value for $G(s)$ [one corresponding point in the $G(s)$-plane], $G(s)$ is said to be a *single-valued function*, and the mapping (correspondence) from points in the s-plane onto points in the $G(s)$-plane is described as *single valued* (Fig. 2-2). However, there are many functions for which the mapping from the function plane to the complex-variable plane is

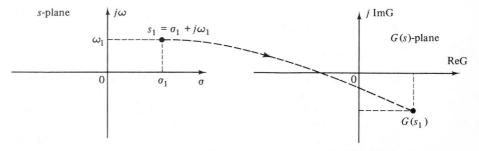

Fig. 2-2. Single-valued mapping from the s-plane to the $G(s)$-plane.

not single valued. For instance, given the function

$$G(s) = \frac{1}{s(s+1)} \tag{2-2}$$

it is apparent that for each value of s there is only one unique corresponding value for $G(s)$. However, the reverse is not true; for instance, the point $G(s) = \infty$ is mapped onto two points, $s = 0$ and $s = -1$, in the s-plane.

Analytic Function

A function $G(s)$ of the complex variable s is called an analytic function in a region of the s-plane if the function and all its derivatives exist in the region. For instance, the function given in Eq. (2-2) is analytic at every point in the s-plane except at the points $s = 0$ and $s = -1$. At these two points the value of the function is infinite. The function $G(s) = s + 2$ is analytic at every point in the finite s-plane.

Singularities and Poles of a Function

The *singularities* of a function are the points in the s-plane at which the function or its derivatives does not exist. A pole is the most common type of singularity and plays a very important role in the studies of the classical control theory.

The definition of a pole can be stated as: *If a function $G(s)$ is analytic and single valued in the neighborhood of s_i, except at s_i, it is said to have a pole of order r at $s = s_i$ if the limit*

$$\lim_{s \to s_i} [(s - s_i)^r G(s)]$$

has a finite, nonzero value. In other words, the denominator of $G(s)$ must include the factor $(s - s_i)^r$, so when $s = s_i$, the function becomes infinite. If $r = 1$, the pole at $s = s_i$ is called a *simple pole*. As an example, the function

$$G(s) = \frac{10(s+2)}{s(s+1)(s+3)^2} \tag{2-3}$$

has a pole of order 2 at $s = -3$ and simple poles at $s = 0$ and $s = -1$. It can also be said that the function is analytic in the s-plane except at these poles.

Zeros of a Function

The definition of a zero of a function can be stated as: *If the function $G(s)$ is analytic at $s = s_i$, it is said to have a zero of order r at $s = s_i$ if the limit*

$$\lim_{s \to s_i} [(s - s_i)^{-r} G(s)] \tag{2-4}$$

has a finite, nonzero value. Or simply, $G(s)$ *has a zero of order r at $s = s_i$ if $1/G(s)$ has an rth-order pole at $s = s_i$.* For example, the function in Eq. (2-3) has a simple zero at $s = -2$.

If the function under consideration is a rational function of s, that is, a quotient of two polynomials of s, the total number of poles equals the total number of zeros, counting the multiple-order poles and zeros, if the poles and zeros at infinity and at zero are taken into account. The function in Eq. (2-3) has four finite poles at $s = 0, -1, -3, -3$; there is one finite zero at $s = -2$, but there are three zeros at infinity, since

$$\lim_{s \to \infty} G(s) = \lim_{s \to \infty} \frac{10}{s^3} = 0 \tag{2-5}$$

Therefore, the function has a total of four poles and four zeros in the entire s-plane.

2.3 Laplace Transform[3-5]

The Laplace transform is one of the mathematical tools used for the solution of ordinary linear differential equations. In comparison with the classical method of solving linear differential equations, the Laplace transform method has the following two attractive features:

1. The homogeneous equation and the particular integral are solved in one operation.
2. The Laplace transform converts the differential equation into an algebraic equation in s. It is then possible to manipulate the algebraic equation by simple algebraic rules to obtain the solution in the s domain. The final solution is obtained by taking the inverse Laplace transform.

Definition of the Laplace Transform

Given the function $f(t)$ which satisfies the condition

$$\int_0^\infty |f(t)e^{-\sigma t}| \, dt < \infty \tag{2-6}$$

for some finite real σ, the Laplace transform of $f(t)$ is defined as

$$F(s) = \int_0^\infty f(t)e^{-st} \, dt \tag{2-7}$$

or

$$F(s) = \mathcal{L}[f(t)] \tag{2-8}$$

The variable s is referred to as the Laplace operator, which is a complex variable; that is, $s = \sigma + j\omega$. The defining equation of Eq. (2-7) is also known as the *one-sided Laplace transform*, as the integration is evaluated from 0 to ∞. This simply means that all information contained in $f(t)$ prior to $t = 0$ is ignored or considered to be zero. This assumption does not place any serious limitation on the applications of the Laplace transform to linear system problems, since in the usual time-domain studies, time reference is often chosen at the instant $t = 0$. Furthermore, for a physical system when an input is applied at $t = 0$, the response of the system does not start sooner than $t = 0$; that is, response does not precede excitation.

The following examples serve as illustrations on how Eq. (2-7) may be used for the evaluation of the Laplace transform of a function $f(t)$.

EXAMPLE 2-1   Let $f(t)$ be a unit step function that is defined to have a constant value of unity for $t > 0$ and a zero value for $t < 0$. Or,

$$f(t) = u_s(t) \tag{2-9}$$

Then the Laplace transform of $f(t)$ is

$$F(s) = \mathcal{L}[u_s(t)] = \int_0^\infty u_s(t)e^{-st}\, dt$$

$$= -\frac{1}{s}e^{-st}\Big|_0^\infty = \frac{1}{s} \tag{2-10}$$

Of course, the Laplace transform given by Eq. (2-10) is valid if

$$\int_0^\infty |u_s(t)e^{-\sigma t}|\, dt = \int_0^\infty |e^{-\sigma t}|\, dt < \infty$$

which means that the real part of s, σ, must be greater than zero. However, in practice, we simply refer to the Laplace transform of the unit step function as $1/s$, and rarely do we have to be concerned about the region in which the transform integral converges absolutely.

EXAMPLE 2-2   Consider the exponential function

$$f(t) = e^{-at}, \qquad t \geq 0$$

where a is a constant.

The Laplace transform of $f(t)$ is written

$$F(s) = \int_0^\infty e^{-at}e^{-st}\, dt = -\frac{e^{-(s+a)t}}{s+a}\Big|_0^\infty = \frac{1}{s+a} \tag{2-11}$$

Inverse Laplace Transformation

The operation of obtaining $f(t)$ from the Laplace transform $F(s)$ is termed the *inverse Laplace transformation*. The inverse Laplace transform of $F(s)$ is denoted by

$$f(t) = \mathcal{L}^{-1}[F(s)] \tag{2-12}$$

and is given by the inverse Laplace transform integral

$$f(t) = \frac{1}{2\pi j} \int_{c-j\infty}^{c+j\infty} F(s)e^{st}\, ds \tag{2-13}$$

where c is a real constant that is greater than the real parts of all the singularities of $F(s)$. Equation (2-13) represents a line integral that is to be evaluated in the s-plane. However, for most engineering purposes the inverse Laplace transform operation can be accomplished simply by referring to the Laplace transform table, such as the one given in Appendix B.

Important Theorems of the Laplace Transform

The applications of the Laplace transform in many instances are simplified by the utilization of the properties of the transform. These properties are presented in the following in the form of theorems, and no proofs are given.

1. *Multiplication by a Constant*
   The Laplace transform of the product of a constant k and a time function $f(t)$ is the constant k multiplied by the Laplace transform of $f(t)$; that is,

$$\mathcal{L}[k f(t)] = kF(s) \tag{2-14}$$

   where $F(s)$ is the Laplace transform of $f(t)$.

2. *Sum and Difference*
   The Laplace transform of the sum (or difference) of two time functions is the sum (or difference) of the Laplace transforms of the time functions; that is,

$$\mathcal{L}[f_1(t) \pm f_2(t)] = F_1(s) \pm F_2(s) \tag{2-15}$$

   where $F_1(s)$ and $F_2(s)$ are the Laplace transforms of $f_1(t)$ and $f_2(t)$, respectively.

3. *Differentiation*
   The Laplace transform of the first derivative of a time function $f(t)$ is s times the Laplace transform of $f(t)$ minus the limit of $f(t)$ as t approaches $0+$; that is,

$$\mathcal{L}\left[\frac{df(t)}{dt}\right] = sF(s) - \lim_{t \to 0+} f(t)$$
$$= sF(s) - f(0+) \tag{2-16}$$

   In general, for higher-order derivatives,

$$\mathcal{L}\left[\frac{d^n f(t)}{dt^n}\right] = s^n F(s) - \lim_{t \to 0+}\left[s^{n-1}f(t) + s^{n-2}\frac{df(t)}{dt} + \ldots + \frac{d^{n-1}f(t)}{dt^{n-1}}\right]$$
$$= s^n F(s) - s^{n-1}f(0+) - s^{n-2}f^{(1)}(0+) - \ldots - f^{(n-1)}(0+) \tag{2-17}$$

4. *Integration*
   The Laplace transform of the first integral of a function $f(t)$ with respect to time is the Laplace transform of $f(t)$ divided by s; that is,

$$\mathcal{L}\left[\int_0^t f(\tau)\, d\tau\right] = \frac{F(s)}{s} \tag{2-18}$$

   In general, for nth-order integration,

$$\mathcal{L}\left[\int_0^{t_1}\int_0^{t_2}\ldots\int_0^{t_n} f(\tau)\, d\tau\, dt_1 \ldots dt_{n-1}\right] = \frac{F(s)}{s^n} \tag{2-19}$$

5. *Shift in Time*
   The Laplace transform of $f(t)$ delayed by time T is equal to the Laplace transform of $f(t)$ multiplied by e^{-Ts}; that is,

$$\mathcal{L}[f(t - T)u_s(t - T)] = e^{-Ts}F(s) \tag{2-20}$$

   where $u_s(t - T)$ denotes the unit step function, which is shifted in time to the right by T.

6. *Initial-Value Theorem*

   If the Laplace transform of $f(t)$ is $F(s)$, then

$$\lim_{t \to 0} f(t) = \lim_{s \to \infty} sF(s) \qquad (2\text{-}21)$$

   if the time limit exists.

7. *Final-Value Theorem*

   If the Laplace transform of $f(t)$ is $F(s)$ and if $sF(s)$ is analytic on the imaginary axis and in the right half of the s-plane, then

$$\lim_{t \to \infty} f(t) = \lim_{s \to 0} sF(s) \qquad (2\text{-}22)$$

The final-value theorem is a very useful relation in the analysis and design of feedback control systems, since it gives the final value of a time function by determining the behavior of its Laplace transform as s tends to zero. However, the final-value theorem is not valid if $sF(s)$ contains any poles whose real part is zero or positive, which is equivalent to the analytic requirement of $sF(s)$ stated in the theorem. The following examples illustrate the care that one must take in applying the final-value theorem.

EXAMPLE 2-3   Consider the function

$$F(s) = \frac{5}{s(s^2 + s + 2)}$$

Since $sF(s)$ is analytic on the imaginary axis and in the right half of the s-plane, the final-value theorem may be applied. Therefore, using Eq. (2-22),

$$\lim_{t \to \infty} f(t) = \lim_{s \to 0} sF(s) = \lim_{s \to 0} \frac{5}{s^2 + s + 2} = \frac{5}{2} \qquad (2\text{-}23)$$

EXAMPLE 2-4   Consider the function

$$F(s) = \frac{\omega}{s^2 + \omega^2} \qquad (2\text{-}24)$$

which is known to be the Laplace transform of $f(t) = \sin \omega t$. Since the function $sF(s)$ has two poles on the imaginary axis, the final-value theorem *cannot* be applied in this case. In other words, although the final-value theorem would yield a value of zero as the final value of $f(t)$, the result is erroneous.

2.4 Inverse Laplace Transform by Partial-Fraction Expansion[7-11]

In a great majority of the problems in control systems, the evaluation of the inverse Laplace transform does not necessitate the use of the inversion integral of Eq. (2-13). The inverse Laplace transform operation involving rational functions can be carried out using a Laplace transform table and partial-fraction expansion.

When the Laplace transform solution of a differential equation is a rational function in s, it can be written

$$X(s) = \frac{P(s)}{Q(s)} \qquad (2\text{-}25)$$

where $P(s)$ and $Q(s)$ are polynomials of s. It is assumed that the order of $Q(s)$ in s is greater than that of $P(s)$. The polynomial $Q(s)$ may be written

$$Q(s) = s^n + a_1 s^{n-1} + \ldots + a_{n-1} s + a_n \qquad (2\text{-}26)$$

where $a_1, \ldots, a_n$ are real coefficients. The zeros of $Q(s)$ are either real or in complex-conjugate pairs, in simple or multiple order. The methods of partial-fraction expansion will now be given for the cases of simple poles, multiple-order poles, and complex poles, of $X(s)$.

Partial-Fraction Expansion When All the Poles of X(s) Are Simple and Real

If all the poles of $X(s)$ are real and simple, Eq. (2-25) can be written

$$X(s) = \frac{P(s)}{Q(s)} = \frac{P(s)}{(s + s_1)(s + s_2) \ldots (s + s_n)} \qquad (2\text{-}27)$$

where the poles $-s_1, -s_2, \ldots, -s_n$ are considered to be real numbers in the present case. Applying the partial-fraction expansion technique, Eq. (2-27) is written

$$X(s) = \frac{K_{s1}}{s + s_1} + \frac{K_{s2}}{s + s_2} + \ldots + \frac{K_{sn}}{s + s_n} \qquad (2\text{-}28)$$

The coefficient, K_{si} $(i = 1, 2, \ldots, n)$, is determined by multiplying both sides of Eq. (2-28) or (2-27) by the factor $(s + s_i)$ and then setting s equal to $-s_i$. To find the coefficient K_{s1}, for instance, we multiply both sides of Eq. (2-27) by $(s + s_1)$ and let $s = -s_1$; that is,

$$K_{s1} = \left[(s + s_1) \frac{P(s)}{Q(s)} \right]_{s=-s_1} = \frac{P(-s_1)}{(s_2 - s_1)(s_3 - s_1) \ldots (s_n - s_1)} \qquad (2\text{-}29)$$

As an illustrative example, consider the function

$$X(s) = \frac{5s + 3}{(s + 1)(s + 2)(s + 3)} \qquad (2\text{-}30)$$

which is written in the partial-fractioned form

$$X(s) = \frac{K_{-1}}{s + 1} + \frac{K_{-2}}{s + 2} + \frac{K_{-3}}{s + 3} \qquad (2\text{-}31)$$

The coefficients K_{-1}, K_{-2}, and K_{-3} are determined as follows:

$$K_{-1} = [(s + 1)X(s)]_{s=-1} = \frac{5(-1) + 3}{(2 - 1)(3 - 1)} = -1 \qquad (2\text{-}32)$$

$$K_{-2} = [(s + 2)X(s)]_{s=-2} = \frac{5(-2) + 3}{(1 - 2)(3 - 2)} = 7 \qquad (2\text{-}33)$$

$$K_{-3} = [(s + 3)X(s)]_{s=-3} = \frac{5(-3) + 3}{(1 - 3)(2 - 3)} = -6 \qquad (2\text{-}34)$$

Therefore, Eq. (2-31) becomes

$$X(s) = \frac{-1}{s + 1} + \frac{7}{s + 2} - \frac{6}{s + 3} \qquad (2\text{-}35)$$

Partial-Fraction Expansion When Some Poles of X(s) Are of Multiple Order

If r of the n poles of $X(s)$ are identical, or say, the pole at $s = -s_i$ is of multiplicity r, $X(s)$ is written

$$X(s) = \frac{P(s)}{Q(s)} = \frac{P(s)}{(s + s_1)(s + s_2) \ldots (s + s_i)^r (s + s_n)} \tag{2-36}$$

Then $X(s)$ can be expanded as

$$X(s) = \underbrace{\frac{K_{s1}}{s + s_1} + \frac{K_{s2}}{s + s_2} + \ldots + \frac{K_{sn}}{s + s_n}}_{(n-r) \text{ terms of simple poles}}$$

$$+ \underbrace{\frac{A_1}{s + s_i} + \frac{A_2}{(s + s_i)^2} + \ldots + \frac{A_r}{(s + s_i)^r}}_{r \text{ terms of repeated poles}} \tag{2-37}$$

The $n - r$ coefficients, which correspond to simple poles, $K_{s1}, K_{s2}, \ldots, K_{sn}$, may be evaluated by the method described by Eq. (2-29). The determination of the coefficients that correspond to the multiple-order poles is described below.

$$A_r = [(s + s_i)^r X(s)]_{s=-s_i} \tag{2-38}$$

$$A_{r-1} = \frac{d}{ds}[(s + s_i)^r X(s)]\bigg|_{s=-s_i} \tag{2-39}$$

$$A_{r-2} = \frac{1}{2!}\frac{d^2}{ds^2}[(s + s_i)^r X(s)]\bigg|_{s=-s_i} \tag{2-40}$$

$$A_1 = \frac{1}{(r-1)!}\frac{d^{r-1}}{ds^{r-1}}[(s + s_i)^r X(s)]\bigg|_{s=-s_i} \tag{2-41}$$

EXAMPLE 2-5   Consider the function

$$X(s) = \frac{1}{s(s + 1)^3(s + 2)} \tag{2-42}$$

Using the format of Eq. (2-37), $X(s)$ is written

$$X(s) = \frac{K_0}{s} + \frac{K_{-2}}{s + 2} + \frac{A_1}{s + 1} + \frac{A_2}{(s + 1)^2} + \frac{A_3}{(s + 1)^3} \tag{2-43}$$

Then the coefficients corresponding to the simple poles are

$$K_0 = [sX(s)]_{s=0} = \tfrac{1}{2} \tag{2-44}$$

$$K_{-2} = [(s + 2)X(s)]_{s=-2} = \tfrac{1}{2} \tag{2-45}$$

and those of the third-order poles are

$$A_3 = [(s + 1)^3 X(s)]|_{s=-1} = -1 \tag{2-46}$$

$$A_2 = \frac{d}{ds}[(s + 1)^3 X(s)]\bigg|_{s=-1} = \frac{d}{ds}\left[\frac{1}{s(s + 2)}\right]\bigg|_{s=-1}$$

$$= \frac{-(2s + 2)}{s^2(s + 2)^2}\bigg|_{s=-1} = 0 \tag{2-47}$$

and

$$A_1 = \frac{1}{2!}\frac{d^2}{ds^2}[(s + 1)^3 X(s)]\bigg|_{s=-1} = \frac{1}{2}\frac{d}{ds}\left[\frac{-2(s + 1)}{s^2(s + 2)^2}\right]\bigg|_{s=-1}$$

$$= \left[\frac{-1}{s^2(s + 2)^2} + \frac{2(s + 1)}{s^2(s + 2)^3} + \frac{2(s + 1)}{s^3(s + 2)^2}\right]\bigg|_{s=-1} = -1 \tag{2-48}$$

The completed partial-fraction expansion is

$$X(s) = \frac{1}{2s} + \frac{1}{2(s+2)} - \frac{1}{s+1} - \frac{1}{(s+1)^3} \tag{2-49}$$

Partial-Fraction Expansion of Simple Complex-Conjugate Poles

The partial-fraction expansion of Eq. (2-28) is valid also for simple complex-conjugate poles. However, since complex-conjugate poles are more difficult to handle and are of special interest in control-systems studies, they deserve separate treatment here.

Suppose that the rational function $X(s)$ of Eq. (2-25) contains a pair of complex poles:

$$s = -\alpha + j\omega \quad \text{and} \quad s = -\alpha - j\omega$$

Then the corresponding coefficients of these poles are

$$K_{-\alpha+j\omega} = (s + \alpha - j\omega)X(s)|_{s=-\alpha+j\omega} \tag{2-50}$$

$$K_{-\alpha-j\omega} = (s + \alpha + j\omega)X(s)|_{s=-\alpha-j\omega} \tag{2-51}$$

EXAMPLE 2-6   Consider the function

$$X(s) = \frac{\omega_n^2}{s(s^2 + 2\zeta\omega_n s + \omega_n^2)} \tag{2-52}$$

Let us assume that the values of ζ and ω_n are such that the nonzero poles of $X(s)$ are complex numbers. Then $X(s)$ is expanded as follows:

$$X(s) = \frac{K_0}{s} + \frac{K_{-\alpha+j\omega}}{s + \alpha - j\omega} + \frac{K_{-\alpha-j\omega}}{s + \alpha + j\omega} \tag{2-53}$$

where

$$\alpha = \zeta\omega_n \tag{2-54}$$

and

$$\omega = \omega_n\sqrt{1 - \zeta^2} \tag{2-55}$$

The coefficients in Eq. (2-53) are determined as

$$K_0 = sX(s)|_{s=0} = 1 \tag{2-56}$$

$$K_{-\alpha+j\omega} = (s + \alpha - j\omega)X(s)|_{s=-\alpha+j\omega}$$

$$= \frac{\omega_n^2}{2j\omega(-\alpha + j\omega)} = \frac{\omega_n}{2\omega}e^{-j(\theta+\pi/2)} \tag{2-57}$$

where

$$\theta = \tan^{-1}\left[-\frac{\omega}{\alpha}\right] \tag{2-58}$$

Also,

$$K_{-\alpha-j\omega} = (s + \alpha + j\omega)X(s)|_{s=-\alpha-j\omega}$$

$$= \frac{\omega_n^2}{-2j\omega(-\alpha - j\omega)} = \frac{\omega_n}{2\omega}e^{j(\theta+\pi/2)} \tag{2-59}$$

The complete expansion is

$$X(s) = \frac{1}{s} + \frac{\omega_n}{2\omega}\left[\frac{e^{-j(\theta+\pi/2)}}{s + \alpha - j\omega} + \frac{e^{j(\theta+\pi/2)}}{s + \alpha + j\omega}\right] \tag{2-60}$$

Taking the inverse Laplace transform on both sides of the last equation gives

$$x(t) = 1 + \frac{\omega_n}{2\omega}(e^{-j(\theta+\pi/2)}e^{(-\alpha+j\omega)t} + e^{j(\theta+\pi/2)}e^{(-\alpha-j\omega)t})$$

$$= 1 + \frac{\omega_n}{\omega}e^{-\alpha t}\sin(\omega t - \theta)$$

(2-61)

or

$$x(t) = 1 + \frac{1}{\sqrt{1-\zeta^2}}e^{-\zeta\omega_n t}\sin(\omega_n\sqrt{1-\zeta^2}\,t - \theta)$$

(2-62)

where θ is given by Eq. (2-58).

2.5 Application of Laplace Transform to the Solution of Linear Ordinary Differential Equations

With the aid of the theorems concerning Laplace transform given in Section 2.3 and a table of transforms, linear ordinary differential equations can be solved by the Laplace transform method. The advantages with the Laplace transform method are that, with the aid of a transform table the steps involved are all algebraic, and the homogeneous solution and the particular integral solution are obtained simultaneously.

Let us illustrate the method by several illustrative examples.

EXAMPLE 2-7   Consider the differential equation

$$\frac{d^2x(t)}{dt^2} + 3\frac{dx(t)}{dt} + 2x(t) = 5u_s(t)$$

(2-63)

where $u_s(t)$ is the unit step function, which is defined as

$$u_s(t) = \begin{cases} 1 & t > 0 \\ 0 & t < 0 \end{cases}$$

(2-64)

The initial conditions are $x(0+) = -1$ and $x^{(1)}(0+) = dx(t)/dt\,|_{t=0+} = 2$. To solve the differential equation we first take the Laplace transform on both sides of Eq. (2-63); we have

$$s^2 X(s) - sx(0+) - x^{(1)}(0+) + 3sX(s) - 3x(0+) + 2X(s) = \frac{5}{s}$$

(2-65)

Substituting the values of $x(0+)$ and $x^{(1)}(0+)$ into Eq. (2-65) and solving for $X(s)$, we get

$$X(s) = \frac{-s^2 - s + 5}{s(s^2 + 3s + 2)} = \frac{-s^2 - s + 5}{s(s+1)(s+2)}$$

(2-66)

Equation (2-66) is expanded by partial-fraction expansion to give

$$X(s) = \frac{5}{2s} - \frac{5}{s+1} + \frac{3}{2(s+2)}$$

(2-67)

Now taking the inverse Laplace transform of the last equation, we get the complete solution as

$$x(t) = \tfrac{5}{2} - 5e^{-t} + \tfrac{3}{2}e^{-2t} \qquad t \geq 0$$

(2-68)

The first term in the last equation is the steady-state solution, and the last two terms are the transient solution. Unlike the classical method, which requires separate steps to give the transient and the steady-state solutions, the Laplace transform method gives the entire solution of the differential equation in one operation.

If only the magnitude of the steady-state solution is of interest, the final-value theorem may be applied. Thus

$$\lim_{t\to\infty} x(t) = \lim_{s\to 0} sX(s) = \lim_{s\to 0} \frac{-s^2 - s + 5}{s^2 + 3s + 2} = \frac{5}{2} \tag{2-69}$$

where we have first checked and found that the function, $sX(s)$, has poles only in the left half of the s-plane.

EXAMPLE 2-8   Consider the linear differential equation

$$\frac{d^2x(t)}{dt^2} + 34.5\frac{dx(t)}{dt} + 1000x(t) = 1000u_s(t) \tag{2-70}$$

where $u_s(t)$ is the unit step function. The initial values of $x(t)$ and $dx(t)/dt$ are assumed to be zero.

Taking the Laplace transform on both sides of Eq. (2-70) and applying zero initial conditions, we have

$$s^2X(s) + 34.5sX(s) + 1000X(s) = \frac{1000}{s} \tag{2-71}$$

Solving $X(s)$ from the last equation, we obtain

$$X(s) = \frac{1000}{s(s^2 + 34.5s + 1000)} \tag{2-72}$$

The poles of $X(s)$ are at $s = 0$, $s = -17.25 + j26.5$, and $s = -17.25 - j26.5$. Therefore, Eq. (2-72) can be written as

$$X(s) = \frac{1000}{s(s + 17.25 - j26.5)(s + 17.25 + j26.5)} \tag{2-73}$$

One way of solving for $x(t)$ is to perform the partial-fraction expansion of Eq. (2-73), giving

$$X(s) = \frac{1}{s} + \frac{31.6}{2(26.5)}\left[\frac{e^{-j(\theta+\pi/2)}}{s + 17.25 - j26.5} + \frac{e^{j(\theta+\pi/2)}}{s + 17.25 + j26.5}\right] \tag{2-74}$$

where

$$\theta = \tan^{-1}\left(\frac{-26.5}{17.25}\right) = -56.9° \tag{2-75}$$

Then, using Eq. (2-61),

$$x(t) = 1 + 1.193e^{-17.25t}\sin(26.5t - \theta) \tag{2-76}$$

Another approach is to compare Eq. (2-72) with Eq. (2-52), so that

$$\omega_n = \pm\sqrt{1000} = \pm 31.6 \tag{2-77}$$

and

$$\zeta = 0.546 \tag{2-78}$$

and the solution to $x(t)$ is given directly by Eq. (2-62).

2.6 Elementary Matrix Theory[1,2,6]

In the study of modern control theory it is often desirable to use matrix notation to simplify complex mathematical expressions. The simplifying matrix notation may not reduce the amount of work required to solve the mathematical equations, but it usually makes the equations much easier to handle and manipulate.

As a motivation to the reason of using matrix notation, let us consider the following set of n simultaneous algebraic equations:

$$a_{11}x_1 + a_{12}x_2 + \ldots + a_{1n}x_n = y_1$$
$$a_{21}x_1 + a_{22}x_2 + \ldots + a_{2n}x_n = y_2$$
$$\cdots\cdots\cdots\cdots\cdots\cdots\cdots\cdots\cdots\cdots\cdots\cdots$$ (2-79)
$$a_{n1}x_1 + a_{n2}x_2 + \ldots + a_{nn}x_n = y_n$$

We may use the matrix equation

$$\mathbf{A}\mathbf{x} = \mathbf{y} \tag{2-80}$$

as a simplified representation for Eq. (2-79).

The symbols **A**, **x**, and **y** are defined as matrices, which contain the coefficients and variables of the original equations as their elements. In terms of matrix algebra, which will be discussed later, Eq. (2-80) can be stated as: *The product of the matrices* **A** *and* **x** *is equal to the matrix* **y**. The three matrices involved here are defined to be

$$\mathbf{A} = \begin{bmatrix} a_{11} & a_{12} & \cdots & a_{1n} \\ a_{21} & a_{22} & \cdots & a_{2n} \\ \cdot & \cdot & & \cdot \\ \cdot & \cdot & & \cdot \\ \cdot & \cdot & & \cdot \\ a_{n1} & a_{n2} & \cdots & a_{nn} \end{bmatrix} \tag{2-81}$$

$$\mathbf{x} = \begin{bmatrix} x_1 \\ x_2 \\ \cdot \\ \cdot \\ \cdot \\ x_n \end{bmatrix} \tag{2-82}$$

$$\mathbf{y} = \begin{bmatrix} y_1 \\ y_2 \\ \cdot \\ \cdot \\ \cdot \\ y_n \end{bmatrix} \tag{2-83}$$

which are simply *bracketed arrays of coefficients and variables*. Thus, we can define a matrix as follows:

Definition of a Matrix

A matrix is a collection of elements arranged in a rectangular or square array. Several ways of representing a matrix are as follows:

$$\mathbf{A} = \begin{bmatrix} 0 & 3 & 10 \\ 1 & -2 & 0 \end{bmatrix}, \qquad \mathbf{A} = \begin{pmatrix} 0 & 3 & 10 \\ 1 & -2 & 0 \end{pmatrix}$$

$$\mathbf{A} = \begin{Vmatrix} 0 & 3 & 10 \\ 1 & -2 & 0 \end{Vmatrix}, \qquad \mathbf{A} = [a_{ij}]_{2,3}$$

In this text we shall use square brackets to represent the matrix.

It is important to distinguish between a matrix and a determinant:

Matrix	*Determinant*
An array of numbers or elements with n rows and m columns. Does not have a value, although a square matrix ($n = m$) has a determinant.	*An array of numbers or elements with n rows and n columns (always square). Has a value.*

Some important definitions of matrices are given in the following.

Matrix elements. When a matrix is written

$$\mathbf{A} = \begin{bmatrix} a_{11} & a_{12} & a_{13} \\ a_{21} & a_{22} & a_{23} \\ a_{31} & a_{32} & a_{33} \end{bmatrix} \tag{2-84}$$

a_{ij} is identified as the element in the *i*th *row* and the *j*th *column* of the matrix. As a rule, we always refer to the row first and the column last.

Order of a matrix. The order of a matrix refers to the total number of rows and columns of the matrix. For example, the matrix in Eq. (2-84) has three rows and three columns and, therefore, is called a 3×3 (three by three) matrix. In general, a matrix with *n* rows and *m* columns is termed "$n \times m$" or "*n* by *m*."

Square matrix. *A square matrix is one that has the same number of rows as columns.*

Column matrix. *A column matrix is one that has one column and more than one row, that is, an $m \times 1$ matrix, $m > 1$.*

Quite often, a column matrix is referred to as a *column vector* or simply an *m-vector* if there are *m* rows. The matrix in Eq. (2-82) is a typical column matrix that is $n \times 1$, or an *n-vector*.

Row matrix. *A row matrix is one that has one row and more than one column, that is, a $1 \times n$ matrix. A row matrix can also be referred to as a row vector.*

Diagonal matrix. *A diagonal matrix is a square matrix with $a_{ij} = 0$ for all $i \neq j$.* Examples of a diagonal matrix are

$$\begin{bmatrix} a_{11} & 0 & 0 \\ 0 & a_{22} & 0 \\ 0 & 0 & a_{33} \end{bmatrix}, \qquad \begin{bmatrix} 5 & 0 \\ 0 & 3 \end{bmatrix} \tag{2-85}$$

Unity matrix (Identity matrix). *A unity matrix is a diagonal matrix with all the elements on the main diagonal ($i = j$) equal to 1.* A unity matrix is often designated by $\mathbf{I}$ or $\mathbf{U}$. An example of a unity matrix is

$$\mathbf{I} = \begin{bmatrix} 1 & 0 & 0 \\ 0 & 1 & 0 \\ 0 & 0 & 1 \end{bmatrix} \tag{2-86}$$

Null matrix. *A null matrix is one whose elements are all equal to zero;* for example,

$$\mathbf{O} = \begin{bmatrix} 0 & 0 & 0 \\ 0 & 0 & 0 \end{bmatrix} \qquad (2\text{-}87)$$

Symmetric matrix. *A symmetric matrix is a square matrix that satisfies the condition*

$$a_{ij} = a_{ji} \qquad (2\text{-}88)$$

for all i and j. A symmetric matrix has the property that if its rows are interchanged with its columns, the same matrix is preserved. Two examples of symmetric matrices are

$$\begin{bmatrix} 6 & 5 & 1 \\ 5 & 0 & 10 \\ 1 & 10 & -1 \end{bmatrix}, \qquad \begin{bmatrix} 1 & -4 \\ -4 & 1 \end{bmatrix} \qquad (2\text{-}89)$$

Determinant of a matrix. With each square matrix a determinant having the same elements and order as the matrix may be defined. *The determinant of a square matrix* $\mathbf{A}$ *is designated by*

$$\det \mathbf{A} = \Delta_A = |\mathbf{A}| \qquad (2\text{-}90)$$

As an illustrative example, consider the matrix

$$\mathbf{A} = \begin{bmatrix} 1 & 0 & -1 \\ 0 & 3 & 2 \\ -1 & 1 & 0 \end{bmatrix} \qquad (2\text{-}91)$$

The determinant of $\mathbf{A}$ is

$$|\mathbf{A}| = \begin{vmatrix} 1 & 0 & -1 \\ 0 & 3 & 2 \\ -1 & 1 & 0 \end{vmatrix} = -5 \qquad (2\text{-}92)$$

Singular matrix. *A square matrix is said to be singular if the value of its determinant is zero.* On the other hand, if a square matrix has a nonzero determinant, it is called a *nonsingular matrix.*

When a matrix is singular, it usually means that not all the rows or not all the columns of the matrix are independent of each other. When the matrix is used to represent a set of algebraic equations, singularity of the matrix means that these equations are not independent of each other. As an illustrative example, let us consider the following set of equations:

$$\begin{align} 2x_1 - 3x_2 + x_3 &= 0 \\ -x_1 + x_2 + x_3 &= 0 \\ x_1 - 2x_2 + 2x_3 &= 0 \end{align} \qquad (2\text{-}93)$$

Note that in Eq. (2-93), the third equation is equal to the sum of the first two equations. Therefore, these three equations are not completely independent.

In matrix form, these equations may be represented by

$$\mathbf{AX} = \mathbf{O}$$

where

$$\mathbf{A} = \begin{bmatrix} 2 & -3 & 1 \\ -1 & 1 & 1 \\ 1 & -2 & 2 \end{bmatrix} \tag{2-94}$$

$$\mathbf{X} = \begin{bmatrix} x_1 \\ x_2 \\ x_3 \end{bmatrix} \tag{2-95}$$

and $\mathbf{O}$ is a 3×1 null matrix. The determinant of $\mathbf{A}$ is

$$|\mathbf{A}| = \begin{vmatrix} 2 & -3 & 1 \\ -1 & 1 & 1 \\ 1 & -2 & 2 \end{vmatrix} = 4 - 3 + 2 - 1 - 6 + 4 = 0 \tag{2-96}$$

Therefore, the matrix $\mathbf{A}$ of Eq. (2-94) is singular. In this case the rows of $\mathbf{A}$ are dependent.

Transpose of a matrix. The transpose of a matrix $\mathbf{A}$ is defined as the matrix that is obtained by interchanging the corresponding rows and columns in $\mathbf{A}$.

Let $\mathbf{A}$ be an $n \times m$ matrix which is represented by

$$\mathbf{A} = [a_{ij}]_{n,m} \tag{2-97}$$

Then the transpose of $\mathbf{A}$, denoted by $\mathbf{A}'$, is given by

$$\mathbf{A}' = \text{transpose of } \mathbf{A} = [a_{ji}]_{m,n} \tag{2-98}$$

Notice that the order of $\mathbf{A}$ is $n \times m$; the transpose of $\mathbf{A}$ has an order $m \times n$.

EXAMPLE 2-9   As an example of the transpose of a matrix, consider the matrix

$$\mathbf{A} = \begin{bmatrix} 3 & 2 & 1 \\ 0 & -1 & 5 \end{bmatrix}$$

The transpose of $\mathbf{A}$ is given by

$$\mathbf{A}' = \begin{bmatrix} 3 & 0 \\ 2 & -1 \\ 1 & 5 \end{bmatrix}$$

Skew-symmetric matrix. A skew-symmetric matrix is a square matrix that equals its negative transpose; that is,

$$\mathbf{A} = -\mathbf{A}' \tag{2-99}$$

Some Operations of a Matrix Transpose

1. $(\mathbf{A}')' = \mathbf{A}$ (2-100)
2. $(k\mathbf{A})' = k\mathbf{A}'$, where k is a scalar   (2-101)
3. $(\mathbf{A} + \mathbf{B})' = \mathbf{A}' + \mathbf{B}'$ (2-102)
4. $(\mathbf{AB})' = \mathbf{B}'\mathbf{A}'$ (2-103)

Adjoint of a matrix. Let $\mathbf{A}$ be a square matrix of order n. *The adjoint matrix of* $\mathbf{A}$*, denoted by* adj $\mathbf{A}$*, is defined as*

$$\text{adj } \mathbf{A} = [ij \text{ cofactor of det } \mathbf{A}]'_{n, n} \qquad (2\text{-}104)$$

where the ij *cofactor of the determinant of* $\mathbf{A}$ *is the determinant obtained by omitting the* ith *row and the* jth *column of* $|\mathbf{A}|$ *and then multiplying it by* $(-1)^{i+j}$.

EXAMPLE 2-10   As an example of determining the adjoint matrix, let us consider first a 2×2 matrix,

$$\mathbf{A} = \begin{bmatrix} a_{11} & a_{12} \\ a_{21} & a_{22} \end{bmatrix}$$

The determinant of $\mathbf{A}$ is

$$|\mathbf{A}| = \begin{vmatrix} a_{11} & a_{12} \\ a_{21} & a_{22} \end{vmatrix}$$

The 1,1 cofactor, or the cofactor of the $(1, 1)$ element of $|\mathbf{A}|$, is a_{22}; the 1,2 cofactor is $-a_{21}$; the 2,1 cofactor is $-a_{12}$; and the 2,2 cofactor is a_{11}. Thus, from Eq. (2-104), the adjoint matrix of $\mathbf{A}$ is

$$\text{adj } \mathbf{A} = \begin{bmatrix} 1,1 \text{ cofactor} & 1,2 \text{ cofactor} \\ 2,1 \text{ cofactor} & 2,2 \text{ cofactor} \end{bmatrix}'$$

$$= \begin{bmatrix} 1,1 \text{ cofactor} & 2,1 \text{ cofactor} \\ 1,2 \text{ cofactor} & 2,2 \text{ cofactor} \end{bmatrix} \qquad (2\text{-}105)$$

$$= \begin{bmatrix} a_{22} & -a_{12} \\ -a_{21} & a_{11} \end{bmatrix}$$

EXAMPLE 2-11   As a second example of the adjoint matrix, consider

$$\mathbf{A} = \begin{bmatrix} a_{11} & a_{12} & a_{13} \\ a_{21} & a_{22} & a_{23} \\ a_{31} & a_{32} & a_{33} \end{bmatrix} \qquad (2\text{-}106)$$

Then

$$\text{adj } \mathbf{A} = \begin{bmatrix} 1,1 \text{ cofactor} & 2,1 \text{ cofactor} & 3,1 \text{ cofactor} \\ 1,2 \text{ cofactor} & 2,2 \text{ cofactor} & 3,2 \text{ cofactor} \\ 1,3 \text{ cofactor} & 2,3 \text{ cofactor} & 3,3 \text{ cofactor} \end{bmatrix}$$

$$= \begin{bmatrix} (a_{22}a_{33} - a_{23}a_{32}) & -(a_{12}a_{33} - a_{13}a_{32}) & (a_{12}a_{23} - a_{13}a_{22}) \\ -(a_{21}a_{33} - a_{23}a_{31}) & (a_{11}a_{33} - a_{13}a_{31}) & -(a_{11}a_{23} - a_{13}a_{21}) \\ (a_{21}a_{32} - a_{22}a_{31}) & -(a_{11}a_{32} - a_{12}a_{31}) & (a_{11}a_{22} - a_{12}a_{21}) \end{bmatrix}$$

$$(2\text{-}107)$$

Conjugate matrix. *Given a matrix* $\mathbf{A}$ *whose elements are represented by* a_{ij}*, the conjugate of* $\mathbf{A}$*, denoted by* $\bar{\mathbf{A}}$*, is obtained by replacing the elements of* $\mathbf{A}$ *by their complex conjugates; that is,*

$$\bar{\mathbf{A}} = \text{conjugate matrix of } \mathbf{A}$$
$$= [\bar{a}_{ij}]$$

(2-108)

where $\bar{a}_{ij} = $ complex conjugate of a_{ij}.

2.7 Matrix Algebra

When carrying out matrix operations it is necessary to define matrix algebra in the form of addition, subtraction, multiplication, division, and other necessary operations. It is important to point out at this stage that matrix operations are defined independently of the algebraic operations for scalar quantities.

Equality of Matrices

Two matrices $\mathbf{A}$ *and* $\mathbf{B}$ *are said to be equal to each other if they satisfy the following conditions:*

1. They are of the same order.
2. The corresponding elements are equal; that is,

$$a_{ij} = b_{ij} \qquad \text{for every } i \text{ and } j$$

For example,

$$\mathbf{A} = \begin{bmatrix} a_{11} & a_{12} \\ a_{21} & a_{22} \end{bmatrix} = \mathbf{B} = \begin{bmatrix} b_{11} & b_{12} \\ b_{21} & b_{22} \end{bmatrix}$$

(2-109)

implies that $a_{11} = b_{11}$, $a_{12} = b_{12}$, $a_{21} = b_{21}$, and $a_{22} = b_{22}$.

Addition of Matrices

Two matrices $\mathbf{A}$ and $\mathbf{B}$ can be added to form $\mathbf{A} + \mathbf{B}$ if they are of the same order. Then

$$\mathbf{A} + \mathbf{B} = [a_{ij}]_{n,m} + [b_{ij}]_{n,m} = \mathbf{C} = [c_{ij}]_{n,m}$$

(2-110)

where

$$c_{ij} = a_{ij} + b_{ij}$$

(2-111)

for all i and j. The order of the matrices is preserved after addition.

EXAMPLE 2-12   As an illustrative example, consider the two matrices

$$\mathbf{A} = \begin{bmatrix} 3 & 2 \\ -1 & 4 \\ 0 & -1 \end{bmatrix}, \qquad \mathbf{B} = \begin{bmatrix} 0 & 3 \\ -1 & 2 \\ 1 & 0 \end{bmatrix}$$

which are of the same order. Then the sum of $\mathbf{A}$ and $\mathbf{B}$ is given by

$$\mathbf{C} = \mathbf{A} + \mathbf{B} = \begin{bmatrix} 3+0 & 2+3 \\ -1-1 & 4+2 \\ 0+1 & -1+0 \end{bmatrix} = \begin{bmatrix} 3 & 5 \\ -2 & 6 \\ 1 & -1 \end{bmatrix}$$

(2-112)

Matrix Subtraction

The rules governing the subtraction of matrices are similar to those of matrix addition. In other words, Eqs. (2-110) and (2-111) are true if all the plus signs are replaced by minus signs. Or,

$$\mathbf{C} = \mathbf{A} - \mathbf{B} = [a_{ij}]_{n,m} - [b_{ij}]_{n,m}$$
$$= [a_{ij}]_{n,m} + [-b_{ij}]_{n,m} \qquad (2\text{-}113)$$
$$= [c_{ij}]_{n,m}$$

where

$$c_{ij} = a_{ij} - b_{ij} \qquad (2\text{-}114)$$

for all i and j.

Associate Law of Matrix (Addition and Subtraction)

The associate law of scalar algebra still holds for matrix addition and subtraction. Therefore,

$$(\mathbf{A} + \mathbf{B}) + \mathbf{C} = \mathbf{A} + (\mathbf{B} + \mathbf{C}) \qquad (2\text{-}115)$$

Commutative Law of Matrix (Addition and Subtraction)

The commutative law for matrix addition and subtraction states that the following matrix relationship is true:

$$\mathbf{A} + \mathbf{B} + \mathbf{C} = \mathbf{B} + \mathbf{C} + \mathbf{A}$$
$$= \mathbf{A} + \mathbf{C} + \mathbf{B} \qquad (2\text{-}116)$$

Matrix Multiplication

The matrices $\mathbf{A}$ and $\mathbf{B}$ may be multiplied together to form the product $\mathbf{AB}$ if they are *conformable*. This means that the number of columns of $\mathbf{A}$ must equal the number of rows of $\mathbf{B}$. In other words, let

$$\mathbf{A} = [a_{ij}]_{n,p}$$
$$\mathbf{B} = [b_{ij}]_{q,m}$$

Then $\mathbf{A}$ and $\mathbf{B}$ are conformable to form the product

$$\mathbf{C} = \mathbf{AB} = [a_{ij}]_{n,p}[b_{ij}]_{q,m} = [c_{ij}]_{n,m} \qquad (2\text{-}117)$$

if and only if $p = q$. The matrix $\mathbf{C}$ will have the same number of rows as $\mathbf{A}$ and the same number of columns as $\mathbf{B}$.

It is important to note that $\mathbf{A}$ and $\mathbf{B}$ may be conformable for $\mathbf{AB}$, but they may not be conformable for the product $\mathbf{BA}$, unless in Eq. (2-117) n also equals m. This points out an important fact that the commutative law is not generally valid for matrix multiplication. It is also noteworthy that even though $\mathbf{A}$ and $\mathbf{B}$ are conformable for both $\mathbf{AB}$ and $\mathbf{BA}$, usually $\mathbf{AB} \neq \mathbf{BA}$. In general, the following references are made with respect to matrix multiplication whenever they exist:

$$\mathbf{AB} = \mathbf{A} \text{ postmultiplied by } \mathbf{B}$$

or

$$\mathbf{AB} = \mathbf{B} \text{ premultiplied by } \mathbf{A}$$

Having established the condition for matrix multiplication, let us now turn to the rule of matrix multiplication. When the matrices $\mathbf{A}$ and $\mathbf{B}$ are conformable to form the matrix $\mathbf{C} = \mathbf{AB}$ as in Eq. (2-117), the ijth element of $\mathbf{C}$, c_{ij}, is given by

$$c_{ij} = \sum_{k=1}^{p} a_{ik}b_{kj} \tag{2-118}$$

for $i = 1, 2, \ldots, n$, and $j = 1, 2, \ldots, m$.

EXAMPLE 2-13   Given the matrices

$$\mathbf{A} = [a_{ij}]_{2,3} \qquad \mathbf{B} = [b_{ij}]_{3,1}$$

we notice that these two matrices are conformable for the product $\mathbf{AB}$ but not for $\mathbf{BA}$. Thus,

$$\mathbf{AB} = \begin{bmatrix} a_{11} & a_{12} & a_{13} \\ a_{21} & a_{22} & a_{23} \end{bmatrix} \begin{bmatrix} b_{11} \\ b_{21} \\ b_{31} \end{bmatrix} \tag{2-119}$$

$$= \begin{bmatrix} a_{11}b_{11} + a_{12}b_{21} + a_{13}b_{31} \\ a_{21}b_{11} + a_{22}b_{21} + a_{23}b_{31} \end{bmatrix}$$

EXAMPLE 2-14   Given the matrices

$$\mathbf{A} = \begin{bmatrix} 3 & -1 \\ 0 & 1 \\ 2 & 0 \end{bmatrix}, \qquad \mathbf{B} = \begin{bmatrix} 1 & 0 & -1 \\ 2 & 1 & 0 \end{bmatrix}$$

we notice that both $\mathbf{AB}$ and $\mathbf{BA}$ are conformable for multiplication.

$$\mathbf{AB} = \begin{bmatrix} 3 & -1 \\ 0 & 1 \\ 2 & 0 \end{bmatrix} \begin{bmatrix} 1 & 0 & -1 \\ 2 & 1 & 0 \end{bmatrix}$$

$$= \begin{bmatrix} (3)(1) + (-1)(2) & (3)(0) + (-1)(1) & (3)(-1) + (-1)(0) \\ (0)(1) + (1)(2) & (0)(0) + (1)(1) & (0)(-1) + (1)(0) \\ (2)(1) + (0)(2) & (2)(0) + (0)(1) & (2)(-1) + (0)(0) \end{bmatrix} \tag{2-120}$$

$$= \begin{bmatrix} 1 & -1 & -3 \\ 2 & 1 & 0 \\ 2 & 0 & -2 \end{bmatrix}$$

$$\mathbf{BA} = \begin{bmatrix} 1 & 0 & -1 \\ 2 & 1 & 0 \end{bmatrix} \begin{bmatrix} 3 & -1 \\ 0 & 1 \\ 2 & 0 \end{bmatrix}$$

$$= \begin{bmatrix} (1)(3) + (0)(0) + (-1)(2) & (1)(-1) + (0)(1) + (-1)(0) \\ (2)(3) + (1)(0) + (0)(2) & (2)(-1) + (1)(1) + (0)(0) \end{bmatrix} \tag{2-121}$$

$$= \begin{bmatrix} 1 & -1 \\ 6 & -1 \end{bmatrix}$$

Therefore, even though $\mathbf{AB}$ and $\mathbf{BA}$ may both exist, in general, they are not equal. In this case the products are not even of the same order.

Although the commutative law does not hold in general for matrix multiplication, the *associative* and the *distributive* laws are valid. For the distributive law, we state that

$$\mathbf{A}(\mathbf{B} + \mathbf{C}) = \mathbf{AB} + \mathbf{AC} \tag{2-122}$$

if the products are conformable.

For the associative law,

$$(\mathbf{AB})\mathbf{C} = \mathbf{A}(\mathbf{BC}) \tag{2-123}$$

if the product is conformable.

Multiplication by a Scalar k

Multiplying a matrix $\mathbf{A}$ by any scalar k is equivalent to multiplying each element of $\mathbf{A}$ by k. Therefore, if $\mathbf{A} = [a_{ij}]_{n,\,m}$

$$k\mathbf{A} = [ka_{ij}]_{n,\,m} \tag{2-124}$$

Inverse of a Matrix (Matrix Division)

In the algebra for scalar quantities, when we write

$$ax = y \tag{2-125}$$

it leads to

$$x = \frac{1}{a}y \tag{2-126}$$

or

$$x = a^{-1}y \tag{2-127}$$

Equations (2-126) and (2-127) are notationally equivalent.

In matrix algebra, if

$$\mathbf{Ax} = \mathbf{y} \tag{2-128}$$

then it *may be possible* to write

$$\mathbf{x} = \mathbf{A}^{-1}\mathbf{y} \tag{2-129}$$

where $\mathbf{A}^{-1}$ denotes the "inverse of $\mathbf{A}$." The conditions that $\mathbf{A}^{-1}$ exists are:

1. $\mathbf{A}$ is a square matrix.
2. $\mathbf{A}$ must be nonsingular.

If $\mathbf{A}^{-1}$ exists, it is given by

$$\mathbf{A}^{-1} = \frac{\text{adj } \mathbf{A}}{|\mathbf{A}|} \tag{2-130}$$

EXAMPLE 2-15   Given the matrix

$$\mathbf{A} = \begin{bmatrix} a_{11} & a_{12} \\ a_{21} & a_{22} \end{bmatrix} \tag{2-131}$$

the inverse of $\mathbf{A}$ is given by

$$\mathbf{A}^{-1} = \frac{\text{adj } \mathbf{A}}{|\mathbf{A}|} = \frac{\begin{bmatrix} a_{22} & -a_{12} \\ -a_{21} & a_{11} \end{bmatrix}}{a_{11}a_{22} - a_{12}a_{21}} \tag{2-132}$$

where for **A** to be nonsingular, $|\mathbf{A}| \neq 0$, or

$$a_{11}a_{22} - a_{12}a_{21} \neq 0 \tag{2-133}$$

If we pay attention to the adjoint matrix of **A**, which is the numerator of $\mathbf{A}^{-1}$, we see that for a 2×2 matrix, adj **A** is obtained by interchanging the two elements on the main diagonal and changing the signs of the elements on the off diagonal of **A**.

EXAMPLE 2-16   Given the matrix

$$\mathbf{A} = \begin{bmatrix} 1 & 1 & 0 \\ -1 & 0 & 2 \\ 1 & 1 & 1 \end{bmatrix} \tag{2-134}$$

the determinant of **A** is

$$|\mathbf{A}| = \begin{vmatrix} 1 & 1 & 0 \\ -1 & 0 & 2 \\ 1 & 1 & 1 \end{vmatrix} = 1 \tag{2-135}$$

Therefore, **A** has an inverse matrix, and is given by

$$\mathbf{A}^{-1} = \begin{bmatrix} -2 & -1 & 2 \\ 3 & 1 & -2 \\ -1 & 0 & 1 \end{bmatrix} \tag{2-136}$$

Some Properties of Matrix Inverse

1. $\mathbf{A}\mathbf{A}^{-1} = \mathbf{A}^{-1}\mathbf{A} = \mathbf{I}$ $\qquad\qquad\qquad\qquad\qquad\qquad\qquad$ (2-137)
2. $(\mathbf{A}^{-1})^{-1} = \mathbf{A}$ $\qquad\qquad\qquad\qquad\qquad\qquad\qquad\qquad$ (2-138)
3. In matrix algebra, in general,

$$\mathbf{AB} = \mathbf{AC} \tag{2-139}$$

does not necessarily imply $\mathbf{B} = \mathbf{C}$. The reader can easily construct an example to illustrate this property. However, if **A** is a square matrix, and is nonsingular, we can premultiply both sides of Eq. (2-139) by $\mathbf{A}^{-1}$. Then

$$\mathbf{A}^{-1}\mathbf{AB} = \mathbf{A}^{-1}\mathbf{AC} \tag{2-140}$$

or

$$\mathbf{IB} = \mathbf{IC} \tag{2-141}$$

which leads to

$$\mathbf{B} = \mathbf{C}$$

4. If **A** and **B** are square matrices and are nonsingular, then

$$(\mathbf{AB})^{-1} = \mathbf{B}^{-1}\mathbf{A}^{-1} \tag{2-142}$$

Rank of a Matrix

The rank of a matrix **A** is the maximum number of linearly independent columns of **A**; or, it is the order of the largest nonsingular matrix contained in **A**. Several examples on the rank of a matrix are as follows:

$$\begin{bmatrix} 0 & 1 \\ 0 & 0 \end{bmatrix} \quad \text{rank} = 1, \qquad \begin{bmatrix} 0 & 5 & 1 & 4 \\ 3 & 0 & 3 & 2 \end{bmatrix} \quad \text{rank} = 2,$$

$$\begin{bmatrix} 3 & 9 & 2 \\ 1 & 3 & 0 \\ 2 & 6 & 1 \end{bmatrix} \quad \text{rank} = 2, \qquad \begin{bmatrix} 3 & 0 & 0 \\ 1 & 2 & 0 \\ 0 & 0 & 1 \end{bmatrix} \quad \text{rank} = 3$$

The following properties on rank are useful in the determination of the rank of a matrix. Given an $n \times m$ matrix $\mathbf{A}$,

1. Rank of $\mathbf{A}$ = Rank of $\mathbf{A}'$.
2. Rank of $\mathbf{A}$ = Rank of $\mathbf{A}'\mathbf{A}$.
3. Rank of $\mathbf{A}$ = Rank of $\mathbf{A}\mathbf{A}'$.

Properties 2 and 3 are useful in the determination of rank; since $\mathbf{A}'\mathbf{A}$ and $\mathbf{A}\mathbf{A}'$ are always square, the rank condition can be checked by evaluating the determinant of these matrices.

Quadratic Forms

Consider the scalar function

$$f(\mathbf{x}) = \sum_{i=1}^{n} \sum_{j=1}^{n} a_{ij} x_i x_j \tag{2-143}$$

which is called the *quadratic form*. We can write this equation as

$$f(\mathbf{x}) = \sum_{i=1}^{n} x_i \sum_{j=1}^{n} a_{ij} x_j \tag{2-144}$$

Let

$$y_i = \sum_{j=1}^{n} a_{ij} x_j \tag{2-145}$$

Then Eq. (2-144) becomes

$$f(\mathbf{x}) = \sum_{i=1}^{n} x_i y_i \tag{2-146}$$

Now if we define

$$\mathbf{x} = \begin{bmatrix} x_1 \\ x_2 \\ \cdot \\ \cdot \\ \cdot \\ x_n \end{bmatrix}, \qquad \mathbf{y} = \begin{bmatrix} y_1 \\ y_2 \\ \cdot \\ \cdot \\ \cdot \\ y_n \end{bmatrix}$$

Eq. (2-146) can be written

$$f(\mathbf{x}) = \mathbf{x}'\mathbf{y} \tag{2-147}$$

and from Eq. (2-145),

$$\mathbf{y} = \mathbf{A}\mathbf{x} \tag{2-148}$$

where

$$\mathbf{A} = [a_{ij}]_{n,n} \tag{2-149}$$

Finally, $f(\mathbf{x})$ becomes

$$f(\mathbf{x}) = \mathbf{x}'\mathbf{A}\mathbf{x} \tag{2-150}$$

Since the coefficient of $x_i x_j$ is $a_{ij} + a_{ji}$ for $i \neq j$, given any quadratic form as in Eq. (2-150), we can always replace $\mathbf{A}$ with a symmetric matrix. In other words, given any $\mathbf{A}$, we can always define a symmetric matrix $\mathbf{B}$ such that

$$b_{ij} = b_{ji} = \frac{a_{ij} + a_{ji}}{2}, \qquad i \neq j \tag{2-151}$$

The quadratic forms are often used as performance indices in control systems design, since they usually lead to mathematical conveniences in the design algorithms.

Definiteness

Positive definite. An $n \times n$ matrix $\mathbf{A}$ is said to be *positive definite* if all the roots of the equation

$$|\lambda\mathbf{I} - \mathbf{A}| = 0 \tag{2-152}$$

are positive. Equation (2-152) is called the *characteristic equation* of $\mathbf{A}$, and the roots are referred to as the *eigenvalues* of $\mathbf{A}$.

Positive semidefinite. The matrix $\mathbf{A}$ ($n \times n$) is *positive semidefinite* if all its eigenvalues are nonnegative and at least one of the eigenvalues is zero.

Negative definite. The matrix $\mathbf{A}$ ($n \times n$) is *negative semidefinite* if all its eigenvalues are nonpositive and at least one of the eigenvalues is zero.

Indefinite. The matrix $\mathbf{A}$ ($n \times n$) is *indefinite* if some of the eigenvalues are negative and some are positive.

An alternative way of testing the definiteness of a square matrix is to check the signs of all the leading principal minors of the matrix. The leading principal minors of an $n \times n$ matrix $\mathbf{A}$ are defined as follows. Given the square matrix

$$\mathbf{A} = \begin{bmatrix} a_{11} & a_{12} & \cdots & a_{1n} \\ a_{21} & a_{22} & \cdots & a_{2n} \\ \cdot & \cdot & & \cdot \\ \cdot & \cdot & & \cdot \\ \cdot & \cdot & & \cdot \\ a_{n1} & a_{n2} & \cdots & a_{nn} \end{bmatrix}$$

the n leading principal minors are the following determinants:

$$a_{11} \qquad \begin{vmatrix} a_{11} & a_{12} \\ a_{21} & a_{22} \end{vmatrix} \qquad \begin{vmatrix} a_{11} & a_{12} & a_{13} \\ a_{21} & a_{22} & a_{23} \\ a_{31} & a_{32} & a_{33} \end{vmatrix} \quad \cdots \quad |\mathbf{A}|$$

Then the definiteness of $\mathbf{A}$ is determined as follows:

$\mathbf{A}$ is positive (negative) definite if all the leading principal minors of $\mathbf{A}$ are positive (negative).

$\mathbf{A}$ is positive semidefinite if $|\mathbf{A}| = 0$ and all the leading principal minors of $\mathbf{A}$ are nonnegative.

A is negative semidefinite if $|\mathbf{A}| = 0$ and all the leading principal minors of $-\mathbf{A}$ are nonnegative.

We may also refer to the definiteness of the quadratic form, $\mathbf{x}'\mathbf{A}\mathbf{x}$.

The quadratic form, $\mathbf{x}'\mathbf{A}\mathbf{x}$ (**A** is symmetric), is positive definite (positive semidefinite, negative definite, negative semidefinite) if the matrix **A** is positive definite (positive semidefinite, negative definite, negative semidefinite).

2.8 z-Transform[12,13]

The Laplace transform is a powerful tool for the analysis and design of linear time-invariant control systems with continuous data. However, for linear systems with sampled or discrete data, we may find that the z-transform is more appropriate.

Let us first consider the analysis of a discrete-data system which is represented by the block diagram of Fig. 2-3. One way of describing the discrete nature of the signals is to consider that the input and the output of the system are sequences of numbers. These numbers are spaced T seconds apart. Thus, the input sequence and the output sequence may be represented by $r(kT)$ and $c(kT)$, respectively, $k = 0, 1, 2, \ldots$. To represent these input and output sequences by time-domain expressions, the numbers are represented by impulse functions in such a way that the strengths of the impulses correspond to the values of these numbers at the corresponding time instants. This way, the input sequence is expressed as a train of impulses,

$$r^*(t) = \sum_{k=0}^{\infty} r(kT)\delta(t - kT) \tag{2-153}$$

A similar expression can be written for the output sequence.

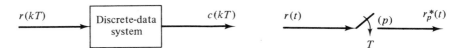

Fig. 2-3. Block diagram of a discrete-data system.

Fig. 2-4. Block diagram of a finite-pulsewidth sampler.

Another type of system that has discontinuous signals is the sampled-data system. A sampled-data system is characterized by having samplers in the system. A sampler is a device that converts continuous data into some form of sampled data. For example, Fig. 2-4 shows the block diagram of a typical sampler that closes for a very short duration of p seconds once every T seconds. This is referred to as a sampler with a uniform *sampling period T* and a finite *sampling duration p*. Figure 2-5 illustrates a set of typical input and output signals of the sampler.

With the notation of Figs. 2-4 and 2-5, the output of the finite-pulse-duration sampler is written

$$r_p^*(t) = r(t) \sum_{k=0}^{\infty} [u_s(t - kT) - u_s(t - kT - p)] \tag{2-154}$$

where $u_s(t)$ is the unit step function.

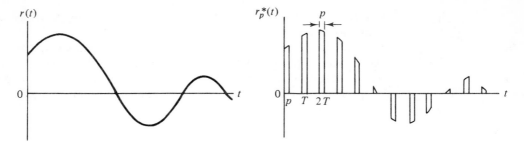

Fig. 2-5. Input and output signals of a finite-pulsewidth sampler.

For small p, that is, $p \ll T$, the narrow-width pulses of Fig. 2-5 may be approximated by flat-topped pulses. In other words, Eq. (2-154) can be written

$$r_p^*(t) \simeq \sum_{k=0}^{\infty} r(kT)[u_s(t - kT) - u_s(t - kT - p)] \qquad (2\text{-}155)$$

Multiplying both sides of Eq. (2-155) by $1/p$ and taking the limit as p approaches zero, we have

$$\lim_{p \to 0} \frac{1}{p} r_p^*(t) \simeq \lim_{p \to 0} \sum_{k=0}^{\infty} \frac{1}{p} r(kT)[u_s(t - kT) - u_s(t - kT - p)]$$

$$= \sum_{k=0}^{\infty} r(kT)\delta(t - kT)$$

or

$$\lim_{p \to 0} \frac{1}{p} r_p^*(t) \simeq r^*(t) \qquad (2\text{-}156)$$

In arriving at this result we have made use of the fact that

$$\delta(t) = \lim_{p \to 0} \frac{1}{p}[u_s(t) - u_s(t - p)] \qquad (2\text{-}157)$$

The significance of Eq. (2-156) is that the output of the finite-pulsewidth sampler can be approximated by a train of impulses if the pulsewidth approaches zero in the limit. A sampler whose output is a train of impulses with the strength of each impulse equal to the magnitude of the input at the corresponding sampling instant is called an *ideal sampler*. Figure 2-6 shows the block diagram of an ideal sampler connected in cascade with a constant factor p so that the combination is an approximation to the finite-pulsewidth sampler of Fig. 2-4 if p is very small. Figure 2-7 illustrates the typical input and output signals of an ideal sampler; the arrows are used to represent impulses with the heights representing the strengths (or areas) of the latter.

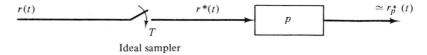

Fig. 2-6. Approximation of a finite-pulsewidth sampler by an ideal sampler and a cascade constant factor.

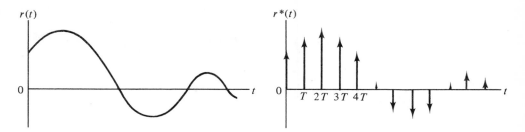

Fig. 2-7. Input and output signals of an ideal sampler.

In view of these considerations we may now use the ideal sampler to represent the discrete data, $r(kT)$. This points to the fact that the signals of the system in Fig. 2-3 can essentially be treated as outputs of ideal samplers.

Now we are ready to investigate the application of transform methods to discrete and sampled-data systems. Taking the Laplace transform on both sides of Eq. (2-153), we have

$$R^*(s) = \sum_{k=0}^{\infty} r(kT)e^{-kTs} \tag{2-158}$$

The fact that Eq. (2-158) contains the exponential term e^{-kTs} reveals the difficulty of using Laplace transform for the general treatment of discrete-data systems, since the transfer function relations will no longer be algebraic as in the continuous-data case. Although it is conceptually simple to perform inverse Laplace transform on algebraic transfer relations, it is not a simple matter to perform inverse Laplace transform on transcendental functions. One simple fact is that the commonly used Laplace transform tables do not have entries with transcendental functions in s. This necessitates the use of the z-transform. Our motivation here for the generation of the z-transform is simply to convert transcendental functions in s into algebraic ones in z. The definition of z-transform is given with this objective in mind.

Definition of the z-Transform

The z-transform is defined as

$$z = e^{Ts} \tag{2-159}$$

where s is the Laplace transform variable and T is the sampling period. Equation (2-159) also leads to

$$s = \frac{1}{T} \ln z \tag{2-160}$$

Using Eq. (2-159), the expression in Eq. (2-158) is written

$$R^*\left(s = \frac{1}{T} \ln z\right) = R(z) = \sum_{k=0}^{\infty} r(kT)z^{-k} \tag{2-161}$$

or

$$R(z) = z\text{-transform of } r^*(t)$$

$$= \mathfrak{z}[r^*(t)] \tag{2-162}$$

$$= [\text{Laplace transform of } r^*(t)]_{s=1/T \ln z}$$

Therefore, we have treated the z-transform as simply a change in variable, $z = e^{Ts}$.

The following examples illustrate some of the simple z-transform operations.

EXAMPLE 2-17   Consider the sequence

$$r(kT) = e^{-akT}, \qquad k = 0, 1, 2, \ldots \qquad (2\text{-}163)$$

where a is a constant.

From Eq. (2-153),

$$r^*(t) = \sum_{k=0}^{\infty} e^{-akT}\delta(t - kT) \qquad (2\text{-}164)$$

Then

$$R^*(s) = \sum_{k=0}^{\infty} e^{-akT}e^{-kTs} \qquad (2\text{-}165)$$

Multiply both sides of Eq. (2-165) by $e^{-(s+a)T}$ and subtract the resulting equation from Eq. (2-165); now we can show easily that $R^*(s)$ can be written in a closed form,

$$R^*(s) = \frac{1}{1 - e^{-(s+a)T}} \qquad (2\text{-}166)$$

for

$$|e^{-(\sigma+a)T}| < 1 \qquad (2\text{-}167)$$

where σ is the real part of s. Then the z-transform of $r^*(t)$ is

$$R(z) = \frac{1}{1 - e^{-aT}z^{-1}} = \frac{z}{z - e^{-aT}} \qquad (2\text{-}168)$$

for $|e^{-aT}z^{-1}| < 1$.

EXAMPLE 2-18   In Example 2-17, if $a = 0$, we have

$$r(kT) = 1, \qquad k = 0, 1, 2, \ldots \qquad (2\text{-}169)$$

which represents a sequence of numbers all equal to unity. Then

$$R^*(s) = \sum_{k=0}^{\infty} e^{-kTs} \qquad (2\text{-}170)$$

$$R(z) = \sum_{k=0}^{\infty} z^{-k} = 1 + z^{-1} + z^{-2} + z^{-3} + \ldots \qquad (2\text{-}171)$$

This expression is written in closed form as

$$R(z) = \frac{1}{1 - z^{-1}} \qquad |z^{-1}| < 1 \qquad (2\text{-}172)$$

or

$$R(z) = \frac{z}{z - 1} \qquad |z^{-1}| < 1 \qquad (2\text{-}173)$$

In general, the z-transforms of more complex functions are obtained by use of the same procedure as described in the preceding two examples. If a time function $r(t)$ is given as the starting point, the procedure of finding its z-transform is to first form the sequence $r(kT)$ and then use Eq. (2-161) to get $R(z)$. An equivalent interpretation of this step is to send the signal $r(t)$ through an ideal sampler whose output is $r^*(t)$. We then take the Laplace transform of $r^*(t)$ to give $R^*(s)$ as in Eq. (2-158), and $R(z)$ is obtained by substituting z for e^{Ts}.

Table 2-1   Table of z-Transforms

Laplace Transform	Time Function	z-Transform
1	Unit impulse $\delta(t)$	1
$\dfrac{1}{s}$	Unit step $u(t)$	$\dfrac{z}{z-1}$
$\dfrac{1}{1-e^{-Ts}}$	$\delta_T(t) = \sum\limits_{n=0}^{\infty} \delta(t-nT)$	$\dfrac{z}{z-1}$
$\dfrac{1}{s^2}$	t	$\dfrac{Tz}{(z-1)^2}$
$\dfrac{1}{s^3}$	$\dfrac{t^2}{2}$	$\dfrac{T^2z(z+1)}{2(z-1)^3}$
$\dfrac{1}{s^{n+1}}$	$\dfrac{t^n}{n!}$	$\lim\limits_{a\to 0} \dfrac{(-1)^n}{n!} \dfrac{\partial^n}{\partial a^n}\left(\dfrac{z}{z-e^{-aT}}\right)$
$\dfrac{1}{s+a}$	e^{-at}	$\dfrac{z}{z-e^{-aT}}$
$\dfrac{1}{(s+a)^2}$	te^{-at}	$\dfrac{Tze^{-aT}}{(z-e^{-aT})^2}$
$\dfrac{a}{s(s+a)}$	$1-e^{-at}$	$\dfrac{(1-e^{-aT})z}{(z-1)(z-e^{-aT})}$
$\dfrac{\omega}{s^2+\omega^2}$	$\sin \omega t$	$\dfrac{z \sin \omega T}{z^2 - 2z \cos \omega T + 1}$
$\dfrac{\omega}{(s+a)^2+\omega^2}$	$e^{-at}\sin \omega t$	$\dfrac{ze^{-aT}\sin \omega T}{z^2 e^{2aT} - 2ze^{aT}\cos \omega T + 1}$
$\dfrac{s}{s^2+\omega^2}$	$\cos \omega t$	$\dfrac{z(z-\cos \omega T)}{z^2 - 2z \cos \omega T + 1}$
$\dfrac{s+a}{(s+a)^2+\omega^2}$	$e^{-at}\cos \omega t$	$\dfrac{z^2 - ze^{-aT}\cos \omega T}{z^2 - 2ze^{-aT}\cos \omega T + e^{-2aT}}$

Table 2-1 gives the *z*-transforms of some of the time functions commonly used in systems analysis. A more extensive table may be found in the literature.[12,13]

Inverse z-Transformation

Just as in the Laplace transformation, one of the major objectives of the *z*-transformation is that algebraic manipulations can be made first in the *z*-domain, and then the final time response is determined by the inverse *z*-transformation. In general, the inverse *z*-transformation of $R(z)$ can yield information only on $r(kT)$, not on $r(t)$. In other words, the *z*-transform carries information only in a discrete fashion. When the time signal $r(t)$ is sampled by the ideal sampler, only information on the signal at the sampling instants, $t = kT$, is retained. With this in mind, the inverse *z*-transformation can be effected by one of the following three methods:

1. The partial-fraction expansion method.
2. The power-series method.
3. The inversion formula.

Partial-fraction expansion method. The *z*-transform function $R(z)$ is expanded by partial-fraction expansion into a sum of simple recognizable terms, and the z-transform table is used to determine the corresponding $r(kT)$. In carrying out the partial-fraction expansion, there is a slight difference between the z-transform and the Laplace transform procedures. With reference to the z-transform table, we note that practically all the transform functions have the term z in the numerator. Therefore, we should expand $R(z)$ into the form

$$R(z) = \frac{K_1 z}{z - e^{-aT}} + \frac{K_2 z}{z - e^{-bT}} + \cdots \qquad (2\text{-}174)$$

For this, we should first expand $R(z)/z$ into fractions and then multiply z across to obtain the final desired expression. The following example will illustrate this recommended procedure.

EXAMPLE 2-19   Given the z-transform function

$$R(z) = \frac{(1 - e^{-aT})z}{(z - 1)(z - e^{-aT})} \qquad (2\text{-}175)$$

it is desired to find the inverse z-transform.

Expanding $R(z)/z$ by partial-fraction expansion, we have

$$\frac{R(z)}{z} = \frac{1}{z - 1} - \frac{1}{z - e^{-aT}} \qquad (2\text{-}176)$$

Thus,

$$R(z) = \frac{z}{z - 1} - \frac{z}{z - e^{-aT}} \qquad (2\text{-}177)$$

From the z-transform table of Table 2-1, the corresponding inverse z-transform of $R(z)$ is found to be

$$r(kT) = 1 - e^{-akT} \qquad (2\text{-}178)$$

Power-series method. The z-transform $R(z)$ is expanded into a power series in powers of z^{-1}. In view of Eq. (2-161), the coefficient of z^{-k} is the value of $r(t)$ at $t = kT$, or simply $r(kT)$. For example, for the $R(z)$ in Eq. (2-175), we expand it into a power series in powers of z^{-1} by long division; then we have

$$R(z) = (1 - e^{-aT})z^{-1} + (1 - e^{-2aT})z^{-2} + (1 - e^{-3aT})z^{-3}$$
$$+ \ldots + (1 - e^{-akT})z^{-k} + \cdots \qquad (2\text{-}179)$$

or

$$R(z) = \sum_{k=0}^{\infty} (1 - e^{-akT})z^{-k} \qquad (2\text{-}180)$$

Thus,

$$r(kT) = 1 - e^{-akT} \qquad (2\text{-}181)$$

which is the same result as in Eq. (2-178).

Inversion formula. The time sequence $r(kT)$ may be determined from $R(z)$ by use of the inversion formula,

$$r(kT) = \frac{1}{2\pi j} \oint_{\Gamma} R(z)z^{k-1} \, dz \qquad (2\text{-}182)$$

which is a contour integration[5] along the path Γ, where Γ is a circle of radius $|z| = \epsilon^{cT}$ centered at the origin in the z-plane, and c is of such a value that all the poles of $R(z)$ are inside the circle.

One way of evaluating the contour integration of Eq. (2-182) is by use of the residue theorem of complex variable theory. Equation (2-182) may be written

$$r(kT) = \frac{1}{2\pi j} \oint_\Gamma R(z)z^{k-1}\, dz$$

$$= \sum \text{Residues of } R(z)z^{k-1} \text{ at the poles of } R(z)z^{k-1} \qquad (2\text{-}183)$$

For simple poles, the residues of $R(z)z^{k-1}$ at the pole $z = z_j$ is obtained as

$$\text{Residue of } R(z)z^{k-1} \text{ at the pole } z_j = (z - z_j)R(z)z^{k-1}|_{z=z_j} \qquad (2\text{-}184)$$

Now let us consider the same function used in Example 2-19. The function $R(z)$ of Eq. (2-175) has two poles: $z = 1$ and $z = e^{-aT}$. Using Eq. (2-183), we have

$$r(kT) = [\text{Residue of } R(z)z^{k-1} \text{ at } z = 1] + [\text{Residue of } R(z)z^{k-1} \text{ at } z = e^{-aT}]$$

$$= \frac{(1 - e^{-aT})z^k}{(z - e^{-aT})}\bigg|_{z=1} + \frac{(1 - e^{-aT})z^k}{(z - 1)}\bigg|_{z=e^{-aT}} \qquad (2\text{-}185)$$

$$= 1 - e^{-akT}$$

which again agrees with the result obtained earlier.

Some Important Theorems of the z-Transformation

Some of the commonly used theorems of the *z*-transform are stated in the following without proof. Just as in the case of the Laplace transform, these theorems are useful in many aspects of the *z*-transform analysis.

1. *Addition and Subtraction*
   If $r_1(kT)$ and $r_2(kT)$ have *z*-transforms $R_1(z)$ and $R_2(z)$, respectively, then

$$\mathfrak{z}[r_1(kT) \pm r_2(kT)] = R_1(z) \pm R_2(z) \qquad (2\text{-}186)$$

2. *Multiplication by a Constant*

$$\mathfrak{z}[ar(kT)] = a\mathfrak{z}[r(kT)] = aR(z) \qquad (2\text{-}187)$$

   where a is a constant.

3. *Real Translation*

$$\mathfrak{z}[r(kT - nT)] = z^{-n}R(z) \qquad (2\text{-}188)$$

   and

$$\mathfrak{z}[r(kT + nT)] = z^n\left[R(z) - \sum_{k=0}^{n-1} r(kT)z^{-k} \right] \qquad (2\text{-}189)$$

   where n is a positive integer. Equation (2-188) represents the *z*-transform of a time sequence that is shifted to the right by nT, and Eq. (2-189) denotes that of a time sequence shifted to the left by nT. The reason the right-hand side of Eq. (2-189) is not $z^n R(z)$ is because the *z*-transform, similar to the Laplace transform, is defined only for

$k \geq 0$. Thus, the second term on the right-hand side of Eq. (2-189) simply represents the sequence that is lost after it is shifted to the left by nT.

4. *Complex Translation*

$$\mathfrak{z}[e^{\mp akT}r(kT)] = R(ze^{\pm aT}) \tag{2-190}$$

5. *Initial-Value Theorem*

$$\lim_{k \to 0} r(kT) = \lim_{z \to \infty} R(z) \tag{2-191}$$

if the limit exists.

6. *Final-Value Theorem*

$$\lim_{k \to \infty} r(kT) = \lim_{z \to 1} (1 - z^{-1})R(z) \tag{2-192}$$

if the function, $(1 - z^{-1})R(z)$, has no poles on or outside the unit circle centered at the origin in the z-plane, $|z| = 1$.

The following examples illustrate the usefulness of these theorems.

EXAMPLE 2-20   Apply the complex translation theorem to find the z-transform of
$f(t) = te^{-at}$, $t \geq 0$.
Let $r(t) = t$, $t \geq 0$; then

$$R(z) = \mathfrak{z}[tu_s(t)] = \mathfrak{z}(kT) = \frac{Tz}{(z - 1)^2} \tag{2-193}$$

Using the complex translation theorem,

$$F(z) = \mathfrak{z}[te^{-at}u_s(t)] = R(ze^{aT}) = \frac{Tze^{-aT}}{(z - e^{-aT})^2} \tag{2-194}$$

EXAMPLE 2-21   Given the function

$$R(z) = \frac{0.792z^2}{(z - 1)(z^2 - 0.416z + 0.208)} \tag{2-195}$$

determine the value of $r(kT)$ as k approaches infinity.
Since

$$(1 - z^{-1})R(z) = \frac{0.792z}{z^2 - 0.416z + 0.208} \tag{2-196}$$

does not have any pole on or outside the unit circle $|z| = 1$ in the z-plane, the final-value theorem of the z-transform can be applied. Hence,

$$\lim_{k \to \infty} r(kT) = \lim_{z \to 1} \frac{0.792z}{z^2 - 0.416z + 0.208} = 1 \tag{2-197}$$

This result is easily checked by expanding $R(z)$ in powers of z^{-1},

$$R(z) = 0.792z^{-1} + 1.121z^{-2} + 1.091z^{-3} + 1.013z^{-4} + 0.986z^{-5}$$
$$+ 0.981z^{-6} + 0.998z^{-7} + \ldots \tag{2-198}$$

It is apparent that the coefficients of this power series converge rapidly to the final value of unity.

REFERENCES

Complex Variables, Laplace Transforms, and Matrix Algebra

1. F. B. HILDEBRAND, *Methods of Applied Mathematics*, Prentice-Hall, Inc., Englewood Cliffs, N.J., 1952.

2. R. BELLMAN, *Introduction to Matrix Analysis*, McGraw-Hill Book Company, New York, 1960.

3. B. C. KUO, *Linear Networks and Systems*, McGraw-Hill Book Company, New York, 1967.

4. R. LEGROS and A. V. J. MARTIN, *Transform Calculus for Electrical Engineers*, Prentice-Hall, Inc., Englewood Cliffs, N.J., 1961.

5. C. R. WYLIE, JR., *Advanced Engineering Mathematics*, 2nd ed., McGraw-Hill Book Company, New York, 1960.

6. S. BARNETT, "Matrices, Polynomials, and Linear Time-Invariant Systems," *IEEE Trans. Automatic Control*, Vol. AC-18, pp. 1–10, Feb. 1973.

Partial Fraction Expansion

7. D. HAZONY and J. RILEY, "Evaluating Residues and Coefficients of High Order Poles," *IRE Trans. Automatic Control*, Vol. AC-4, pp. 132–136, Nov. 1959.

8. C. POTTLE, "On the Partial Fraction Expansion of a Rational Function with Multiple Poles by Digital Computer," *IEEE Trans. Circuit Theory*, Vol. CT-11, pp. 161–162, Mar. 1964.

9. M. I. YOUNIS, "A Quick Check on Partial Fraction Expansion Coefficients," *IEEE Trans. Automatic Control*, Vol. AC-11, pp. 318–319, Apr. 1966.

10. N. AHMED and K. R. Rao, "Partial Fraction Expansion of Rational Functions with One High-Order Pole," *IEEE Trans. Automatic Control*, Vol. AC-13, p. 133, Feb. 1968.

11. B. O. WATKINS, "A Partial Fraction Algorithm," *IEEE Trans. Automatic Control*, Vol. AC-16, pp. 489–491, Oct. 1971.

Sampled-Data and Discrete-Data Control Systems

12. B. C. KUO, *Analysis and Synthesis of Sampled-Data Control Systems*, Prentice-Hall, Inc., Englewood Cliffs, N.J., 1963.

13. B. C. KUO, *Discrete Data Control Systems*, Science-Tech, Box 2277 Station A, Champaign, Illinois, 1970.

PROBLEMS

2.1. Find the poles and zeros of the following functions (include the ones at infinity):

(a) $G(s) = \dfrac{5(s + 1)}{s^2(s + 2)(s + 5)}$

(b) $G(s) = \dfrac{s^2(s + 1)}{(s + 2)(s^2 + 3s + 2)}$

(c) $G(s) = \dfrac{K(s + 2)}{s(s^2 + s + 1)}$

(d) $G(s) = \dfrac{Ke^{-2s}}{(s + 1)(s + 2)}$

2.2. Find the Laplace transforms of the following functions:

(a) $g(t) = te^{-2t}$

(b) $g(t) = t \cos 5t$

(c) $g(t) = e^{-t} \sin \omega t$

(d) $g(t) = \sum\limits_{k=0}^{\infty} g(kT)\delta(t - kT);\ \delta(t) =$ unit impulse function

2.3. Find the Laplace transforms of the functions shown in Fig. P2-3.

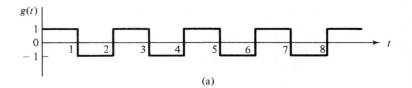

(a)

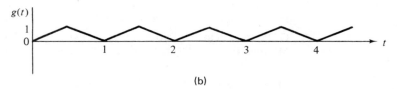

(b)

Figure P2-3.

2.4. Find the Laplace transform of the following function:

$$f(t) = \begin{cases} 0 & t < 1 \\ t + 1 & 1 \le t < 3 \\ 4 & 3 \le t \le 4 \\ 0 & 4 < t \end{cases}$$

2.5. Solve the following differential equation by means of the Laplace transformation:

$$\frac{d^2 f(t)}{dt^2} + 5\frac{df(t)}{dt} + 4f(t) = e^{-t}u_s(t)$$

Assume that all the initial conditions are zero.

2.6. Find the inverse Laplace transforms of the following functions:

(a) $G(s) = \dfrac{1}{(s + 2)(s + 3)}$

(b) $G(s) = \dfrac{1}{(s + 1)^2(s + 4)}$

(c) $G(s) = \dfrac{10}{s(s^2 + 4)(s + 1)}$

(d) $G(s) = \dfrac{2(s + 1)}{s(s^2 + s + 2)}$

2.7. Carry out the following matrix sums and differences:

(a) $\begin{bmatrix} 3 & 6 \\ 0 & -5 \end{bmatrix} + \begin{bmatrix} 7 & 0 \\ -3 & 10 \end{bmatrix}$

(b) $\begin{bmatrix} 15 \\ -1 \\ 3 \end{bmatrix} - \begin{bmatrix} 20 \\ -4 \\ 5 \end{bmatrix}$

(c) $\begin{bmatrix} \dfrac{1}{s} & 0 & s+1 \\ 5 & \dfrac{1}{s-3} & s^2 \end{bmatrix} + \begin{bmatrix} 0 & 10 & -s \\ s & \dfrac{1}{s} & 0 \end{bmatrix}$

2.8. Determine if the following matrices are conformable for the products **AB** and **BA**. Find the valid products.

(a) $\mathbf{A} = \begin{bmatrix} 1 \\ 0 \\ 3 \end{bmatrix}$ $\mathbf{B} = [6 \quad 0 \quad 1]$

(b) $\mathbf{A} = \begin{bmatrix} 2 & -1 \\ 3 & 0 \end{bmatrix}$ $\mathbf{B} = \begin{bmatrix} 10 & 0 & 9 \\ -1 & -1 & 0 \end{bmatrix}$

2.9. Express the following set of algebraic equations in matrix form:

$$5x_1 + x_2 - x_3 = 1$$
$$-x_1 + 3x_2 - x_3 = 1$$
$$3x_1 - 7x_2 - 2x_3 = 0$$

2.10. Express the following set of differential equations in the form $\dot{\mathbf{x}}(t) = \mathbf{A}\mathbf{x}(t) + \mathbf{B}\mathbf{u}(t)$:

$$\dot{x}_1(t) = -x_1(t) + x_2(t)$$
$$\dot{x}_2(t) = -2x_2(t) - 3x_3(t) + u_1(t)$$
$$\dot{x}_3(t) = -x_1(t) - 5x_2(t) - 3x_3(t) + u_2(t)$$

2.11. Find the inverse of the following matrices:

(a) $\mathbf{A} = \begin{bmatrix} 2 & 5 \\ 10 & -1 \end{bmatrix}$

(b) $\mathbf{A} = \begin{bmatrix} 3 & 0 & -1 \\ -2 & 1 & 2 \\ 0 & 1 & -1 \end{bmatrix}$

(c) $\mathbf{A} = \begin{bmatrix} 1 & 3 & 4 \\ -1 & 1 & 0 \\ -1 & 0 & -1 \end{bmatrix}$

2.12. Determine the ranks of the following matrices:

(a) $\begin{bmatrix} 3 & 2 \\ 7 & 1 \\ 0 & 3 \end{bmatrix}$

(b) $\begin{bmatrix} 2 & 4 & 0 & 8 \\ 1 & 2 & 6 & 3 \end{bmatrix}$

2.13. Determine the definiteness of the following matrices:

(a) $\begin{bmatrix} 2 & 3 \\ -1 & 2 \end{bmatrix}$

(b) $\begin{bmatrix} 1 & 5 & -1 \\ -2 & 0 & 0 \\ 3 & 1 & 1 \end{bmatrix}$

2.14. The following signals are sampled by an ideal sampler with a sampling period of T seconds. Determine the output of the sampler, $f^*(t)$, and find the Laplace transform of $f^*(t)$, $F^*(s)$. Express $F^*(s)$ in closed form.

(a) $f(t) = te^{-at}$

(b) $f(t) = e^{-at} \sin \omega t$

2.15. Determine the z-transform of the following functions:

(a) $G(s) = \dfrac{1}{(s + a)^n}$

(b) $G(s) = \dfrac{1}{s(s + 5)^2}$

(c) $G(s) = \dfrac{1}{s^3(s + 2)}$

(d) $g(t) = t^2 e^{-2t}$

(e) $g(t) = t \sin \omega t$

2.16. Find the inverse z-transform of

$$G(z) = \frac{10z(z + 1)}{(z - 1)(z^2 + z + 1)}$$

by means of the following methods:

(a) the inversion formula

(b) partial-fraction expansion

3

Transfer Function and
Signal Flow Graphs

3.1 Introduction

One of the most important steps in the analysis of a physical system is the mathematical description and modeling of the system. A mathematical model of a system is essential because it allows one to gain a clear understanding of the system in terms of cause-and-effect relationships among the system components.

In general, a physical system can be represented by a schematic diagram that portrays the relationships and interconnections among the system components. From the mathematical standpoint, algebraic and differential or difference equations can be used to describe the dynamic behavior of a system.

In systems theory, the block diagram is often used to portray systems of all types. For linear systems, transfer functions and signal flow graphs are valuable tools for analysis as well as for design.

In this chapter we give the definition of transfer function of a linear system, and demonstrate the power of the signal-flow-graph technique in the analysis of linear systems.

3.2 Transfer Functions of Linear Systems

Transfer function plays an important role in the characterization of linear time-invariant systems. Together with block diagram and signal flow graph, transfer function forms the basis of representing the input–output relationships of a linear time-invariant system in classical control theory.

The starting point of defining the transfer function is the differential equation of a dynamic system. Consider that a linear time-invariant system is described by the following nth-order differential equation:

$$a_0 \frac{d^n c(t)}{dt^n} + a_1 \frac{d^{n-1} c(t)}{dt^{n-1}} + \ldots + a_{n-1} \frac{dc(t)}{dt} + a_n c(t)$$
$$= b_0 \frac{d^m r(t)}{dt^m} + b_1 \frac{d^{m-1} r(t)}{dt^{m-1}} + \ldots + b_{m-1} \frac{dr(t)}{dt} + b_m r(t) \qquad (3\text{-}1)$$

where $c(t)$ is the output variable and $r(t)$ is the input variable. The coefficients, $a_0, a_1, \ldots, a_n$ and $b_0, b_1, \ldots, b_m$ are constants, and $n \geq m$.

The differential equation in Eq. (3-1) represents a complete description of the system between the input $r(t)$ and the output $c(t)$. Once the input and the initial conditions of the system are specified, the output response may be obtained by solving Eq. (3-1). However, it is apparent that the differential equation method of describing a system is, although essential, a rather cumbersome one, and the higher-order differential equation of Eq. (3-1) is of little practical use in design. More important is the fact that although efficient subroutines are available on digital computers for the solution of high-order differential equations, the important development in linear control theory relies on analysis and design techniques without actual solutions of the system differential equations.

A convenient way of describing linear systems is made possible by the use of *transfer function* and *impulse response*. To obtain the transfer function of the linear system that is represented by Eq. (3-1), we take the Laplace transform on both sides of the equation, and assuming zero initial conditions, we have

$$(a_0 s^n + a_1 s^{n-1} + \ldots + a_{n-1} s + a_n) C(s)$$
$$= (b_0 s^m + b_1 s^{m-1} + \ldots + b_{m-1} s + b_m) R(s) \qquad (3\text{-}2)$$

The transfer function of the system is defined as the ratio of $C(s)$ to $R(s)$; therefore,

$$G(s) = \frac{C(s)}{R(s)} = \frac{b_0 s^m + b_1 s^{m-1} + \ldots + b_{m-1} s + b_m}{a_0 s^n + a_1 s^{n-1} + \ldots + a_{n-1} s + a_n} \qquad (3\text{-}3)$$

Summarizing over the properties of a transfer function we state:

1. A transfer function is defined only for a linear system, and, strictly, only for time-invariant systems.
2. A transfer function between an input variable and an output variable of a system is defined as the ratio of the Laplace transform of the output to the Laplace transform of the input.
3. All initial conditions of the system are assumed to be zero.
4. A transfer function is independent of input excitation.

The following example is given to illustrate how transfer functions for a linear system are derived.

EXAMPLE 3-1   A series *RLC* network is shown in Fig. 3-1. The input voltage is designated by $e_i(t)$. The output variable in this case can be defined as the voltage across any one of the three network elements, or the current

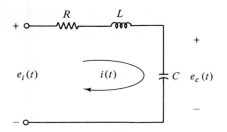

Fig. 3-1. *RLC* network.

$i(t)$. The loop equation of the network is written

$$e_i(t) = Ri(t) + L\frac{di(t)}{dt} + \frac{1}{C}\int i(t)\,dt \qquad (3\text{-}4)$$

Taking the Laplace transform on both sides of Eq. (3-4) and assuming zero initial conditions, we have

$$E_i(s) = \left(R + Ls + \frac{1}{Cs}\right)I(s) \qquad (3\text{-}5)$$

If we regard the current $i(t)$ as an output variable, the transfer function between $e_i(t)$ and $i(t)$ is simply

$$\frac{I(s)}{E_i(s)} = \frac{1}{R + Ls + (1/Cs)} = \frac{Cs}{1 + RCs + LCs^2} \qquad (3\text{-}6)$$

If the voltage across the capacitor $e_c(t)$ is considered as an output, the transfer function between $e_i(t)$ and $e_c(t)$ is obtained by substituting

$$E_c(s) = \frac{1}{Cs}I(s) \qquad (3\text{-}7)$$

into Eq. (3-5). Therefore,

$$\frac{E_c(s)}{E_i(s)} = \frac{1}{1 + RCs + LCs^2} \qquad (3\text{-}8)$$

The definition of transfer function is easily extended to a system with a multiple number of inputs and outputs. A system of this type is often referred to as a *multivariable system*. In a multivariable system, a differential equation of the form of Eq. (3-1) may be used to describe the relationship between a pair of input and output. When dealing with the relationship between one input and one output, it is assumed that all other inputs are set to zero. Since the principle of superposition is valid for linear systems, the total effect on any output variable due to all the inputs acting simultaneously can be obtained by adding the individual effects.

As a simple illustrative example of the transfer functions of a multivariable system, let us consider the control of a turbopropeller engine. In this case the input variables are the fuel rate and the propeller blade angle. The output variables are the speed of rotation of the engine and the turbine-inlet tempera-ture. In general, either one of the outputs is affected by the changes in both inputs. For instance, when the blade angle of the propeller is increased, the speed of rotation of the engine will decrease and the temperature usually increases. The following transfer relations may be written from steady-state tests performed on the system:

$$C_1(s) = G_{11}(s)R_1(s) + G_{12}(s)R_2(s) \qquad (3\text{-}9)$$
$$C_2(s) = G_{21}(s)R_1(s) + G_{22}(s)R_2(s) \qquad (3\text{-}10)$$

where

$C_1(s) =$ transformed variable of speed of rotation
$C_2(s) =$ transformed variable of turbine-inlet temperature
$R_1(s) =$ transformed variable of fuel rate
$R_2(s) =$ transformed variable of propeller blade angle

All these variables are assumed to be measured from some reference levels.

Since Eqs. (3-9) and (3-10) are written with the assumption that the system is linear, the principle of superposition holds. Therefore, $G_{11}(s)$ represents the transfer function between the fuel rate and the speed of rotation of the engine with the propeller blade angle held at the reference value; that is, $R_2(s) = 0$. Similar statements can be made for the other transfer functions.

In general, if a linear system has p inputs and q outputs, the transfer function between the ith output and the jth input is defined as

$$G_{ij}(s) = \frac{C_i(s)}{R_j(s)} \tag{3-11}$$

with $R_k(s) = 0$, $k = 1, 2, \ldots, p$, $k \neq j$. Note that Eq. (3-11) is defined with only the jth input in effect, while the other inputs are set to zero. The ith output transform of the system is related to all the input transforms by

$$C_i(s) = G_{i1}(s)R_1(s) + G_{i2}(s)R_2(s) + \ldots + G_{ip}(s)R_p(s)$$
$$= \sum_{j=1}^{p} G_{ij}(s)R_j(s) \qquad (i = 1, 2, \ldots, q) \tag{3-12}$$

where $G_{ij}(s)$ is defined in Eq. (3-11).

It is convenient to represent Eq. (3-12) by a matrix equation

$$\mathbf{C}(s) = \mathbf{G}(s)\mathbf{R}(s) \tag{3-13}$$

where

$$\mathbf{C}(s) = \begin{bmatrix} C_1(s) \\ C_2(s) \\ \cdot \\ \cdot \\ \cdot \\ C_q(s) \end{bmatrix}, \tag{3-14}$$

is a $q \times 1$ matrix, called the *transformed output vector*;

$$\mathbf{R}(s) = \begin{bmatrix} R_1(s) \\ R_2(s) \\ \cdot \\ \cdot \\ \cdot \\ R_p(s) \end{bmatrix} \tag{3-15}$$

is a $p \times 1$ matrix, called the *transformed input vector*;

$$\mathbf{G}(s) = \begin{bmatrix} G_{11}(s) & G_{12}(s) & \ldots & G_{1p}(s) \\ G_{21}(s) & G_{22}(s) & \ldots & G_{2p}(s) \\ \cdot \\ \cdot \\ \cdot \\ G_{q1}(s) & G_{q2}(s) & \ldots & G_{qp}(s) \end{bmatrix} \tag{3-16}$$

is a $q \times p$ matrix, called the *transfer function matrix*.

3.3 Impulse Response of Linear Systems

The impulse response of a linear system is defined as the output response of the system when the input is a unit impulse function. Therefore, for a system with a single input and a single output, if $r(t) = \delta(t)$, the Laplace transform of the system output is simply the transfer function of the system, that is,

$$C(s) = G(s) \tag{3-17}$$

since the Laplace transform of the unit impulse function is unity.

Taking the inverse Laplace transform on both sides of Eq. (3-17) yields

$$c(t) = g(t) \tag{3-18}$$

where $g(t)$ is the inverse Laplace transform of $G(s)$ and is the *impulse response* (sometimes also called the *weighing function*) of a linear system. Therefore, we can state that *the Laplace transform of the impulse response is the transfer function.* Since the transfer function is a powerful way of characterizing linear systems, this means that if a linear system has zero initial conditions, theoretically, the system can be described or identified by exciting it with a unit impulse response and measuring the output. In practice, although a true impulse cannot be generated physically, a pulse with a very narrow pulsewidth usually provides a suitable approximation.

For a multivariable system, an *impulse response matrix* must be defined and is given by

$$\mathbf{g}(t) = \mathcal{L}^{-1}[\mathbf{G}(s)] \tag{3-19}$$

where the inverse Laplace transform of $\mathbf{G}(s)$ implies the transform operation on each term of the matrix.

The derivation of $G(s)$ in Eq. (3-3) is based on the knowledge of the system differential equation, and the solution of $C(s)$ from Eq. (3-3) also assumes that $R(s)$ and $G(s)$ are all available in analytical forms. This is not always possible, for quite often the input signal $r(t)$ is not Laplace transformable or is available only in the form of experimental data. Under such conditions, to analyze the system we would have to work with the time function $r(t)$ and $g(t)$.

Let us consider that the input signal $r(\tau)$ shown in Fig. 3-2(a) is applied to a linear system whose impulse response is $g(t)$. The output response $c(t)$ is to be determined. In this case we have denoted the input signal as a function of τ, which is the time variable; this is necessary since t is reserved as a fixed time quantity in the analysis. For all practical purposes, $r(\tau)$ is assumed to extend from minus infinity to plus infinity in time.

Now consider that the input $r(\tau)$ is approximated by a sequence of pulses of pulsewidth $\Delta\tau$, as shown in Fig. 3-2(b). In the limit, as $\Delta\tau$ approaches zero, these pulses become impulses, and the impulse at time $k\Delta\tau$ has a strength or area equal to $\Delta\tau \cdot r(k\Delta\tau)$, which is the area of the pulse at $k\Delta\tau$. Also, when $\Delta\tau$ decreases, k has to be increased proportionally, so the value of $k\Delta\tau$ remains constant and equals τ, which is a particular point on the time axis. We now compute the output response of the linear system, using the impulse-approxi-

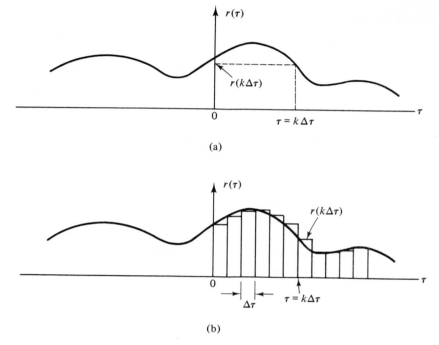

Fig. 3-2. (a) Input signal of a linear system. (b) Input signal represented by sum of rectangular pulses.

mated signal. When only the impulse at time $\tau = k\Delta\tau$ is considered, the system response is given by

$$\Delta\tau \cdot r(k\Delta\tau)g(t - k\Delta\tau) \tag{3-20}$$

which is the system impulse response delayed by $k\Delta\tau$, multiplied by the impulse strength $\Delta\tau \cdot r(k\Delta\tau)$. By use of the superposition principle, the total response due to $r(\tau)$ is obtained by adding up the responses due to each of the impulses from $-\infty$ to $+\infty$. Therefore,

$$c(t) = \lim_{\substack{\Delta\tau \to 0 \\ N \to \infty}} \sum_{k=-N}^{N} r(k\Delta\tau)g(t - k\Delta\tau)\,\Delta\tau \tag{3-21}$$

or

$$c(t) = \int_{-\infty}^{\infty} r(\tau)g(t - \tau)\,d\tau \tag{3-22}$$

For all physical systems, output response does not precede excitation. Thus

$$g(t) = 0 \tag{3-23}$$

for $t < 0$, since the impulse function is applied at $t = 0$. Or

$$g(t - \tau) = 0 \qquad t < \tau \tag{3-24}$$

The output response of the system is now written

$$c(t) = \int_{-\infty}^{t} r(\tau)g(t - \tau)\, d\tau \tag{3-25}$$

Further, if $r(\tau) = 0$ for $\tau < 0$, Eq. (3-25) becomes

$$c(t) = \int_{0}^{t} r(\tau)g(t - \tau)\, d\tau \tag{3-26}$$

The expressions of Eqs. (3-25) and (3-26) are called the *convolution integral*. The convolution operation is denoted by the symbol $*$, so

$$c(t) = r(t) * g(t) \tag{3-27}$$

is interpreted as

$$c(t) = r(t) \text{ convolves into } g(t) \tag{3-28}$$

The positions of $r(t)$ and $g(t)$ in the convolution operation may be interchanged, since basically there is no difference between the two functions. Therefore, the convolution integral can also be written as

$$\begin{aligned} c(t) &= \int_{0}^{t} g(\tau)r(t - \tau)\, d\tau \\ &= g(t) * r(t) \\ &= g(t) \text{ convolves into } r(t) \end{aligned} \tag{3-29}$$

The evaluation of the impulse response of a linear system is sometimes an important step in the analysis and design of a class of systems known as the *adaptive control systems*. In real life the dynamic characteristics of most systems vary to some extent over an extended period of time. This may be caused by simple deterioration of components due to wear and tear, drift in operating environments, and the like. Some systems simply have parameters that vary with time in a predictable or unpredictable fashion. For instance, the transfer characteristic of a guided missile in flight will vary in time because of the change of mass of the missile and the change of atmospheric conditions. On the other hand, for a simple mechanical system with mass and friction, the latter may be subject to unpredictable variation either due to "aging" or surface conditions. Thus the control system designed under the assumption of known and fixed parameters may fail to yield satisfactory response should the system parameters vary. In order that the system may have the ability of self-correction or self-adjustment in accordance with varying parameters and environment, it is necessary that the system's transfer characteristics be identified continuously or at appropriate intervals during the operation of the system. One of the methods of identification is to measure the impulse response of the system so that design parameters may be adjusted accordingly to attain optimal control at all times.

In the two preceding sections, definitions of transfer function and impulse response of a linear system have been presented. The two functions are directly related through the Laplace transformation, and they represent essentially the same information about the system. However, it must be reiterated that transfer function and impulse response are defined only for linear systems and that the initial conditions are assumed to be zero.

3.4 Block Diagrams[1]

Because of its simplicity and versatility, *block diagram* is often used by control engineers to portray systems of all types. A block diagram can be used simply to represent the composition and interconnection of a system. Or, it can be used, together with transfer functions, to represent the cause-and-effect relationships throughout the system. For instance, the block diagram of Fig. 3-3 represents a turbine-driven hydraulic power system for an aircraft. The main components of the system include a pressure-compensated hydraulic pump, an air-driven pump, an electronic speed controller, and a control valve. The block diagram in the figure depicts how these components are interconnected.

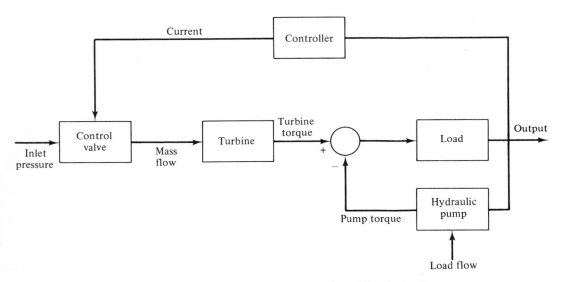

Fig. 3-3. Block diagram of a turbine-driven hydraulic power system.

If the mathematical and functional relationships of all the system elements are known, the block diagram can be used as a reference for the analytical or the computer solution of the system. Furthermore, if all the system elements are assumed to be linear, the transfer function for the overall system can be obtained by means of block-diagram algebra.

The essential point is that block diagram can be used to portray nonlinear as well as linear systems. For example, Fig. 3-4(a) shows the block diagram of a simple control system which includes an amplifier and a motor. In the figure the nonlinear characteristic of the amplifier is depicted by its nonlinear

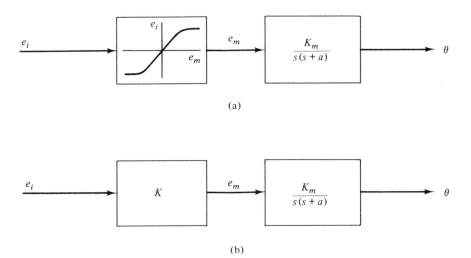

(a)

(b)

Fig. 3-4. Block diagram of a simple control system. (a) Amplifier shown with a nonlinear gain characteristic. (b) Amplifier shown with a linear gain characteristic.

input–output relation. The motor is assumed to be linear and its dynamics are represented by a transfer function between the input voltage and the output displacement. Figure 3-4(b) illustrates the same system but with the amplifier characteristic approximated by a constant gain. In this case the overall system is linear, and it is now possible to write the transfer function for the overall system as

$$\frac{\theta(s)}{E_i(s)} = \frac{\theta(s)}{E_m(s)} \frac{E_m(s)}{E_i(s)} = \frac{KK_m}{s(s+a)} \qquad (3\text{-}30)$$

Block Diagrams of Control Systems

We shall now define some block-diagram elements used frequently in control systems and the block-diagram algebra. One of the important components of a feedback control system is the sensing device that acts as a junction point for signal comparisons. The physical components involved are the potentiometer, synchros, resolvers, differential amplifiers, multipliers, and so on. In general, the operations of the sensing devices are addition, subtraction, multiplication, and sometimes combinations of these. The block-diagram elements of these operations are illustrated as shown in Fig. 3-5. It should be pointed out that the signals shown in the diagram of Fig. 3-5 can be functions of time t or functions of the Laplace transform variable s.

In Fig. 3-4 we have already used block-diagram elements to represent input–output relationships of linear and nonlinear elements. It simply shows that the block-diagram notation can be used to represent practically any input–output relation as long as the relation is defined. For instance, the block diagram of

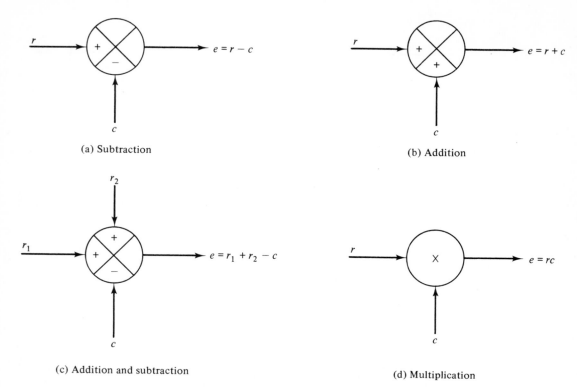

(a) Subtraction

(b) Addition

(c) Addition and subtraction

(d) Multiplication

Fig. 3-5. Block-diagram elements of typical sensing devices of control systems. (a) Subtraction. (b) Addition. (c) Addition and subtraction. (d) Multiplication.

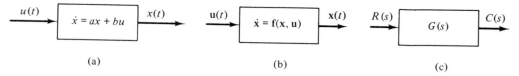

(a)

(b)

(c)

Fig. 3-6. Block-diagram representations of input–output relationships of systems.

Fig. 3-6(a) represents a system that is described by the linear differential equation

$$\dot{x}(t) = ax(t) + bu(t) \qquad (3\text{-}31)$$

Figure 3-6(b) illustrates the input–output relation of a system described by the vector-matrix differential equation

$$\dot{\mathbf{x}}(t) = \mathbf{f}[\mathbf{x}(t), \mathbf{u}(t)] \qquad (3\text{-}32)$$

where $\mathbf{x}(t)$ is an $n \times 1$ vector and $\mathbf{u}(t)$ is an $r \times 1$ vector. As another example, Fig. 3-6(c) shows a block diagram which represents the transfer function of a linear system; that is,

$$C(s) = G(s)R(s) \qquad (3\text{-}33)$$

where $G(s)$ is the transfer function.

Figure 3-7 shows the block diagram of a linear feedback control system. The following terminology often used in control systems is defined with reference to the block diagram:

$$r(t), R(s) = \text{reference input}$$
$$c(t), C(s) = \text{output signal (controlled variable)}$$
$$b(t), B(s) = \text{feedback signal}$$
$$\epsilon(t), \mathcal{E}(s) = \text{actuating signal}$$
$$e(t), E(s) = R(s) - C(s) = \text{error signal}$$
$$G(s) = \frac{C(s)}{\mathcal{E}(s)} = \text{open-loop transfer function or forward-}$$
$$\text{path transfer function}$$
$$M(s) = \frac{C(s)}{R(s)} = \text{closed-loop transfer function}$$
$$H(s) = \text{feedback-path transfer function}$$
$$G(s)H(s) = \text{loop transfer function}$$

The closed-loop transfer function, $M(s) = C(s)/R(s)$, can be expressed as a function of $G(s)$ and $H(s)$. From Fig. 3-7 we write

$$C(s) = G(s)\mathcal{E}(s) \qquad (3\text{-}34)$$

and

$$B(s) = H(s)C(s) \qquad (3\text{-}35)$$

The actuating signal is written

$$\mathcal{E}(s) = R(s) - B(s) \qquad (3\text{-}36)$$

Substituting Eq. (3-36) into Eq. (3-34) yields

$$C(s) = G(s)R(s) - G(s)B(s) \qquad (3\text{-}37)$$

Substituting Eq. (3-35) into Eq. (3-37) gives

$$C(s) = G(s)R(s) - G(s)H(s)C(s) \qquad (3\text{-}38)$$

Solving $C(s)$ from the last equation, the closed-loop transfer function of the

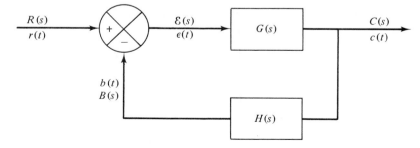

Fig. 3-7. Basic block diagram of a feedback control system.

system is given by

$$M(s) = \frac{C(s)}{R(s)} = \frac{G(s)}{1 + G(s)H(s)} \tag{3-39}$$

In general, a practical control system may contain many feedback loops, and the evaluation of the transfer function from the block diagram by means of the algebraic method described above may be tedious. In principle at least, the block diagram of a system with one input and one output can always be reduced to the basic single-loop form of Fig. 3-7. However, the steps involved in the reduction process may again be quite involved. We shall show later that the transfer function of any linear system can be obtained directly from its block diagram by use of the signal-flow-graph gain formula.

Block Diagram and Transfer Function of Multivariable Systems

A *multivariable system* is defined as one that has a multiple number of inputs and outputs. Two block-diagram representations of a multiple-variable system with p inputs and q outputs are shown in Fig. 3-8(a) and (b). In Fig. 3-8(a) the individual input and output signals are designated, whereas in the block diagram of Fig. 3-8(b), the multiplicity of the inputs and outputs is denoted by vectors. The case of Fig. 3-8(b) is preferable in practice because of its simplicity.

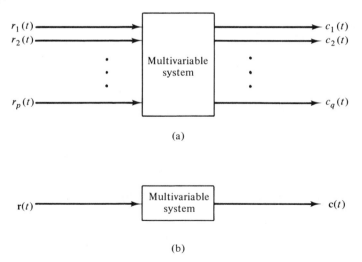

(a)

(b)

Fig. 3-8. Block-diagram representations of a multivariable system.

Figure 3-9 shows the block diagram of a multivariable feedback control system. The transfer function relationship between the input and the output of the system is obtained by using matrix algebra:

$$\mathbf{C}(s) = \mathbf{G}(s)\boldsymbol{\mathcal{E}}(s) \tag{3-40}$$

$$\boldsymbol{\mathcal{E}}(s) = \mathbf{R}(s) - \mathbf{B}(s) \tag{3-41}$$

$$\mathbf{B}(s) = \mathbf{H}(s)\mathbf{C}(s) \tag{3-42}$$

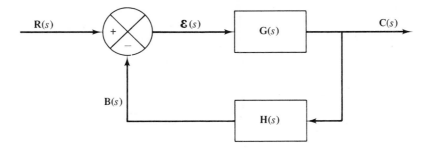

Fig. 3-9. Block diagram of a multivariable feedback control system.

Substituting Eq. (3-42) into Eq. (3-41) and then from Eq. (3-41) into Eq. (3-40) yields

$$\mathbf{C}(s) = \mathbf{G}(s)\mathbf{R}(s) - \mathbf{G}(s)\mathbf{H}(s)\mathbf{C}(s) \tag{3-43}$$

Solving for $\mathbf{C}(s)$ from Eq. (3-43) gives

$$\mathbf{C}(s) = [\mathbf{I} + \mathbf{G}(s)\mathbf{H}(s)]^{-1}\mathbf{G}(s)\mathbf{R}(s) \tag{3-44}$$

provided that $\mathbf{I} + \mathbf{G}(s)\mathbf{H}(s)$ is nonsingular.

It should be mentioned that although the development of the input–output relationship here is similar to that of the single input–output case, in the present situation it is improper to speak of the ratio $C(s)/R(s)$ since $\mathbf{C}(s)$ and $\mathbf{R}(s)$ are matrices. However, it is still possible to define the closed-loop transfer matrix as

$$\mathbf{M}(s) = [\mathbf{I} + \mathbf{G}(s)\mathbf{H}(s)]^{-1}\mathbf{G}(s) \tag{3-45}$$

Then Eq. (3-44) is written

$$\mathbf{C}(s) = \mathbf{M}(s)\mathbf{R}(s) \tag{3-46}$$

EXAMPLE 3-2   Consider that the forward-path transfer function matrix and the feedback-path transfer function matrix of the system shown in Fig. 3-9 are

$$\mathbf{G}(s) = \begin{bmatrix} \dfrac{1}{s+1} & -\dfrac{1}{s} \\ 2 & \dfrac{1}{s+2} \end{bmatrix} \tag{3-47}$$

and

$$\mathbf{H}(s) = \begin{bmatrix} 1 & 0 \\ 0 & 1 \end{bmatrix}$$

respectively.

The closed-loop transfer matrix of the system is given by Eq. (3-45) and is evaluated as follows:

$$\mathbf{I} + \mathbf{G}(s)\mathbf{H}(s) = \begin{bmatrix} 1 + \dfrac{1}{s+1} & -\dfrac{1}{s} \\ 2 & 1 + \dfrac{1}{s+2} \end{bmatrix} = \begin{bmatrix} \dfrac{s+2}{s+1} & -\dfrac{1}{s} \\ 2 & \dfrac{s+3}{s+2} \end{bmatrix} \tag{3-48}$$

The closed-loop transfer matrix is

$$\mathbf{M}(s) = [\mathbf{I} + \mathbf{G}(s)\mathbf{H}(s)]^{-1}\mathbf{G}(s) = \frac{1}{\Delta} \begin{bmatrix} \dfrac{s+3}{s+2} & \dfrac{1}{s} \\ -2 & \dfrac{s+2}{s+1} \end{bmatrix} \begin{bmatrix} \dfrac{1}{s+1} & -\dfrac{1}{s} \\ 2 & \dfrac{1}{s+2} \end{bmatrix} \tag{3-49}$$

where

$$\Delta = \frac{s+2}{s+1}\frac{s+3}{s+2} + \frac{2}{s} = \frac{s^2+5s+2}{s(s+1)} \tag{3-50}$$

Thus

$$\mathbf{M}(s) = \frac{s(s+1)}{s^2+5s+2} \begin{bmatrix} \dfrac{3s^2+9s+4}{s(s+1)(s+2)} & -\dfrac{1}{s} \\ 2 & \dfrac{3s+2}{s(s+1)} \end{bmatrix} \tag{3-51}$$

3.5 Signal Flow Graphs[2]

A signal flow graph may be regarded as a simplified notation for a block diagram, although it was originally introduced by S. J. Mason[2] as a cause-and-effect representation of linear systems. In general, besides the difference in the physical appearances of the signal flow graph and the block diagram, we may regard the signal flow graph to be constrained by more rigid mathematical relationships, whereas the rules of using the block-diagram notation are far more flexible and less stringent.

A signal flow graph may be defined as a graphical means of portraying the input–output relationships between the variables of a set of linear algebraic equations.

Consider that a linear system is described by the set of N algebraic equations

$$y_j = \sum_{k=1}^{N} a_{kj} y_k \qquad j = 1, 2, \ldots, N \tag{3-52}$$

It should be pointed out that these N equations are written in the form of cause-and-effect relations:

$$j\text{th effect} = \sum_{k=1}^{N} (\text{gain from } k \text{ to } j)(k\text{th cause}) \tag{3-53}$$

or simply

$$\text{output} = \sum (\text{gain})(\text{input}) \tag{3-54}$$

This is the single most important axiom in the construction of the set of algebriac equations from which a signal flow graph is drawn.

In the case when a system is represented by a set of integrodifferential equations, we must first transform them into Laplace transform equations and then rearrange the latter into the form of Eq. (3-52), or

$$Y_j(s) = \sum_{k=1}^{N} G_{kj}(s) Y_k(s) \qquad j = 1, 2, \ldots, N \tag{3-55}$$

When constructing a signal flow graph, junction points or *nodes* are used to represent the variables y_j and y_k. The nodes are connected together by line segments called *branches*, according to the cause-and-effect equations. The branches have associated branch gains and directions. A signal can transmit through a branch only in the direction of the arrow. In general, given a set of equations such as those of Eq. (3-52) or Eq. (3-55), the construction of the signal flow graph is basically a matter of following through the cause-and-effect relations relating each variable in terms of itself and the other variables. For instance, consider that a linear system is represented by the simple equation

$$y_2 = a_{12}y_1 \qquad (3\text{-}56)$$

where y_1 is the input variable, y_2 the output variable, and a_{12} the gain or trans-mittance between the two variables. The signal-flow-graph representation of Eq. (3-56) is shown in Fig. 3-10. Notice that the branch directing from node y_1 to node y_2 expresses the depend-ence of y_2 upon y_1. It should be reiterated that Eq. (3-56) and Fig. 3-10 represent only the dependence of the out-put variable upon the input variable, not the reverse. An important consideration in the application of signal flow graphs is that the branch between the two nodes y_1

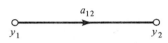

Fig. 3-10. Signal flow graph of y_2 = $a_{12}y_1$.

and y_2 should be integrated as a unilateral amplifier with gain a_{12}, so that when a signal of one unit is applied at the input y_1, the signal is multiplied by a_{12} and a signal of strength a_{12} is delivered at node y_2. Although algebraically Eq. (3-56) can be rewritten

$$y_1 = \frac{1}{a_{12}}y_2 \qquad (3\text{-}57)$$

the signal flow graph of Fig. 3-10 does not imply this relationship. If Eq. (3-57) is valid as a cause-and-effect equation in the physical sense, a new signal flow graph must be drawn.

As another illustrative example, consider the following set of algebraic equations:

$$
\begin{aligned}
y_2 &= a_{12}y_1 + a_{32}y_3 \\
y_3 &= a_{23}y_2 + a_{43}y_4 \\
y_4 &= a_{24}y_2 + a_{34}y_3 + a_{44}y_4 \\
y_5 &= a_{25}y_2 + a_{45}y_4
\end{aligned}
\qquad (3\text{-}58)
$$

The signal flow graph for these equations is constructed step by step as shown in Fig. 3-11, although the indicated sequence of steps is not unique. The nodes representing the variables y_1, y_2, y_3, y_4, and y_5 are located in order from left to right. The first equation states that y_2 depends upon two signals, $a_{12}y_1$ and $a_{32}y_3$; the signal flow graph representing this equation is drawn as shown in Fig. 3-11(a). The second equation states that y_3 depends upon $a_{23}y_2$ and $a_{43}y_4$; therefore, on the signal flow graph of Fig. 3-11(a), a branch of gain a_{23} is drawn from node y_2 to y_3, and a branch of gain a_{43} is drawn from y_4 to y_3 with the directions of the branches indicated by the arrows, as shown in Fig. 3-11(b). Similarly, with the consideration of the third equation, Fig. 3-11(c) is obtained. Finally, when the last equation of Eq. (3-58) is portrayed, the complete signal flow graph is shown in Fig. 3-11(d). The branch that begins from the node y_4

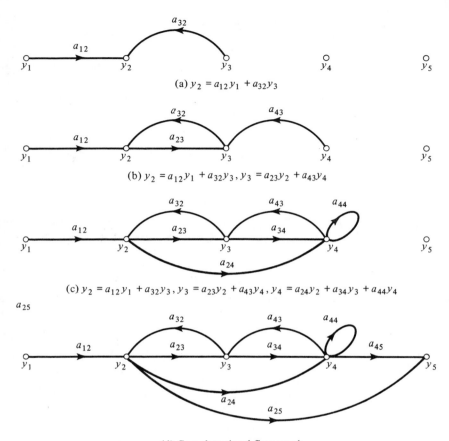

(a) $y_2 = a_{12}y_1 + a_{32}y_3$

(b) $y_2 = a_{12}y_1 + a_{32}y_3, y_3 = a_{23}y_2 + a_{43}y_4$

(c) $y_2 = a_{12}y_1 + a_{32}y_3, y_3 = a_{23}y_2 + a_{43}y_4, y_4 = a_{24}y_2 + a_{34}y_3 + a_{44}y_4$

(d) Complete signal flow graph

Fig. 3-11. Step-by-step construction of the signal flow graph for Eq.
(3-58). (a) $y_2 = a_{12}y_1 + a_{32}y_3$. (b) $y_2 = a_{12}y_1 + a_{32}y_3, y_3 = a_{23}y_2$
$+ a_{43}y_4$. (c) $y_2 = a_{12}y_1 + a_{32}y_3, y_3 = a_{23}y_2 + a_{43}y_4, y_4 = a_{24}y_2$
$+ a_{34}y_3 + a_{44}y_4$. (d) Complete signal flow graph.

and ends at y_4 is called a loop, and with a gain a_{44}, represents the dependence
of y_4 upon itself.

3.6 Summary of Basic Properties of Signal Flow Graphs

At this point it is best to summarize some of the important properties of the
signal flow graph.

1. A signal flow graph applies only to linear systems.
2. The equations based on which a signal flow graph is drawn must be
   algebraic equations in the form of effects as functions of causes.
3. Nodes are used to represent variables. Normally, the nodes are

arranged from left to right, following a succession of causes and effects through the system.

4. Signals travel along branches only in the direction described by the arrows of the branches.
5. The branch directing from node y_k to y_j represents the dependence of the variable y_j upon y_k, but not the reverse.
6. A signal y_k traveling along a branch between nodes y_k and y_j is multiplied by the gain of the branch, a_{kj}, so that a signal $a_{kj}y_k$ is delivered at node y_j.

3.7 Definitions for Signal Flow Graphs

In addition to the branches and nodes defined earlier for the signal flow graph, the following terms are useful for the purposes of identification and reference.

Input node (source). An input node is a node that has only outgoing branches. (Example: node y_1 in Fig. 3-11.)

Output node (sink). An output node is a node which has only incoming branches. (Example: node y_5 in Fig. 3-11.) However, this condition is not always readily met by an output node. For instance, the signal flow graph shown in Fig. 3-12(a) does not have any node that satisfies the condition of an output node. However, it may be necessary to regard nodes y_2 and/or y_3 as output nodes. In order to meet the definition requirement, we may simply introduce branches with unity gains and additional variables y_2 and y_3, as shown in Fig. 3-12(b). Notice that in the modified signal flow graph it is equivalent that the

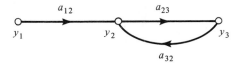

(a) Original signal flow graph

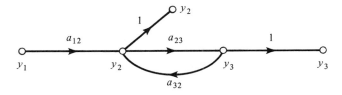

(b) Modified signal flow graph

Fig. 3-12. Modification of a signal flow graph so that y_2 and y_3 satisfy the requirement as output nodes. (a) Original signal flow graph. (b) Modified signal flow graph.

equations $y_2 = y_2$ and $y_3 = y_3$ are added. In general, we can state that any noninput node of a signal flow graph can always be made an output node by the aforementioned operation. However, we cannot convert a noninput node into an input node by using a similar operation. For instance, node y_2 of the signal flow graph of Fig. 3-12(a) does not satisfy the definition of an input node. If we attempt to convert it into an input node by adding an incoming branch of unity gain from another identical node y_2, the signal flow graph of Fig. 3-13 would result. However, the equation that portrays the relationship at node y_2 now reads

$$y_2 = y_2 + a_{12}y_1 + a_{32}y_3 \tag{3-59}$$

which is different from the original equation, as written from Fig. 3-12(a),

$$y_2 = a_{12}y_1 + a_{32}y_3 \tag{3-60}$$

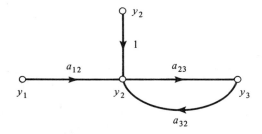

Fig. 3-13. Erroneous way to make the node y_2 an input node.

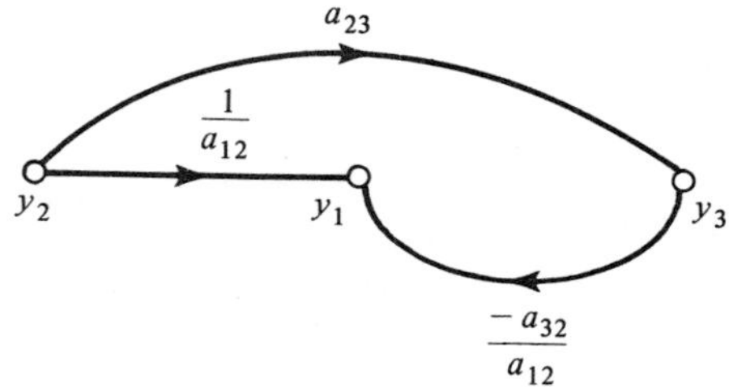

Fig. 3-14. Signal flow graph with y_2 as an input node.

Since the only proper way that a signal flow graph can be drawn is from a set of cause-and-effect equations, that is, with the causes on the right side of the equation and the effects on the left side of the equation, we must transfer y_2 to the right side of Eq. (3-60) if it were to be an input. Rearranging Eq. (3-60), the two equations originally for the signal flow graph of Fig. 3-12 now become

$$y_1 = \frac{1}{a_{12}}y_2 - \frac{a_{32}}{a_{12}}y_3 \tag{3-61}$$

$$y_3 = a_{23}y_2 \tag{3-62}$$

The signal flow graph for these two equations is shown in Fig. 3-14, with y_2 as an input node.

Path. A path is any collection of continuous succession of branches traversed in the same direction. The definition of a path is entirely general since it does not prevent any node to be traversed more than once. Therefore, as simple as the signal flow graph of Fig. 3-12(a) is, it may have numerous paths.

Forward path. A forward path is a path that starts at an input node and ends at an output node and along which no node is traversed more than once. For example, in the signal flow graph of Fig. 3-11(d), y_1 is the input node, and there are four possible output nodes in y_2, y_3, y_4, and y_5. The forward path between y_1 and y_2 is simply the branch connected between y_1 and y_2. There are two forward paths between y_1 and y_3; one contains the branches from y_1 to y_2 to y_3, and the other one contains the branches from y_1 to y_2 to y_4 (through the branch with gain a_{24}) and then back to y_3 (through the branch with gain a_{43}). The reader may determine the two forward paths between y_1 and y_4. Similarly, there are also two forward paths between y_1 and y_5.

Loop. A loop is a path that originates and terminates on the same node and along which no other node is encountered more than once. For example, there are four loops in the signal flow graph of Fig. 3-11(d). These are shown in Fig. 3-15.

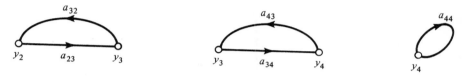

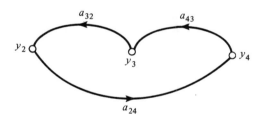

Fig. 3-15. Four loops in the signal flow graph of Fig. 3-11(d).

Path gain. The product of the branch gains encountered in traversing a path is called the path gain. For example, the path gain for the path $y_1 - y_2 - y_3 - y_4$ in Fig. 3-11(d) is $a_{12}a_{23}a_{34}$.

Forward-path gain. Forward-path gain is defined as the path gain of a forward path.

Loop gain. Loop gain is defined as the path gain of a loop. For example, the loop gain of the loop $y_2 - y_4 - y_3 - y_2$ in Fig. 3-15 is $a_{24}a_{43}a_{32}$.

3.8 Signal-Flow-Graph Algebra

Based on the properties of the signal flow graph, we can state the following manipulation and algebra of the signal flow graph.

    1. The value of the variable represented by a node is equal to the sum of all the signals entering the node. Therefore, for the signal flow

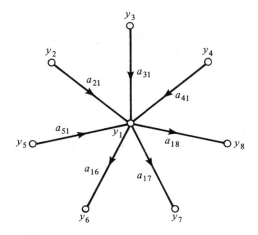

Fig. 3-16. Node as a summing point and as a transmitting point.

graph of Fig. 3-16, the value of y_1 is equal to the sum of the signals transmitted through all the incoming branches; that is,

$$y_1 = a_{21}y_2 + a_{31}y_3 + a_{41}y_4 + a_{51}y_5 \qquad (3\text{-}63)$$

2. The value of the variable represented by a node is transmitted through all branches leaving the node. In the signal flow graph of Fig. 3-16, we have

$$y_6 = a_{16}y_1$$
$$y_7 = a_{17}y_1 \qquad (3\text{-}64)$$
$$y_8 = a_{18}y_1$$

3. Parallel branches in the same direction connected between two nodes can be replaced by a single branch with gain equal to the sum of the gains of the parallel branches. An example of this case is illustrated in Fig. 3-17.

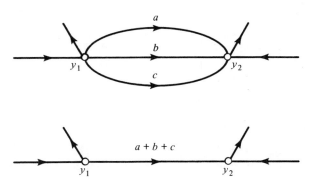

Fig. 3-17. Signal flow graph with parallel paths replaced by one with a single branch.

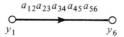

Fig. 3-18. Signal flow graph with cascaded unidirectional branches replaced by a single branch.

4. A series connection of unidirectional branches, as shown in Fig. 3-18, can be replaced by a single branch with gain equal to the product of the branch gains.

5. *Signal flow graph of a feedback control system.* Figure 3-19 shows the signal flow graph of a feedback control system whose block diagram is given in Fig. 3-7. Therefore, the signal flow graph may be regarded as a simplified notation for the block diagram. Writing the equations for the signals at the nodes $\mathcal{E}(s)$ and $C(s)$, we have

$$\mathcal{E}(s) = R(s) - H(s)C(s) \tag{3-65}$$

and

$$C(s) = G(s)\mathcal{E}(s) \tag{3-66}$$

The closed-loop transfer function is obtained from these two equations,

$$\frac{C(s)}{R(s)} = \frac{G(s)}{1 + G(s)H(s)} \tag{3-67}$$

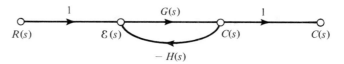

Fig. 3-19. Signal flow graph of a feedback control system.

For complex signal flow graphs we do not need to rely on algebraic manipulation to determine the input–output relation. In Section 3.10 a general gain formula will be introduced which allows the determination of the gain between an input node and an output node by mere inspection.

3.9 Examples of the Construction of Signal Flow Graphs

It was emphasized earlier that the construction of a signal flow graph of a physical system depends upon first writing the equations of the system in the cause-and-effect form. In this section we shall give two simple illustrative examples.

Owing to the lack of background on systems at this early stage, we are using two electric networks as examples. More elaborate cases will be discussed in Chapter 5, where the modeling of systems is formally covered.

EXAMPLE 3-3   The passive network shown in Fig. 3-20(a) is considered to consist of R, L, and C elements so that the network elements can be represented by impedance functions, $Z(s)$, and admittance functions, $Y(s)$. The Laplace transform of the input voltage is denoted by $E_{in}(s)$ and that of the output voltage is $E_o(s)$. In this case it is more convenient to use the branch currents and node voltages designated as shown in Fig. 3-20(a). Then one set of independent equations representing cause-and-effect relation is

$$I_1(s) = [E_{in}(s) - E_2(s)]Y_1(s) \tag{3-68}$$

$$E_2(s) = [I_1(s) - I_3(s)]Z_2(s) \tag{3-69}$$

$$I_3(s) = [E_2(s) - E_o(s)]Y_3(s) \tag{3-70}$$

$$E_o(s) = Z_4(s)I_3(s) \tag{3-71}$$

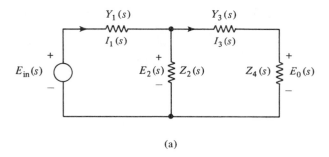

(a)

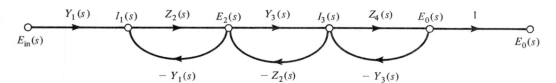

Fig. 3-20. (a) Passive ladder network. (b) A signal flow graph for the network.

With the variables $E_{in}(s)$, $I_1(s)$, $E_2(s)$, $I_3(s)$, and $E_o(s)$ arranged from left to right in order, the signal flow graph of the network is constructed as shown in Fig. 3-20(b).

It is noteworthy that in the case of network analysis, the cause-and-effect equations that are most convenient for the construction of a signal flow graph are neither the loop equations nor the node equations. Of course, this does not mean that we cannot construct a signal flow graph using the loop or the node equations. For instance, in Fig. 3-20(a), if we let $I_1(s)$ and $I_3(s)$ be the loop currents of the two loops, the loop equations are

$$E_{in}(s) = [Z_1(s) + Z_2(s)]I_1(s) - Z_2(s)I_3(s) \tag{3-72}$$

$$0 = -Z_2(s)I_1(s) + [Z_2(s) + Z_3(s) + Z_4(s)]I_3(s) \tag{3-73}$$

$$E_o(s) = Z_4(s)I_3(s) \tag{3-74}$$

However, Eqs. (3-72) and (3-73) should be rearranged, since only effect variables can appear on the left-hand sides of the equations. Therefore, solving for $I_1(s)$ from Eq. (3-72) and $I_3(s)$ from Eq. (3-73), we get

$$I_1(s) = \frac{1}{Z_1(s) + Z_2(s)}E_{in}(s) + \frac{Z_2(s)}{Z_1(s) + Z_2(s)}I_3(s) \tag{3-75}$$

$$I_3(s) = \frac{Z_2(s)}{Z_2(s) + Z_3(s) + Z_4(s)}I_1(s) \tag{3-76}$$

Now, Eqs. (3-74), (3-75), and (3-76) are in the form of cause-and-effect equations. The signal flow graph portraying these equations is drawn as shown in Fig. 3-21. This exercise also illustrates that the signal flow graph of a system is not unique.

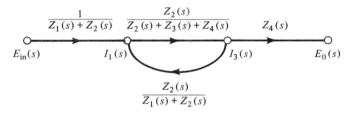

Fig. 3-21. Signal flow graph of the network in Fig. 3-20(a) using the loop equations as a starting point.

EXAMPLE 3-4   Let us consider the *RLC* network shown in Fig. 3-22(a). We shall define the current $i(t)$ and the voltage $e_c(t)$ as the dependent variables of the network. Writing the voltage across the inductance and the current in the capacitor, we have the following differential equations:

$$L\frac{di(t)}{dt} = e_1(t) - Ri(t) - e_c(t) \tag{3-77}$$

$$C\frac{de_c(t)}{dt} = i(t) \tag{3-78}$$

However, we cannot construct a signal flow graph using these two equations since they are differential equations. In order to arrive at algebraic equations, we divide Eqs. (3-77) and (3-78) by L and C, respectively. When we take the Laplace transform, we have

$$sI(s) = i(0+) + \frac{1}{L}E_1(s) - \frac{R}{L}I(s) - \frac{1}{L}E_c(s) \tag{3-79}$$

$$sE_c(s) = e_c(0+) + \frac{1}{C}I(s) \tag{3-80}$$

where $i(0+)$ is the initial current and $e_c(0+)$ is the initial voltage at $t = 0+$. In these last two equations, $e_c(0+)$, $i(0+)$, and $E_1(s)$ are the input variables. There are several possible ways of constructing the signal flow graph for these equations. One way is to solve for $I(s)$ from Eq. (3-79) and $E_c(s)$ from Eq. (3-80); we get

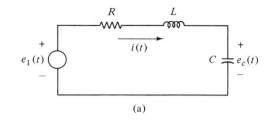

(a)

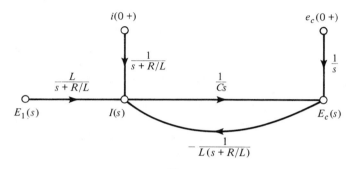

(b)

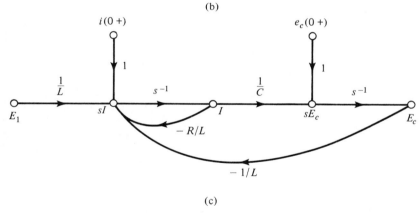

(c)

Fig. 3-22. (a) *RLC* network. (b) Signal flow graph. (c) Alternative signal flow graph.

$$I(s) = \frac{1}{s + (R/L)}i(0+) + \frac{1}{L[s + (R/L)]}E_1(s) - \frac{1}{L[s + (R/L)]}E_c(s) \qquad (3\text{-}81)$$

$$E_c(s) = \frac{1}{s}e_c(0+) + \frac{1}{Cs}I(s) \qquad (3\text{-}82)$$

The signal flow graph using the last equations is drawn as shown in Fig. 3-22(b).

The signal flow graph in Fig. 3-22(b) is of analytical value only. In other words, we can solve for $I(s)$ and $E_c(s)$ from the signal flow graph in terms of the inputs, $e_c(0+)$, $i(0+)$, and $E_1(s)$, but the value of the signal flow graph would probably end here. As an alternative, we can use Eqs. (3-79) and (3-80) directly, and define $I(s)$, $E_c(s)$, $sI(s)$, and $sE_c(s)$ as the noninput variables. These four variables are related by the equations

$$I(s) = s^{-1}[sI(s)] \qquad (3\text{-}83)$$

$$E_c(s) = s^{-1}[sE_c(s)] \qquad (3\text{-}84)$$

The significance of using s^{-1} is that it represents pure integration in the time domain. Now, a signal flow graph using Eqs. (3-79), (3-80), (3-83), and (3-84) is constructed as shown in Fig. 3-22(c). Notice that in this signal flow graph the Laplace transform variable appears only in the form of s^{-1}. Therefore, this signal flow graph may be used as a basis for analog or digital computer solution of the problem. Signal flow graphs in this form are defined in Chapter 4 as the *state diagrams*.[5]

3.10 General Gain Formula for Signal Flow Graphs[3]

Given a signal flow graph or a block diagram, it is usually a tedious task to solve for its input–output relationships by analytical means. Fortunately, there is a general gain formula available which allows the determination of the input–output relationship of a signal flow graph by mere inspection. The general gain formula is

$$M = \frac{y_{\text{out}}}{y_{\text{in}}} = \sum_{k=1}^{N} \frac{M_k \, \Delta_k}{\Delta} \qquad (3\text{-}85)$$

where

$M =$ gain between y_{in} and y_{out}

$y_{\text{out}} =$ output node variable

$y_{\text{in}} =$ input node variable

$N =$ total number of forward paths

$M_k =$ gain of the kth forward path

$$\Delta = 1 - \sum_m P_{m1} + \sum_m P_{m2} - \sum_m P_{m3} + \cdots \qquad (3\text{-}86)$$

$P_{mr} =$ gain product of the mth possible combination of r nontouching* loops

or

$\Delta = 1 -$ (sum of all individual loop gains) + (sum of gain products of all possible combinations of two nontouching loops) − (sum of the gain products of all possible combinations of three nontouching (3-87) loops) + . . .

$\Delta_k =$ the Δ for that part of the signal flow graph which is nontouching with the kth forward path

This general formula may seem formidable to use at first glance. However, the only complicated term in the gain formula is Δ; but in practice, systems having a large number of nontouching loops are rare. An error that is frequently made with regard to the gain formula is the condition under which it is valid. It must be emphasized that *the gain formula can be applied only between an input node and an output node.*

*Two parts of a signal flow graph are said to be nontouching if they do not share a common node.

EXAMPLE 3-5   Consider the signal flow graph of Fig. 3-19. We wish to find the transfer function $C(s)/R(s)$ by use of the gain formula, Eq. (3-85). The following conclusions are obtained by inspection from the signal flow graph:

1. There is only one forward path between $R(s)$ and $C(s)$, and the forward-path gain is

$$M_1 = G(s) \qquad (3\text{-}88)$$

2. There is only one loop; the loop gain is

$$P_{11} = -G(s)H(s) \qquad (3\text{-}89)$$

3. There are no nontouching loops since there is only one loop. Furthermore, the forward path is in touch with the only loop. Thus $\Delta_1 = 1$, and $\Delta = 1 - P_{11} = 1 + G(s)H(s)$.

By use of Eq. (3-85), the transfer function of the system is obtained as

$$\frac{C(s)}{R(s)} = \frac{M_1 \Delta_1}{\Delta} = \frac{G(s)}{1 + G(s)H(s)} \qquad (3\text{-}90)$$

which agrees with the result obtained in Eq. (3-67).

EXAMPLE 3-6   Consider, in Fig. 3-20(b) that the functional relation between E_{in} and E_o is to be determined by use of the general gain formula. The signal flow graph is redrawn in Fig. 3-23(a). The following conclusions are obtained by inspection from the signal flow graph:

1. There is only one forward path between E_{in} and E_o, as shown in Fig. 3-23(b). The forward-path gain is

$$M_1 = Y_1 Z_2 Y_3 Z_4 \qquad (3\text{-}91)$$

2. There are three individual loops, as shown in Fig. 3-23(c); the loop gains are

$$P_{11} = -Z_2 Y_1 \qquad (3\text{-}92)$$
$$P_{21} = -Z_2 Y_3 \qquad (3\text{-}93)$$
$$P_{31} = -Z_4 Y_3 \qquad (3\text{-}94)$$

3. There is one pair of nontouching loops, as shown in Fig. 3-23(d); the loop gains of these two loops are

$$-Z_2 Y_1 \qquad \text{and} \qquad -Z_4 Y_3$$

Thus

$$P_{12} = \text{product of gains of the first (and only) possible combination of two nontouching loops} = Z_2 Z_4 Y_1 Y_3 \qquad (3\text{-}95)$$

4. There are no three nontouching loops, four nontouching loops, and so on; thus

$$P_{m3} = 0, \quad P_{m4} = 0, \ldots$$

From Eq. (3-86),

$$\Delta = 1 - (P_{11} + P_{21} + P_{31}) + P_{12}$$
$$= 1 + Z_2 Y_1 + Z_2 Y_3 + Z_4 Y_3 + Z_2 Z_4 Y_1 Y_3 \qquad (3\text{-}96)$$

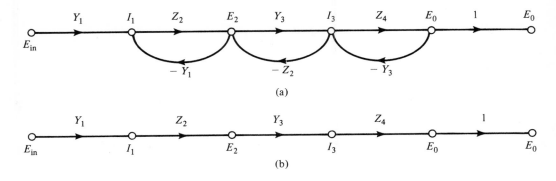

(a)

(b)

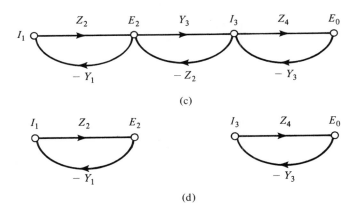

(c)

(d)

Fig. 3-23. (a) Signal flow graph of the passive network in Fig. 3-20(a). (b) Forward path between E_{in} and E_o. (c) Three individual loops. (d) Two nontouching loops.

5. All the three feedback loops are in touch with the forward path; thus

$$\Delta_1 = 1 \qquad\qquad (3\text{-}97)$$

Substituting the quantities in Eqs. (3-91) through (3-97) into Eq. (3-85), we obtain

$$\frac{E_o}{E_{in}} = \frac{M_1 \Delta_1}{\Delta} = \frac{Y_1 Y_3 Z_2 Z_4}{1 + Z_2 Y_1 + Z_2 Y_3 + Z_4 Y_3 + Z_2 Z_4 Y_1 Y_3} \qquad (3\text{-}98)$$

EXAMPLE 3-7   Consider the signal flow graph of Fig. 3-22(c). It is desired to find the relationships between I and the three inputs, E_1, $i(0+)$, and $e_c(0+)$.

Similar relationship is desired for E_c. Since the system is linear, the principle of superposition applies. The gain between one input and one output is determined by applying the gain formula to the two variables while setting the rest of the inputs to zero.

The signal flow graph is redrawn as shown in Fig. 3-24(a). Let us first consider I as the output variable. The forward paths between each inputs and I are shown in Fig. 3-24(b), (c), and (d), respectively.

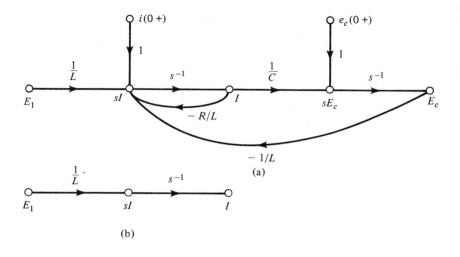

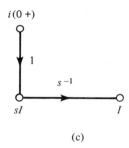

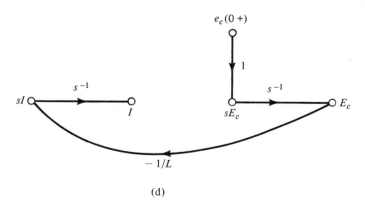

Fig. 3-24. (a) Signal flow graph of the *RLC* network in Fig. 3-22(a). (b) Forward path between E_1 and I. (c) Forward path between $i(0+)$ and I. (d) Forward path between $e_c(0+)$ and I.

The signal flow graph has two loops; the Δ is given by

$$\Delta = 1 + \frac{R}{L}s^{-1} + \frac{1}{LC}s^{-2} \qquad (3\text{-}99)$$

All the forward paths are in touch with the two loops; thus $\Delta_1 = 1$ for all cases. Considering each input separately, we have

$$\frac{I}{E_1} = \frac{(1/L)s^{-1}}{\Delta} \qquad i(0+) = 0, \qquad e_c(0+) = 0 \qquad (3\text{-}100)$$

$$\frac{I}{i(0+)} = \frac{s^{-1}}{\Delta} \qquad E_1 = 0, \qquad e_c(0+) = 0 \qquad (3\text{-}101)$$

$$\frac{I}{e_c(0+)} = \frac{-(1/L)s^{-2}}{\Delta} \qquad i(0+) = 0, \qquad E_1 = 0 \qquad (3\text{-}102)$$

When all three inputs are applied simultaneously, we write

$$I = \frac{1}{\Delta}\left[\frac{1}{L}s^{-1}E_1 + s^{-1}i(0+) - \frac{1}{L}s^{-2}e_c(0+)\right] \qquad (3\text{-}103)$$

In a similar fashion, the reader should verify that when E_c is considered as the output variable, we have

$$E_c = \frac{1}{\Delta}\left[\frac{1}{LC}s^{-2}E_1 + \frac{1}{C}s^{-2}i(0+) + s^{-1}\left(1 + \frac{R}{L}s^{-1}\right)e_c(0+)\right] \qquad (3\text{-}104)$$

Notice that the loop between the nodes sI and I is not in touch with the forward path between $e_c(0+)$ and E_c.

EXAMPLE 3-8   Consider the signal flow graph of Fig. 3-25. The following input–output relations are obtained by use of the general gain formula:

$$\frac{y_2}{y_1} = \frac{a(1 + d)\cdot}{\Delta} \qquad (3\text{-}105)$$

$$\frac{y_3}{y_1} = \frac{ae(1 + d) + abc}{\Delta} \qquad (3\text{-}106)$$

where

$$\Delta = 1 + eg + d + bcg + deg \qquad (3\text{-}107)$$

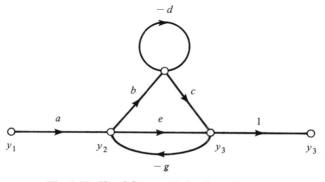

Fig. 3-25. Signal flow graph for Example 3-8.

3.11 Application of the General Gain Formula to Block Diagrams

Because of the similarity between the block diagram and the signal flow graph, the general gain formula in Eq. (3-85) can be used to determine the input–output relationships of either. In general, given a block diagram of a linear system we can apply the gain formula directly to it. However, in order to be able to identify all the loops and nontouching parts clearly, sometimes it may be helpful if an equivalent signal flow graph is drawn for a block diagram before applying the gain formula.

To illustrate how the signal flow graph and the block diagram are related, the equivalent models of a control system are shown in Fig. 3-26. Note that since a node on the signal flow graph is interpreted as a summing point of all

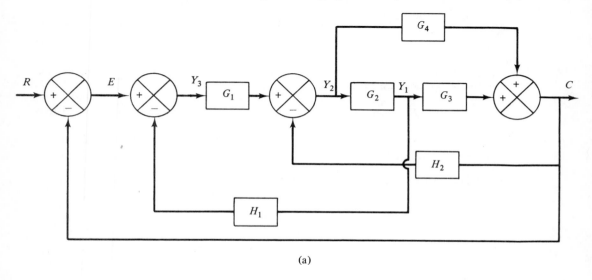

(a)

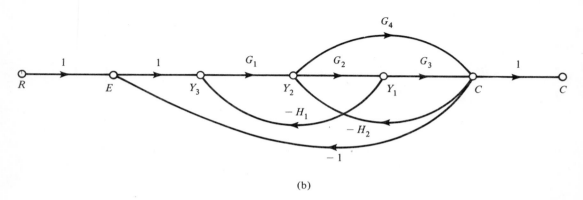

(b)

Fig. 3-26. (a) Block diagram of a control system. (b) Equivalent signal flow graph.

incoming signals to the node, the negative feedback paths in this case are represented by assigning negative gains to the feedback paths.

The closed-loop transfer function of the system is obtained by applying Eq. (3-85) to either the block diagram or the signal flow graph:

$$\frac{C(s)}{R(s)} = \frac{G_1 G_2 G_3 + G_1 G_4}{1 + G_1 G_2 H_1 + G_2 G_3 H_2 + G_1 G_2 G_3 + G_4 H_2 + G_1 G_4} \tag{3-108}$$

Similarly,

$$\frac{E(s)}{R(s)} = \frac{1 + G_1 G_2 H_1 + G_2 G_3 H_2 + G_4 H_2}{\Delta} \tag{3-109}$$

$$\frac{Y_3(s)}{R(s)} = \frac{1 + G_2 G_3 H_2 + G_4 H_2}{\Delta} \tag{3-110}$$

where

$$\Delta = 1 + G_1 G_2 H_1 + G_2 G_3 H_2 + G_1 G_2 G_3 + G_4 H_2 + G_1 G_4 \tag{3-111}$$

3.12 Transfer Functions of Discrete-Data Systems[7,8]

It is shown in Chapter 2 that the signals in a discrete-data or sampled-data system are in the form of pulse trains. Therefore, the Laplace transform and the transfer functions defined for continuous-data systems, in the s-domain, cannot be used adequately to describe these systems.

Figure 3-27(a) illustrates a linear system with transfer function $G(s)$ whose input is the output of a finite-pulsewidth sampler. As described in Section 2.8, the finite-pulsewidth sampler closes for a short duration of p seconds once every T seconds, and a typical set of input and output signals of the sampler is shown in Fig. 2-5. Since for a very small pulse duration p, as compared with the sampling period T, the finite-pulsewidth sampler can be approximated by an ideal sampler connected in cascade with a constant attenuation p, the system of Fig. 3-27(a) may be approximated by the system shown in Fig. 3-27(b).

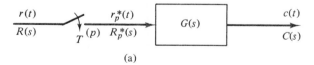

(a)

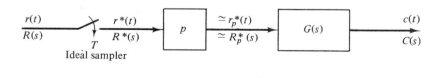

(b)

Fig. 3-27. (a) Discrete-data system with a finite-pulsewidth sampler. (b) Discrete-data system with an ideal sampler that approximates the system in (a).

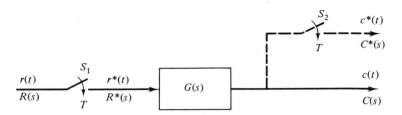

Fig. 3-28. Discrete-data system with an ideal sampler.

Normally, for convenience, it is assumed that the attenuation factor p is included in the transfer function of the process, $G(s)$. Therefore, the block diagram of Fig. 3-28 is considered as that of a typical open-loop discrete-data or sampled-data system.

There are several ways of deriving the transfer function representation of the system of Fig. 3-28. In the following we shall show two different representations of the transfer function of the system. Let us assume that $r^*(t)$, the output of the ideal sampler S_1, is a unit impulse function. This may be obtained by sampling a unit step function $u_s(t)$ once at $t = 0$ or if $r(t)$ is a unit impulse function.* Unless stated otherwise, the samplers considered in the remaining portion of this text are ideal samplers.

The output of $G(s)$ is the impulse $g(t)$. If a fictitious ideal sampler S_2, which is synchronized with S_1 and has the same sampling period as that of S_1, is placed at the output of the system as shown in Fig. 3-28, the output of the sampler S_2 may be written as

$$c^*(t) = g^*(t) = \sum_{k=0}^{\infty} g(kT)\delta(t - kT) \tag{3-112}$$

where $c(kT) = g(kT)$ is defined as the *weighting sequence* of the linear process $G(s)$. In other words, the sampled version of the impulse response or weighting function is the weighting sequence.

Taking the Laplace transform on both sides of Eq. (3-112) yields

$$C^*(s) = G^*(s) = \mathcal{L}[g^*(t)]$$
$$= \sum_{k=0}^{\infty} g(kT)e^{-kTs} \tag{3-113}$$

which is defined as the pulse transfer function of the linear process.

At this point we can summarize our findings about the description of the discrete-data system of Fig. 3-28 as follows. *When a unit impulse function is applied to the linear process, the output is simply the impulse response of the process; the impulse response is sampled by a fictitious ideal sampler S_2, and the output of the sampler is the weighting sequence of the process. The Laplace transform of the weighting sequence impulse train gives the pulse transfer function $G^*(s)$.*

*Although from a mathematical standpoint the meaning of sampling an impulse function is questionable and difficult to define, physically, we may argue that sending a pulse through a finite-pulsewidth sampler will retain the same identity of the pulse.

Once the weighting sequence of a linear system is defined, the output of the system, $c(t)$, and the sampled output, $c^*(t)$, which is due to any arbitrary input $r(t)$, can be obtained by means of the principle of superposition.

Consider that an arbitrary input $r(t)$ is applied to the system of Fig. 3-28 at $t = 0$. The output of the ideal sampler is the impulse train,

$$r^*(t) = \sum_{k=0}^{\infty} r(kT)\delta(t - kT) \tag{3-114}$$

By means of superposition, the output of the process, which is due to $r^*(t)$, is

$$c(t) = r(0)g(t) + r(T)g(t - T) + \ldots + r(kT)g(t - kT) + \ldots \tag{3-115}$$

At $t = kT$, the last equation becomes

$$c(kT) = r(0)g(kT) + r(T)g[(k - 1)T] + \ldots + r[(k - 1)T]g(T) + r(kT)g(0) \tag{3-116}$$

where it is assumed that $g(t)$ is zero for all $t < 0$, since the process is a physical system so that its output does not precede the input.

Multiplying both sides of Eq. (3-116) by e^{-kTs}, and taking the summation from $k = 0$ to $k = \infty$, we have

$$\sum_{k=0}^{\infty} c(kT)e^{-kTs} = \sum_{k=0}^{\infty} r(0)g(kT)e^{-kTs} + \sum_{k=0}^{\infty} r(T)g[(k - 1)T]e^{-kTs} + \ldots$$
$$+ \sum_{k=0}^{\infty} r[(k - 1)T]g(T)e^{-kTs} + \sum_{k=0}^{\infty} r(kT)g(0)e^{-kTs} \tag{3-117}$$

Again, using the fact that $g(t)$ is zero for negative time, Eq. (3-117) is simplified to

$$\sum_{k=0}^{\infty} c(kT)e^{-kTs} = [r(0) + r(T)e^{-Ts} + r(2T)e^{-2Ts} + \ldots] \sum_{k=0}^{\infty} g(kT)e^{-kTs} \tag{3-118}$$

or

$$\sum_{k=0}^{\infty} c(kT)e^{-kTs} = \sum_{k=0}^{\infty} r(kT)e^{-kTs} \sum_{k=0}^{\infty} g(kT)e^{-kTs} \tag{3-119}$$

Therefore, using the definition of the pulse transfer function, the last equation is written

$$C^*(s) = R^*(s)G^*(s) \tag{3-120}$$

which is the input–output transfer relationship of the discrete-data system shown in Fig. 3-28. The z-transform relationship is obtained directly from the definition of the z-transform. Since $z = e^{Ts}$, Eq. (3-119) is also written

$$\sum_{k=0}^{\infty} c(kT)z^{-k} = \sum_{k=0}^{\infty} r(kT)z^{-k} \sum_{k=0}^{\infty} g(kT)z^{-k} \tag{3-121}$$

Therefore, defining the z-*transfer function* of the process as

$$G(z) = \sum_{k=0}^{\infty} g(kT)z^{-k} \tag{3-122}$$

which implies that the z-transfer function of a linear system, $G(z)$, is the z-transform of the weighting sequence, $g(kT)$, of the system. Equation (3-121) is written

$$C(z) = R(z)G(z) \tag{3-123}$$

It is important to point out that the output of the discrete-data system is continuous with respect to time. However, the pulse transform of the output, $C^*(s)$, and the z-transform of the output, $C(z)$, specify the values of $c(t)$ only at the sampling instants. If $c(t)$ is a well-behaved function between sampling instants, $c^*(t)$ or $C(z)$ may give an accurate description of the true output $c(t)$. However, if $c(t)$ has wide fluctuations between the sampling instants, the z-transform method, which gives information only at the sampling instants, will yield misleading or inaccurate results.

The pulse transfer relation of Eq. (3-120) can also be obtained by use of the following relation between $C^*(s)$ and $C(s)$, which is given in the literature[7]:

$$C^*(s) = \frac{1}{T} \sum_{n=-\infty}^{\infty} C(s + jn\omega_s) \qquad (3\text{-}124)$$

where ω_s is the sampling frequency in radians per second and $\omega_s = 2\pi/T$.

From Fig. 3-28, the Laplace transform of the continuous-data output $c(t)$ is

$$C(s) = G(s)R^*(s) \qquad (3\text{-}125)$$

Substituting Eq. (3-125) into Eq. (3-124) gives

$$C^*(s) = \frac{1}{T} \sum_{k=-\infty}^{\infty} G(s + jn\omega_s)R^*(s + jn\omega_s) \qquad (3\text{-}126)$$

We can write

$$R^*(s + jn\omega_s) = \sum_{k=0}^{\infty} r(kT)e^{-kT(s+jn\omega_s)}$$

$$= \sum_{k=0}^{\infty} r(kT)e^{-kTs} \qquad (3\text{-}127)$$

since for integral k and n,

$$e^{-jnkT\omega_s} = e^{-j2\pi nk} = 1 \qquad (3\text{-}128)$$

Thus Eq. (3-127) becomes

$$R^*(s + jn\omega_s) = R^*(s) \qquad (3\text{-}129)$$

Using this identity, Eq. (3-126) is simplified to

$$C^*(s) = R^*(s)\frac{1}{T} \sum_{n=-\infty}^{\infty} G(s + jn\omega_s) \qquad (3\text{-}130)$$

or

$$C^*(s) = R^*(s)G^*(s) \qquad (3\text{-}131)$$

where

$$G^*(s) = \frac{1}{T} \sum_{n=-\infty}^{\infty} G(s + jn\omega_s) \qquad (3\text{-}132)$$

The transfer function in z of Eq. (3-123) can again be obtained directly from Eq. (3-131) by use of $z = e^{Ts}$.

In conclusion, we note that when the input to a linear system is sampled but the output is unsampled, the Laplace transform of the continuous output is given by

$$C(s) = G(s)R^*(s) \qquad (3\text{-}133)$$

If the continuous-data output is sampled by a sampler that is synchronized with and has the same sampling period as the input, the Laplace transform of the discrete-data output is given by

$$C^*(s) = G^*(s)R^*(s) \qquad (3\text{-}134)$$

The result in Eq. (3-133) is natural, since it is in line with the well-established transfer relation for linear time-invariant systems. The expression in Eq. (3-134) is obtained by use of Eqs. (3-124) and (3-132). However, it can be interpreted as being obtained directly from Eq. (3-133) by taking the pulse transform on both sides of the equation. In other words, in view of Eq. (3-129), we can write, from Eq. (3-133),

$$C^*(s) = [G(s)R^*(s)]^* \qquad (3\text{-}135)$$
$$= G^*(s)[R^*(s)]^*$$

where Eq. (3-134) implies that

$$[R^*(s)]^* = R^*(s) \qquad (3\text{-}136)$$

Transfer Functions of Discrete-Data Systems with Cascaded Elements

The transfer function representation of discrete-data systems with cascaded elements is slightly more involved than that for continuous-data systems, because of the variation of having or not having any samplers in between the elements. Figure 3-29 illustrates two different situations of a discrete-data system which contains two cascaded elements. In the system of Fig. 3-29(a), the two elements are separated by a sampler S_2 which is synchronized to and has

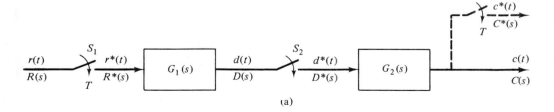

(a)

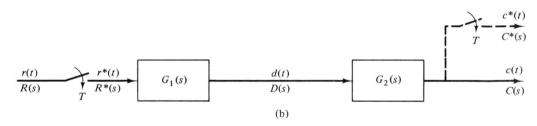

(b)

Fig. 3-29. (a) Discrete-data system with cascaded elements and sampler separates the two elements. (b) Discrete-data system with cascaded elements and no sampler in between.

the same period as the sampler S_1. The two elements with transfer functions $G_1(s)$ and $G_2(s)$ of the system in Fig. 3-29(b) are connected directly together. In discrete-data systems, it is important to distinguish these two cases when deriving the pulse transfer functions.

Let us consider first the system of Fig. 3-29(a). The output of $G_1(s)$ is written

$$D(s) = G_1(s)R^*(s) \tag{3-137}$$

and the system output is

$$C(s) = G_2(s)D^*(s) \tag{3-138}$$

Taking the pulse transform on both sides of Eq. (3-137) and substituting the result into Eq. (3-138) yields

$$C(s) = G_2(s)G_1^*(s)R^*(s) \tag{3-139}$$

Then, taking the pulse transform on both sides of the last equation gives

$$C^*(s) = G_2^*(s)G_1^*(s)R^*(s) \tag{3-140}$$

where we have made use of the relation in Eq. (3-136). The corresponding z-transform expression of the last equation is

$$C(z) = G_2(z)G_1(z)R(z) \tag{3-141}$$

We conclude that the z-transform of two linear elements separated by a sampler is equal to the product of the z-transforms of the two individual transfer functions.

The Laplace transform of the output of the system in Fig. 3-29(b) is

$$C(s) = G_1(s)G_2(s)R^*(s) \tag{3-142}$$

The pulse transform of the last equation is

$$C^*(s) = [G_1(s)G_2(s)]^*R^*(s) \tag{3-143}$$

where

$$[G_1(s)G_2(s)]^* = \frac{1}{T} \sum_{n=-\infty}^{\infty} G_1(s + jn\omega_s)G_2(s + jn\omega_s) \tag{3-144}$$

Notice that since $G_1(s)$ and $G_2(s)$ are not separated by a sampler, they have to be treated as one element when taking the pulse transform. For simplicity, we define the following notation:

$$\begin{aligned} [G_1(s)G_2(s)]^* &= G_1G_2^*(s) \\ &= G_2G_1^*(s) \end{aligned} \tag{3-145}$$

Then Eq. (3-143) becomes

$$C^*(s) = G_1G_2^*(s)R^*(s) \tag{3-146}$$

Taking the z-transform on both sides of Eq. (3-146) gives

$$C(z) = G_1G_2(z)R(z) \tag{3-147}$$

where $G_1G_2(z)$ is defined as the z-transform of the product of $G_1(s)$ and $G_2(s)$, and it should be treated as a single function.

It is important to note that, in general,

$$G_1 G_2^*(s) \neq G_1^*(s) G_2^*(s) \tag{3-148}$$

and

$$G_1 G_2(z) \neq G_1(z) G_2(z) \tag{3-149}$$

Therefore, we conclude that the z-transform of two cascaded elements with no sampler in between is equal to the z-transform of the product of the transfer functions of the two elements.

Transfer Functions of Closed-Loop Discrete-Data Systems

In this section the transfer functions of simple closed-loop discrete-data systems are derived by algebraic means. Consider the closed-loop system shown in Fig. 3-30. The output transform is

$$C(s) = G(s)E^*(s) \tag{3-150}$$

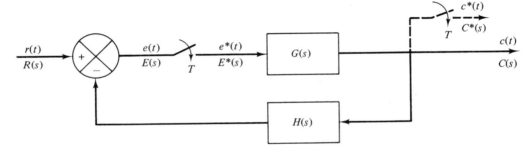

Fig. 3-30. Closed-loop discrete-data system.

The Laplace transform of the continuous error function is

$$E(s) = R(s) - H(s)C(s) \tag{3-151}$$

Substituting Eq. (3-150) into Eq. (3-151) yields

$$E(s) = R(s) - G(s)H(s)E^*(s) \tag{3-152}$$

Taking the pulse transform on both sides of the last equation and solving for $E^*(s)$ gives

$$E^*(s) = \frac{R^*(s)}{1 + GH^*(s)} \tag{3-153}$$

The output transform $C(s)$ is obtained by substituting $E^*(s)$ from Eq. (3-153) into Eq. (3-150); we have

$$C(s) = \frac{G(s)}{1 + GH^*(s)} R^*(s) \tag{3-154}$$

Now taking the pulse transform on both sides of Eq. (3-154) gives

$$C^*(s) = \frac{G^*(s)}{1 + GH^*(s)} R^*(s) \tag{3-155}$$

In this case it is possible to define the pulse transfer function between the input and the output of the closed-loop system as

$$\frac{C^*(s)}{R^*(s)} = \frac{G^*(s)}{1 + GH^*(s)} \qquad (3\text{-}156)$$

The z-transfer function of the system is

$$\frac{C(z)}{R(z)} = \frac{G(z)}{1 + GH(z)} \qquad (3\text{-}157)$$

We shall show in the following that although it is possible to define a transfer function for the closed-loop system of Fig. 3-30, in general, this may not be possible for all discrete-data systems. Let us consider the system shown in Fig. 3-31. The output transforms, $C(s)$ and $C(z)$, are derived as follows:

$$C(s) = G(s)E(s) \qquad (3\text{-}158)$$
$$E(s) = R(s) - H(s)C^*(s) \qquad (3\text{-}159)$$

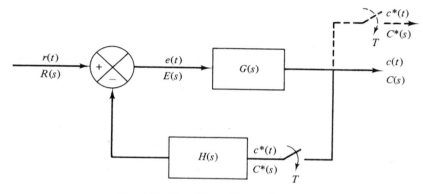

Fig. 3-31. Closed-loop discrete-data system.

Substituting Eq. (3-159) into Eq. (3-158) yields

$$C(s) = G(s)R(s) - G(s)H(s)C^*(s) \qquad (3\text{-}160)$$

Taking the pulse transform on both sides of the last equation and solving for $C^*(s)$, we have

$$C^*(s) = \frac{GR^*(s)}{1 + GH^*(s)} \qquad (3\text{-}161)$$

Note that the input and the transfer function $G(s)$ are now combined as one function, $GR^*(s)$, and the two cannot be separated. In this case we cannot define a transfer function in the form of $C^*(s)/R^*(s)$.

The z-transform of the output is determined directly from Eq. (3-161) to be

$$C(z) = \frac{GR(z)}{1 + GH(z)} \qquad (3\text{-}162)$$

where it is important to note that

$$GR(z) = \mathfrak{z}[G(s)R(s)] \qquad (3\text{-}163)$$

and

$$GH(z) = \mathfrak{z}[G(s)H(s)] \qquad (3\text{-}164)$$

To determine the transform of the continuous output, $C(s)$, we substitute $C^*(s)$ from Eq. (3-161) into Eq. (3-160). We have

$$C(s) = G(s)R(s) - \frac{G(s)H(s)}{1 + GH^*(s)}GR^*(s) \qquad (3\text{-}165)$$

Although we have been able to arrive at the input–output transfer function and transfer relation of the systems of Figs. 3-30 and 3-31 by algebraic means without difficulty, for more complex system configurations, the algebraic method may become tedious. The signal-flow-graph method is extended to the analysis of discrete-data systems; the reader may refer to the literature.[7,8]

REFERENCES

Block Diagram and Signal Flow Graphs

1. T. D. GRAYBEAL, "Block Diagram Network Transformation," *Elec. Eng.*, Vol. 70, pp. 985–990, 1951.

2. S. J. MASON, "Feedback Theory—Some Properties of Signal Flow Graphs," *Proc. IRE*, Vol. 41, No. 9, pp. 1144–1156, Sept. 1953.

3. S. J. MASON, "Feedback Theory—Further Properties of Signal Flow Graphs," *Proc. IRE*, Vol. 44, No. 7, pp. 920–926, July 1956.

4. L. P. A. ROBICHAUD, M. BOISVERT, and J. ROBERT, *Signal Flow Graphs and Applications*, Prentice-Hall, Inc., Englewood Cliffs, N.J., 1962.

5. B. C. KUO, *Linear Networks and Systems*, McGraw-Hill Book Company, New York, 1967.

6. N. AHMED, "On Obtaining Transfer Functions from Gain-Function Derivatives," *IEEE Trans. Automatic Control*, Vol. AC-12, p. 229, Apr. 1967.

Signal Flow Graphs of Sampled-Data Systems

7. B. C. KUO, *Analysis and Synthesis of Sampled-Data Control Systems*, Prentice-Hall, Inc., Englewood Cliffs, N.J., 1963.

8. B. C. KUO, *Discrete Data Control Systems*, Science-Tech, Box 2277, Station A, Champaign, Illinois, 1970.

PROBLEMS

3.1. The following differential equations represent linear time-invariant systems, where $r(t)$ denotes the input and $c(t)$ denotes the output. Find the transfer function of each of the systems.

(a) $\dfrac{d^3c(t)}{dt^3} + 3\dfrac{d^2c(t)}{dt^2} + 4\dfrac{dc(t)}{dt} + c(t) = 2\dfrac{dr(t)}{dt} + r(t)$

(b) $\dfrac{d^2c(t)}{dt^2} + 10\dfrac{dc(t)}{dt} + 2c(t) = r(t-2)$

3.2. The block diagram of a multivariable feedback control system is shown in Fig. P3-2. The transfer function matrices of the system are

$$G(s) = \begin{bmatrix} \dfrac{1}{s} & \dfrac{1}{s+2} \\[2mm] 5 & \dfrac{1}{s+1} \end{bmatrix}$$

$$H(s) = \begin{bmatrix} 1 & 0 \\ 0 & 1 \end{bmatrix}$$

Find the closed-loop transfer function matrix for the system.

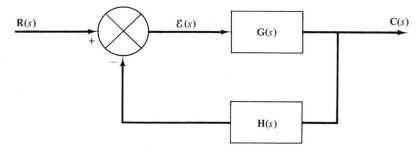

Figure P3-2.

3.3. A multivariable system with two inputs and two outputs is shown in Fig. P3-3. Determine the following transfer function relationships:

$$\left.\frac{C_1(s)}{R_1(s)}\right|_{R_2=0} \qquad \left.\frac{C_2(s)}{R_1(s)}\right|_{R_2=0} \qquad \left.\frac{C_1(s)}{R_2(s)}\right|_{R_1=0} \qquad \left.\frac{C_2(s)}{R_2(s)}\right|_{R_1=0}$$

Write the transfer function relation of the system in the form

$$C(s) = G(s)R(s)$$

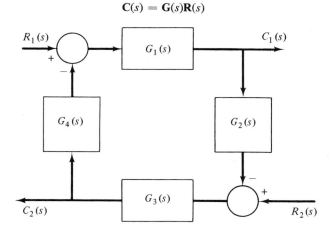

Figure P3-3.

3.4. Draw a signal flow graph for the following set of algebraic equations:

$$3x_1 + x_2 + 5x_3 = 0$$
$$x_1 + 2x_2 - 4x_3 = 2$$
$$-x_2 - x_3 = 0$$

3.5. Draw an equivalent signal flow graph for the block diagram in Fig. P3-5. Find the transfer function $C(s)/R(s)$.

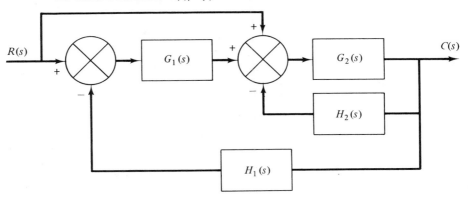

Figure P3-5.

3.6. Find the gains, y_6/y_1, y_3/y_1, and y_5/y_2 for the signal flow graph shown in Fig. P3-6.

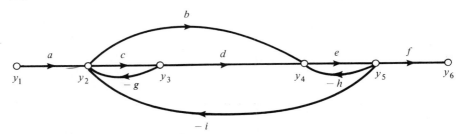

Figure P3-6.

3.7. Find the gains y_5/y_1 and y_2/y_1 for the signal flow graph shown in Fig. P3-7.

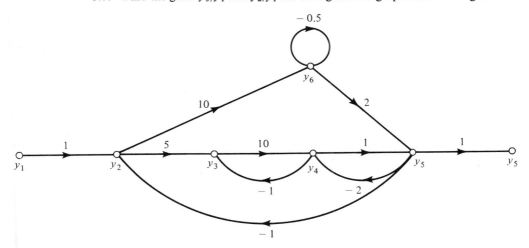

Figure P3-7.

3.8. In the circuit of Fig. P3-8, $e_s(t)$, $e_d(t)$, and $i_s(t)$ are ideal sources. Find the value of a so that the voltage $e_0(t)$ is not affected by the source $e_d(t)$.

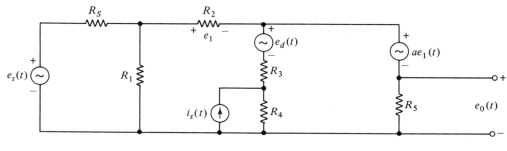

Figure P3-8.

3.9. Are the two systems shown in Fig. P3-9(a) and (b) equivalent? Explain.

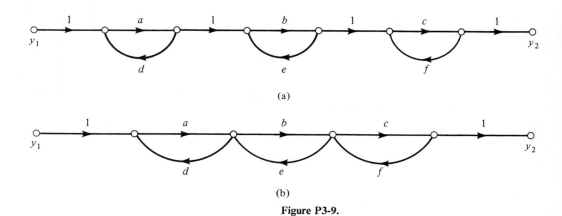

Figure P3-9.

3.10. Given the signal flow graph of Fig. P3-10(a) and the transfer functions G_1, G_2, G_3, G_4, and G_5, find the transfer functions G_A, G_B, and G_C so that the three systems shown in Fig. P3-10 are all equivalent.

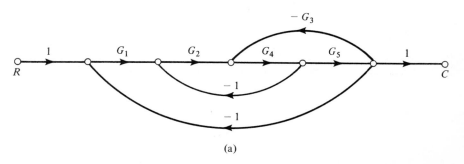

Figure P3-10.

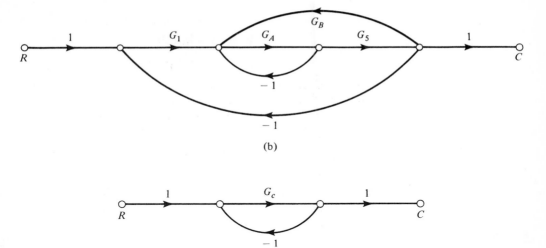

(b)

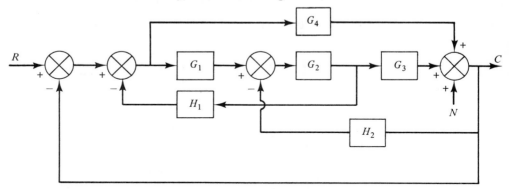

(c)

Figure P3-10. (Cont.)

3.11. Construct an equivalent signal flow graph for the block diagram of Fig. P3-11.
(a) Evaluate the transfer function C/R when $N = 0$. (b) Determine the relation
among the transfer functions G_1, G_2, G_3, G_4, H_1, and H_2 so that the output C
is not affected by the disturbance signal N.

Figure P3-11.

3.12. A multivariable system is described by the following matrix transfer function
relations:

$$\mathbf{C}(s) = \mathbf{G}(s)\mathbf{\mathcal{E}}(s)$$
$$\mathbf{\mathcal{E}}(s) = \mathbf{R}(s) - \mathbf{H}(s)\mathbf{C}(s)$$

where

$$\mathbf{C}(s) = \begin{bmatrix} C_1(s) \\ C_2(s) \end{bmatrix} \qquad \mathbf{R}(s) = \begin{bmatrix} R_1(s) \\ R_2(s) \end{bmatrix}$$

$$\mathbf{G}(s) = \begin{bmatrix} \dfrac{1}{s} & \dfrac{5}{s+1} \\ 1 & \dfrac{1}{s} \end{bmatrix} \qquad \mathbf{H}(s) = \begin{bmatrix} 1 & 0 \\ 0 & 0 \end{bmatrix}$$

(a) Derive the closed-loop transfer function relationship

$$C(s) = M(s)R(s)$$

by using

$$M(s) = [I + G(s)H(s)]^{-1}G(s)$$

(b) Draw a signal flow graph for the system and find $M(s)$ from the signal flow graph using Mason's gain formula.

3.13. Find the transfer function relations $C(s)/R(s)$ and $C(s)/E(s)$ for the system shown in Fig. P3-13.

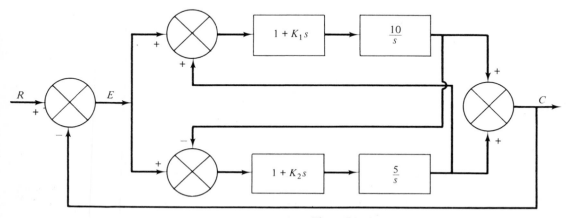

Figure P3-13.

3.14. Find the transfer function $C(z)/R(z)$ of the discrete-data system shown in Fig. P3-14. The sampling period is 1 sec.

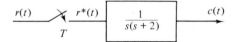

Figure P3-14.

3.15. Find the z-transfer functions $C(z)/R(z)$ of the discrete-data systems shown in Fig. P3-15.

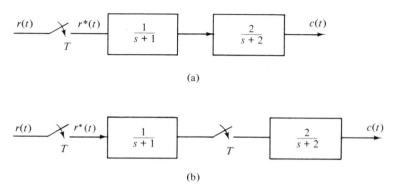

(a)

(b)

Figure P3-15.

4

State-Variable Characterization

of Dynamic Systems

4.1 Introduction to the State Concept

In Chapter 3 the classical methods of describing a linear system by transfer function, impulse response, block diagram, and signal flow graph have been presented. An important feature of this type of representation is that the system dynamics are described by the input–output relations. For instance, the transfer function describes the input–output relation in the Laplace transform domain. However, the transform method suffers from the disadvantage that all the initial conditions of the system are neglected. Therefore, when one is interested in a time-domain solution, which depends to a great deal on the past history of the system, the transfer function does not carry all the necessary information. Transfer function is valuable for frequency-domain analysis and design, as well as for stability studies. The greatest advantage of transfer function is in its compactness and the ease that we can obtain qualitative information on the system from the poles and zeros of the transfer function.

An alternative to the transfer function method of describing a linear system is the *state-variable method*. The state-variable representation is not limited to linear systems and time-invariant systems. It can be applied to nonlinear as well as time-varying systems.

The state-variable method is often referred to as a modern approach. However, in reality, the state equations are simply first-order differential equations, which have been used for the characterization of dynamic systems for many years by physicists and mathematicians.

To begin with the state-variable approach, we should first begin by defining the state of a system. As the word implies, the *state* of a system refers to the *past*, *present*, and *future* conditions of the system. It is interesting to note that an easily understood example is the "State of the Union" speech given by the President of the United States every year. In this case, the entire system encompasses all elements of the government, society, economy, and so on. In general, the state can be described by a set of numbers, a curve, an equation, or something that is more abstract in nature. From a mathematical sense it is convenient to define a set of *state variables* and *state equations* to portray systems. There are some basic ground rules regarding the definition of a state variable and what constitutes a state equation. Consider that the set of variables, $x_1(t), x_2(t), \ldots, x_n(t)$ is chosen to describe the dynamic characteristics of a system. Let us define these variables as the state variables of the system. Then these state variables must satisfy the following conditions:

1. At any time $t = t_0$, the state variables, $x_1(t_0), x_2(t_0), \ldots, x_n(t_0)$ define the *initial states* of the system at the selected initial time.
2. Once the inputs of the system for $t \geq t_0$ and the initial states defined above are specified, the state variables should completely define the future behavior of the system.

Therefore, we may define the state variables as follows:

Definition of state variables. *The state variables of a system are defined as a minimal set of variables, $x_1(t), x_2(t), \ldots, x_n(t)$ such that knowledge of these variables at any time t_0, plus information on the input excitation subsequently applied, are sufficient to determine the state of the system at any time $t > t_0$.*

One should not confuse the state variables with the outputs of a system. An output of a system is a variable that can be *measured*, but a state variable does not always, and often does not, satisfy this requirement. However, an output variable is usually defined as a function of the state variables.

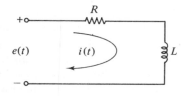

Fig. 4-1. *RL* network.

EXAMPLE 4-1   As a simple illustrative example of state variables, let us consider the *RL* network shown in Fig. 4-1. The *history* of the network is completely specified by the initial current of the inductance, $i(0+)$, at $t = 0$. At $t = 0$, a constant input voltage of amplitude E_1 is applied to the network. The loop equation of the network for $t \geq 0$ is

$$e(t) = Ri(t) + L\frac{di(t)}{dt} \tag{4-1}$$

Taking the Laplace transform on both sides of the last equation, we get

$$E(s) = \frac{E_1}{s} = (R + Ls)I(s) - Li(0+) \tag{4-2}$$

Solving for $I(s)$ from the last equation yields

$$I(s) = \frac{E_1}{s(R + Ls)} + \frac{Li(0+)}{R + Ls} \tag{4-3}$$

The current $i(t)$ for $t \geq 0$ is obtained by taking the inverse Laplace transform of both sides of Eq. (4-3). We have

$$i(t) = \frac{E_1}{R}(1 - e^{-Rt/L}) + i(0+)e^{-Rt/L} \qquad (4\text{-}4)$$

Once the current $i(t)$ is determined for $t \ge 0$, the behavior of the entire network is defined for the same time interval. Therefore, it is apparent that the current $i(t)$ in this case satisfies the basic requirements as a state variable. This is not surprising since an inductor is an electric element that stores kinetic energy, and it is the energy storage capability that holds the information on the history of the system. Similarly, it is easy to see that the voltage across a capacitor also qualifies as a state variable.

4.2   State Equations and the Dynamic Equations

The first-order differential equation of Eq. (4-1), which gives the relationship between the state variable $i(t)$ and the input $e(t)$, can be rearranged to give

$$\frac{di(t)}{dt} = \frac{-R}{L}i(t) + \frac{1}{L}e(t) \qquad (4\text{-}5)$$

This first-order differential equation is referred to as a *state equation*.

For a system with p inputs and q outputs, the system may be linear or nonlinear, time varying or time invariant, the state equations of the system are written as

$$\frac{dx_i(t)}{dt} = f_i[x_1(t), x_2(t), \ldots, x_n(t), r_1(t), r_2(t), \ldots, r_p(t)] \qquad (4\text{-}6)$$
$$i = 1, 2, \ldots, n$$

where $x_1(t), x_2(t), \ldots, x_n(t)$ are the state variables; $r_1(t), r_2(t), \ldots, r_p(t)$ are the input variables; and f_i denotes the ith functional relationship.

The outputs of the system $c_k(t), k = 1, 2, \ldots, q$, are related to the state variables and the inputs through the *output equation*,

$$c_k(t) = g_k[x_1(t), x_2(t), \ldots, x_n(t), r_1(t), r_2(t), \ldots, r_p(t)] \qquad (4\text{-}7)$$
$$k = 1, 2, \ldots, q$$

where g_k denotes the kth functional relationship. The state equations and the output equations together form the set of equations which are often called the *dynamic equations* of the system.

Notice that for the state equations, the left side of the equation should contain only the first derivatives of the state variables, while the right side should have only the state variables and the inputs.

EXAMPLE 4-2   Consider the *RLC* network shown in Fig. 4-2. Using the conventional network approach, the loop equation of the network is written

$$e(t) = Ri(t) + L\frac{di(t)}{dt} + \frac{1}{C}\int i(t)\, dt \qquad (4\text{-}8)$$

We notice that this equation is not in the form of a state equation, since the last term is a time integral. One method of writing the state equations of the network, starting with Eq. (4-8), is to let the state variables be defined as

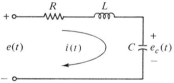

Fig. 4-2. *RLC* network.

$$x_1(t) = i(t) \tag{4-9}$$

$$x_2(t) = \int i(t)\, dt \tag{4-10}$$

Substituting the last two equations into Eq. (4-8), we have

$$e(t) = Rx_1(t) + L\frac{dx_1(t)}{dt} + \frac{1}{C}x_2(t) \tag{4-11}$$

Rearranging the terms in Eq. (4-11) and taking the time derivative on both sides of Eq. (4-10), we have the two state equations of the network,

$$\frac{dx_1(t)}{dt} = -\frac{R}{L}x_1(t) - \frac{1}{LC}x_2(t) + \frac{1}{L}e(t) \tag{4-12}$$

$$\frac{dx_2(t)}{dt} = x_1(t) \tag{4-13}$$

which are linear first-order differential equations.

We have demonstrated how the state equations of the *RLC* network may be written from the loop equations by defining the state variables in a specific way. The objective, of course, is to replace Eq. (4-8) by two first-order differential equations. An alternative approach is to start with the network and define the state variables according to the elements of the network. As stated in Section 4.1, we may assign the current through an inductor and the voltage across a capacitor as state variables. Therefore, with reference to Fig. 4-2, the state variables are defined as

$$x_1(t) = i(t) \tag{4-14}$$

$$x_2(t) = e_c(t) \tag{4-15}$$

Then, knowing that the state equations would have to contain the first derivatives of $x_1(t)$ and $x_2(t)$ on the left side of the equations, we can write the equations directly by inspection from Fig. 4-2:

$$\text{Voltage across } L: \quad L\frac{di(t)}{dt} = -Ri(t) - e_c(t) + e(t) \tag{4-16}$$

$$\text{Current in } C: \quad C\frac{de_c(t)}{dt} = i(t) \tag{4-17}$$

Using Eqs. (4-14) and (4-15), and dividing both sides of the last two equations by L and C, respectively, we have

$$\frac{dx_1(t)}{dt} = -\frac{R}{L}x_1(t) - \frac{1}{L}x_2(t) + \frac{1}{L}e(t) \tag{4-18}$$

$$\frac{dx_2(t)}{dt} = \frac{1}{C}x_1(t) \tag{4-19}$$

which are the state equations of the *RLC* network. We notice that using the two independent methods, the only difference in the results is in the definition of the second state variable $x_2(t)$. Equation (4-19) differs from Eq. (4-13) by the factor of the capacitance C.

From these two simple examples we see that for linear time-invariant systems, the state equations can generally be written as

$$\frac{dx_i(t)}{dt} = \sum_{j=1}^{n} a_{ij}x_j(t) + \sum_{k=1}^{p} b_{ik}r_k(t) \qquad i = 1, 2, \ldots, n \qquad (4\text{-}20)$$

where a_{ij} and b_{ik} are constant coefficients. The output equations are written

$$c_k(t) = \sum_{j=1}^{n} d_{kj}x_j(t) + \sum_{m=1}^{p} e_{km}r_m(t) \qquad k = 1, 2, \ldots, q \qquad (4\text{-}21)$$

where d_{kj} and e_{km} are constant coefficients.

For a linear system with time-varying parameters, the coefficients of Eqs. (4-20) and (4-21) become time dependent.

4.3 Matrix Representation of State Equations

The dynamic equations are more conveniently expressed in matrix form. Let us define the following column matrices:

$$\mathbf{x}(t) = \begin{bmatrix} x_1(t) \\ x_2(t) \\ \cdot \\ \cdot \\ \cdot \\ x_n(t) \end{bmatrix} \qquad (n \times 1) \qquad (4\text{-}22)$$

where $\mathbf{x}(t)$ is defined as the *state vector;*

$$\mathbf{r}(t) = \begin{bmatrix} r_1(t) \\ r_2(t) \\ \cdot \\ \cdot \\ \cdot \\ r_p(t) \end{bmatrix} \qquad (p \times 1) \qquad (4\text{-}23)$$

where $\mathbf{r}(t)$ is defined as the *input vector;* and

$$\mathbf{c}(t) = \begin{bmatrix} c_1(t) \\ c_2(t) \\ \cdot \\ \cdot \\ \cdot \\ c_q(t) \end{bmatrix} \qquad (q \times 1) \qquad (4\text{-}24)$$

where $\mathbf{c}(t)$ is defined as the *output vector.* Then the state equations of Eq. (4-6) can be written

$$\frac{d\mathbf{x}(t)}{dt} = \mathbf{f}[\mathbf{x}(t), \mathbf{r}(t)] \qquad (4\text{-}25)$$

where $\mathbf{f}$ denotes an $n \times 1$ column matrix that contains the functions $f_1, f_2, \ldots,$ f_n as elements, and the output equations of Eq. (4-7) become

$$\mathbf{c}(t) = \mathbf{g}[\mathbf{x}(t), \mathbf{r}(t)] \qquad (4\text{-}26)$$

where $\mathbf{g}$ denotes a $q \times 1$ column matrix that contains the functions $g_1, g_2, \ldots,$ g_q as elements.

For a linear time-invariant system, the dynamic equations are written

State equation: $\dfrac{d\mathbf{x}(t)}{dt} = \mathbf{A}\mathbf{x}(t) + \mathbf{B}\mathbf{r}(t)$ (4-27)

Output equation: $\mathbf{c}(t) = \mathbf{D}\mathbf{x}(t) + \mathbf{E}\mathbf{r}(t)$ (4-28)

where $\mathbf{A}$ is an $n \times n$ coefficient matrix given by

$$\mathbf{A} = \begin{bmatrix} a_{11} & a_{12} & \cdots & a_{1n} \\ a_{21} & a_{22} & \cdots & a_{2n} \\ \cdot & \cdot & & \cdot \\ \cdot & \cdot & & \cdot \\ \cdot & \cdot & & \cdot \\ a_{n1} & a_{n2} & \cdots & a_{nn} \end{bmatrix}$$ (4-29)

$\mathbf{B}$ is an $n \times p$ matrix given by

$$\mathbf{B} = \begin{bmatrix} b_{11} & b_{12} & \cdots & b_{1p} \\ b_{21} & b_{22} & \cdots & b_{2p} \\ \cdot & \cdot & & \cdot \\ \cdot & \cdot & & \cdot \\ \cdot & \cdot & & \cdot \\ b_{n1} & b_{n2} & \cdots & b_{np} \end{bmatrix}$$ (4-30)

$\mathbf{D}$ is a $q \times n$ matrix given by

$$\mathbf{D} = \begin{bmatrix} d_{11} & d_{12} & \cdots & d_{1n} \\ d_{21} & d_{22} & \cdots & d_{2n} \\ \cdot & \cdot & & \cdot \\ \cdot & \cdot & & \cdot \\ \cdot & \cdot & & \cdot \\ d_{q1} & d_{q2} & \cdots & d_{qn} \end{bmatrix}$$ (4.31)

and $\mathbf{E}$ is a $q \times p$ matrix,

$$\mathbf{E} = \begin{bmatrix} e_{11} & e_{12} & \cdots & e_{1p} \\ e_{21} & e_{22} & \cdots & e_{2p} \\ \cdot & \cdot & & \cdot \\ \cdot & \cdot & & \cdot \\ \cdot & \cdot & & \cdot \\ e_{q1} & e_{q2} & \cdots & e_{qp} \end{bmatrix}$$ (4-32)

EXAMPLE 4-3   The state equations of Eqs. (4-18) and (4-19) are expressed in matrix-vector form as follows:

$$\begin{bmatrix} \dfrac{dx_1(t)}{dt} \\ \dfrac{dx_2(t)}{dt} \end{bmatrix} = \begin{bmatrix} -\dfrac{R}{L} & -\dfrac{1}{L} \\ \dfrac{1}{C} & 0 \end{bmatrix} \begin{bmatrix} x_1(t) \\ x_2(t) \end{bmatrix} + \begin{bmatrix} \dfrac{1}{L} \\ 0 \end{bmatrix} e(t)$$ (4-33)

Thus the coefficient matrices $\mathbf{A}$ and $\mathbf{B}$ are identified to be

$$\mathbf{A} = \begin{bmatrix} -\dfrac{R}{L} & -\dfrac{1}{L} \\ \dfrac{1}{C} & 0 \end{bmatrix}$$ (4-34)

$$\mathbf{B} = \begin{bmatrix} \dfrac{1}{L} \\ 0 \end{bmatrix} \tag{4-35}$$

4.4 State Transition Matrix

The *state transition matrix* is defined as a matrix that satisfies the linear homogeneous state equation

$$\frac{d\mathbf{x}(t)}{dt} = \mathbf{A}\mathbf{x}(t) \tag{4-36}$$

Let $\boldsymbol{\phi}(t)$ be an $n \times n$ matrix that represents the state transition matrix; then it must satisfy the equation

$$\frac{d\boldsymbol{\phi}(t)}{dt} = \mathbf{A}\boldsymbol{\phi}(t) \tag{4-37}$$

Furthermore, let $\mathbf{x}(0+)$ denote the initial state at $t = 0$; then $\boldsymbol{\phi}(t)$ is also defined by the matrix equation

$$\mathbf{x}(t) = \boldsymbol{\phi}(t)\mathbf{x}(0+) \tag{4-38}$$

which is the solution of the homogeneous state equation for $t \geq 0$.

One way of determining $\boldsymbol{\phi}(t)$ is by taking the Laplace transform on both sides of Eq. (4-36); we have

$$s\mathbf{X}(s) - \mathbf{x}(0+) = \mathbf{A}\mathbf{X}(s) \tag{4-39}$$

Solving for $\mathbf{X}(s)$ from the last equation, we get

$$\mathbf{X}(s) = (s\mathbf{I} - \mathbf{A})^{-1}\mathbf{x}(0+) \tag{4-40}$$

where it is assumed that the matrix $(s\mathbf{I} - \mathbf{A})$ is nonsingular. Taking the inverse Laplace transform on both sides of the last equation yields

$$\mathbf{x}(t) = \mathcal{L}^{-1}[(s\mathbf{I} - \mathbf{A})^{-1}]\mathbf{x}(0+) \qquad t \geq 0 \tag{4-41}$$

Comparing Eq. (4-41) with Eq. (4-38), the state transition matrix is identified to be

$$\boldsymbol{\phi}(t) = \mathcal{L}^{-1}[(s\mathbf{I} - \mathbf{A})^{-1}] \tag{4-42}$$

An alternative way of solving the homogeneous state equation is to assume a solution, as in the classical method of solving differential equations. We let the solution to Eq. (4-36) be

$$\mathbf{x}(t) = e^{\mathbf{A}t}\mathbf{x}(0+) \tag{4-43}$$

for $t \geq 0$, where $e^{\mathbf{A}t}$ represents a power series of the matrix $\mathbf{A}t$ and

$$e^{\mathbf{A}t} = \mathbf{I} + \mathbf{A}t + \frac{1}{2!}\mathbf{A}^2 t^2 + \frac{1}{3!}\mathbf{A}^3 t^3 + \ldots \tag{4-44}*$$

It is easy to show that Eq. (4-43) is a solution of the homogeneous state equation, since, from Eq. (4-44),

$$\frac{de^{\mathbf{A}t}}{dt} = \mathbf{A}e^{\mathbf{A}t} \tag{4-45}$$

---

*It can be proved that this power series is uniformly convergent.

Therefore, in addition to Eq. (4-42), we have obtained another expression for the state transition matrix in

$$\boldsymbol{\phi}(t) = e^{\mathbf{A}t} = \mathbf{I} + \mathbf{A}t + \frac{1}{2!}\mathbf{A}^2 t^2 + \frac{1}{3!}\mathbf{A}^3 t^3 + \dots \tag{4-46}$$

Equation (4-46) can also be obtained directly from Eq. (4-42). This is left as an exercise for the reader (Problem 4-3).

EXAMPLE 4-4   Consider the *RL* network of Fig. 4-1 with the input short circuited; that is, $e(t) = 0$. The homogeneous state equation is written

$$\frac{di(t)}{dt} = -\frac{R}{L}i(t) \tag{4-47}$$

The solution of the last equation for $t \geq 0$ is obtained from Eq. (4-4) by setting $E_1 = 0$. Thus

$$i(t) = e^{-Rt/L}i(0+) \tag{4-48}$$

The state transition matrix in this case is a scalar and is given by

$$\boldsymbol{\phi}(t) = e^{-Rt/L} \qquad t \geq 0 \tag{4-49}$$

which is a simple exponential decay function.

Significance of the State Transition Matrix

Since the state transition matrix satisfies the homogeneous state equation, it represents the *free response* of the system. In other words, it governs the response that is excited by the initial conditions only. In view of Eqs. (4-42) and (4-46), the state transition matrix is dependent only upon the matrix $\mathbf{A}$. As the name implies, the state transition matrix $\boldsymbol{\phi}(t)$ completely defines the *transition* of the states from the initial time $t = 0$ to any time t.

Properties of the State Transition Matrix

The state transition matrix $\boldsymbol{\phi}(t)$ possesses the following properties:

1. $\boldsymbol{\phi}(0) = \mathbf{I}$ the identity matrix   (4-50)

   Proof: Equation (4-50) follows directly from Eq. (4-46) by setting $t = 0$.

2. $\boldsymbol{\phi}^{-1}(t) = \boldsymbol{\phi}(-t)$ (4-51)

   Proof: Postmultiplying both sides of Eq. (4-46) by $e^{-\mathbf{A}t}$, we get

$$\boldsymbol{\phi}(t)e^{-\mathbf{A}t} = e^{\mathbf{A}t}e^{-\mathbf{A}t} = \mathbf{I} \tag{4-52}$$

Then premultiplying both sides of Eq. (4-52) by $\boldsymbol{\phi}^{-1}(t)$, we get

$$e^{-\mathbf{A}t} = \boldsymbol{\phi}^{-1}(t) \tag{4-53}$$

Thus

$$\boldsymbol{\phi}(-t) = \boldsymbol{\phi}^{-1}(t) = e^{-\mathbf{A}t} \tag{4-54}$$

An interesting result from this property of $\boldsymbol{\phi}(t)$ is that Eq. (4-43) can be rearranged to read

$$\mathbf{x}(0+) = \boldsymbol{\phi}(-t)\mathbf{x}(t) \tag{4-55}$$

which means that the state transition process can be considered as bilateral in time. That is, the transition in time can take place in either direction.

    3. $\boldsymbol{\phi}(t_2 - t_1)\boldsymbol{\phi}(t_1 - t_0) = \boldsymbol{\phi}(t_2 - t_0)$ for any t_0, t_1, t_2 (4-56)

    Proof:

$$\boldsymbol{\phi}(t_2 - t_1)\boldsymbol{\phi}(t_1 - t_0) = e^{\mathbf{A}(t_2 - t_1)} e^{\mathbf{A}(t_1 - t_0)}$$
$$= e^{\mathbf{A}(t_2 - t_0)} \tag{4-57}$$
$$= \boldsymbol{\phi}(t_2 - t_0)$$

This property of the state transition matrix is important since it implies that a state transition process can be divided into a number of sequential transitions. Figure 4-3 illustrates that the transition from $t = t_0$ to $t = t_2$ is

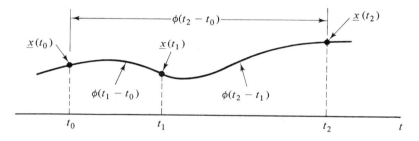

Fig. 4-3. Property of the state transition matrix.

equal to the transition from t_0 to t_1, and then from t_1 to t_2. In general, of course, the transition process can be broken up into any number of parts.

    Another way of proving Eq. (4-56) is to write

$$\mathbf{x}(t_2) = \boldsymbol{\phi}(t_2 - t_1)\mathbf{x}(t_1) \tag{4-58}$$
$$\mathbf{x}(t_1) = \boldsymbol{\phi}(t_1 - t_0)\mathbf{x}(t_0) \tag{4-59}$$
$$\mathbf{x}(t_2) = \boldsymbol{\phi}(t_2 - t_0)\mathbf{x}(t_0) \tag{4-60}$$

The proper result is obtained by substituting Eq. (4-59) into Eq. (4-58) and comparing the result with Eq. (4-60).

    4. $[\boldsymbol{\phi}(t)]^k = \boldsymbol{\phi}(kt)$ for $k = $ integer   (4-61)

    Proof:

$$[\boldsymbol{\phi}(t)]^k = e^{\mathbf{A}t} e^{\mathbf{A}t} \dots e^{\mathbf{A}t} \quad (k \text{ terms})$$
$$= e^{k\mathbf{A}t} \tag{4-62}$$
$$= \boldsymbol{\phi}(kt)$$

4.5 State Transition Equation

The *state transition equation* is defined as the solution of the linear nonhomogeneous state equation. For example, Eq. (4-5) is a state equation of the *RL* network of Fig. 4-1. Then Eq. (4-4) is the state transition equation when the input voltage is constant of amplitude E_1 for $t \geq 0$.

In general, the linear time-invariant state equation

$$\frac{d\mathbf{x}(t)}{dt} = \mathbf{A}\mathbf{x}(t) + \mathbf{B}\mathbf{r}(t) \tag{4-63}$$

can be solved by using either the classical method of solving differential equations or the Laplace transform method. The Laplace transform method is presented in the following.

Taking the Laplace transform on both sides of Eq. (4-63), we have

$$s\mathbf{X}(s) - \mathbf{x}(0+) = \mathbf{A}\mathbf{X}(s) + \mathbf{B}\mathbf{R}(s) \tag{4-64}$$

where $\mathbf{x}(0+)$ denotes the initial state vector evaluated at $t = 0+$. Solving for $\mathbf{X}(s)$ in Eq. (4-64) yields

$$\mathbf{X}(s) = (s\mathbf{I} - \mathbf{A})^{-1}\mathbf{x}(0+) + (s\mathbf{I} - \mathbf{A})^{-1}\mathbf{B}\mathbf{R}(s) \tag{4-65}$$

The state transition equation of Eq. (4-63) is obtained by taking the inverse Laplace transform on both sides of Eq. (4-65),

$$\mathbf{x}(t) = \mathcal{L}^{-1}[(s\mathbf{I} - \mathbf{A})^{-1}]\mathbf{x}(0+) + \mathcal{L}^{-1}[(s\mathbf{I} - \mathbf{A})^{-1}\mathbf{B}\mathbf{R}(s)] \tag{4-66}$$

Using the definition of the state transition matrix of Eq. (4-42), and the convolution integral, Eq. (3-26), Eq. (4-66) is written

$$\mathbf{x}(t) = \boldsymbol{\phi}(t)\mathbf{x}(0+) + \int_0^t \boldsymbol{\phi}(t - \tau)\mathbf{B}\mathbf{r}(\tau)\, d\tau \qquad t \geq 0 \tag{4-67}$$

The state transition equation in Eq. (4-67) is useful only when the initial time is defined to be at $t = 0$. In the study of control systems, especially discrete-data control systems, it is often desirable to break up a state transition process into a sequence of transitions, so that a more flexible initial time must be chosen. Let the initial time be represented by t_0 and the corresponding initial state by $\mathbf{x}(t_0)$, and assume that an input $\mathbf{r}(t)$ is applied for $t \geq t_0$.

We start with Eq. (4-67) by setting $t = t_0$, and solving for $\mathbf{x}(0+)$, we get

$$\mathbf{x}(0+) = \boldsymbol{\phi}(-t_0)\mathbf{x}(t_0) - \boldsymbol{\phi}(-t_0)\int_0^{t_0} \boldsymbol{\phi}(t_0 - \tau)\mathbf{B}\mathbf{r}(\tau)\, d\tau \tag{4-68}$$

where the property on $\boldsymbol{\phi}(t)$ of Eq. (4-51) has been used.

Substituting Eq. (4-68) into Eq. (4-67) yields

$$\begin{aligned}
\mathbf{x}(t) = \boldsymbol{\phi}(t)\boldsymbol{\phi}(-t_0)\mathbf{x}(t_0) - \boldsymbol{\phi}(t)\boldsymbol{\phi}(-t_0)\int_0^{t_0} \boldsymbol{\phi}(t_0 - \tau)\mathbf{B}\mathbf{r}(\tau)\, d\tau \\
+ \int_0^t \boldsymbol{\phi}(t - \tau)\mathbf{B}\mathbf{r}(\tau)\, d\tau
\end{aligned} \tag{4-69}$$

Now using the property of Eq. (4-56), and combining the last two integrals, Eq. (4-69) becomes

$$\mathbf{x}(t) = \boldsymbol{\phi}(t - t_0)\mathbf{x}(t_0) + \int_{t_0}^t \boldsymbol{\phi}(t - \tau)\mathbf{B}\mathbf{r}(\tau)\, d\tau \tag{4-70}$$

It is apparent that Eq. (4-70) reverts to Eq. (4-67) when $t_0 = 0$.

Once the state transition equation is determined, the output vector can be expressed as a function of the initial state and the input vector simply by substituting $\mathbf{x}(t)$ from Eq. (4-70) into Eq. (4-28). Thus the output vector is written

$$\mathbf{c}(t) = \mathbf{D}\boldsymbol{\phi}(t - t_0)\mathbf{x}(t_0) + \int_{t_0}^t \mathbf{D}\boldsymbol{\phi}(t - \tau)\mathbf{B}\mathbf{r}(\tau)\, d\tau + \mathbf{E}\mathbf{r}(t) \tag{4-71}$$

   The following example illustrates the application of the state transition equation.

EXAMPLE 4-5   Consider the state equation

$$
\begin{bmatrix} \dfrac{dx_1(t)}{dt} \\[2mm] \dfrac{dx_2(t)}{dt} \end{bmatrix} = \begin{bmatrix} 0 & 1 \\ -2 & -3 \end{bmatrix} \begin{bmatrix} x_1(t) \\ x_2(t) \end{bmatrix} + \begin{bmatrix} 0 \\ 1 \end{bmatrix} r(t) \tag{4-72}
$$

The problem is to determine the state vector $\mathbf{x}(t)$ for $t \geq 0$ when the input $r(t) = 1$ for $t \geq 0$; that is, $r(t) = u_s(t)$. The coefficient matrices $\mathbf{A}$ and $\mathbf{B}$ are identified to be

$$
\mathbf{A} = \begin{bmatrix} 0 & 1 \\ -2 & -3 \end{bmatrix} \qquad \mathbf{B} = \begin{bmatrix} 0 \\ 1 \end{bmatrix} \tag{4-73}
$$

Therefore,

$$
s\mathbf{I} - \mathbf{A} = \begin{bmatrix} s & 0 \\ 0 & s \end{bmatrix} - \begin{bmatrix} 0 & 1 \\ -2 & -3 \end{bmatrix} = \begin{bmatrix} s & -1 \\ 2 & s+3 \end{bmatrix}
$$

The matrix inverse of $(s\mathbf{I} - \mathbf{A})$ is

$$
(s\mathbf{I} - \mathbf{A})^{-1} = \frac{1}{s^2 + 3s + 2} \begin{bmatrix} s+3 & 1 \\ -2 & s \end{bmatrix} \tag{4-74}
$$

The state transition matrix of $\mathbf{A}$ is found by taking the inverse Laplace transform of the last equation. Thus

$$
\boldsymbol{\phi}(t) = \mathcal{L}^{-1}[(s\mathbf{I} - \mathbf{A})^{-1}] = \begin{bmatrix} 2e^{-t} - e^{-2t} & e^{-t} - e^{-2t} \\ -2e^{-t} + 2e^{-2t} & -e^{-t} + 2e^{-2t} \end{bmatrix} \tag{4-75}
$$

The state transition equation for $t \geq 0$ is obtained by substituting Eq. (4-75), $\mathbf{B}$, and $r(t)$ into Eq. (4-67). We have

$$
\mathbf{x}(t) = \begin{bmatrix} 2e^{-t} - e^{-2t} & e^{-t} - e^{-2t} \\ -2e^{-t} + e^{-2t} & -e^{-t} + 2e^{-2t} \end{bmatrix} \mathbf{x}(0+)
$$
$$
+ \int_0^t \begin{bmatrix} 2e^{-(t-\tau)} - e^{-2(t-\tau)} & e^{-(t-\tau)} - e^{-2(t-\tau)} \\ -2e^{-(t-\tau)} + 2e^{-2(t-\tau)} & -e^{-(t-\tau)} + 2e^{-2(t-\tau)} \end{bmatrix} \begin{bmatrix} 0 \\ 1 \end{bmatrix} d\tau \tag{4-76}
$$

or

$$
\mathbf{x}(t) = \begin{bmatrix} 2e^{-t} - e^{-2t} & e^{-t} - e^{-2t} \\ -2e^{-t} + 2e^{-2t} & -e^{-t} + 2e^{-2t} \end{bmatrix} \mathbf{x}(0+)
$$
$$
+ \begin{bmatrix} \frac{1}{2} - e^{-t} + \frac{1}{2}e^{-2t} \\ e^{-t} - e^{-2t} \end{bmatrix} \qquad t \geq 0 \tag{4-77}
$$

As an alternative, the second term of the state transition equation can be obtained by taking the inverse Laplace transform of $(s\mathbf{I} - \mathbf{A})^{-1}\mathbf{B}R(s)$. Therefore,

$$
\mathcal{L}^{-1}[(s\mathbf{I} - \mathbf{A})^{-1}\mathbf{B}R(s)] = \mathcal{L}^{-1}\frac{1}{s^2 + 3s + 2} \begin{bmatrix} s+3 & 1 \\ -2 & s \end{bmatrix} \begin{bmatrix} 0 \\ 1 \end{bmatrix} \frac{1}{s}
$$
$$
= \mathcal{L}^{-1}\frac{1}{s^2 + 3s + 2} \begin{bmatrix} \dfrac{1}{s} \\ 1 \end{bmatrix} \tag{4-78}
$$
$$
= \begin{bmatrix} \frac{1}{2} - e^{-t} + \frac{1}{2}e^{-2t} \\ e^{-t} - e^{-2t} \end{bmatrix} \qquad t \geq 0
$$

EXAMPLE 4-6   In this example we shall illustrate the utilization of the state transition method to a system with input discontinuity. Let us consider that the input voltage to the RL network of Fig. 4-1 is as shown in Fig. 4-4.

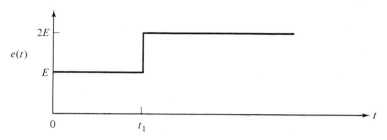

Fig. 4-4. Input voltage waveform for the network in Fig. 4-1.

The state equation of the network is

$$\frac{di(t)}{dt} = -\frac{R}{L}\,i(t) + \frac{1}{L}\,e(t) \tag{4-79}$$

Thus

$$A = -\frac{R}{L} \qquad B = \frac{1}{L} \tag{4-80}$$

The state transition matrix is

$$\phi(t) = e^{-Rt/L} \tag{4-81}$$

One approach to the problem of solving for $i(t)$ for $t \geq 0$ is to express the input voltage as

$$e(t) = Eu_s(t) + Eu_s(t - t_1) \tag{4-82}$$

where $u_s(t)$ is the unit step function. The Laplace transform of $e(t)$ is

$$E(s) = \frac{E}{s}(1 + e^{-t_1 s}) \tag{4-83}$$

Then

$$(s\mathbf{I} - \mathbf{A})^{-1}\mathbf{B}R(s) = \frac{1}{s + R/L}\frac{1}{L}\frac{E}{s}(1 + e^{-t_1 s})$$

$$= \frac{E}{Rs[1 + (L/R)s]}(1 + e^{-t_1 s}) \tag{4-84}$$

Substituting Eq. (4-84) into Eq. (4-66), the current for $t \geq 0$ is obtained:

$$i(t) = e^{-Rt/L}i(0+)u_s(t) + \frac{E}{R}(1 - e^{-Rt/L})u_s(t)$$

$$+ \frac{E}{R}[1 - e^{-R(t-t_1)/L}]u_s(t - t_1) \tag{4-85}$$

Using the state transition approach we can divide the transition period into two parts: $t = 0$ to $t = t_1$, and $t = t_1$ to $t = \infty$. First for the time interval, $0 \leq t \leq t_1$, the input is

$$e(t) = Eu_s(t) \qquad 0 \leq t < t_1 \tag{4-86}$$

Then

$$(s\mathbf{I} - \mathbf{A})^{-1}\mathbf{B}R(s) = \frac{1}{s + R/L}\frac{1}{L}\frac{E}{s}$$

$$= \frac{1}{Rs[1 + (L/R)s]} \tag{4-87}$$

Thus the state transition equation for the time interval $0 \leq t \leq t_1$ is

$$i(t) = \left[e^{-Rt/L} i(0+) + \frac{E}{R} (1 - e^{-Rt/L}) \right] u_s(t) \qquad (4\text{-}88)$$

Substituting $t = t_1$ into this equation, we get

$$i(t_1) = e^{-Rt_1/L} i(0+) + \frac{E}{R} (1 - e^{-Rt_1/L}) \qquad (4\text{-}89)$$

The value of $i(t)$ at $t = t_1$ is now used as the initial state for the next transition period of $t_1 \leq t < \infty$. The magnitude of the input for this interval is $2E$. Therefore, the state transition equation for the second transition period is

$$i(t) = e^{-R(t-t_1)/L} \, i(t_1) + \frac{2E}{R} [1 - e^{-R(t-t_1)/L}] \qquad t \geq t_1 \qquad (4\text{-}90)$$

where $i(t_1)$ is given by Eq. (4-89).

This example illustrates two possible ways of solving a state transition problem. In the first approach, the transition is treated as one continuous process, whereas in the second, the transition period is divided into parts over which the input can be more easily represented. Although the first approach requires only one operation, the second method yields relatively simple results to the state transition equation, and it often presents computational advantages. Notice that in the second method the state at $t = t_1$ is used as the initial state for the next transition period, which begins at t_1.

4.6 Relationship Between State Equations and High-Order Differential Equations

In preceding sections we defined the state equations and their solutions for linear time-invariant systems. In general, although it is always possible to write the state equations from the schematic diagram of a system, in practice the system may have been described by a high-order differential equation or transfer function. Therefore, it is necessary to investigate how state equations can be written directly from the differential equation or the transfer function. The relationship between a high-order differential equation and the state equations is discussed in this section.

Let us consider that a single-variable, linear time-invariant system is described by the following nth-order differential equation:

$$\frac{d^n c(t)}{dt^n} + a_1 \frac{d^{n-1} c(t)}{dt^{n-1}} + a_2 \frac{d^{n-2} c(t)}{dt^{n-2}} + \ldots + a_{n-1} \frac{dc(t)}{dt} + a_n c(t) = r(t) \qquad (4\text{-}91)$$

where $c(t)$ is the output variable and $r(t)$ is the input.

The problem is to represent Eq. (4-91) by n state equations and an output equation. This simply involves the defining of the n state variables in terms of the output $c(t)$ and its derivatives. We have shown earlier that the state variables of a given system are not unique. Therefore, in general, we seek the most convenient way of assigning the state variables as long as the definition of state variables stated in Section 4.1 is met.

For the present case it is convenient to define the state variables as

$$x_1(t) = c(t)$$

$$x_2(t) = \frac{dc(t)}{dt}$$

$$\vdots \tag{4-92}$$

$$x_n(t) = \frac{d^{n-1}c(t)}{dt^{n-1}}$$

Then the state equations are

$$\frac{dx_1(t)}{dt} = x_2(t)$$

$$\frac{dx_2(t)}{dt} = x_3(t)$$

$$\vdots \tag{4-93}$$

$$\frac{dx_{n-1}(t)}{dt} = x_n(t)$$

$$\frac{dx_n(t)}{dt} = -a_n x_1(t) - a_{n-1}x_2(t) - \ldots - a_2 x_{n-1}(t) - a_1 x_n(t) + r(t)$$

where the last state equation is obtained by equating the highest-ordered derivative term to the rest of Eq. (4-91). The output equation is simply

$$c(t) = x_1(t) \tag{4-94}$$

In vector-matrix form, Eq. (4-93) is written

$$\frac{d\mathbf{x}(t)}{dt} = \mathbf{A}\mathbf{x}(t) + \mathbf{B}r(t) \tag{4-95}$$

where $\mathbf{x}(t)$ is the $n \times 1$ state vector and $r(t)$ is the scalar input. The coefficient matrices are

$$\mathbf{A} = \begin{bmatrix} 0 & 1 & 0 & 0 & 0 & \cdots & 0 \\ 0 & 0 & 1 & 0 & 0 & \cdots & 0 \\ 0 & 0 & 0 & 1 & 0 & \cdots & 0 \\ \multicolumn{7}{c}{\cdots\cdots\cdots\cdots\cdots\cdots\cdots} \\ 0 & 0 & 0 & 0 & 0 & \cdots & 1 \\ -a_n & -a_{n-1} & -a_{n-2} & -a_{n-3} & -a_{n-4} & \cdots & -a_1 \end{bmatrix} \quad (n \times n) \tag{4-96}$$

$$\mathbf{B} = \begin{bmatrix} 0 \\ 0 \\ \vdots \\ 1 \end{bmatrix} \quad (n \times 1) \tag{4-97}$$

The output equation in vector-matrix form is

$$c(t) = \mathbf{D}x(t) \tag{4-98}$$

where

$$\mathbf{D} = [1 \quad 0 \quad 0 \quad . \quad . \quad . \quad 0] \qquad (1 \times n) \tag{4-99}$$

The state equation of Eq. (4-95) with the matrices $\mathbf{A}$ and $\mathbf{B}$ defined as in Eqs. (4-96) and (4-97) is called the *phase-variable canonical form* in the next section.

EXAMPLE 4-7   Consider the differential equation

$$\frac{d^3c(t)}{dt^3} + 5\frac{d^2c(t)}{dt^2} + \frac{dc(t)}{dt} + 2c(t) = r(t) \tag{4-100}$$

Rearranging the last equation so that the highest-order derivative term is equated to the rest of the terms, we have

$$\frac{d^3c(t)}{dt^3} = -5\frac{d^2c(t)}{dt^2} - \frac{dc(t)}{dt} - 2c(t) + r(t) \tag{4-101}$$

The state variables are defined as

$$x_1(t) = c(t)$$
$$x_2(t) = \frac{dc(t)}{dt} \tag{4-102}$$
$$x_3(t) = \frac{d^2c(t)}{dt^2}$$

Then the state equations are represented by the vector-matrix equation of Eq. (4-95) with

$$\mathbf{A} = \begin{bmatrix} 0 & 1 & 0 \\ 0 & 0 & 1 \\ -2 & -1 & -5 \end{bmatrix} \tag{4-103}$$

and

$$\mathbf{B} = \begin{bmatrix} 0 \\ 0 \\ 1 \end{bmatrix} \tag{4-104}$$

The output equation is

$$c(t) = x_1(t) \tag{4-105}$$

4.7 Transformation to Phase-Variable Canonical Form

In general, when the coefficient matrices $\mathbf{A}$ and $\mathbf{B}$ are given by Eqs. (4-96) and (4-97), respectively, the state equation of Eq. (4-95) is called the *phase-variable canonical form*. It is shown in the following that any linear time-invariant system with single input and satisfying a certain condition of *controllability* (see section 4.15) can always be represented in the phase-variable canonical form.

Theorem 4-1. Let the state equation of a linear time-invariant system be given by

$$\frac{d\mathbf{x}(t)}{dt} = \mathbf{A}\mathbf{x}(t) + \mathbf{B}r(t) \tag{4-106}$$

where $\mathbf{x}(t)$ *is an* $n \times 1$ *state vector,* $\mathbf{A}$ *an* $n \times n$ *coefficient matrix,* $\mathbf{B}$ *an* $n \times 1$ *coefficient matrix, and* $r(t)$ *a scalar input. If the matrix*

$$\mathbf{S} = [\mathbf{B} \quad \mathbf{AB} \quad \mathbf{A^2B} \quad \ldots \quad \mathbf{A^{n-1}B}] \tag{4-107}$$

is nonsingular, then there exists a nonsingular transformation

$$\mathbf{y}(t) = \mathbf{P}\mathbf{x}(t) \tag{4-108}$$

or

$$\mathbf{x}(t) = \mathbf{P^{-1}}\mathbf{y}(t) \tag{4-109}$$

which transforms Eq. (4-106) to the phase-variable canonical form

$$\dot{\mathbf{y}}(t) = \mathbf{A_1}\mathbf{y}(t) + \mathbf{B_1}r(t) \tag{4-110}$$

where

$$\mathbf{A_1} = \begin{bmatrix} 0 & 1 & 0 & 0 & \ldots & 0 \\ 0 & 0 & 1 & 0 & \ldots & 0 \\ 0 & 0 & 0 & 1 & \ldots & 0 \\ \cdots & \cdots & \cdots & \cdots & \cdots & \cdots \\ 0 & 0 & 0 & 0 & \ldots & 1 \\ -a_n & -a_{n-1} & -a_{n-2} & -a_{n-3} & \ldots & -a_1 \end{bmatrix} \tag{4-111}$$

and

$$\mathbf{B_1} = \begin{bmatrix} 0 \\ 0 \\ \cdot \\ \cdot \\ \cdot \\ 1 \end{bmatrix} \tag{4-112}$$

The transforming matrix $\mathbf{P}$ *is given by*

$$\mathbf{P} = \begin{bmatrix} \mathbf{P_1} \\ \mathbf{P_1A} \\ \cdot \\ \cdot \\ \cdot \\ \mathbf{P_1A^{n-1}} \end{bmatrix} \tag{4-113}$$

where

$$\mathbf{P_1} = [0 \quad 0 \quad \ldots \quad 1][\mathbf{B} \quad \mathbf{AB} \quad \mathbf{A^2B} \quad \ldots \quad \mathbf{A^{n-1}B}]^{-1} \tag{4-114}$$

Proof: Let

$$\mathbf{x}(t) = \begin{bmatrix} x_1(t) \\ x_2(t) \\ \cdot \\ \cdot \\ \cdot \\ x_n(t) \end{bmatrix} \tag{4-115}$$

$$\mathbf{y}(t) = \begin{bmatrix} y_1(t) \\ y_2(t) \\ \cdot \\ \cdot \\ \cdot \\ y_n(t) \end{bmatrix} \tag{4-116}$$

and

$$\mathbf{P} = \begin{bmatrix} p_{11} & p_{12} & \cdots & p_{1n} \\ p_{21} & p_{22} & \cdots & p_{2n} \\ \cdots\cdots\cdots\cdots\cdots\cdots \\ p_{n1} & p_{n2} & \cdots & p_{nn} \end{bmatrix} = \begin{bmatrix} \mathbf{P}_1 \\ \mathbf{P}_2 \\ \cdot \\ \cdot \\ \cdot \\ \mathbf{P}_n \end{bmatrix} \tag{4-117}$$

where

$$\mathbf{P}_i = [p_{i1} \quad p_{i2} \cdots p_{in}] \qquad i = 1, 2, \ldots, n \tag{4-118}$$

Then, from Eq. (4-108),

$$\begin{aligned} y_1(t) &= p_{11}x_1(t) + p_{12}x_2(t) + \ldots + p_{1n}x_n(t) \\ &= \mathbf{P}_1\mathbf{x}(t) \end{aligned} \tag{4-119}$$

Taking the time derivative on both sides of the last equation and in view of Eqs. (4-110) and (4-111),

$$\dot{y}_1(t) = y_2(t) = \mathbf{P}_1\dot{\mathbf{x}}(t) = \mathbf{P}_1\mathbf{A}\mathbf{x}(t) + \mathbf{P}_1\mathbf{B}r(t) \tag{4-120}$$

Since Eq. (4-108) states that $\mathbf{y}(t)$ is a function of $\mathbf{x}(t)$ only, in Eq. (4-120), $\mathbf{P}_1\mathbf{B} = \mathbf{0}$. Therefore,

$$\dot{y}_1(t) = y_2(t) = \mathbf{P}_1\mathbf{A}\mathbf{x}(t) \tag{4-121}$$

Taking the time derivative of the last equation once again leads to

$$\dot{y}_2(t) = y_3(t) = \mathbf{P}_1\mathbf{A}^2\mathbf{x}(t) \tag{4-122}$$

with $\mathbf{P}_1\mathbf{A}\mathbf{B} = \mathbf{0}$.

Repeating the procedure leads to

$$\dot{y}_{n-1}(t) = y_n(t) = \mathbf{P}_1\mathbf{A}^{n-1}\mathbf{x}(t) \tag{4-123}$$

with $\mathbf{P}_1\mathbf{A}^{n-2}\mathbf{B} = \mathbf{0}$. Therefore, using Eq. (4-108), we have

$$\mathbf{y}(t) = \mathbf{P}\mathbf{x}(t) = \begin{bmatrix} \mathbf{P}_1 \\ \mathbf{P}_1\mathbf{A} \\ \cdot \\ \cdot \\ \cdot \\ \mathbf{P}_1\mathbf{A}^{n-1} \end{bmatrix} \mathbf{x}(t) \tag{4-124}$$

or

$$\mathbf{P} = \begin{bmatrix} \mathbf{P}_1 \\ \mathbf{P}_1\mathbf{A} \\ \cdot \\ \cdot \\ \cdot \\ \mathbf{P}_1\mathbf{A}^{n-1} \end{bmatrix} \tag{4-125}$$

and $\mathbf{P}_1$ should satisfy the condition

$$\mathbf{P}_1\mathbf{B} = \mathbf{P}_1\mathbf{AB} = \ldots = \mathbf{P}_1\mathbf{A}^{n-2}\mathbf{B} = 0 \tag{4-126}$$

Now taking the derivative of Eq. (4-108) with respect to time, we get

$$\dot{\mathbf{y}}(t) = \mathbf{P}\dot{\mathbf{x}}(t) = \mathbf{PAx}(t) + \mathbf{PB}r(t) \tag{4-127}$$

Comparing Eq. (4-127) with Eq. (4-110), we obtain

$$\mathbf{A}_1 = \mathbf{PAP}^{-1} \tag{4-128}$$

and

$$\mathbf{B}_1 = \mathbf{PB} \tag{4-129}$$

Then, from Eq. (4-125),

$$\mathbf{PB} = \begin{bmatrix} \mathbf{P}_1\mathbf{B} \\ \mathbf{P}_1\mathbf{AB} \\ \cdot \\ \cdot \\ \cdot \\ \mathbf{P}_1\mathbf{A}^{n-1}\mathbf{B} \end{bmatrix} = \begin{bmatrix} 0 \\ 0 \\ \cdot \\ \cdot \\ \cdot \\ 1 \end{bmatrix} \tag{4-130}$$

Since $\mathbf{P}_1$ is an $1 \times n$ row matrix, Eq. (4-130) can be written

$$\mathbf{P}_1[\mathbf{B} \quad \mathbf{AB} \quad \mathbf{A}^2\mathbf{B} \quad \ldots \quad \mathbf{A}^{n-1}\mathbf{B}] = [0 \quad 0 \quad \ldots \quad 1] \tag{4-131}$$

Thus $\mathbf{P}_1$ is obtained as

$$\begin{aligned} \mathbf{P}_1 &= [0 \quad 0 \quad \ldots \quad 1][\mathbf{B} \quad \mathbf{AB} \quad \mathbf{A}^2\mathbf{B} \quad \ldots \quad \mathbf{A}^{n-1}\mathbf{B}]^{-1} \\ &= [0 \quad 0 \quad \ldots \quad 1]\mathbf{S}^{-1} \end{aligned} \tag{4-132}$$

if the matrix $\mathbf{S} = [\mathbf{B} \quad \mathbf{AB} \quad \mathbf{A}^2\mathbf{B} \quad \ldots \quad \mathbf{A}^{n-1}\mathbf{B}]$ is nonsingular. This is the condition of complete state controllability. Once $\mathbf{P}_1$ is determined from Eq. (4-132), the transformation matrix $\mathbf{P}$ is given by Eq. (4-125).

EXAMPLE 4-8   Let a linear time-invariant system be described by Eq. (4-95) with

$$\mathbf{A} = \begin{bmatrix} 1 & -1 \\ 0 & -1 \end{bmatrix} \qquad \mathbf{B} = \begin{bmatrix} 1 \\ 1 \end{bmatrix} \tag{4-133}$$

It is desired to transform the state equation into the phase-variable canonical form. Since the matrix

$$\mathbf{S} = [\mathbf{B} \quad \mathbf{AB}] = \begin{bmatrix} 1 & 0 \\ 1 & -1 \end{bmatrix} \tag{4-134}$$

is nonsingular, the system may be expressed in the phase-variable canonical form. Therefore, $\mathbf{P}_1$ is obtained as a row matrix which contains the elements of the last row of $\mathbf{S}^{-1}$; that is,

$$\mathbf{P}_1 = [1 \quad -1] \tag{4-135}$$

Using Eq. (4-125),

$$\mathbf{P} = \begin{bmatrix} \mathbf{P}_1 \\ \mathbf{P}_1\mathbf{A} \end{bmatrix} = \begin{bmatrix} 1 & -1 \\ 1 & 0 \end{bmatrix} \tag{4-136}$$

Thus

$$A_1 = PAP^{-1} = \begin{bmatrix} 0 & 1 \\ 1 & 0 \end{bmatrix} \tag{4-137}$$

$$B_1 = PB = \begin{bmatrix} 0 \\ 1 \end{bmatrix} \tag{4-138}$$

The method of defining state variables by inspection as described earlier with reference to Eq. (4-91) is inadequate when the right-hand side of the differential equation also includes the derivatives of $r(t)$. To illustrate the point we consider the following example.

EXAMPLE 4-9   Given the differential equation

$$\frac{d^3c(t)}{dt^3} + 5\frac{d^2c(t)}{dt^2} + \frac{dc(t)}{dt} + 2c(t) = \frac{dr(t)}{dt} + 2r(t) \tag{4-139}$$

it is desired to represent the equation by three state equations. Since the right side of the state equations cannot include any derivatives of the input $r(t)$, it is necessary to include $r(t)$ when defining the state variables. Let us rewrite Eq. (4-139) as

$$\frac{d^3c(t)}{dt^3} - \frac{dr(t)}{dt} = -5\frac{d^2c(t)}{dt^2} - \frac{dc(t)}{dt} - 2c(t) + 2r(t) \tag{4-140}$$

The state variables are now defined as

$$x_1(t) = c(t) \tag{4-141}$$

$$x_2(t) = \frac{dc(t)}{dt} \tag{4-142}$$

$$x_3(t) = \frac{d^2c(t)}{dt^2} - r(t) \tag{4-143}$$

Using these last three equations and Eq. (4-140), the state equations are written

$$\frac{dx_1(t)}{dt} = x_2(t)$$

$$\frac{dx_2(t)}{dt} = x_3(t) + r(t) \tag{4-144}$$

$$\frac{dx_3(t)}{dt} = -2x_1(t) - x_2(t) - 5x_3(t) - 3r(t)$$

In general, it can be shown that for the nth-order differential equation

$$\frac{d^nc(t)}{dt^n} + a_1\frac{d^{n-1}c(t)}{dt^{n-1}} + \ldots + a_{n-1}\frac{dc(t)}{dt} + a_nc(t)$$
$$= b_0\frac{d^nr(t)}{dt^n} + b_1\frac{d^{n-1}r(t)}{dt^{n-1}} + \ldots + b_{n-1}\frac{dr(t)}{dt} + b_nr(t) \tag{4-145}$$

the state variables should be defined as

$$x_1(t) = c(t) - b_0 r(t)$$

$$x_2(t) = \frac{dx_1(t)}{dt} - h_1 r(t)$$

$$x_3(t) = \frac{dx_2(t)}{dt} - h_2 r(t)$$

$$\cdot$$
$$\cdot$$ (4-146)
$$\cdot$$

$$x_n(t) = \frac{dx_{n-1}(t)}{dt} - h_{n-1} r(t)$$

where

$$h_1 = b_1 - a_1 b_0$$

$$h_2 = (b_2 - a_2 b_0) - a_1 h_1$$

$$h_3 = (b_3 - a_3 b_0) - a_2 h_1 - a_1 h_2$$ (4-147)

$$\cdot$$
$$\cdot$$
$$\cdot$$

$$h_n = (b_n - a_n b_0) - a_{n-1} h_1 - a_{n-2} h_2 - \ldots - a_2 h_{n-2} - a_1 h_{n-1}$$

Using Eqs. (4-146) and (4-147), we resolve the nth-order differential equation in Eq. (4-145) into the following n state equations:

$$\frac{dx_1(t)}{dt} = x_2(t) + h_1 r(t)$$

$$\frac{dx_2(t)}{dt} = x_3(t) + h_2 r(t)$$

$$\cdot$$
$$\cdot$$ (4-148)
$$\cdot$$

$$\frac{dx_{n-1}(t)}{dt} = x_n(t) + h_{n-1} r(t)$$

$$\frac{dx_n(t)}{dt} = -a_n x_1(t) - a_{n-1} x_2(t) - \ldots - a_2 x_{n-1}(t) - a_1 x_n(t) + h_n r(t)$$

The output equation is obtained by rearranging the first equation of Eq. (4-146):

$$c(t) = x_1(t) + b_0 r(t) \qquad (4\text{-}149)$$

Now if we apply these equations to the case of Example 4-9, we have

$$a_1 = 5 \qquad b_0 = 0 \qquad b_3 = 2$$
$$a_2 = 1 \qquad b_1 = 0$$
$$a_3 = 2 \qquad b_2 = 1$$
$$h_1 = b_1 - a_1 b_0 = 0$$
$$h_2 = (b_2 - a_2 b_0) - a_1 h_1 = 1$$
$$h_3 = (b_3 - a_3 b_0) - a_2 h_1 - a_1 h_2 = -3$$

When we substitute these parameters into Eqs. (4-146) and (4-147), we have the same results for the state variables and the state equations as obtained in Example 4-9.

The disadvantage with the method of Eqs. (4-146), (4-147), and (4-148) is that these equations are difficult and impractical to memorize. It is not expected that one will always have these equations available for reference. However, we shall later describe a more convenient method using the transfer function.

4.8 Relationship Between State Equations and Transfer Functions

We have presented the methods of describing a linear time-invariant system by transfer functions and by dynamic equations. It is interesting to investigate the relationship between these two representations.

In Eq. (3-3), the transfer function of a linear single-variable system is defined in terms of the coefficients of the system's differential equation. Similarly, Eq. (3-16) gives the matrix transfer function relation for a multivariable system that has p inputs and q outputs. Now we shall investigate the transfer function matrix relation using the dynamic equation notation.

Consider that a linear time-invariant system is described by the dynamic equations

$$\frac{d\mathbf{x}(t)}{dt} = \mathbf{A}\mathbf{x}(t) + \mathbf{B}\mathbf{r}(t) \qquad (4\text{-}150)$$

$$\mathbf{c}(t) = \mathbf{D}\mathbf{x}(t) + \mathbf{E}\mathbf{r}(t) \qquad (4\text{-}151)$$

where

$$\mathbf{x}(t) = n \times 1 \text{ state vector}$$
$$\mathbf{r}(t) = p \times 1 \text{ input vector}$$
$$\mathbf{c}(t) = q \times 1 \text{ output vector}$$

and $\mathbf{A}, \mathbf{B}, \mathbf{D}$, and $\mathbf{E}$ are matrices of appropriate dimensions.

Taking the Laplace transform on both sides of Eq. (4-150) and solving for $\mathbf{X}(s)$, we have

$$\mathbf{X}(s) = (s\mathbf{I} - \mathbf{A})^{-1}\mathbf{x}(0+) + (s\mathbf{I} - \mathbf{A})^{-1}\mathbf{B}\mathbf{R}(s) \qquad (4\text{-}152)$$

The Laplace transform of Eq. (4-151) is

$$\mathbf{C}(s) = \mathbf{D}\mathbf{X}(s) + \mathbf{E}\mathbf{R}(s) \qquad (4\text{-}153)$$

Substituting Eq. (4-152) into Eq. (4-153), we have

$$\mathbf{C}(s) = \mathbf{D}(s\mathbf{I} - \mathbf{A})^{-1}\mathbf{x}(0+) + \mathbf{D}(s\mathbf{I} - \mathbf{A})^{-1}\mathbf{B}\mathbf{R}(s) + \mathbf{E}\mathbf{R}(s) \qquad (4\text{-}154)$$

Since the definition of transfer function requires that the initial conditions be set to zero, $\mathbf{x}(0+) = \mathbf{0}$; thus Eq. (4-154) becomes

$$\mathbf{C}(s) = [\mathbf{D}(s\mathbf{I} - \mathbf{A})^{-1}\mathbf{B} + \mathbf{E}]\mathbf{R}(s) \qquad (4\text{-}155)$$

Therefore, the transfer function is defined as

$$\mathbf{G}(s) = \mathbf{D}(s\mathbf{I} - \mathbf{A})^{-1}\mathbf{B} + \mathbf{E} \qquad (4\text{-}156)$$

which is a $q \times p$ matrix. Of course, the existence of $\mathbf{G}(s)$ requires that the matrix $(s\mathbf{I} - \mathbf{A})$ is nonsingular.

EXAMPLE 4-10   Consider that a multivariable system is described by the differential equations

$$\frac{d^2 c_1}{dt^2} + 4\frac{dc_1}{dt} - 3c_2 = r_1 \tag{4-157}$$

$$\frac{dc_2}{dt} + \frac{dc_1}{dt} + c_1 + 2c_2 = r_2 \tag{4-158}$$

The state variables of the system are assigned as follows:

$$x_1 = c_1 \tag{4-159}$$

$$x_2 = \frac{dc_1}{dt} \tag{4-160}$$

$$x_3 = c_2 \tag{4-161}$$

These state variables have been defined by mere inspection of the two differential equations, as no particular reasons for the definitions are given other than that these are the most convenient.

Now equating the first term of each of the equations of Eqs. (4-157) and (4-158) to the rest of the terms and using the state-variable relations of Eqs. (4-159) through (4-161), we arrive at the following state equation and output equation in matrix form:

$$\begin{bmatrix} \dfrac{dx_1}{dt} \\[2mm] \dfrac{dx_2}{dt} \\[2mm] \dfrac{dx_3}{dt} \end{bmatrix} = \begin{bmatrix} 0 & 1 & 0 \\ 0 & -4 & 3 \\ -1 & -1 & -2 \end{bmatrix} \begin{bmatrix} x_1 \\ x_2 \\ x_3 \end{bmatrix} + \begin{bmatrix} 0 & 0 \\ 1 & 0 \\ 0 & 1 \end{bmatrix} \begin{bmatrix} r_1 \\ r_2 \end{bmatrix} \tag{4-162}$$

$$\begin{bmatrix} c_1 \\ c_2 \end{bmatrix} = \begin{bmatrix} 1 & 0 & 0 \\ 0 & 0 & 1 \end{bmatrix} \begin{bmatrix} x_1 \\ x_2 \\ x_3 \end{bmatrix} = \mathbf{Dx} \tag{4-163}$$

To determine the transfer function matrix of the system using the state-variable formulation, we substitute the **A**, **B**, **D**, and **E** matrices into Eq. (4-156). First, we form the matrix $(s\mathbf{I} - \mathbf{A})$,

$$(s\mathbf{I} - \mathbf{A}) = \begin{bmatrix} s & -1 & 0 \\ 0 & s+4 & -3 \\ 1 & 1 & s+2 \end{bmatrix} \tag{4-164}$$

The determinant of $(s\mathbf{I} - \mathbf{A})$ is

$$|s\mathbf{I} - \mathbf{A}| = s^3 + 6s^2 + 11s + 3 \tag{4-165}$$

Thus

$$(s\mathbf{I} - \mathbf{A})^{-1} = \frac{1}{|s\mathbf{I} - \mathbf{A}|} \begin{bmatrix} s^2 + 6s + 11 & s+2 & 3 \\ -3 & s(s+2) & 3s \\ -(s+4) & -(s+1) & s(s+4) \end{bmatrix} \tag{4-166}$$

The transfer function matrix is, in this case,

$$\mathbf{G}(s) = \mathbf{D}(s\mathbf{I} - \mathbf{A})^{-1}\mathbf{B}$$

$$= \frac{1}{s^3 + 6s^2 + 11s + 3} \begin{bmatrix} s+2 & 3 \\ -(s+1) & s(s+4) \end{bmatrix} \tag{4-167}$$

Using the conventional approach, we take the Laplace transform on both sides of Eqs. (4-157) and (4-158) and assume zero initial conditions. The resulting transformed equations are written in matrix form as

$$\begin{bmatrix} s(s+4) & -3 \\ s+1 & s+2 \end{bmatrix}\begin{bmatrix} C_1(s) \\ C_2(s) \end{bmatrix} = \begin{bmatrix} R_1(s) \\ R_2(s) \end{bmatrix} \tag{4-168}$$

Solving for $\mathbf{C}(s)$ from Eq. (4-168), we obtain

$$\mathbf{C}(s) = \mathbf{G}(s)\mathbf{R}(s) \tag{4-169}$$

where

$$\mathbf{G}(s) = \begin{bmatrix} s(s+4) & -3 \\ s+1 & s+2 \end{bmatrix}^{-1} \tag{4-170}$$

and the same result as in Eq. (4-167) is obtained when the matrix inverse is carried out.

4.9 Characteristic Equation, Eigenvalues, and Eigenvectors

The characteristic equation plays an important part in the study of linear systems. It can be defined from the basis of the differential equation, the transfer function, or the state equations.

Consider that a linear time-invariant system is described by the differential equation

$$\frac{d^n c}{dt^n} + a_1 \frac{d^{n-1} c}{dt^{n-1}} + a_2 \frac{d^{n-2} c}{dt^{n-2}} + \ldots + a_{n-1} \frac{dc}{dt} + a_n c$$

$$= b_0 \frac{d^n r}{dt^n} + b_1 \frac{d^{n-1} r}{dt^{n-1}} + \ldots + b_{n-1} \frac{dr}{dt} + b_n r \tag{4-171}$$

By defining the operator p as

$$p^k = \frac{d^k}{dt^k} \qquad k = 1, 2, \ldots, n \tag{4-172}$$

Eq. (4-171) is written

$$(p^n + a_1 p^{n-1} + a_2 p^{n-2} + \ldots + a_{n-1} p + a_n)c$$
$$= (b_0 p^n + b_1 p^{n-1} + \ldots + b^{n-1} p + b_n)r \tag{4-173}$$

Then the characteristic equation of the system is defined as

$$s^n + a_1 s^{n-1} + a_2 s^{n-2} + \ldots + a_{n-1} s + a_n = 0 \tag{4-174}$$

which is setting the homogeneous part of Eq. (4-173) to zero. Furthermore, the operator p is replaced by the Laplace transform variable s.

The transfer function of the system is

$$G(s) = \frac{C(s)}{R(s)} = \frac{b_0 s^n + b_1 s^{n-1} + \ldots + b_{n-1} s + b_n}{s^n + a_1 s^{n-1} + \ldots + a_{n-1} s + a_n} \tag{4-175}$$

Therefore, *the characteristic equation is obtained by equating the denominator of the transfer function to zero.*

From the state-variable approach, we can write Eq. (4-156) as

$$\mathbf{G}(s) = \mathbf{D}\frac{\text{adj}(s\mathbf{I} - \mathbf{A})}{|s\mathbf{I} - \mathbf{A}|}\mathbf{B} + \mathbf{E}$$

$$= \frac{\mathbf{D}[\text{adj}(s\mathbf{I} - \mathbf{A})]\mathbf{B} + |s\mathbf{I} - \mathbf{A}|\mathbf{E}}{|s\mathbf{I} - \mathbf{A}|} \tag{4-176}$$

Setting the denominator of the transfer function matrix $\mathbf{G}(s)$ to zero, we get the characteristic equation expressed as

$$|s\mathbf{I} - \mathbf{A}| = 0 \qquad (4\text{-}177)$$

which is an alternative form of Eq. (4-174).

Eigenvalues

The roots of the characteristic equation are often referred to as the eigenvalues of the matrix $\mathbf{A}$. It is interesting to note that if the state equations are represented in the phase-variable canonical form, the coefficients of the characteristic equation are readily given by the elements in the last row of the elements of the $\mathbf{A}$ matrix. That is, if $\mathbf{A}$ is given by Eq. (4-96), the characteristic equation is readily given by Eq. (4-174).

Another important property of the characteristic equation and the eigenvalues is that they are invariant under a nonsingular transformation. In other words, when the $\mathbf{A}$ matrix is transformed by a nonsingular transformation $\mathbf{x} = \mathbf{P}\mathbf{y}$, so that

$$\hat{\mathbf{A}} = \mathbf{P}^{-1}\mathbf{A}\mathbf{P} \qquad (4\text{-}178)$$

then the characteristic equation and the eigenvalues of $\hat{\mathbf{A}}$ are identical to those of $\mathbf{A}$. This is proved by writing

$$s\mathbf{I} - \hat{\mathbf{A}} = s\mathbf{I} - \mathbf{P}^{-1}\mathbf{A}\mathbf{P} \qquad (4\text{-}179)$$

or

$$s\mathbf{I} - \hat{\mathbf{A}} = s\mathbf{P}^{-1}\mathbf{P} - \mathbf{P}^{-1}\mathbf{A}\mathbf{P} \qquad (4\text{-}180)$$

The characteristic equation of $\hat{\mathbf{A}}$ is

$$|s\mathbf{I} - \hat{\mathbf{A}}| = |s\mathbf{P}^{-1}\mathbf{P} - \mathbf{P}^{-1}\mathbf{A}\mathbf{P}|$$
$$= |\mathbf{P}^{-1}(s\mathbf{I} - \mathbf{A})\mathbf{P}| \qquad (4\text{-}181)$$

Since the determinant of a product is equal to the product of the determinants, Eq. (4-181) becomes

$$|s\mathbf{I} - \hat{\mathbf{A}}| = |\mathbf{P}^{-1}||s\mathbf{I} - \mathbf{A}||\mathbf{P}|$$
$$= |s\mathbf{I} - \mathbf{A}| \qquad (4\text{-}182)$$

Eigenvectors

The $n \times 1$ vector $\mathbf{p}_i$ which satisfies the matrix equation

$$(\lambda_i\mathbf{I} - \mathbf{A})\mathbf{p}_i = \mathbf{0} \qquad (4\text{-}183)$$

where λ_i is the ith eigenvalue of $\mathbf{A}$, is called the *eigenvector* of $\mathbf{A}$ associated with the eigenvalue λ_i. Illustrative examples of how the eigenvectors of a matrix are determined are given in the following section.

4.10 Diagonalization of the A Matrix (Similarity Transformation)

One of the motivations for diagonalizing the $\mathbf{A}$ matrix is that if $\mathbf{A}$ is a diagonal matrix, the eigenvalues of $\mathbf{A}$, $\lambda_1, \lambda_2, \ldots, \lambda_n$, all assumed to be distinct, are located on the main diagonal; then the state transition matrix $e^{\mathbf{A}t}$ will also be diagonal, with its nonzero elements given by $e^{\lambda_1 t}, e^{\lambda_2 t}, \ldots, e^{\lambda_n t}$. There are other reasons for wanting to diagonalize the **A** matrix, such as the controllability of a system (Section 4.15). We have to assume that all the eigenvalues of **A** are distinct, since, unless it is real and symmetric, **A** cannot always be diagonalized if it has multiple-order eigenvalues.

The problem can be stated as, given the linear system

$$\dot{\mathbf{x}}(t) = \mathbf{Ax}(t) + \mathbf{Bu}(t) \tag{4-184}$$

where $\mathbf{x}(t)$ is an n-vector, $\mathbf{u}(t)$ an r-vector, and **A** has distinct eigenvalues $\lambda_1, \lambda_2, \ldots, \lambda_n$, it is desired to find a nonsingular matrix **P** such that the transformation

$$\mathbf{x}(t) = \mathbf{Py}(t) \tag{4-185}$$

transforms Eq. (4-184) into

$$\dot{\mathbf{y}}(t) = \mathbf{\Lambda y}(t) + \mathbf{\Gamma u}(t) \tag{4-186}$$

with $\mathbf{\Lambda}$ given by the diagonal matrix

$$\mathbf{\Lambda} = \begin{bmatrix} \lambda_1 & 0 & 0 & \ldots & 0 \\ 0 & \lambda_2 & 0 & \ldots & 0 \\ 0 & 0 & \lambda_3 & \ldots & 0 \\ \ldots & \ldots & \ldots & \ldots & 0 \\ 0 & 0 & 0 & \ldots & \lambda_n \end{bmatrix} \qquad (n \times n) \tag{4-187}$$

This transformation is also known as the *similarity transformation*. The state equation of Eq. (4-186) is known as the *canonical form*.

Substituting Eq. (4-185) into Eq. (4-184) it is easy to see that

$$\mathbf{\Lambda} = \mathbf{P}^{-1}\mathbf{AP} \tag{4-188}$$

and

$$\mathbf{\Gamma} = \mathbf{P}^{-1}\mathbf{B} \qquad (n \times r) \tag{4-189}$$

In general, there are several methods of determining the matrix **P**. We show in the following that **P** can be formed by use of the eigenvectors of **A**; that is,

$$\mathbf{P} = [\mathbf{p}_1 \quad \mathbf{p}_2 \quad \mathbf{p}_3 \cdots \mathbf{p}_n] \tag{4-190}$$

where $\mathbf{p}_i$ $(i = 1, 2, \ldots, n)$ denotes the eigenvector that is associated with the eigenvalue λ_i. This is proved by use of Eq. (4-183), which is written

$$\lambda_i \mathbf{p}_i = \mathbf{Ap}_i \qquad i = 1, 2, \ldots, n \tag{4-191}$$

Now forming the $n \times n$ matrix,

$$\begin{aligned} [\lambda_1 \mathbf{p}_1 \quad \lambda_2 \mathbf{p}_2 \ldots \lambda_n \mathbf{p}_n] &= [\mathbf{Ap}_1 \quad \mathbf{Ap}_2 \ldots \mathbf{Ap}_n] \\ &= \mathbf{A}[\mathbf{p}_1 \quad \mathbf{p}_2 \ldots \mathbf{p}_n] \end{aligned} \tag{4-192}$$

or

$$[\mathbf{p}_1 \quad \mathbf{p}_2 \ldots \mathbf{p}_n]\mathbf{\Lambda} = \mathbf{A}[\mathbf{p}_1 \quad \mathbf{p}_2 \ldots \mathbf{p}_n] \tag{4-193}$$

Therefore, if we let

$$\mathbf{P} = [\mathbf{p}_1 \quad \mathbf{p}_2 \cdots \mathbf{p}_n] \tag{4-194}$$

Eq. (4-193) gives

$$\mathbf{P\Lambda} = \mathbf{AP} \tag{4-195}$$

or

$$\mathbf{\Lambda} = \mathbf{P}^{-1}\mathbf{AP} \tag{4-196}$$

which is the desired transformation.

If the matrix $\mathbf{A}$ is of the phase-variable canonical form, it can be shown that the $\mathbf{P}$ matrix which diagonalizes $\mathbf{A}$ may be the Vandermonde matrix,

$$\mathbf{P} = \begin{bmatrix} 1 & 1 & \ldots & 1 \\ \lambda_1 & \lambda_2 & & \lambda_n \\ \lambda_1^2 & \lambda_2^2 & & \lambda_n^2 \\ \cdot & \cdot & \ldots & \cdot \\ \cdot & \cdot & \ldots & \cdot \\ \cdot & \cdot & \ldots & \cdot \\ \lambda_1^{n-1} & \lambda_2^{n-1} & & \lambda_n^{n-1} \end{bmatrix} \tag{4-197}$$

where $\lambda_1, \lambda_2, \ldots, \lambda_n$ are the eigenvalues of $\mathbf{A}$.

Since it has been proven that $\mathbf{P}$ contains as its columns the eigenvectors of $\mathbf{A}$, we shall show that the ith column of the matrix in Eq. (4-197) is the eigenvector of $\mathbf{A}$ that is associated with λ_i, $i = 1, 2, \ldots, n$.

Let

$$\mathbf{p}_i = \begin{bmatrix} p_{i1} \\ p_{i2} \\ \cdot \\ \cdot \\ \cdot \\ p_{in} \end{bmatrix} \tag{4-198}$$

be the ith eigenvector of $\mathbf{A}$. Then

$$(\lambda_i\mathbf{I} - \mathbf{A})\mathbf{p}_i = \mathbf{0} \tag{4-199}$$

or

$$\begin{bmatrix} \lambda_i & -1 & 0 & 0 & \ldots & 0 \\ 0 & \lambda_i & -1 & 0 & \ldots & 0 \\ 0 & 0 & \lambda_i & -1 & \ldots & 0 \\ \hdotsfor{6} \\ 0 & 0 & 0 & 0 & \ldots & -1 \\ a_n & a_{n-1} & a_{n-2} & a_{n-3} & \ldots & \lambda_i + a_1 \end{bmatrix} \begin{bmatrix} p_{i1} \\ p_{i2} \\ \cdot \\ \cdot \\ \cdot \\ p_{in} \end{bmatrix} = \mathbf{0} \tag{4-200}$$

This equation implies that

$$\lambda_i p_{i1} - p_{i2} = 0$$
$$\lambda_i p_{i2} - p_{i3} = 0$$
$$\cdot$$
$$\cdot \tag{4-201}$$
$$\cdot$$
$$\lambda_i p_{i,n-1} - p_{in} = 0$$
$$a_n p_{i1} + a_{n-1} p_{i2} + \ldots + (\lambda_i + a_1)p_{in} = 0$$

Now we arbitrarily let $p_{i1} = 1$. Then Eq. (4-201) gives

$$p_{i2} = \lambda_i$$
$$p_{i3} = \lambda_i^2$$
$$\cdot$$
$$\cdot \qquad (4\text{-}202)$$
$$\cdot$$
$$p_{i,n-1} = \lambda_i^{n-2}$$
$$p_{in} = \lambda_i^{n-1}$$

which represent the elements of the ith column of the matrix in Eq. (4-197). Substitution of these elements of $\mathbf{p}_i$ into the last equation of Eq. (4-201) simply verifies that the characteristic equation is satisfied.

EXAMPLE 4-11   Given the matrix

$$\mathbf{A} = \begin{bmatrix} 0 & 1 & 0 \\ 0 & 0 & 1 \\ -6 & -11 & -6 \end{bmatrix} \qquad (4\text{-}203)$$

which is the phase-variable canonical form, the eigenvalues of $\mathbf{A}$ are $\lambda_1 = -1$, $\lambda_2 = -2$, $\lambda_3 = -3$. The similarity transformation may be carried out by use of the Vandermonde matrix of Eq. (4-197). Therefore,

$$\mathbf{P} = \begin{bmatrix} 1 & 1 & 1 \\ \lambda_1 & \lambda_2 & \lambda_3 \\ \lambda_1^2 & \lambda_2^2 & \lambda_3^2 \end{bmatrix} = \begin{bmatrix} 1 & 1 & 1 \\ -1 & -2 & -3 \\ 1 & 4 & 9 \end{bmatrix} \qquad (4\text{-}204)$$

The canonical-form state equation is given by Eq. (4-186) with

$$\mathbf{\Lambda} = \mathbf{P}^{-1}\mathbf{A}\mathbf{P} = \begin{bmatrix} -1 & 0 & 0 \\ 0 & -2 & 0 \\ 0 & 0 & -3 \end{bmatrix} = \begin{bmatrix} \lambda_1 & 0 & 0 \\ 0 & \lambda_2 & 0 \\ 0 & 0 & \lambda_3 \end{bmatrix} \qquad (4\text{-}205)$$

EXAMPLE 4-12   Given the matrix

$$\mathbf{A} = \begin{bmatrix} 0 & 1 & -1 \\ -6 & -11 & 6 \\ -6 & -11 & 5 \end{bmatrix} \qquad (4\text{-}206)$$

it can be shown that the eigenvalues of $\mathbf{A}$ are $\lambda_1 = -1$, $\lambda_2 = -2$, and $\lambda_3 = -3$. It is desired to find a nonsingular matrix $\mathbf{P}$ that will transform $\mathbf{A}$ into a diagonal matrix $\mathbf{\Lambda}$, such that $\mathbf{\Lambda} = \mathbf{P}^{-1}\mathbf{A}\mathbf{P}$.

We shall follow the guideline that $\mathbf{P}$ contains the eigenvectors of $\mathbf{A}$. Since $\mathbf{A}$ is not of the phase-variable canonical form, we cannot use the Vandermonde matrix.

Let the eigenvector associated with $\lambda_1 = -1$ be represented by

$$\mathbf{p}_1 = \begin{bmatrix} p_{11} \\ p_{21} \\ p_{31} \end{bmatrix} \qquad (4\text{-}207)$$

Then $\mathbf{p}_1$ must satisfy

$$(\lambda_1 \mathbf{I} - \mathbf{A})\mathbf{p}_1 = 0 \qquad (4\text{-}208)$$

or

$$\begin{bmatrix} \lambda_1 & -1 & 1 \\ 6 & \lambda_1 + 11 & -6 \\ 6 & 11 & \lambda_1 - 5 \end{bmatrix} \begin{bmatrix} p_{11} \\ p_{21} \\ p_{31} \end{bmatrix} = \mathbf{0} \tag{4-209}$$

The last matrix equation leads to

$$-p_{11} - p_{21} + p_{31} = 0$$
$$6p_{11} + 10p_{21} - 6p_{31} = 0 \tag{4-210}$$
$$6p_{11} + 11p_{21} - 6p_{31} = 0$$

from which we get $p_{21} = 0$ and $p_{11} = p_{31}$. Therefore, we can let $p_{11} = p_{31} = 1$, and get

$$\mathbf{p}_1 = \begin{bmatrix} 1 \\ 0 \\ 1 \end{bmatrix} \tag{4-211}$$

For the eigenvector associated with $\lambda_2 = -2$, the following matrix equation must be satisfied:

$$\begin{bmatrix} \lambda_2 & -1 & 1 \\ 6 & \lambda_2 + 11 & -6 \\ 6 & 11 & \lambda_2 - 5 \end{bmatrix} \begin{bmatrix} p_{12} \\ p_{22} \\ p_{32} \end{bmatrix} = \mathbf{0} \tag{4-212}$$

or

$$-2p_{12} - p_{22} + p_{32} = 0$$
$$6p_{12} + 9p_{22} - 6p_{32} = 0 \tag{4-213}$$
$$6p_{12} + 11p_{22} - 7p_{32} = 0$$

In these three equations we let $p_{12} = 1$; then $p_{22} = 2$ and $p_{32} = 4$. Thus

$$\mathbf{p}_2 = \begin{bmatrix} 1 \\ 2 \\ 4 \end{bmatrix} \tag{4-214}$$

Finally, for the eigenvector $\mathbf{p}_3$, we have

$$\begin{bmatrix} \lambda_3 & -1 & 1 \\ 6 & \lambda_3 + 11 & -6 \\ 6 & 11 & \lambda_3 - 5 \end{bmatrix} \begin{bmatrix} p_{13} \\ p_{23} \\ p_{33} \end{bmatrix} = \mathbf{0} \tag{4-215}$$

or

$$-3p_{13} - p_{23} + p_{33} = 0$$
$$6p_{13} + 8p_{23} - 6p_{33} = 0 \tag{4-216}$$
$$6p_{13} + 11p_{23} - 8p_{33} = 0$$

Now if we arbitrarily let $p_{13} = 1$, the last three equations give $p_{23} = 6$ and $p_{33} = 9$. Therefore,

$$\mathbf{p}_3 = \begin{bmatrix} 1 \\ 6 \\ 9 \end{bmatrix} \tag{4-217}$$

The matrix $\mathbf{P}$ is now given by

$$\mathbf{P} = [\mathbf{p}_1 \quad \mathbf{p}_2 \quad \mathbf{p}_3] = \begin{bmatrix} 1 & 1 & 1 \\ 0 & 2 & 6 \\ 1 & 4 & 9 \end{bmatrix} \tag{4-218}$$

It is easy to show that

$$\Lambda = \mathbf{P}^{-1}\mathbf{A}\mathbf{P} = \begin{bmatrix} \lambda_1 & 0 & 0 \\ 0 & \lambda_2 & 0 \\ 0 & 0 & \lambda_3 \end{bmatrix} = \begin{bmatrix} -1 & 0 & 0 \\ 0 & -2 & 0 \\ 0 & 0 & -3 \end{bmatrix} \qquad (4\text{-}219)$$

4.11. Jordan Canonical Form

In general when the **A** matrix has multiple-order eigenvalues, unless the matrix is symmetric and has real elements, it cannot be diagonalized. However, there exists a similarity transformation

$$\Lambda = \mathbf{P}^{-1}\mathbf{A}\mathbf{P} \qquad (n \times n) \qquad (4\text{-}220)$$

such that the matrix Λ is *almost* a diagonal matrix. The matrix Λ is called the *Jordan canonical form.* Typical Jordan canonical forms are shown in the following examples:

$$\Lambda = \begin{bmatrix} \lambda_1 & 1 & 0 & 0 & 0 \\ 0 & \lambda_1 & 1 & 0 & 0 \\ 0 & 0 & \lambda_1 & 0 & 0 \\ 0 & 0 & 0 & \lambda_2 & 0 \\ 0 & 0 & 0 & 0 & \lambda_3 \end{bmatrix} \qquad (4\text{-}221)$$

$$\Lambda = \begin{bmatrix} \lambda_1 & 1 & 0 & 0 & 0 \\ 0 & \lambda_1 & 0 & 0 & 0 \\ 0 & 0 & \lambda_2 & 0 & 0 \\ 0 & 0 & 0 & \lambda_3 & 0 \\ 0 & 0 & 0 & 0 & \lambda_4 \end{bmatrix} \qquad (4\text{-}222)$$

The Jordan canonical form generally has the following properties:

1. The elements on the main diagonal of Λ are the eigenvalues of the matrix.
2. All the elements below the main diagonal of Λ are zero.
3. Some of the elements immediately above the multiple-ordered eigenvalues on the main diagonal are 1s, such as the cases illustrated by Eqs. (4-221) and (4-222).
4. The 1s, together with the eigenvalues, form typical blocks which are called *Jordan blocks.* In Eqs. (4-221) and (4-222) the Jordan blocks are enclosed by dotted lines.
5. When the nonsymmetrical **A** matrix has multiple-order eigenvalues, its eigenvectors are not linearly independent. For an $n \times n$ **A**, there is only r $(r < n)$ linearly independent eigenvectors.
6. The number of Jordan blocks is equal to the number of independent eigenvectors, r. There is one and only one linearly independent eigenvector associated with each Jordan block.
7. The number of 1s above the main diagonal is equal to $n - r$.

The matrix $\mathbf{P}$ is determined with the following considerations. Let us assume that $\mathbf{A}$ has q distinct eigenvalues among n eigenvalues. In the first place, the eigenvectors that correspond to the first-order eigenvalues are determined in the usual manner from

$$(\lambda_i \mathbf{I} - \mathbf{A})\mathbf{p}_i = \mathbf{0} \tag{4-223}$$

where λ_i denotes the ith distinct eigenvalue, $i = 1, 2, \ldots, q$.

The eigenvectors associated with an mth-order Jordan block are determined by referring to the Jordan block being written as

$$\begin{bmatrix} \lambda_j & 1 & 0 & \ldots & 0 \\ 0 & \lambda_j & 1 & \ldots & 0 \\ \ldots & \ldots & \ldots & \ldots & \cdot \\ 0 & 0 & \ldots & \lambda_j & 1 \\ 0 & 0 & \ldots & \ldots & \lambda_j \end{bmatrix} \quad (m \times m) \tag{4-224}$$

where λ_j denotes the jth eigenvalue.

Then the following transformation must hold:

$$[\mathbf{p}_1 \quad \mathbf{p}_2 \quad \ldots \quad \mathbf{p}_m]\begin{bmatrix} \lambda_j & 1 & 0 & \ldots & 0 \\ 0 & \lambda_j & 1 & & 0 \\ \ldots & \ldots & \ldots & \ldots & \cdot \\ 0 & 0 & \ldots & \lambda_j & 1 \\ 0 & 0 & \ldots & \ldots & \lambda_j \end{bmatrix} = \mathbf{A}[\mathbf{p}_1 \quad \mathbf{p}_2 \quad \ldots \quad \mathbf{p}_m] \tag{4-225}$$

or

$$\lambda_j \mathbf{p}_1 = \mathbf{A}\mathbf{p}_1$$
$$\mathbf{p}_1 + \lambda_j \mathbf{p}_2 = \mathbf{A}\mathbf{p}_2$$
$$\mathbf{p}_2 + \lambda_j \mathbf{p}_3 = \mathbf{A}\mathbf{p}_3 \tag{4-226}$$
$$\cdot$$
$$\cdot$$
$$\mathbf{p}_{m-1} + \lambda_j \mathbf{p}_m = \mathbf{A}\mathbf{p}_m$$

The vectors $\mathbf{p}_1, \mathbf{p}_2, \ldots, \mathbf{p}_m$ are determined from these equations, which can also be written

$$(\lambda_j \mathbf{I} - \mathbf{A})\mathbf{p}_1 = \mathbf{0}$$
$$(\lambda_j \mathbf{I} - \mathbf{A})\mathbf{p}_2 = -\mathbf{p}_1$$
$$(\lambda_j \mathbf{I} - \mathbf{A})\mathbf{p}_3 = -\mathbf{p}_2 \tag{4-227}$$
$$\cdot$$
$$\cdot$$
$$(\lambda_j \mathbf{I} - \mathbf{A})\mathbf{p}_m = -\mathbf{p}_{m-1}$$

EXAMPLE 4-13   Given the matrix

$$\mathbf{A} = \begin{bmatrix} 0 & 6 & -5 \\ 1 & 0 & 2 \\ 3 & 2 & 4 \end{bmatrix} \tag{4-228}$$

the determinant of $\lambda \mathbf{I} - \mathbf{A}$ is

$$|\lambda \mathbf{I} - \mathbf{A}| = \begin{vmatrix} \lambda & -6 & 5 \\ -1 & \lambda & -2 \\ -3 & -2 & \lambda - 4 \end{vmatrix} = \lambda^3 - 4\lambda^2 + 5\lambda - 2 \tag{4-229}$$

$$= (\lambda - 2)(\lambda - 1)^2$$

Therefore, $\mathbf{A}$ has a simple eigenvalue at $\lambda_1 = 2$ and a double eigenvalue at $\lambda_2 = 1$.

To find the Jordan canonical form of $\mathbf{A}$ involves the determination of the matrix $\mathbf{P}$ such that $\mathbf{\Lambda} = \mathbf{P}^{-1}\mathbf{A}\mathbf{P}$. The eigenvector that is associated with $\lambda_1 = 2$ is determined from

$$(\lambda_1 \mathbf{I} - \mathbf{A})\mathbf{p}_1 = 0 \tag{4-230}$$

Thus

$$\begin{bmatrix} 2 & -6 & 5 \\ -1 & 2 & -2 \\ -3 & -2 & -2 \end{bmatrix} \begin{bmatrix} p_{11} \\ p_{21} \\ p_{31} \end{bmatrix} = 0 \tag{4-231}$$

Setting $p_{11} = 2$ arbitrarily, the last equation gives $p_{21} = -1$ and $p_{31} = -2$. Therefore,

$$\mathbf{p}_1 = \begin{bmatrix} 2 \\ -1 \\ -2 \end{bmatrix} \tag{4-232}$$

For the eigenvector associated with the second-order eigenvalue, we turn to Eq. (4-227). We have (the two remaining eigenvectors are $\mathbf{p}_2$ and $\mathbf{p}_3$)

$$(\lambda_2 \mathbf{I} - \mathbf{A})\mathbf{p}_2 = 0 \tag{4-233}$$

and

$$(\lambda_2 \mathbf{I} - \mathbf{A})\mathbf{p}_3 = -\mathbf{p}_2 \tag{4-234}$$

Equation (4-233) leads to

$$\begin{bmatrix} 1 & -6 & 5 \\ -1 & 1 & -2 \\ -3 & -2 & -3 \end{bmatrix} \begin{bmatrix} p_{12} \\ p_{22} \\ p_{32} \end{bmatrix} = 0 \tag{4-235}$$

Setting $p_{12} = 1$ arbitrarily, we have $p_{22} = -\frac{3}{7}$ and $p_{32} = -\frac{5}{7}$. Thus

$$\mathbf{p}_2 = \begin{bmatrix} 1 \\ -\frac{3}{7} \\ -\frac{5}{7} \end{bmatrix} \tag{4-236}$$

Equation (4-234), when expanded, gives

$$\begin{bmatrix} 1 & -6 & -5 \\ -1 & 1 & -2 \\ -3 & -2 & -3 \end{bmatrix} \begin{bmatrix} p_{13} \\ p_{23} \\ p_{33} \end{bmatrix} = \begin{bmatrix} -1 \\ \frac{3}{7} \\ \frac{5}{7} \end{bmatrix} \tag{4-237}$$

from which we have

$$\mathbf{p}_3 = \begin{bmatrix} p_{13} \\ p_{23} \\ p_{33} \end{bmatrix} = \begin{bmatrix} 1 \\ -\frac{22}{49} \\ -\frac{46}{49} \end{bmatrix} \tag{4-238}$$

Thus

$$\mathbf{P} = \begin{bmatrix} 2 & 1 & 1 \\ -1 & -\frac{3}{7} & -\frac{22}{49} \\ -2 & -\frac{5}{7} & -\frac{46}{49} \end{bmatrix} \tag{4-239}$$

The Jordan canonical form is now obtained as

$$\mathbf{\Lambda} = \mathbf{P}^{-1}\mathbf{AP} = \begin{bmatrix} 2 & 0 & 0 \\ 0 & 1 & 1 \\ 0 & 0 & 1 \end{bmatrix} \tag{4-240}$$

Note that in this case there are two Jordan blocks and there is one element of unity above the main diagonal of $\mathbf{\Lambda}$.

4.12 State Diagram

The signal flow graph discussed in Section 3.5 applies only to algebraic equations. In this section we introduce the methods of the *state diagram*, which represents an extension of the signal flow graph to portray state equations and differential equations. The important significance of the state diagram is that it forms a close relationship among the state equations, state transition equation, computer simulation, and transfer functions. A state diagram is constructed following all the rules of the signal flow graph. Therefore, the state diagram may be used for solving linear systems either analytically or by computers.

Basic Analog Computer Elements

Before taking up the subject of state diagrams, it is useful to discuss the basic elements of an analog computer. The fundamental linear operations that can be performed on an analog computer are *multiplication by a constant*, *addition*, and *integration*. These are discussed separately in the following.

Multiplication by a constant. Multiplication of a machine variable by a constant is done by potentiometers and amplifiers. Let us consider the operation

$$x_2(t) = ax_1(t) \tag{4-241}$$

where a is a constant. If a lies between zero and unity, a potentiometer is used to realize the operation of Eq. (4-241). An operational amplifier is used to simulate Eq. (4-241) if a is a negative integer less than -1. The negative value of a considered is due to the fact that there is always an $180°$ phase shift between the output and the input of an operational amplifier. The computer block diagram symbols of the potentiometer and the operational amplifier are shown in Figs. 4-5 and 4-6, respectively.

Fig. 4-5. Analog-computer block-diagram symbol of a potentiometer.

Fig. 4-6. Analog-computer block diagram of an operational amplifier.

Algebraic sum of two or more variables. The algebraic sum of two or more machine variables may be obtained by means of the operational amplifier. Amplification may be accompanied by algebraic sum. For example, Fig. 4-7

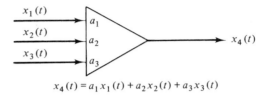

$$x_4(t) = a_1 x_1(t) + a_2 x_2(t) + a_3 x_3(t)$$

Fig. 4-7. Operational amplifier used as a summer.

illustrates the analog computer block diagram of a summing operational amplifier which portrays the following equation:

$$x_4(t) = a_1 x_1(t) + a_2 x_2(t) + a_3 x_3(t) \qquad (4\text{-}242)$$

Integration. The integration of a machine variable on an analog computer is achieved by means of a computer element called the *integrator*. If $x_1(t)$ is the output of the integrator with initial condition $x_1(t_0)$ given at $t = t_0$ and $x_2(t)$ is the input, the integrator performs the following operations:

$$x_1(t) = \int_{t_0}^{t} a x_2(\tau) \, d\tau + x_1(t_0) \qquad a \leq 1 \qquad (4\text{-}243)$$

The block diagram symbol of the integrator is shown in Fig. 4-8. The integrator can also serve simultaneously as a summing and amplification device.

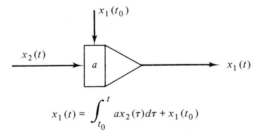

$$x_1(t) = \int_{t_0}^{t} a x_2(\tau) d\tau + x_1(t_0)$$

Fig. 4-8. Analog computer block diagram of an integrator.

We shall now show that these analog computer operations can be portrayed by signal flow graphs which are called state diagrams because the state variables are involved.

First consider the multiplication of a variable by a constant; we take the Laplace transform on both sides of Eq. (4-241). We have

$$X_2(s) = a X_1(s) \qquad (4\text{-}244)$$

The signal flow graph of Eq. (4-244) is shown in Fig. 4-9.

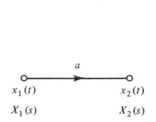

Fig. 4-9. Signal-flow-graph representation of $x_2(t) = ax_1(t)$ or $X_2(s) = aX_1(s)$.

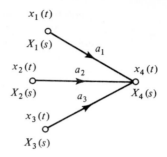

Fig. 4-10. Signal-flow-graph representation of $x_4(t) = a_1x_1(t) + a_2x_2(t) + a_3x_3(t)$ or $X_4(s) = a_1X_1(s) + a_2X_2(s) + a_3X_3(s)$.

For the summing operation of Eq. (4-242), the Laplace transform equation is

$$X_4(s) = a_1 X_1(s) + a_2 X_2(s) + a_3 X_3(s) \qquad (4\text{-}245)$$

The signal-flow-graph representation of the last equation is shown in Fig. 4-10. It is important to note that the variables in the signal flow graphs of Figs. 4-9 and 4-10 may be in the time domain or the Laplace transform domain. Since the branch gains are constants in these cases, the equations are algebraic in both domains.

For the integration operation, we take the Laplace transform on both sides of Eq. (4-243). In this case the transform operation is necessary, since the signal-flow-graph algebra does not handle integration in the time domain. We have

$$X_1(s) = a\mathscr{L}\left(\int_{t_0}^{t} x_2(\tau)\, d\tau\right) + \frac{x_1(t_0)}{s}$$

$$= a\mathscr{L}\left(\int_{0}^{t} x_2(\tau)\, d\tau - \int_{0}^{t_0} x_2(\tau)\, d\tau\right) + \frac{x_1(t_0)}{s} \qquad (4\text{-}246)$$

$$= \frac{aX_2(s)}{s} + \frac{x_1(t_0)}{s} - a\mathscr{L}\left(\int_{0}^{t_0} x_2(\tau)\, d\tau\right)$$

However, since the past history of the integrator is represented by $x_2(t_0)$, and the state transition starts from $\tau = t_0$, $x_2(\tau) = 0$ for $0 < \tau < t_0$. Thus Eq. (4-246) becomes

$$X_1(s) = \frac{aX_2(s)}{s} + \frac{x_1(t_0)}{s} \qquad \tau \geq t_0 \qquad (4\text{-}247)$$

It should be emphasized that Eq. (4-247) is defined only for the period $\tau \geq t_0$. Therefore, the inverse Laplace transform of $X_1(s)$ in Eq. (4-247) will lead to $x_1(t)$ of Eq. (4-243).

Equation (4-247) is now algebraic and can be represented by a signal flow graph as shown in Fig. 4-11. An alternative signal flow graph for Eq. (4-247) is shown in Fig. 4-12.

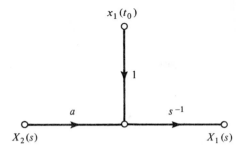

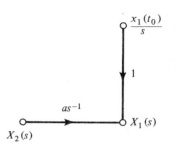

Fig. 4-11. Signal-flow-graph representation of $X_1(s) = [aX_2(s)/s] + [x_1(t_0)/s]$

Fig. 4-12. Signal-flow-graph representation of $X_1(s) = [aX_2(s)/s] + [x_1(t_0)/s]$.

Thus we have established a correspondence between the simple analog computer operations and the signal-flow-graph representations. Since, as shown in Fig. 4-12, these signal-flow-graph elements may include initial conditions and can be used to solve state transition problems, they form the basic elements of the state diagram.

Before embarking on several illustrative examples on state diagrams, let us point out the important usages of the state diagrams.

1. A state diagram can be constructed directly from the system's differential equation. This allows the determination of the state variables and the state equations once the differential equation of the system is given.
2. A state diagram can be constructed from the system's transfer function. This step is defined as the decomposition of transfer functions (Section 4.13).
3. The state diagram can be used for the programming of the system on an analog computer.
4. The state diagram can be used for the simulation of the system on a digital computer.
5. The state transition equation in the Laplace transform domain may be obtained from the state diagram by means of the signal-flow-graph gain formula.
6. The transfer functions of a system can be obtained from the state diagram.
7. The state equations and the output equations can be determined from the state diagram.

The details of these techniques are given below.

From Differential Equation to the State Diagram

When a linear system is described by a high-order differential equation, a state diagram can be constructed from these equations, although a direct approach is not always the most convenient. Consider the following differential equation:

$$\frac{d^n c}{dt^n} + a_1 \frac{d^{n-1}c}{dt^{n-1}} + \ldots + a_{n-1}\frac{dc}{dt} + a_n c = r \qquad (4\text{-}248)$$

In order to construct a state diagram using this equation, we rearrange the equation to read

$$\frac{d^n c}{dt^n} = -a_1 \frac{d^{n-1}c}{dt^{n-1}} - \ldots - a_{n-1}\frac{dc}{dt} - a_n c + r \qquad (4\text{-}249)$$

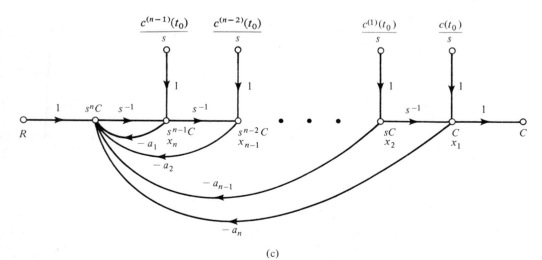

Fig. 4-13. State diagram representation of the differential equation of Eq. (4-248).

Let us use the following symbols to simplify the representation of the derivatives of c:

$$c^{(i)} = \frac{d^i c}{dt^i} \qquad i = 1, 2, \ldots, n \tag{4-250}$$

Frequently, $\dot{c}$ in the literature, is used to represent dc/dt.

Now the variables $r, c, c^{(1)}, c^{(2)}, \ldots, c^{(n)}$ are represented by nodes arranged as shown in Fig. 4-13(a). In terms of Laplace transform, these variables are denoted by $R(s)$, $C(s)$, $sC(s)$, $s^2C(s)$, $\ldots$, $s^nC(s)$, respectively.

As the next step, the nodes in Fig. 4-13(a) are connected by branches to portray Eq. (4-249). Since the variables $c^{(i)}$ and $c^{(i-1)}$ are related through integration with respect to time, they can be interconnected by a branch with gain s^{-1} and the elements of Figs. 4-11 and 4-12 can be used. Therefore, the complete state diagram is drawn as shown in Fig. 4-13(c).

When the differential equation is that of Eq. (4-145), with derivatives of the input on the right side, the problem of drawing the state diagram directly is not so straightforward. We shall show later that, in general, it is more convenient to obtain the transfer function from the differential equation first and then obtain the state diagram through decomposition (Section 4.13).

EXAMPLE 4-14   Consider the following differential equation:

$$\frac{d^2 c}{dt^2} + 3\frac{dc}{dt} + 2c = r \tag{4-251}$$

Equating the highest-ordered term of Eq. (4-251) to the rest of the terms, we have

$$\frac{d^2 c}{dt^2} = -2c - 3\frac{dc}{dt} + r \tag{4-252}$$

Following the procedure outlined above, the state diagram of the system is shown in Fig. 4-14.

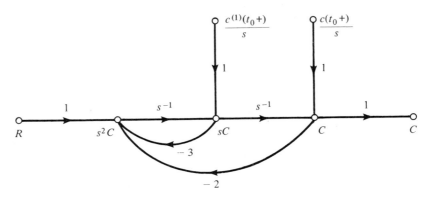

Fig. 4-14. State diagram for Eq. (4-251).

From State Diagram to Analog Computer Block Diagram

It was mentioned earlier that the state diagram is essentially a block diagram for the programming of an analog computer, except for the phase reversal through amplification and integration.

EXAMPLE 4-15   An analog computer block diagram of the system described by Eq. (4-251) is shown in Fig. 4-15. The final practical version of the computer block diagram for programming may be somewhat different from shown, since amplitude and time scaling may be necessary.

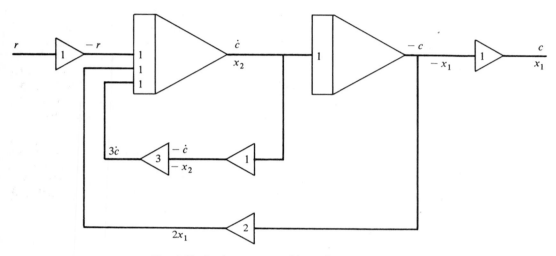

Fig. 4-15. Analog-computer block diagram for the system described by Eq. (4-251).

From State Diagram to Digital Computer Simulation

The solution of differential equations by FORTRAN on the digital computer has been well established. However, from the standpoint of programming, a convenient way of modeling a system on the digital computer is by CSMP (Continuous System Modeling Program).[32] In many respects CSMP serves the same purpose as an analog computer program, except that the scaling problem is practically eliminated. The state diagram or the state equations form a natural basis for the solution by CSMP. The following examples illustrate typical CSMP statements for the mathematical equations listed:

Mathematical Equations	*CSMP Statements*
$c = a_1 x_1 + a_2 x_2$	$C = A1 * X1 + A2 * X2$
$y = \dfrac{x_1}{2}$	$Y = X1/2.$
$x_1 = \displaystyle\int_0^t x_2(\tau)\, d\tau + x_1(0)$	$X1 = \text{INTGRL}(X2, X10)$

EXAMPLE 4-16   From the state diagram of Fig. 4-14, the following equations are written:

$$c = \int \dot{c}\, dt \tag{4-253}$$

$$\dot{c} = \int \ddot{c}\, dt \tag{4-254}$$

$$\ddot{c} = r - 3\dot{c} - 2c \tag{4-255}$$

Let the variables of these equations be denoted by

$$c = C \qquad c(0) = CO$$
$$\dot{c} = C1 \qquad \dot{c}(0) = C10$$
$$\ddot{c} = C2$$
$$r = R$$

on the CSMP. Then the main program of the CSMP representation of the system is given as follows:

$$C = \text{INTGRL}\ (C1,\ C0) \tag{4-256}$$

$$C1 = \text{INTGRL}\ (C2,\ C10) \tag{4-257}$$

$$C2 = R - 3.*C1 - 2.*C \tag{4-258}$$

From State Diagram to the State Transition Equation

We have shown earlier that the Laplace transform method of solving the state equation requires the carrying out of the matrix inverse of $(s\mathbf{I} - \mathbf{A})$. With the state diagram, the equivalence of the matrix inverse operation is carried out by use of the signal-flow-graph formula.

The state transition equation in the Laplace transform domain is

$$\mathbf{X}(s) = (s\mathbf{I} - \mathbf{A})^{-1}\mathbf{x}(t_0^+) + (s\mathbf{I} - \mathbf{A})^{-1}\mathbf{BR}(s) \qquad t \geq t_0 \tag{4-259}$$

Therefore, the last equation can be written directly from the state diagram by use of the gain formula, with $X_i(s)$, $i = 1, 2, \ldots, n$, as the output nodes, and $x_i(t_0^+)$, $i = 1, 2, \ldots, n$, and $R_j(s), j = 1, 2, \ldots, p$, as the input nodes.

EXAMPLE 4-17   Consider the state diagram of Fig. 4-14. The outputs of the integrators are assigned as state variables and the state diagram is redrawn as shown in Fig. 4-16.

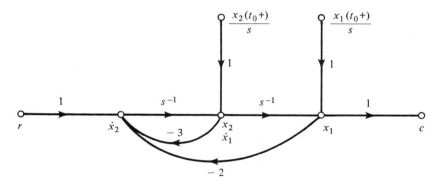

Fig. 4-16. State diagram for Eq. (4-251).

Applying the gain formula to the state diagram in Fig. 4-16, with $X_1(s)$ and $X_2(s)$ as output nodes, $x_1(t_0^+)$, $x_2(t_0^+)$, and $R(s)$ as input nodes, we have

$$X_1(s) = \frac{s^{-1}(1 + 3s^{-1})}{\Delta} x_1(t_0^+) + \frac{s^{-2}}{\Delta} x_2(t_0^+) + \frac{s^{-2}}{\Delta} R(s) \tag{4-260}$$

$$X_2(s) = \frac{-2s^{-2}}{\Delta} x_1(t_0^+) + \frac{s^{-1}}{\Delta} x_2(t_0^+) + \frac{s^{-1}}{\Delta} R(s) \tag{4-261}$$

where
$$\Delta = 1 + 3s^{-1} + 2s^{-2} \qquad (4\text{-}262)$$

After simplification, Eqs. (4-260) and (4-261) are presented in matrix form:

$$\begin{bmatrix} X_1(s) \\ X_2(s) \end{bmatrix} = \frac{1}{(s+1)(s+2)} \begin{bmatrix} s+3 & 1 \\ -2 & s \end{bmatrix} \begin{bmatrix} x_1(t_0^+) \\ x_2(t_0^+) \end{bmatrix} + \begin{bmatrix} \dfrac{1}{(s+1)(s+2)} \\ \dfrac{s}{(s+1)(s+2)} \end{bmatrix} R(s) \qquad (4\text{-}263)$$

The state transition equation for $t \geq t_0$ is obtained by taking the inverse Laplace transform on both sides of Eq. (4-263).

Consider that the input $r(t)$ is a unit step function applied at $t = t_0$. Then the following inverse Laplace transform relationships are identified:

$$\mathcal{L}^{-1}\left(\frac{1}{s}\right) = u_s(t - t_0) \qquad t \geq t_0 \qquad (4\text{-}264)$$

$$\mathcal{L}^{-1}\left(\frac{1}{s+a}\right) = e^{-a(t-t_0)}u_s(t - t_0) \qquad t \geq t_0 \qquad (4\text{-}265)$$

The inverse Laplace transform of Eq. (4-263) is

$$\begin{bmatrix} x_1(t) \\ x_2(t) \end{bmatrix} = \begin{bmatrix} 2e^{-(t-t_0)} - e^{-2(t-t_0)} & e^{-(t-t_0)} - e^{-2(t-t_0)} \\ -2e^{-(t-t_0)} + 2e^{-2(t-t_0)} & -e^{-(t-t_0)} + 2e^{-2(t-t_0)} \end{bmatrix} \begin{bmatrix} x_1(t_0^+) \\ x_2(t_0^+) \end{bmatrix}$$
$$+ \begin{bmatrix} \frac{1}{2}u_s(t - t_0) - e^{-(t-t_0)} + \frac{1}{2}e^{-2(t-t_0)} \\ e^{-(t-t_0)} - e^{-2(t-t_0)} \end{bmatrix} \qquad t \geq t_0 \qquad (4\text{-}266)$$

The reader should compare this result with that of Eq. (4-77), obtained for $t \geq 0$.

From State Diagram to Transfer Function

The transfer function between an input and an output is obtained from the state diagram by setting all other inputs and all initial states to zero.

EXAMPLE 4-18   Consider the state diagram of Fig. 4-16. The transfer function $C(s)/R(s)$ is obtained by applying the gain formula between these two nodes and setting $x_1(t_0^+) = 0$ and $x_2(t_0^+) = 0$. Therefore,

$$\frac{C(s)}{R(s)} = \frac{1}{s^2 + 3s + 2} \qquad (4\text{-}267)$$

The characteristic equation of the system is

$$s^2 + 3s + 2 = 0 \qquad (4\text{-}268)$$

From State Diagram to the State Equations

When the state diagram of a system is already given, the state equations and the output equations can be obtained directly from the state diagram by use of the gain formula. Some clarification seems necessary here, since the state transition equations are determined from the state diagram by use of the gain formula. However, when writing the state transition equations from the state diagram, the state variables, $X_i(s)$, $i = 1, 2, \ldots, n$, are regarded as the output nodes, and the inputs, $R_j(s)$, $j = 1, 2, \ldots, p$, and the initial states, $x_i(t_0)$, $i = 1, 2, \ldots, n$, are regarded as the input nodes. Furthermore, the state transition equations, as written directly from the state diagram, are necessarily in the Laplace transform domain. The state transition equations in the time domain are subsequently obtained by taking the inverse Laplace transform.

The left side of the state equation contains the first-order time derivative of the state variable $\dot{x}_i(t)$. The right side of the equation contains the state variables and the input variables. There are no Laplace operator s or initial state variables in a state equation. Therefore, to obtain state equations from the state diagram, we should disregard all the initial states and all the integrator branches with gains s^{-1}. To avoid confusion, the initial states and the branches with the gain s^{-1} can actually be eliminated from the state diagram. The state diagram of Fig. 4-16 is simplified as described above, and the result is shown in Fig. 4-17. Then, using $\dot{x}_1$ and $\dot{x}_2$ as output nodes and x_1, x_2, and r as input nodes,

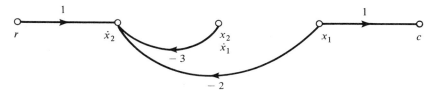

Fig. 4-17. State diagram of Fig. 4-16 with the initial states and the integrator branches eliminated.

and applying the gain formula between these nodes, the state equations are written directly:

$$\frac{dx_1}{dt} = x_2$$

$$\frac{dx_2}{dt} = -2x_1 - 3x_2 + r$$

(4-269)

EXAMPLE 4-19   As another example of illustrating the determination of the state equations from the state diagram, consider the state diagram shown in Fig. 4-18(a). This illustration will emphasize the importance of using the gain formula. Figure 4-18(b) shows the state diagram with the initial states

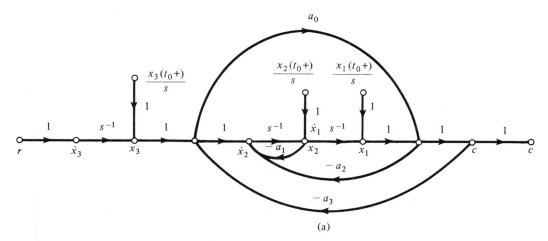

(a)

Fig. 4-18. (a) State diagram.

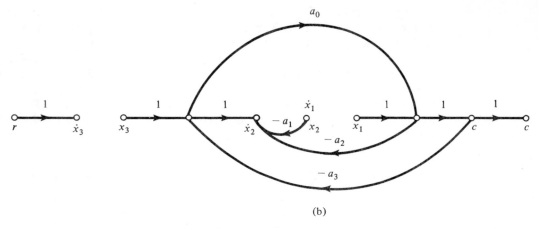

(b)

Fig. 4-18 (Cont.). (b) State diagram in (a) with all initial states and integrators eliminated.

and the integrators being eliminated. Notice that in this case the state diagram in Fig. 4-18(b) still contains a loop. Applying the gain formula to the diagram of Fig. 4-18(b) with $\dot{x}_1$, $\dot{x}_2$, and $\dot{x}_3$ as the output node variables and r, x_1, x_2, and x_3 as the input nodes, the state equations are determined as follows:

$$
\begin{bmatrix} \dfrac{dx_1}{dt} \\[2mm] \dfrac{dx_2}{dt} \\[2mm] \dfrac{dx_3}{dt} \end{bmatrix} = \begin{bmatrix} 0 & 1 & 0 \\[2mm] \dfrac{-(a_2 + a_3)}{1 + a_0 a_3} & -a_1 & \dfrac{1 - a_0 a_2}{1 + a_0 a_3} \\[2mm] 0 & 0 & 0 \end{bmatrix} \begin{bmatrix} x_1 \\[2mm] x_2 \\[2mm] x_3 \end{bmatrix} + \begin{bmatrix} 0 \\[2mm] 0 \\[2mm] 1 \end{bmatrix} r \qquad (4\text{-}270)
$$

4.13 Decomposition of Transfer Functions

Up until this point, various methods of characterizing a linear system have been presented. It will be useful to summarize briefly and gather thoughts at this point, before proceeding to the main topics of this section.

It has been shown that the starting point of the description of a linear system may be the system's differential equation, transfer function, or dynamic equations. It is demonstrated that all these methods are closely related. Further, the state diagram is shown to be a useful tool which not only can lead to the solutions of the state equations but also serves as a vehicle of translation from one type of description to the others. A block diagram is drawn as shown in Fig. 4-19 to illustrate the interrelationships between the various loops of describing a linear system. The block diagram shows that starting, for instance, with the differential equation of a system, one can get to the solution by use of the transfer function method or the state equation method. The block diagram also shows that the majority of the relationships are bilateral, so a great deal of flexibility exists between the methods.

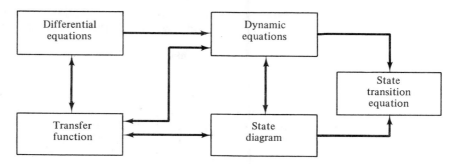

Fig. 4-19. Block diagram showing the relationships among various methods of describing linear systems.

One step remains to be discussed. This involves the construction of the state diagram from the transfer function. In general, it is necessary to establish a better method than using Eqs. (4-146) through (4-148) in getting from a high-order differential equation to the state equations.

The process of going from the transfer function to the state diagram or the state equations is called the *decomposition* of the transfer function. In general, there are three basic ways of decomposing a transfer function: *direct decomposition*, *cascade decomposition*, and *parallel decomposition*. Each of these three schemes of decomposition has its own advantage and is best suited for a particular situation.

Direct Decomposition

The direct decomposition scheme is applied to a transfer function that is not in factored form. Without loss of generality, the method of direct decomposition can be described by the following transfer function:

$$\frac{C(s)}{R(s)} = \frac{a_0 s^2 + a_1 s + a_2}{b_0 s^2 + b_1 s + b_2} \tag{4-271}$$

The objective is to obtain the state diagram and the state equations. The following steps are outlined for the direct decomposition:

1. Alter the transfer function so that it has only negative powers of s. This is accomplished by multiplying the numerator and the denominator of the transfer function by the inverse of its highest power in s. For the transfer function of Eq. (4-271), we multiply the numerator and the denominator of $C(s)/R(s)$ by s^{-2}.

2. Multiply the numerator and the denominator of the transfer function by a dummy variable $X(s)$. Implementing steps 1 and 2, Eq. (4-271) becomes

$$\frac{C(s)}{R(s)} = \frac{a_0 + a_1 s^{-1} + a_2 s^{-2}}{b_0 + b_1 s^{-1} + b_2 s^{-2}} \frac{X(s)}{X(s)} \tag{4-272}$$

3. The numerators and the denominators on both sides of the transfer

function resulting from steps 1 and 2 are equated to each other, respectively. From Eq. (4-272) this step results in

$$C(s) = (a_0 + a_1 s^{-1} + a_2 s^{-2})X(s) \tag{4-273}$$

$$R(s) = (b_0 + b_1 s^{-1} + b_2 s^{-2})X(s) \tag{4-274}$$

4. In order to construct a state diagram using these two equations, they must first be in the proper cause-and-effect relation. It is apparent that Eq. (4-273) already satisfies this prerequisite. However, Eq. (4-274) has the input on the left side and must be rearranged. Dividing both sides of Eq. (4-274) by b_0 and writing $X(s)$ in terms of the other terms, we have

$$X(s) = \frac{1}{b_0} R(s) - \frac{b_1}{b_0} s^{-1}X(s) - \frac{b_2}{b_0} s^{-2}X(s) \tag{4-275}$$

The state diagram is now drawn in Fig. 4-20 using the expressions in Eqs. (4-273) and (4-275). For simplicity, the initial states are not drawn on the diagram. As usual, the state variables are defined as the outputs of the integrators.

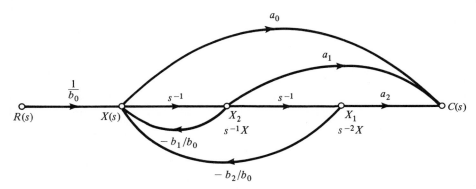

Fig. 4-20. State diagram for the transfer function of Eq. (4-271) by direct decomposition.

Following the method described in the last section, the state equations are written directly from the state diagram:

$$\begin{bmatrix} \dfrac{dx_1}{dt} \\[2mm] \dfrac{dx_2}{dt} \end{bmatrix} = \begin{bmatrix} 0 & 1 \\[2mm] -\dfrac{b_2}{b_0} & -\dfrac{b_1}{b_0} \end{bmatrix} \begin{bmatrix} x_1 \\[2mm] x_2 \end{bmatrix} + \begin{bmatrix} 0 \\[2mm] \dfrac{1}{b_0} \end{bmatrix} r \tag{4-276}$$

The output equation is obtained from Fig. 4-20 by applying the gain formula with $c(t)$ as the output node and $x_1(t)$, $x_2(t)$, and $r(t)$ as the input nodes.

$$c = \left(a_2 - \frac{a_0 b_2}{b_0}\right)x_1 + \left(a_1 - \frac{a_0 b_1}{b_0}\right)x_2 + \frac{a_0}{b_0} r \tag{4-277}$$

Cascade Decomposition

Cascade decomposition may be applied to a transfer function that is in the factored form. Consider that the transfer function of Eq. (4-271) may be factored in the following form (of course, there are other possible combinations of factoring):

$$\frac{C(s)}{R(s)} = \frac{a_0}{b_0} \frac{s + z_1}{s + p_1} \frac{s + z_2}{s + p_2} \tag{4-278}$$

where z_1, z_2, p_1, and p_2 are real constants. Then it is possible to treat the function as the product of two first-order transfer functions. The state diagram of each of the first-order transfer functions is realized by using the direct decomposition method. The complete state diagram is obtained by cascading the two first-order diagrams as shown in Fig. 4-21. As usual, the outputs of the

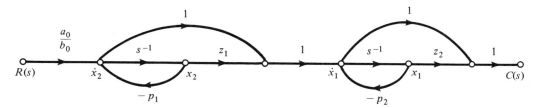

Fig. 4-21. State diagram of the transfer function of Eq. (4-278) by cascade decomposition.

integrators on the state diagram are assigned as the state variables. The state equations are written in matrix form:

$$\begin{bmatrix} \dfrac{dx_1}{dt} \\[2mm] \dfrac{dx_2}{dt} \end{bmatrix} = \begin{bmatrix} -p_2 & z_1 - p_1 \\[2mm] 0 & -p_1 \end{bmatrix} \begin{bmatrix} x_1 \\[2mm] x_2 \end{bmatrix} + \begin{bmatrix} \dfrac{a_0}{b_0} \\[2mm] \dfrac{a_0}{b_0} \end{bmatrix} r \tag{4-279}$$

The output equation is

$$c = (z_2 - p_2)x_1 + (z_1 - p_1)x_2 + \frac{a_0}{b_0}r \tag{4-280}$$

The cascade decomposition has the advantage that the poles and zeros of the transfer function appear as isolated branch gains on the state diagram. This facilitates the study of the effects on the system when the poles and zeros are varied.

Parallel Decomposition

When the denominator of a transfer function is in factored form, it is possible to expand the transfer function by partial fractions. Consider that a second-order system is represented by the following transfer function:

$$\frac{C(s)}{R(s)} = \frac{P(s)}{(s + p_1)(s + p_2)} \tag{4-281}$$

where $P(s)$ is a polynomial of order less than 2. We assume that the poles p_1

and p_2 may be complex conjugate for analytical purposes, but it is difficult to implement complex coefficients on the computer.

In this case if p_1 and p_2 are equal it would not be possible to carry out a partial-fraction expansion of the transfer function of Eq. (4-281). With p_1 and p_2 being distinct, Eq. (4-281) is written

$$\frac{C(s)}{R(s)} = \frac{K_1}{s + p_1} + \frac{K_2}{s + p_2} \tag{4-282}$$

where K_1 and K_2 are constants.

The state diagram for the system is formed by the parallel combination of the state diagram representation of each of the first-order terms on the right side of Eq. (4-282), as shown in Fig. 4-22. The state equations of the system are written

$$\begin{bmatrix} \dfrac{dx_1}{dt} \\ \dfrac{dx_2}{dt} \end{bmatrix} = \begin{bmatrix} -p_1 & 0 \\ 0 & -p_2 \end{bmatrix} \begin{bmatrix} x_1 \\ x_2 \end{bmatrix} + \begin{bmatrix} 1 \\ 1 \end{bmatrix} r \tag{4-283}$$

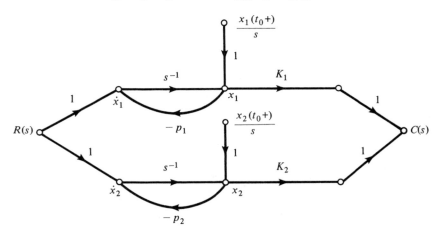

Fig. 4-22. State diagram of the transfer function of Eq. (4-281) by parallel decomposition.

The output equation is

$$c = [K_1 \quad K_2] \begin{bmatrix} x_1 \\ x_2 \end{bmatrix} \tag{4-284}$$

One of the advantages of the parallel decomposition is that for transfer functions with simple poles, the resulting **A** matrix is always a diagonal matrix. Therefore, we can consider that parallel decomposition may be used for the diagonalization of the **A** matrix.

When a transfer function has multiple-order poles, care must be taken that the state diagram, as obtained through the parallel decomposition, contain the minimum number of integrators. To further clarify the point just made, consider the following transfer function and its partial-fraction expansion:

$$\frac{C(s)}{R(s)} = \frac{2s^2 + 6s + 5}{(s+1)^2(s+2)} = \frac{1}{(s+1)^2} + \frac{1}{s+1} + \frac{1}{s+2} \qquad (4\text{-}285)$$

Note that the transfer function is of the third order, and although the total order of the terms on the right side of Eq. (4-285) is four, only three integrators should be used in the state diagram. The state diagram for the system is drawn as shown in Fig. 4-23. The minimum number of three integrators are used, with one in-

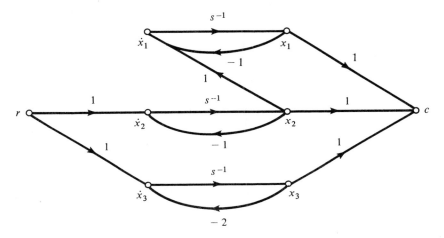

Fig. 4-23. State diagram of the transfer function of Eq. (4-285) by paralled decomposition.

tegrator being shared by two channels. The state equations of the system are written

$$\begin{bmatrix} \dfrac{dx_1}{dt} \\[2ex] \dfrac{dx_2}{dt} \\[2ex] \dfrac{dx_3}{dt} \end{bmatrix} = \begin{bmatrix} -1 & 1 & 0 \\ 0 & -1 & 0 \\ 0 & 0 & -2 \end{bmatrix} \begin{bmatrix} x_1 \\ x_2 \\ x_3 \end{bmatrix} + \begin{bmatrix} 0 \\ 1 \\ 1 \end{bmatrix} r \qquad (4\text{-}286)$$

Therefore, the **A** matrix is of the Jordan canonical form.

4.14 Transformation into Modal Form

When the **A** matrix has complex eigenvalues it may not be possible to transform it into a diagonal matrix with real elements. To facilitate computer computation, it is desirable to avoid matrices with complex elements. When **A** has complex eigenvalues, in general, the matrix can be transformed into a nondiagonal matrix which is called the *modal form* by the transformation

$$\mathbf{\Lambda} = \mathbf{P}^{-1}\mathbf{A}\mathbf{P} \qquad (4\text{-}287)$$

Let us assume that **A** is 2×2 and has eigenvalues $\lambda_1 = \sigma + j\omega$ and $\lambda_2 = \sigma - j\omega$. Then the modal-form matrix is given by

$$\mathbf{\Lambda} = \begin{pmatrix} \sigma & \omega \\ -\omega & \sigma \end{pmatrix} \tag{4-288}$$

The elements of the **P** matrix may be determined by brute force using Eqs. (4-287) and (4-288). If **A** has m real distinct eigenvalues in $\lambda_1, \lambda_2, \ldots, \lambda_m$, and n sets of complex-conjugate eigenvalues in $\lambda_i = \sigma_i \pm j\omega_i, i = 1, 2, \ldots, n$, the modal-form matrix is given by

$$\mathbf{\Lambda} = \begin{bmatrix} \lambda_1 & 0 & \ldots & 0 & 0 & 0 & \ldots & 0 \\ 0 & \lambda_2 & \ldots & 0 & 0 & 0 & \ldots & 0 \\ \ldots & \ldots & \ldots & \ldots & \ldots & \ldots & \ldots & \ldots \\ 0 & . & \ldots & \lambda_m & 0 & 0 & \ldots & 0 \\ 0 & 0 & \ldots & 0 & \mathbf{\Lambda}_1 & 0 & \ldots & 0 \\ 0 & 0 & \ldots & 0 & 0 & \mathbf{\Lambda}_2 & \ldots & 0 \\ \ldots & \ldots & \ldots & \ldots & \ldots & \ldots & \ldots & \ldots \\ 0 & 0 & \ldots & 0 & 0 & 0 & \ldots & \mathbf{\Lambda}_n \end{bmatrix} \tag{4-289}$$

where

$$\mathbf{\Lambda}_i = \begin{pmatrix} \sigma_i & \omega_i \\ -\omega_i & \sigma_i \end{pmatrix} \tag{4-290}$$

If the jth complex eigenvalue pair is of multiplicity m, then $\mathbf{\Lambda}_j$ is written

$$\mathbf{\Lambda}_j = \begin{bmatrix} \mathbf{\Gamma}_j & \mathbf{I} & \mathbf{0} & \ldots & \mathbf{0} \\ \mathbf{0} & \mathbf{\Gamma}_j & \mathbf{I} & \ldots & \mathbf{0} \\ \mathbf{0} & \mathbf{0} & \mathbf{\Gamma}_j & \ldots & \mathbf{0} \\ \ldots & \ldots & \ldots & \ldots & \ldots \\ \mathbf{0} & \mathbf{0} & \mathbf{0} & \ldots & \mathbf{\Gamma}_j \end{bmatrix} \quad (m \times m \text{ blocks}) \tag{4-291}$$

where

$$\mathbf{\Gamma}_j = \begin{bmatrix} \sigma_j & \omega_j \\ -\omega_j & \sigma_j \end{bmatrix}. \tag{4-292}$$

$$\mathbf{I} = \begin{bmatrix} 1 & 0 \\ 0 & 1 \end{bmatrix} \tag{4-293}$$

The modal-form matrix in Eq. (4-289) is easily modified for real and multiple-order eigenvalues by use of the Jordan canonical form.

Although the modal-form matrix is not diagonal and does not represent decoupling of the states from the standpoint of the state diagram, it still has the components of the eigenvalues as its matrix elements.

To determine the transformation matrix **P** for the matrix **Λ** of Eq. (4-288), we let

$$\mathbf{P} = [\mathbf{p}_1 \quad \mathbf{p}_2] \tag{4-294}$$

where $\mathbf{p}_1$ and $\mathbf{p}_2$ are 2×1 vectors. Equation (4-287) is written

$$[\mathbf{p}_1 \quad \mathbf{p}_2] \begin{bmatrix} \sigma & \omega \\ -\omega & \sigma \end{bmatrix} = \mathbf{A}[\mathbf{p}_1 \quad \mathbf{p}_2] \tag{4-295}$$

or

$$\sigma\mathbf{p}_1 - \omega\mathbf{p}_2 = \mathbf{A}\mathbf{p}_1 \tag{4-296}$$

$$\omega\mathbf{p}_1 + \sigma\mathbf{p}_2 = \mathbf{A}\mathbf{p}_2 \tag{4-297}$$

These two equations are expressed in matrix equation form,

$$\begin{bmatrix} \sigma\mathbf{I} & -\omega\mathbf{I} \\ \omega\mathbf{I} & \sigma\mathbf{I} \end{bmatrix}\begin{bmatrix} \mathbf{p}_1 \\ \mathbf{p}_2 \end{bmatrix} = \begin{bmatrix} \mathbf{A} & 0 \\ 0 & \mathbf{A} \end{bmatrix}\begin{bmatrix} \mathbf{p}_1 \\ \mathbf{p}_2 \end{bmatrix} \tag{4-298}$$

where $\mathbf{I}$ denotes a 2×2 identity matrix.

Let $\mathbf{q}_1$ and $\mathbf{q}_2$ denote the eigenvectors that are associated with the two complex-conjugate eigenvalues, $\lambda_1 = \sigma + j\omega$ and $\lambda_2 = \sigma - j\omega$, respectively. Then, according to the definition of eigenvectors, $\mathbf{q}_1$ and $\mathbf{q}_2$ must satisfy

$$(\sigma + j\omega)\mathbf{q}_1 = \mathbf{A}\mathbf{q}_1 \tag{4-299}$$

$$(\sigma - j\omega)\mathbf{q}_2 = \mathbf{A}\mathbf{q}_2 \tag{4-300}$$

Let

$$\mathbf{q}_1 = \boldsymbol{\alpha}_1 + j\boldsymbol{\beta}_1 \tag{4-301}$$

$$\mathbf{q}_2 = \boldsymbol{\alpha}_2 + j\boldsymbol{\beta}_2 \tag{4-302}$$

Then Eqs. (4-299) and (4-300) become

$$(\sigma + j\omega)(\boldsymbol{\alpha}_1 + j\boldsymbol{\beta}_1) = \mathbf{A}(\boldsymbol{\alpha}_1 + j\boldsymbol{\beta}_1) \tag{4-303}$$

$$(\sigma - j\omega)(\boldsymbol{\alpha}_2 + j\boldsymbol{\beta}_2) = \mathbf{A}(\boldsymbol{\alpha}_2 + j\boldsymbol{\beta}_2) \tag{4-304}$$

Equating the real and imaginary parts in the last two equations, we have

$$\begin{bmatrix} \sigma\mathbf{I} & -\omega\mathbf{I} \\ \omega\mathbf{I} & \sigma\mathbf{I} \end{bmatrix}\begin{bmatrix} \boldsymbol{\alpha}_1 \\ \boldsymbol{\beta}_1 \end{bmatrix} = \begin{bmatrix} \mathbf{A} & 0 \\ 0 & \mathbf{A} \end{bmatrix}\begin{bmatrix} \boldsymbol{\alpha}_1 \\ \boldsymbol{\beta}_1 \end{bmatrix} \tag{4-305}$$

and

$$\begin{bmatrix} \sigma\mathbf{I} & \omega\mathbf{I} \\ -\omega\mathbf{I} & \sigma\mathbf{I} \end{bmatrix}\begin{bmatrix} \boldsymbol{\alpha}_2 \\ \boldsymbol{\beta}_2 \end{bmatrix} = \begin{bmatrix} \mathbf{A} & 0 \\ 0 & \mathbf{A} \end{bmatrix}\begin{bmatrix} \boldsymbol{\alpha}_2 \\ \boldsymbol{\beta}_2 \end{bmatrix} \tag{4-306}$$

Comparing Eq. (4-305) with Eq. (4-298), we have the identity

$$\mathbf{P} = [\mathbf{p}_1 \quad \mathbf{p}_2] = [\boldsymbol{\alpha}_1 \quad \boldsymbol{\beta}_1] \tag{4-307}$$

The significance of this result is that the transformation matrix $\mathbf{P}$ is formed by taking the real and imaginary components of the eigenvector of $\mathbf{A}$ associated with $\lambda_1 = \sigma + j\omega$.

EXAMPLE 4-20   Consider the state equation

$$\dot{\mathbf{x}} = \mathbf{A}\mathbf{x} + \mathbf{B}r \tag{4-308}$$

where

$$\mathbf{A} = \begin{bmatrix} 0 & 1 \\ -2 & -2 \end{bmatrix} \quad \mathbf{B} = \begin{bmatrix} 0 \\ 1 \end{bmatrix}$$

The eigenvalues of $\mathbf{A}$ are $\lambda_1 = -1 + j$ and $\lambda_2 = -1 - j$. The eigenvectors are

$$\mathbf{q}_1 = \begin{bmatrix} 1 \\ -1 + j \end{bmatrix} \quad \mathbf{q}_2 = \begin{bmatrix} 1 \\ -1 - j \end{bmatrix}$$

or

$$\mathbf{q}_1 = \boldsymbol{\alpha}_1 + j\boldsymbol{\beta}_1 = \begin{bmatrix} 1 \\ -1 \end{bmatrix} + j\begin{bmatrix} 0 \\ 1 \end{bmatrix} \quad \mathbf{q}_2 = \boldsymbol{\alpha}_2 + j\boldsymbol{\beta}_2 = \begin{bmatrix} 1 \\ -1 \end{bmatrix} + j\begin{bmatrix} 0 \\ -1 \end{bmatrix}$$

Therefore,

$$\mathbf{P} = [\boldsymbol{\alpha}_1 \quad \boldsymbol{\beta}_1] = \begin{bmatrix} 1 & 0 \\ -1 & 1 \end{bmatrix} \tag{4-309}$$

$$\boldsymbol{\Lambda} = \mathbf{P}^{-1}\mathbf{AP} = \begin{bmatrix} -1 & 1 \\ -1 & -1 \end{bmatrix} \tag{4-310}$$

$$\boldsymbol{\Gamma} = \mathbf{P}^{-1}\mathbf{B} = \begin{bmatrix} 0 \\ 1 \end{bmatrix} \tag{4-311}$$

The original state equation of Eq. (4-308) is transformed to

$$\dot{\mathbf{y}} = \boldsymbol{\Lambda}\mathbf{y} + \boldsymbol{\Gamma}r \tag{4-312}$$

Then the state transition matrix is given by

$$\boldsymbol{\phi}(t) = e^{\boldsymbol{\Lambda}t} = e^{-t}\begin{bmatrix} \cos t & \sin t \\ -\sin t & \cos t \end{bmatrix} \tag{4-313}$$

4.15 Controllability of Linear Systems

The concepts of controllability and observability introduced first by Kalman[24] play an important role in both theoretical and practical aspects of modern control theory. The conditions on controllability and observability often govern the existence of a solution to an optimal control problem, particularly in multivariable systems. However, one should not associate these conditions with the concept of stability. Furthermore, not all optimal control problems require that the system be controllable and/or observable in order to achieve the control objectives. These points will be clarified as the subjects are developed. In this section we shall first treat the subject of controllability.

General Concept of Controllability

The concept of controllability can be stated with reference to the block diagram of Fig. 4-24. The process **G** is said to be completely controllable if

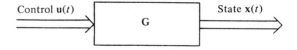

Fig. 4-24. Linear time-invariant system.

every state variable of **G** can be affected or controlled to reach a certain objective in finite time by some unconstrained control $\mathbf{u}(t)$. Intuitively, it is simple to understand that if any one of the state variables is independent of the control $\mathbf{u}(t)$, there would be no way of driving this particular state variable to a desired state in finite time by means of a control effort. Therefore, this particular state is said to be uncontrollable, and as long as there is at least one uncontrollable state, the system is said to be not completely controllable, or simply uncontrollable.

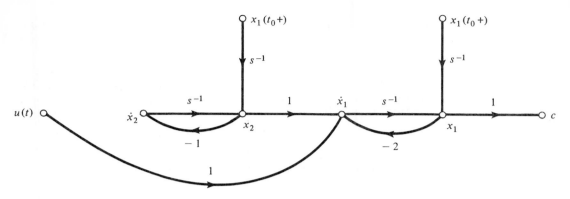

Fig. 4-25. State diagram of a system that is not state controllable.

As a simple example of an uncontrollable system, Fig. 4-25 illustrates the state diagram of a linear system with two state variables. Since the control $u(t)$ affects only the state $x_1(t)$, $x_2(t)$ is uncontrollable. In other words, it would be impossible to drive $x_2(t)$ from an initial state $x_2(t_0)$ to a desired state $x_2(t_f)$ in a finite time interval $t_f - t_0$ by any control $u(t)$. Therefore, the entire system is said to be uncontrollable.

The concept of controllability given above refers to the states and is sometimes referred to as the *state controllability*. Controllability can also be defined for the outputs of a system, so there is a difference between state controllability and output controllability.

Definition of Controllability (State Controllability)

Consider that a linear time-invariant system is described by the following dynamic equations:

$$\dot{\mathbf{x}}(t) = \mathbf{A}\mathbf{x}(t) + \mathbf{B}\mathbf{u}(t) \qquad (4\text{-}314)$$

$$\mathbf{c}(t) = \mathbf{D}\mathbf{x}(t) + \mathbf{E}\mathbf{u}(t) \qquad (4\text{-}315)$$

where

$\mathbf{x}(t) = n \times 1$ state vector

$\mathbf{u}(t) = r \times 1$ input vector

$\mathbf{c}(t) = p \times 1$ output vector

$\mathbf{A} = n \times n$ coefficient matrix

$\mathbf{B} = n \times r$ coefficient matrix

$\mathbf{D} = p \times n$ coefficient matrix

$\mathbf{E} = p \times r$ coefficient matrix

The state $\mathbf{x}(t)$ *is said to be controllable at* $t = t_0$ *if there exists a piecewise continuous input* $\mathbf{u}(t)$ *that will drive the state to any final state* $\mathbf{x}(t_f)$ *for a finite time* $(t_f - t_0) \geq 0$. *If every state* $\mathbf{x}(t_0)$ *of the system is controllable in a finite time interval, the system is said to be completely state controllable or simply state controllable.*

The following theorem shows that the condition of controllability depends on the coefficient matrices $\mathbf{A}$ and $\mathbf{B}$ of the system. The theorem also gives one way of testing state controllability.

Theorem 4-2. For the system described by the state equation of Eq. (4-314) to be completely state controllable, it is necessary and sufficient that the following $n \times nr$ matrix has a rank of n:

$$\mathbf{S} = [\mathbf{B} \quad \mathbf{AB} \quad \mathbf{A}^2\mathbf{B} \ldots \mathbf{A}^{n-1}\mathbf{B}] \tag{4-316}$$

Since the matrices $\mathbf{A}$ and $\mathbf{B}$ are involved, sometimes we say that the pair $[\mathbf{A}, \mathbf{B}]$ is controllable, which implies that $\mathbf{S}$ is of rank n.

Proof: The solution of Eq. (4-314) is

$$\mathbf{x}(t) = \boldsymbol{\phi}(t - t_0)\mathbf{x}(t_0) + \int_{t_0}^{t} \boldsymbol{\phi}(t - \tau)\mathbf{Bu}(\tau)\,d\tau \tag{4-317}$$

for $t \geq t_0$. Without losing any generality we can assume that the desired final state for some finite $t_f \geq t_0$ is $\mathbf{x}(t_f) = \mathbf{0}$. Then Eq. (4-317) gives

$$\mathbf{x}(t_0) = -\int_{t_0}^{t_f} \boldsymbol{\phi}(t_0 - \tau)\mathbf{Bu}(\tau)\,d\tau \tag{4-318}$$

From the Caley–Hamilton's theorem,[21]

$$\mathbf{A}^k = \sum_{m=0}^{n-1} \alpha_{km}\mathbf{A}^m \qquad \text{for any } k \tag{4-319}$$

Then the state transition matrix can be written

$$\boldsymbol{\phi}(t) = e^{\mathbf{A}t} = \sum_{k=0}^{\infty} \frac{\mathbf{A}^k t^k}{k!}$$

$$= \sum_{k=0}^{\infty} \frac{t^k}{k!} \sum_{m=0}^{n-1} \alpha_{km}\mathbf{A}^m \tag{4-320}$$

or

$$\boldsymbol{\phi}(t) = \sum_{m=0}^{n-1} \mathbf{A}^m \sum_{k=0}^{\infty} \alpha_{km}\frac{t^k}{k!} \tag{4-321}$$

Thus $\boldsymbol{\phi}(t)$ can be written in the form

$$\boldsymbol{\phi}(t) = \sum_{m=0}^{n-1} \alpha_m(t)\mathbf{A}^m \tag{4-322}$$

Substituting Eq. (4-322) into Eq. (4-318) and rearranging, we have

$$\mathbf{x}(t_0) = -\sum_{m=0}^{n-1} \mathbf{A}^m\mathbf{B}\int_{t_0}^{t_f} \alpha_m(t_0 - \tau)\mathbf{u}(\tau)\,d\tau \tag{4-323}$$

Let

$$\mathbf{U}_m = \int_{t_0}^{t_f} \alpha_m(t_0 - \tau)\mathbf{u}(\tau)\,d\tau \tag{4-324}$$

Then Eq. (4-323) becomes

$$\mathbf{x}(t_0) = -\sum_{m=0}^{n-1} \mathbf{A}^m\mathbf{BU}_m \tag{4-325}$$

which is written in matrix form:

$$\mathbf{x}(t_0) = -[\mathbf{B} \quad \mathbf{AB} \quad \mathbf{A}^2\mathbf{B} \ldots \mathbf{A}^{n-1}\mathbf{B}]\mathbf{U} \qquad (4\text{-}326)$$

$$= -\mathbf{SU}$$

where

$$\mathbf{U} = [\mathbf{U}_0 \quad \mathbf{U}_1 \ldots \mathbf{U}_{n-1}]' \qquad (4\text{-}327)$$

Equation (4-325) represents n equations with nr unknowns, and the controllability problem may be interpreted as: Given any initial state $\mathbf{x}(t_0)$, find the control vector $\mathbf{u}(t)$ so that the final state is $\mathbf{x}(t_f) = \mathbf{0}$ for finite $t_f - t_0$. This implies that given $\mathbf{x}(t_0)$ and the matrix $\mathbf{S}$, solve $\mathbf{U}$ from Eq. (4-326). Therefore, the system is completely state controllable if and only if there exists a set of n linearly independent column vectors in $\mathbf{S}$. For a system with a scalar input, $r = 1$, the matrix $\mathbf{S}$ is square; then the condition of state controllability is that $\mathbf{S}$ must be nonsingular.

Although the criterion of state controllability given by Theorem 4-2 is quite straightforward, it is not very easy to implement for multiple-input systems. Even with $r = 2$, there are $2n$ columns in $\mathbf{S}$, and there would be a large number of possible combinations of $n \times n$ matrices. A practical way may be to use one column of $\mathbf{B}$ at a time, each time giving an $n \times n$ matrix for $\mathbf{S}$. However, failure to find an $\mathbf{S}$ with a rank for n this way does not mean that the system is uncontrollable, until all the columns of $\mathbf{B}$ are used. An easier way would be to form the matrix $\mathbf{SS}'$, which is $n \times n$; then if $\mathbf{SS}'$ is nonsingular, $\mathbf{S}$ has rank n.

EXAMPLE 4-21   Consider the system shown in Fig. 4-25, which was reasoned earlier to be uncontrollable. Let us investigate the same problem using the condition of Eq. (4-316).

The state equations of the system are written, from Fig. 4-25,

$$\begin{bmatrix} \dfrac{dx_1(t)}{dt} \\ \dfrac{dx_2(t)}{dt} \end{bmatrix} = \begin{bmatrix} -2 & 1 \\ 0 & -1 \end{bmatrix} \begin{bmatrix} x_1(t) \\ x_2(t) \end{bmatrix} + \begin{bmatrix} 1 \\ 0 \end{bmatrix} u(t) \qquad (4\text{-}328)$$

Therefore, from Eq. (4-316),

$$\mathbf{S} = [\mathbf{B} \quad \mathbf{AB}] = \begin{bmatrix} 1 & -2 \\ 0 & 0 \end{bmatrix} \qquad (4\text{-}329)$$

which is singular, and the system is not state controllable.

EXAMPLE 4-22   Determine the state controllability of the system described by the state equation

$$\begin{bmatrix} \dfrac{dx_1(t)}{dt} \\ \dfrac{dx_2(t)}{dt} \end{bmatrix} = \begin{bmatrix} 0 & 1 \\ -1 & 0 \end{bmatrix} \begin{bmatrix} x_1(t) \\ x_2(t) \end{bmatrix} + \begin{bmatrix} 0 \\ 1 \end{bmatrix} u(t) \qquad (4\text{-}330)$$

From Eq. (4-316),

$$\mathbf{S} = [\mathbf{B} \quad \mathbf{AB}] = \begin{bmatrix} 0 & 1 \\ 1 & 0 \end{bmatrix} \qquad (4\text{-}331)$$

which is nonsingular. Therefore, the system is completely state controllable.

Alternative Definition of Controllability

Consider that a linear time invariant system is described by the state equation

$$\dot{\mathbf{x}}(t) = \mathbf{A}\mathbf{x}(t) + \mathbf{B}\mathbf{u}(t) \tag{4-332}$$

If the eigenvalues of $\mathbf{A}$ are distinct and are denoted by λ_i, $i = 1, 2, \ldots, n$, then there exists an nth-order nonsingular matrix $\mathbf{P}$ which transforms $\mathbf{A}$ into a diagonal matrix $\mathbf{\Lambda}$ such that

$$\mathbf{\Lambda} = \mathbf{P}^{-1}\mathbf{A}\mathbf{P} = \begin{bmatrix} \lambda_1 & 0 & 0 & \ldots & 0 \\ 0 & \lambda_2 & 0 & \ldots & 0 \\ 0 & 0 & \lambda_3 & \ldots & 0 \\ \cdot & \cdot & \cdot & \ldots & \cdot \\ \cdot & \cdot & \cdot & \ldots & \cdot \\ \cdot & \cdot & \cdot & \ldots & \cdot \\ 0 & 0 & 0 & \ldots & \lambda_n \end{bmatrix} \tag{4-333}$$

Let the new state variable be

$$\mathbf{y} = \mathbf{P}^{-1}\mathbf{x} \tag{4-334}$$

Then the state equation transformed through $\mathbf{P}$ is

$$\dot{\mathbf{y}} = \mathbf{\Lambda}\mathbf{y} + \mathbf{\Gamma}\mathbf{u} \tag{4-335}$$

where

$$\mathbf{\Gamma} = \mathbf{P}^{-1}\mathbf{B} \tag{4-336}$$

The motivation for the use of the similarity transformation is that the states of the system of Eq. (4-335) are decoupled from each other, and the only way the states are controllable is through the inputs directly. Thus, for state controllability, each state should be controlled by at least one input. Therefore, an alternative definition of state controllability for a system with distinct eigenvalues is: *The system is completely state controllable if $\mathbf{\Gamma}$ has no rows that are all zeros.*

It should be noted that the prerequisite on distinct eigenvalues precedes the condition of diagonalization of $\mathbf{A}$. In other words, all square matrices with distinct eigenvalues can be diagonalized. However, certain matrices with multiple-order eigenvalues can also be diagonalized. The natural question is: Does the alternative definition apply to a system with multiple-order eigenvalues but whose $\mathbf{A}$ matrix can be diagonalized? The answer is *no*. We must not lose sight of the original definition on state controllability that any state $\mathbf{x}(t_0)$ is brought to any state $\mathbf{x}(t_f)$ in finite time. Thus the question of independent control must enter the picture. In other words, consider that we have two states which are uncoupled and are related by the following state equations:

$$\frac{dx_1(t)}{dt} = ax_1(t) + b_1u(t) \tag{4-337}$$

$$\frac{dx_2(t)}{dt} = ax_2(t) + b_2u(t) \tag{4-338}$$

This system is apparently uncontrollable, since

$$S = [B \quad AB] = \begin{bmatrix} b_1 & ab_1 \\ b_2 & ab_2 \end{bmatrix} \tag{4-339}$$

is singular. Therefore, just because **A** is diagonal, and **B** has no rows which are zeros does not mean that the system is controllable. The reason in this case is that **A** has multiple-order eigenvalues.

When **A** has multiple-order eigenvalues and cannot be diagonalized, there is a nonsingular matrix **P** which transforms **A** into a Jordan canonical form $\Lambda = P^{-1}AP$. *The condition of state controllability is that all the elements of* $\Gamma = P^{-1}B$ *that correspond to the last row of each Jordan block are nonzero.* The reason behind this is that the last row of each Jordan block corresponds to a state equation that is completely uncoupled from the other state equations.

The elements in the other rows of Γ need not all be nonzero, since the corresponding states are all coupled. For instance, if the matrix **A** has four eigenvalues, $\lambda_1, \lambda_1, \lambda_1, \lambda_2$, three of which are equal, then there is a nonsingular **P** which transforms **A** into the Jordan canonical form:

$$\Lambda = P^{-1}AP = \begin{bmatrix} \lambda_1 & 1 & 0 & 0 \\ 0 & \lambda_1 & 1 & 0 \\ 0 & 0 & \lambda_1 & 0 \\ 0 & 0 & 0 & \lambda_2 \end{bmatrix} \tag{4-340}$$

Then the condition given above becomes self-explanatory.

EXAMPLE 4-23   Consider the system of Example 4-21. The **A** and **B** matrices are, respectively,

$$A = \begin{bmatrix} -2 & 1 \\ 0 & -1 \end{bmatrix} \qquad B = \begin{bmatrix} 1 \\ 0 \end{bmatrix}$$

Let us check the controllability of the system by checking the rows of the matrix Γ. It can be shown that **A** is diagonalized by the matrix

$$P = \begin{bmatrix} 1 & 1 \\ 0 & 1 \end{bmatrix}$$

Therefore,

$$\Gamma = P^{-1}B = \begin{bmatrix} 1 & -1 \\ 0 & 1 \end{bmatrix}\begin{bmatrix} 1 \\ 0 \end{bmatrix} = \begin{bmatrix} 1 \\ 0 \end{bmatrix} \tag{4-341}$$

The transformed state equation is

$$\dot{y}(t) = \begin{bmatrix} -2 & 0 \\ 0 & -1 \end{bmatrix}y(t) + \begin{bmatrix} 1 \\ 0 \end{bmatrix}u(t) \tag{4-342}$$

Since the second row of Γ is zero, the state variable $y_2(t)$, or $x_2(t)$, is uncontrollable, and the system is uncontrollable.

EXAMPLE 4-24   Consider that a third-order system has the coefficient matrices

$$A = \begin{bmatrix} 1 & 2 & -1 \\ 0 & 1 & 0 \\ 1 & -4 & 3 \end{bmatrix} \qquad B = \begin{bmatrix} 0 \\ 0 \\ 1 \end{bmatrix}$$

Then

$$S = [\mathbf{B} \quad \mathbf{AB} \quad \mathbf{A}^2\mathbf{B}] = \begin{bmatrix} 0 & -1 & -4 \\ 0 & 0 & 0 \\ 1 & 3 & 8 \end{bmatrix} \tag{4-343}$$

Since **S** is singular, the system is not state controllable.

Using the alternative method, the eigenvalues of **A** are found to be $\lambda_1 = 2$, $\lambda_2 = 2$, and $\lambda_3 = 1$. The Jordan canonical form of **A** is obtained with

$$\mathbf{P} = \begin{bmatrix} 1 & 0 & 0 \\ 0 & 0 & 1 \\ -1 & 1 & 2 \end{bmatrix} \tag{4-344}$$

Then

$$\mathbf{\Lambda} = \mathbf{P}^{-1}\mathbf{AP} = \begin{bmatrix} 2 & 1 & 0 \\ 0 & 2 & 0 \\ 0 & 0 & 1 \end{bmatrix} \tag{4-345}$$

$$\mathbf{\Gamma} = \mathbf{P}^{-1}\mathbf{B} = \begin{bmatrix} 0 \\ -1 \\ 0 \end{bmatrix} \tag{4-346}$$

Since the last row of **Γ** is zero, the state variable y_3 is uncontrollable. Since $x_2 = y_3$, this corresponds to x_2 being uncontrollable.

EXAMPLE 4-25   Determine the controllability of the system described by the state equation

$$\dot{\mathbf{x}}(t) = \begin{bmatrix} 0 & 1 \\ -1 & 0 \end{bmatrix}\mathbf{x}(t) + \begin{bmatrix} 0 \\ 1 \end{bmatrix}u(t) \tag{4-347}$$

We form the matrix

$$S = [\mathbf{B} \quad \mathbf{AB}] = \begin{bmatrix} 0 & 1 \\ 1 & 0 \end{bmatrix} \tag{4-348}$$

which is nonsingular. The system is completely controllable.

Let us now check the controllability of the system from the rows of **Γ**. The eigenvalues of **A** are complex and are $\lambda_1 = j$ and $\lambda_2 = -j$. With the similarity transformation,

$$\mathbf{P} = \begin{bmatrix} 1 & 1 \\ j & -j \end{bmatrix}$$

$$\mathbf{\Lambda} = \mathbf{P}^{-1}\mathbf{AP} = \begin{bmatrix} j & 0 \\ 0 & -j \end{bmatrix}$$

and

$$\mathbf{\Gamma} = \mathbf{P}^{-1}\mathbf{B} = \begin{bmatrix} \dfrac{1}{2j} \\ \dfrac{-1}{2j} \end{bmatrix}$$

Since all the rows of **Γ** are nonzero, the system is controllable.

In general, when the eigenvalues are complex, which occurs quite frequently in control systems, it is more difficult to work with complex numbers. However, we may use the modal form so that only real matrices are dealt with. In the present problem $\mathbf{A}$ may be transformed to the modal form

$$\boldsymbol{\Lambda} = \begin{bmatrix} \sigma & \omega \\ -\omega & \sigma \end{bmatrix} = \begin{bmatrix} 0 & 1 \\ -1 & 0 \end{bmatrix} \tag{4-349}$$

by the transform matrix

$$\mathbf{P} = \begin{bmatrix} 1 & -1 \\ 1 & 1 \end{bmatrix}$$

Then

$$\boldsymbol{\Gamma} = \mathbf{P}^{-1}\mathbf{B} = \begin{bmatrix} \frac{1}{2} \\ \frac{1}{2} \end{bmatrix}$$

Since the modal form $\boldsymbol{\Lambda}$ implies that the states are coupled, the condition of controllability is that not *all* the rows of $\boldsymbol{\Gamma}$ are zeros.

Output Controllability[28]

The condition of controllability defined in the preceding sections is referred only to the states of a system. Essentially, a system is controllable if every desired transition of the states can be effected in finite time by an unconstrained control. However, controllability defined in terms of the states is neither necessary nor sufficient for the existence of a solution of the problem of controlling the outputs of the system.

Definition of output controllability. A system is said to be completely *output controllable* if there exists a piecewise continuous function $\mathbf{u}(t)$ that will drive the output $\mathbf{y}(t_0)$ at $t = t_0$ to any final output $\mathbf{y}(t_f)$ for a finite time $(t_f - t_0) \geq 0$.

Theorem 4-3. Consider that an nth-order linear time-invariant system is described by the dynamic equations of Eqs. (4-314) and (4-315). The system is completely output controllable if and only if the $p \times (n + 1)r$ matrix

$$\mathbf{T} = [\mathbf{DB} \quad \mathbf{DAB} \quad \mathbf{DA^2B} \ldots \mathbf{DA^{n-1}B} \quad \mathbf{E}] \tag{4-350}$$

is of rank p. Or, $\mathbf{T}$ has a set of p linearly independent columns.

The proof of this theorem is similar to that of Theorem 4-2.

EXAMPLE 4-26   Consider a linear system whose input–output relationship is described by the differential equation

$$\frac{d^2c(t)}{dt^2} + 2\frac{dc(t)}{dt} + c(t) = \frac{du(t)}{dt} + u(t) \tag{4-351}$$

The state controllability and the output controllability of the system will be investigated. We shall show that the state controllability of the system depends upon how the state variables are defined.

Let the state variables be defined as

$$x_1 = c$$
$$x_2 = \dot{c} - u$$

The state equations of the system are expressed in matrix form as

$$\begin{bmatrix} \dot{x}_1 \\ \dot{x}_2 \end{bmatrix} = \begin{bmatrix} 0 & 1 \\ -1 & -2 \end{bmatrix} \begin{bmatrix} x_1 \\ x_2 \end{bmatrix} + \begin{bmatrix} 1 \\ -1 \end{bmatrix} u \tag{4-352}$$

The output equation is

$$c = x_1 \tag{4-353}$$

The state controllability matrix is

$$\mathbf{S} = [\mathbf{B} \quad \mathbf{AB}] = \begin{bmatrix} 1 & -1 \\ -1 & 1 \end{bmatrix} \tag{4-354}$$

which is singular. The system is *not state controllable*.

From the output equation, $\mathbf{D} = [1 \quad 0]$ and $E = 0$. The output controllability matrix is written

$$\mathbf{T} = [\mathbf{DB} \quad \mathbf{DAB} \quad \mathbf{E}] = [1 \quad -1 \quad 0] \tag{4-355}$$

which is of rank 1, the same as the number of output. Thus the system is *output controllable*.

Now let us define the state variables of the system in a different way. By the method of direct decomposition, the state equations are written in matrix form:

$$\begin{bmatrix} \dot{x}_1 \\ \dot{x}_2 \end{bmatrix} = \begin{bmatrix} 0 & 1 \\ -1 & -2 \end{bmatrix} \begin{bmatrix} x_1 \\ x_2 \end{bmatrix} + \begin{bmatrix} 0 \\ 1 \end{bmatrix} u \tag{4-356}$$

The output equation is

$$c = x_1 + x_2 \tag{4-357}$$

The system is now *completely state controllable* since

$$\mathbf{S} = [\mathbf{B} \quad \mathbf{AB}] = \begin{bmatrix} 0 & 1 \\ 1 & -2 \end{bmatrix} \tag{4-358}$$

which is nonsingular.

The system is still *output controllable* since

$$\mathbf{T} = [\mathbf{DB} \quad \mathbf{DAB} \quad \mathbf{E}] = [1 \quad -1 \quad 0] \tag{4-359}$$

which is of rank 1.

We have demonstrated through this example that given a linear system, state controllability depends on how the state variables are defined. Of course, the output controllability is directly dependent upon the assignment of the output variable. The two types of controllability are not at all related to each other.

4.16 Observability of Linear Systems

The concept of observability is quite similar to that of controllability. Essentially, a system is completely observable if every state variable of the system affects some of the outputs. In other words, it is often desirable to obtain information on the state variables from measurements of the outputs and the inputs. If any one of the states cannot be observed from the measurements of the outputs, the state is said to be unobservable, and the system is not completely observable, or is simply unobservable. Figure 4-26 shows the state diagram of a linear system in which the state x_2 is not connected to the output c in any way. Once we have measured c, we can observe the state x_1, since $x_1 = c$. However, the state x_2 cannot be observed from the information on c. Thus the system is described as not completely observable, or simply unobservable.

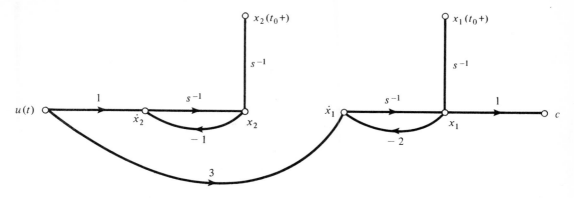

Fig. 4-26. State diagram of a system that is not observable.

Definition of observability. *Given a linear time-invariant system that is described by the dynamic equations of Eqs.* (4-314) *and* (4-315), *the state* $\mathbf{x}(t_0)$ *is said to be observable if given any input* $\mathbf{u}(t)$, *there exists a finite time* $t_f \geq t_0$ *such that the knowledge of* $\mathbf{u}(t)$ *for* $t_0 \leq t < t_f$; *the matrices* $\mathbf{A}, \mathbf{B}, \mathbf{D},$ *and* $\mathbf{E}$; *and the output* $\mathbf{c}(t)$ *for* $t_0 \leq t < t_f$ *are sufficient to determine* $\mathbf{x}(t_0)$. *If every state of the system is observable for a finite* t_f, *we say that the system is completely observable, or simply observable.*

The following theorem shows that the condition of observability depends on the coefficient matrices $\mathbf{A}$ and $\mathbf{D}$ of the system. The theorem also gives one method of testing observability.

Theorem 4-4. For the system described by the dynamic equation of Eqs. (4-314) *and* (4-315) *to be completely observable, it is necessary and sufficient that the following* $n \times np$ *matrix has a rank of* n:

$$\mathbf{V} = [\mathbf{D}' \quad \mathbf{A}'\mathbf{D}' \quad (\mathbf{A}')^2\mathbf{D}' \dots (\mathbf{A}')^{n-1}\mathbf{D}'] \qquad (4\text{-}360)$$

The condition is also referred to as the pair [$\mathbf{A}, \mathbf{D}$] *being observable. In particular, if the system has only one output,* $\mathbf{D}$ *is an* $l \times n$ *matrix;* $\mathbf{V}$ *of Eq.* (4-360) *is an* $n \times n$ *square matrix. Then the system is completely observable if* $\mathbf{V}$ *is nonsingular.*

Proof: Substituting Eq. (4-317) into Eq. (4-315), we have

$$\mathbf{c}(t) = \mathbf{D}\boldsymbol{\phi}(t - t_0)\mathbf{x}(t_0) + \mathbf{D}\int_{t_0}^{t} \boldsymbol{\phi}(t - \tau)\mathbf{B}\mathbf{u}(\tau)\,d\tau + \mathbf{E}\mathbf{u}(t) \qquad (4\text{-}361)$$

Based on the definition of observability, it is apparent that the observability of $\mathbf{x}(t_0)$ depends essentially on the first term of the right side of Eq. (4-361). With $\mathbf{u}(t) = \mathbf{0}$, Eq. (4-361) becomes

$$\mathbf{c}(t) = \mathbf{D}\boldsymbol{\phi}(t - t_0)\mathbf{x}(t_0) \qquad (4\text{-}362)$$

Making use of Eq. (4-322), Eq. (4-362) becomes

$$\mathbf{c}(t) = \sum_{m=0}^{n-1} \alpha_m(t)\mathbf{D}\mathbf{A}^m\mathbf{x}(t_0) \qquad (4\text{-}363)$$

or

$$\mathbf{c}(t) = (\alpha_0\mathbf{I} \quad \alpha_1\mathbf{I} \quad \ldots \quad \alpha_{n-1}\mathbf{I}) \begin{bmatrix} \mathbf{D} \\ \mathbf{DA} \\ \mathbf{DA}^2 \\ \cdot \\ \cdot \\ \cdot \\ \mathbf{DA}^{n-1} \end{bmatrix} \mathbf{x}(t_0) \tag{4-364}$$

Therefore, knowing the output $\mathbf{c}(t)$ over the time interval $t_0 \le t < t_f$, $\mathbf{x}(t_0)$ is uniquely determined from Eq. (4-364) if and only if the matrix

$$\begin{bmatrix} \mathbf{D} \\ \mathbf{DA} \\ \mathbf{DA}^2 \\ \cdot \\ \cdot \\ \cdot \\ \mathbf{DA}^{n-1} \end{bmatrix} \quad (np \times n)$$

has rank n. Or the matrix

$$\mathbf{V} = [\mathbf{D}' \quad \mathbf{A}'\mathbf{D}' \quad (\mathbf{A}')^2\mathbf{D} \ldots (\mathbf{A}')^{n-1}\mathbf{D}'] \tag{4-365}$$

has a rank of n.

Comparing Eq. (4-360) with Eq. (4-316) and the rank condition, the following observations may be made:

1. *Controllability of the pair* $[\mathbf{A}, \mathbf{B}]$ *implies observability of the pair* $[\mathbf{A}', \mathbf{B}']$.
2. *Observability of the pair* $[\mathbf{A}, \mathbf{B}]$ *implies controllability of the pair* $[\mathbf{A}', \mathbf{B}']$.

EXAMPLE 4-27   Consider the system shown in Fig. 4-26, which was earlier defined to be unobservable. The dynamic equations of the system are written directly from the state diagram.

$$\begin{bmatrix} \dot{x}_1 \\ \dot{x}_2 \end{bmatrix} = \begin{bmatrix} -2 & 0 \\ 0 & -1 \end{bmatrix} \begin{bmatrix} x_1 \\ x_2 \end{bmatrix} + \begin{bmatrix} 3 \\ 1 \end{bmatrix} u \tag{4-366}$$

$$c = [1 \quad 0] \begin{bmatrix} x_1 \\ x_2 \end{bmatrix} \tag{4-367}$$

Therefore,

$$\mathbf{D} = [1 \quad 0] \qquad \mathbf{D}' = \begin{bmatrix} 1 \\ 0 \end{bmatrix}$$

$$\mathbf{A}'\mathbf{D}' = \begin{bmatrix} -2 & 0 \\ 0 & -1 \end{bmatrix} \begin{bmatrix} 1 \\ 0 \end{bmatrix} = \begin{bmatrix} -2 \\ 0 \end{bmatrix}$$

and, from Eq. (4-360),

$$\mathbf{V} = [\mathbf{D}' \quad \mathbf{A}'\mathbf{D}'] = \begin{bmatrix} 1 & -2 \\ 0 & 0 \end{bmatrix} \tag{4-368}$$

Since $\mathbf{V}$ is singular, the system is unobservable.

EXAMPLE 4-28   Consider the linear system described by the following dynamic equations:

$$\begin{bmatrix} \dot{x}_1 \\ \dot{x}_2 \end{bmatrix} = \begin{bmatrix} 1 & -1 \\ 1 & 1 \end{bmatrix} \begin{bmatrix} x_1 \\ x_2 \end{bmatrix} + \begin{bmatrix} 2 & -1 \\ 1 & 0 \end{bmatrix} \begin{bmatrix} u_1 \\ u_2 \end{bmatrix} \tag{4-369}$$

$$\begin{bmatrix} c_1 \\ c_2 \end{bmatrix} = \begin{bmatrix} 1 & 0 \\ -1 & 1 \end{bmatrix} \begin{bmatrix} x_1 \\ x_2 \end{bmatrix} \tag{4-370}$$

For the test of observability, we evaluate

$$\mathbf{A'D'} = \begin{bmatrix} 1 & 1 \\ -1 & 1 \end{bmatrix} \begin{bmatrix} 1 & -1 \\ 0 & 1 \end{bmatrix} = \begin{bmatrix} 1 & 0 \\ -1 & 2 \end{bmatrix} \tag{4-371}$$

The observability matrix becomes

$$\mathbf{V} = [\mathbf{D'} \quad \mathbf{A'D'}] = \begin{bmatrix} 1 & -1 & 1 & 0 \\ 0 & 1 & -1 & 2 \end{bmatrix} \tag{4-372}$$

Since $\mathbf{V}$ has a rank of 2, which is the number of inputs, the system is completely observable.

EXAMPLE 4-29   Let us consider the system described by the differential equation of Eq. (4-351), Example 4-26. In Example 4-26 we have shown that state controllability of a system depends on how the state variables are defined. We shall now show that the observability also depends on the definition of the state variables. Let the dynamic equations of the system be defined as in Eqs. (4-352) and (4-353),

$$\mathbf{A} = \begin{bmatrix} 0 & 1 \\ -1 & -2 \end{bmatrix} \qquad \mathbf{D} = [1 \quad 0]$$

Then

$$\mathbf{V} = [\mathbf{D'} \quad \mathbf{A'D'}] = \begin{bmatrix} 1 & 0 \\ 0 & 1 \end{bmatrix} \tag{4-373}$$

and thus the system is completely observable.

Let the dynamic equations of the system be given by Eqs. (4-356) and (4-357). Then

$$\mathbf{A} = \begin{bmatrix} 0 & 1 \\ -1 & -2 \end{bmatrix} \qquad \mathbf{D} = [1 \quad 1]$$

Then

$$\mathbf{V} = [\mathbf{D'} \quad \mathbf{A'D'}] = \begin{bmatrix} 1 & -1 \\ 1 & -1 \end{bmatrix}$$

which is singular. Thus the system is unobservable, and we have shown that given the input–output relation of a linear system, the observability of the system depends on how the state variables are defined. It should be noted that for the system of Eq. (4-351), one method of state variable assignment, Eqs. (4-352) and (4-353) yields a system that is observable but not state controllable. On the other hand, if the dynamic equations of Eqs. (4-356) and (4-357) are used, the system is completely state controllable but not observable. There are definite reasons behind these results, and we shall investigate these phenomena further in the following discussions.

Alternative definition of observability. If the matrix $\mathbf{A}$ has distinct eigenvalues, it can be diagonalized as in Eq. (4-333). The new state variable is

$$\mathbf{y} = \mathbf{P}^{-1}\mathbf{x} \tag{4-374}$$

The new dynamic equations are

$$\dot{\mathbf{y}} = \mathbf{\Lambda}\mathbf{y} + \mathbf{\Gamma}\mathbf{u} \tag{4-375}$$

$$\mathbf{c} = \mathbf{F}\mathbf{y} + \mathbf{E}\mathbf{u} \tag{4-376}$$

where

$$\mathbf{F} = \mathbf{D}\mathbf{P} \tag{4-377}$$

Then the system is completely observable if $\mathbf{F}$ *has no zero columns.*

The reason behind the above condition is that if the jth ($j = 1, 2, \ldots, n$) column of $\mathbf{F}$ contains all zeros, the state variable y_j will not appear in Eq. (4-376) and is not related to the output $\mathbf{c}(t)$. Therefore, y_j will be unobservable. In general, the states that correspond to zero columns of $\mathbf{F}$ are said to be unobservable, and the rest of the state variables are observable.

EXAMPLE 4-30　Consider the system of Example 4-27, which was found to be unobservable. Since the A matrix, as shown in Eq. (4-366), is already a diagonal matrix, the alternative condition of observability stated above requires that the matrix $\mathbf{D} = [1 \quad 0]$ must not contain any zero columns. Since the second column of $\mathbf{D}$ is indeed zero, the state x_2 is unobservable, and the system is unobservable.

4.17　Relationship Among Controllability, Observability, and Transfer Functions

In the classical analysis of control systems, transfer functions are often used for the modeling of linear time-invariant systems. Although controllability and observability are concepts of modern control theory, they are closely related to the properties of the transfer function.

Let us focus our attention on the system considered in Examples 4-26 and 4-29. It was demonstrated in these two examples that the system is either not state controllable or not observable, depending on the ways the state variables are defined. These phenomena can be explained by referring to the transfer function of the system, which is obtained from Eq. (4-351). We have

$$\frac{C(s)}{U(s)} = \frac{s+1}{s^2 + 2s + 1} = \frac{s+1}{(s+1)^2} = \frac{1}{s+1} \tag{4-378}$$

which has an identical pole and zero at $s = -1$. The following theorem gives the relationship between controllability and observability and the pole–zero cancellation of a transfer function.

Theorem 4-5. If the input–output transfer function of a linear system has pole–zero cancellation, the system will be either not state controllable or unobservable, depending on how the state variables are defined. If the input–output transfer function of a linear system does not have pole–zero cancellation, the system can always be represented by dynamic equations as a completely controllable and observable system.

Proof: Consider that an nth-order system with a single input and single output and distinct eigenvalues is represented by the dynamic equations

$$\dot{\mathbf{x}}(t) = \mathbf{A}\mathbf{x}(t) + \mathbf{B}u(t) \tag{4-379}$$

$$c(t) = \mathbf{D}\mathbf{x}(t) \tag{4-380}$$

Let the $\mathbf{A}$ matrix be diagonalized by an $n \times n$ Vandermonde matrix $\mathbf{P}$,

$$
\mathbf{P} = \begin{bmatrix}
1 & 1 & 1 & \cdots & 1 \\
\lambda_1 & \lambda_2 & \lambda_3 & \cdots & \lambda_n \\
\lambda_1^2 & \lambda_2^2 & \lambda_3^2 & \cdots & \lambda_n^2 \\
\cdots\cdots\cdots\cdots\cdots\cdots\cdots \\
\lambda_1^{n-1} & \lambda_2^{n-1} & \lambda_3^{n-1} & \cdots & \lambda_n^{n-1}
\end{bmatrix} \tag{4-381}
$$

The new state equation in canonical form is

$$
\dot{\mathbf{y}}(t) = \mathbf{\Lambda}\mathbf{y}(t) + \mathbf{\Gamma}u(t) \tag{4-382}
$$

where $\mathbf{\Lambda} = \mathbf{P}^{-1}\mathbf{A}\mathbf{P}$. The output equation is transformed into

$$
c(t) = \mathbf{F}\mathbf{y}(t) \tag{4-383}
$$

where $\mathbf{F} = \mathbf{D}\mathbf{P}$. The state vectors $\mathbf{x}(t)$ and $\mathbf{y}(t)$ are related by

$$
\mathbf{x}(t) = \mathbf{P}\mathbf{y}(t) \tag{4-384}
$$

Since $\mathbf{\Lambda}$ is a diagonal matrix, the ith equation of Eq. (4-382) is

$$
\dot{y}_i(t) = \lambda_i y_i(t) + \gamma_i u(t) \tag{4-385}
$$

where λ_i is the ith eigenvalue of $\mathbf{A}$ and γ_i is the ith element of $\mathbf{\Gamma}$, where $\mathbf{\Gamma}$ is an $n \times 1$ matrix in the present case. Taking the Laplace transform on both sides of Eq. (4-385) and assuming zero initial conditions, we obtain the transfer function relation between $Y_i(s)$ and $U(s)$ as

$$
Y_i(s) = \frac{\gamma_i}{s - \lambda_i}U(s) \tag{4-386}
$$

The Laplace transform of Eq. (4-383) is

$$
C(s) = \mathbf{F}\mathbf{Y}(s) = \mathbf{D}\mathbf{P}\mathbf{Y}(s) \tag{4-387}
$$

Now if it is assumed that

$$
\mathbf{D} = [d_1 \quad d_2 \quad \cdots \quad d_n] \tag{4-388}
$$

then

$$
\mathbf{F} = \mathbf{D}\mathbf{P} = [f_1 \quad f_2 \quad \cdots \quad f_n] \tag{4-389}
$$

where

$$
f_i = d_1 + d_2\lambda_i + \cdots + d_n\lambda_i^{n-1} \tag{4-390}
$$

for $i = 1, 2, \ldots, n$. Equation (4-387) is written as

$$
\begin{aligned}
C(s) &= [f_1 \quad f_2 \quad \cdots \quad f_n]\mathbf{Y}(s) \\[4pt]
&= [f_1 \quad f_2 \quad \cdots \quad f_n]\begin{bmatrix}
\dfrac{\gamma_1}{s - \lambda_1} \\[6pt]
\dfrac{\gamma_2}{s - \lambda_2} \\[4pt]
\cdot \\ \cdot \\ \cdot \\[4pt]
\dfrac{\gamma_n}{s - \lambda_n}
\end{bmatrix} U(s) \\[6pt]
&= \sum_{i=1}^{n} \frac{f_i\gamma_i}{s - \lambda_i}U(s)
\end{aligned} \tag{4-391}
$$

For the nth-order system with distinct eigenvalues, let us assume that the input–output transfer function is of the form

$$\frac{C(s)}{U(s)} = \frac{K(s - a_1)(s - a_2)\ldots(s - a_m)}{(s - \lambda_1)(s - \lambda_2)\ldots(s - \lambda_n)} \qquad n > m \qquad (4\text{-}392)$$

which is expanded by partial fraction into

$$\frac{C(s)}{U(s)} = \sum_{i=1}^{n} \frac{\sigma_i}{s - \lambda_i} \qquad (4\text{-}393)$$

where σ_i denotes the residue of $C(s)/U(s)$ at $s = \lambda_i$.

It was established earlier that for the system described by Eq. (4-382) to be state controllable, all the rows of $\mathbf{\Gamma}$ must be nonzero; that is, $\gamma_i \neq 0$ for $i = 1, 2, \ldots, n$. If $C(s)/U(s)$ has one or more pairs of identical pole and zero, for instance in Eq. (4-392), $a_1 = \lambda_1$, then in Eq. (4-393), $\sigma_1 = 0$. Comparing Eq. (4-391) with Eq. (4-393), we see that in general

$$\sigma_i = f_i \gamma_i \qquad (4\text{-}394)$$

Therefore, when $\sigma_i = 0$, γ_i will be zero if $f_i \neq 0$, and the state y_i is uncontrollable.

For observability, it was established earlier that $\mathbf{F}$ must not have columns containing zeros. Or, in the present case, $f_i \neq 0$ for $i = 1, 2, \ldots, n$. However, from Eq. (4-394),

$$f_i = \frac{\sigma_i}{\gamma_i} \qquad (4\text{-}395)$$

When the transfer function has an identical pair of pole and zero at $a_i = \lambda_i$, $\sigma_i = 0$. Thus, from Eq. (4-395), $f_i = 0$ if $\gamma_i \neq 0$.

4.18 Nonlinear State Equations and Their Linearization

When a dynamic system has nonlinear characteristics, the state equations of the system can be represented by the following vector-matrix form:

$$\frac{d\mathbf{x}(t)}{dt} = \mathbf{f}[\mathbf{x}(t), \mathbf{r}(t)] \qquad (4\text{-}396)$$

where $\mathbf{x}(t)$ represents the $n \times 1$ state vector, $\mathbf{r}(t)$ the $p \times 1$ input vector, and $\mathbf{f}[\mathbf{x}(t), \mathbf{r}(t)]$ denotes an $n \times 1$ function vector. In general, $\mathbf{f}$ is a function of the state vector and the input vector.

Being able to represent a nonlinear and/or time-varying system by state equations is a distinct advantage of the state-variable approach over the transfer function method, since the latter is defined strictly only for linear time-invariant systems.

As a simple illustrative example, the following state equations are nonlinear:

$$\frac{dx_1(t)}{dt} = x_1(t) + x_2^2(t)$$

$$\frac{dx_2(t)}{dt} = x_1(t) + r(t) \qquad (4\text{-}397)$$

Since nonlinear systems are usually difficult to analyze and design, it would be desirable to perform a linearization whenever the situation justifies.

A linearization process that depends on expanding the nonlinear state equation into a Taylor series about a nominal operating point or trajectory is now described. All the terms of the Taylor series of order higher than 1 are discarded, and linear approximation of the nonlinear state equation at the nominal point results.

Let the nominal operating trajectory be denoted by $\mathbf{x}_0(t)$, which corresponds to the nominal input $\mathbf{r}_0(t)$ and some fixed initial states. Expanding the nonlinear state equation of Eq. (4-396) into a Taylor series about $\mathbf{x}(t) = \mathbf{x}_0(t)$ and neglecting all the higher-order terms yields

$$\dot{x}_i(t) = f_i(\mathbf{x}_0, \mathbf{r}_0) + \sum_{j=1}^{n} \left. \frac{\partial f_i(\mathbf{x}, \mathbf{r})}{\partial x_j} \right|_{\mathbf{x}_0, \mathbf{r}_0} (x_j - x_{0j})$$
$$+ \sum_{j=1}^{p} \left. \frac{\partial f_i(\mathbf{x}, \mathbf{r})}{\partial r_j} \right|_{\mathbf{x}_0, \mathbf{r}_0} (r_j - r_{0j}) \tag{4-398}$$

$i = 1, 2, \ldots, n$. Let

$$\Delta x_i = x_i - x_{0i} \tag{4-399}$$

and

$$\Delta r_j = r_j - r_{0j} \tag{4-400}$$

Then

$$\Delta \dot{x}_i = \dot{x}_i - \dot{x}_{0i} \tag{4-401}$$

Since

$$\dot{x}_{0i} = f_i(\mathbf{x}_0, \mathbf{r}_0) \tag{4-402}$$

Equation (4-398) is written

$$\Delta \dot{x}_i = \sum_{j=1}^{n} \left. \frac{\partial f_i(\mathbf{x}, \mathbf{r})}{\partial x_j} \right|_{\mathbf{x}_0, \mathbf{r}_0} \Delta x_j + \sum_{j=1}^{p} \left. \frac{\partial f_i(\mathbf{x}, \mathbf{r})}{\partial r_j} \right|_{\mathbf{x}_0, \mathbf{r}_0} \Delta r_j \tag{4-403}$$

The last equation may be written in the vector-matrix form

$$\Delta \dot{\mathbf{x}} = \mathbf{A}^* \Delta \mathbf{x} + \mathbf{B}^* \Delta \mathbf{r} \tag{4-404}$$

where

$$\mathbf{A}^* = \begin{bmatrix} \dfrac{\partial f_1}{\partial x_1} & \dfrac{\partial f_1}{\partial x_2} & \cdots & \dfrac{\partial f_1}{\partial x_n} \\[2mm] \dfrac{\partial f_2}{\partial x_1} & \dfrac{\partial f_2}{\partial x_2} & \cdots & \dfrac{\partial f_2}{\partial x_n} \\[2mm] \cdots & \cdots & \cdots & \cdots \\[2mm] \dfrac{\partial f_n}{\partial x_1} & \dfrac{\partial f_n}{\partial x_2} & \cdots & \dfrac{\partial f_n}{\partial x_n} \end{bmatrix} \tag{4-405}$$

$$\mathbf{B}^* = \begin{bmatrix} \dfrac{\partial f_1}{\partial r_1} & \dfrac{\partial f_1}{\partial r_2} & \cdots & \dfrac{\partial f_1}{\partial r_p} \\[2mm] \dfrac{\partial f_2}{\partial r_1} & \dfrac{\partial f_2}{\partial r_2} & \cdots & \dfrac{\partial f_2}{\partial r_p} \\[2mm] \cdots & \cdots & \cdots & \cdots \\[2mm] \dfrac{\partial f_n}{\partial r_1} & \dfrac{\partial f_n}{\partial r_2} & \cdots & \dfrac{\partial f_n}{\partial r_p} \end{bmatrix} \tag{4-406}$$

where it should be reiterated that $\mathbf{A}^*$ and $\mathbf{B}^*$ are evaluated at the nominal point. Thus we have linearized the nonlinear system of Eq. (4-396) at a nominal operating point. However, in general, although Eq. (4-404) is linear, the elements of $\mathbf{A}^*$ and $\mathbf{B}^*$ may be time varying.

The following examples serve to illustrate the linearization procedure just described.

EXAMPLE 4-31   Figure 4-27 shows the block diagram of a control system with a saturation nonlinearity. The state equations of the system are

$$\dot{x}_1 = f_1 = x_2 \tag{4-407}$$

$$\dot{x}_2 = f_2 = u \tag{4-408}$$

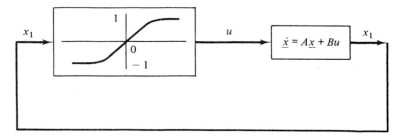

Fig. 4-27. Nonlinear control system.

where the input–output relation of the saturation nonlinearity is represented by

$$u = (1 - e^{-K|x_1|})\, \text{SGN}\, x_1 \tag{4-409}$$

where

$$\text{SGN}\, x_1 = \begin{cases} +1 & x_1 > 0 \\ -1 & x_1 < 0 \end{cases} \tag{4-410}$$

Substituting Eq. (4-409) into Eq. (4-408) and using Eq. (4-403), we have the linearized state equation

$$\Delta \dot{x}_1 = \frac{\partial f_1}{\partial x_2} \Delta x_2 = \Delta x_2 \tag{4-411}$$

$$\Delta \dot{x}_2 = \frac{\partial f_2}{\partial x_1} \Delta x_1 = K e^{-K|x_{01}|} \Delta x_1 \tag{4-412}$$

where x_{01} denotes a nominal value of x_1. Notice that the last two equations are linear and are valid only for small signals. In vector-matrix form, these linearized state equations are written as

$$\begin{bmatrix} \Delta \dot{x}_1 \\ \Delta \dot{x}_2 \end{bmatrix} = \begin{bmatrix} 0 & 1 \\ a & 0 \end{bmatrix} \begin{bmatrix} \Delta x_1 \\ \Delta x_2 \end{bmatrix} \tag{4-413}$$

where

$$a = K e^{-K|x_{01}|} = \text{constant} \tag{4-414}$$

It is of interest to check the significance of the linearization. If x_{01} is chosen to be at the origin of the nonlinearity, $x_{01} = 0$, then $a = K$; Eq. (4-412) becomes

$$\Delta \dot{x}_2 = K \Delta x_1 \tag{4-415}$$

Thus the linearized model is equivalent to having a linear amplifier with a constant gain K. On the other hand, if x_{01} is a large number, the nominal operating point will lie on the saturated portion of the nonlinearity, and $a = 0$. This means that any small variation in x_1 (small Δx_1) will give rise to practically no change in $\Delta \dot{x}_2$.

EXAMPLE 4-32   In the last example the linearized system turns out to be time invariant. In general, linearization of a nonlinear system often leads to a linear time-varying system. Consider the following nonlinear system:

$$\dot{x}_1 = \frac{-1}{x_2^2} \tag{4-416}$$

$$\dot{x}_2 = ux_1 \tag{4-417}$$

We would like to linearize these equations about the nominal trajectory $[x_{01}(t), x_{02}(t)]$, which is the solution of the equations with the initial conditions $x_1(0) = x_2(0) = 1$ and the input $u(t) = 0$.

Integrating both sides of Eq. (4-417), we have

$$x_2 = x_2(0) = 1 \tag{4-418}$$

Then Eq. (4-416) gives

$$x_1 = -t + 1 \tag{4-419}$$

Therefore, the nominal trajectory about which Eqs. (4-416) and (4-417) are to be linearized is described by

$$x_{01}(t) = -t + 1 \tag{4-420}$$

$$x_{02}(t) = 1 \tag{4-421}$$

Now evaluating the coefficients of Eq. (4-403), we get

$$\frac{\partial f_1}{\partial x_1} = 0 \qquad \frac{\partial f_1}{\partial x_2} = \frac{2}{x_2^3} \qquad \frac{\partial f_2}{\partial x_1} = u \qquad \frac{\partial f_2}{\partial u} = x_1$$

Equation (4-403) gives

$$\Delta \dot{x}_1 = \frac{2}{x_{02}^3} \Delta x_2 \tag{4-422}$$

$$\Delta \dot{x}_2 = u_0 \Delta x_1 + x_{01} \Delta u \tag{4-423}$$

Substituting Eqs. (4-420) and (4-421) into Eqs. (4-422) and (4-423), the linearized equations are written as

$$\begin{bmatrix} \Delta \dot{x}_1 \\ \Delta \dot{x}_2 \end{bmatrix} = \begin{bmatrix} 0 & 2 \\ 0 & 0 \end{bmatrix} \begin{bmatrix} \Delta x_1 \\ \Delta x_2 \end{bmatrix} + \begin{bmatrix} 0 \\ 1 - t \end{bmatrix} \Delta u \tag{4-424}$$

which is a set of linear state equations with time-varying coefficients.

4.19 State Equations of Linear Discrete-Data Systems

Similar to the continuous-data systems case, a modern way of modeling a discrete-data system is by means of discrete state equations. As described earlier, when dealing with discrete-data systems, we often encounter two different situations. The first one is that the components of the system are continuous-data elements, but the signals at certain points of the system are discrete or discontinuous with respect to time, because of the sample-and-hold operations. In this case the components of the system are still described by differential equations, but because of the discrete data, a set of difference equations may be generated from the original differential equations. The second situation involves systems that are completely discrete with respect to time in the sense that they receive and send out discrete data only, such as in the case of a digital controller or digital computer. Under this condition, the system dynamics should be described by difference equations.

Let us consider the open-loop discrete-data control system with a sample-and-hold device, as shown in Fig. 4-28. Typical signals that appear at various points in the system are also shown in the figure. The output signal, $c(t)$,

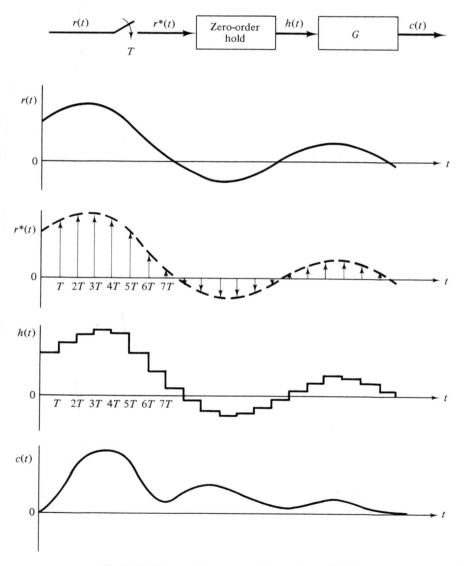

Fig. 4-28. Discrete-data system with sample-and-hold.

ordinarily is a continuous-data signal. The output of the sample-and-hold, $h(t)$, is a train of steps. Therefore, we can write

$$h(kT) = r(kT) \qquad k = 0, 1, 2, \ldots \tag{4-425}$$

Now we let the linear process G be described by the state equation and output equation:

$$\frac{d\mathbf{x}(t)}{dt} = \mathbf{A}\mathbf{x}(t) + \mathbf{B}h(t) \tag{4-426}$$

$$c(t) = \mathbf{D}\mathbf{x}(t) + \mathbf{E}h(t) \tag{4-427}$$

where $\mathbf{x}(t)$ is the state vector and $h(t)$ and $c(t)$ are the scalar input and output signals, respectively. The matrices $\mathbf{A}$, $\mathbf{B}$, $\mathbf{D}$, and $\mathbf{E}$ are coefficient matrices which have been defined earlier. Using Eq. (4-70), the state transition equation of the system is written

$$\mathbf{x}(t) = \boldsymbol{\phi}(t - t_0)\mathbf{x}(t_0) + \int_{t_0}^{t} \boldsymbol{\phi}(t - \tau)\mathbf{B}h(\tau)\, d\tau \tag{4-428}$$

for $t \geq t_0$.

If we are interested only in the responses at the sampling instants, just as in the case of the z-transform solution, we let $t = (k + 1)T$ and $t_0 = kT$. Then Eq. (4-428) becomes

$$\mathbf{x}[(k + 1)T] = \boldsymbol{\phi}(T)\mathbf{x}(kT) + \int_{kT}^{(k+1)T} \boldsymbol{\phi}[(k + 1)T - \tau]\mathbf{B}h(\tau)\, d\tau \tag{4-429}$$

where $\boldsymbol{\phi}(t)$ is the state transition matrix as defined in Section 4.4.

Since $h(t)$ is piecewise constant, that is, $h(kT) = r(kT)$ for $kT \leq t < (k + 1)T$, the input function $h(\tau)$ in Eq. (4-429) can be taken outside the integral sign. Equation (4-429) is written

$$\mathbf{x}[(k + 1)T] = \boldsymbol{\phi}(T)\mathbf{x}(kT) + \int_{kT}^{(k+1)T} \boldsymbol{\phi}[(k + 1)T - \tau]\mathbf{B}\, d\tau\, r(kT) \tag{4-430}$$

or

$$\mathbf{x}[(k + 1)T] = \boldsymbol{\phi}(T)\mathbf{x}(kT) + \boldsymbol{\theta}(T)r(kT) \tag{4-431}$$

where

$$\boldsymbol{\theta}(T) = \int_{kT}^{(k+1)T} \boldsymbol{\phi}[(k + 1)\,T - \tau]\mathbf{B}\, d\tau \tag{4-432}$$

Equation (4-431) is of the form of a linear difference equation in vector-matrix form. Since it represents a set of first-order difference equations, it is referred to as the *vector-matrix discrete state equation*.

The discrete state equation in Eq. (4-431) can be solved by means of a simple recursion procedure. Setting $k = 0, 1, 2, \ldots$ in Eq. (4-431), we find that the following equations result:

$$k = 0: \qquad \mathbf{x}(T) = \boldsymbol{\phi}(T)\mathbf{x}(0) + \boldsymbol{\theta}(T)r(0) \tag{4-433}$$

$$k = 1: \qquad \mathbf{x}(2T) = \boldsymbol{\phi}(T)\mathbf{x}(T) + \boldsymbol{\theta}(T)r(T) \tag{4-434}$$

$$k = 2: \qquad \mathbf{x}(3T) = \boldsymbol{\phi}(T)\mathbf{x}(2T) + \boldsymbol{\theta}(T)r(2T) \tag{4-435}$$

$$\vdots$$

$$k = k - 1: \qquad \mathbf{x}(kT) = \boldsymbol{\phi}(T)\mathbf{x}[(k - 1)T] + \boldsymbol{\theta}(T)r[(k - 1)T] \tag{4-436}$$

Substituting Eq. (4-433) into Eq. (4-434), and then Eq. (4-434) into Eq. (4-435), . . . , and so on, we obtain the following solution for Eq. (4-431):

$$\mathbf{x}(kT) = \boldsymbol{\phi}^k(T)\mathbf{x}(0) + \sum_{i=0}^{k-1} \boldsymbol{\phi}^{k-i-1}(T)\boldsymbol{\theta}(T)r(iT) \qquad (4\text{-}437)$$

Equation (4-437) is defined as the *discrete state transition equation* of the discrete-data system. It is interesting to note that Eq. (4-437) is analogous to its continuous counterpart in Eq. (4-67). In fact, the state transition equation of Eq. (4-67) describes the state of the system of Fig. 4-28 with or without sampling. The discrete state transition equation of Eq. (4-437) is more restricted in that it describes the state only at $t = kT$ ($k = 0, 1, 2, \ldots$), and only if the system has a sample-and-hold device such as in Fig. 4-28.

With kT considered as the initial time, a discrete state transition equation similar to that of Eq. (4-70) can be obtained as

$$\mathbf{x}[(k+N)T] = \boldsymbol{\phi}^N(T)\mathbf{x}(kT) + \sum_{i=0}^{N-1} \boldsymbol{\phi}^{N-i-1}(T)\boldsymbol{\theta}(T)r[(k+i)T] \quad (4\text{-}438)$$

where N is a positive integer. The derivation of Eq. (4-438) is left as an exercise for the reader.

The output of the system of Fig. 4-28 at the sampling instants is obtained by substituting $t = kT$ and Eq. (4-437) into Eq. (4-427), yielding

$$\begin{aligned} c(kT) &= \mathbf{D}\mathbf{x}(kT) + \mathbf{E}h(kT) \\ &= \mathbf{D}\boldsymbol{\phi}^k(T)\mathbf{x}(0) + \mathbf{D}\sum_{i=0}^{k-1} \boldsymbol{\phi}^{k-i-1}(T)\boldsymbol{\theta}(T)r(iT) + \mathbf{E}h(kT) \end{aligned} \qquad (4\text{-}439)$$

An important advantage of the state-variable method over the z-transform method is that it can be modified easily to describe the states and the output between sampling instants. In Eq. (4-428) if we let $t = (k + \Delta)T$, where $0 < \Delta \leq 1$ and $t_0 = kT$, we get

$$\begin{aligned} \mathbf{x}[(k+\Delta)T] &= \boldsymbol{\phi}(\Delta T)\mathbf{x}(kT) + \int_{kT}^{(k+\Delta)T} \boldsymbol{\phi}[(k+\Delta)T - \tau]\mathbf{B}\,d\tau\,r(kT) \\ &= \boldsymbol{\phi}(\Delta T)\mathbf{x}(kT) + \boldsymbol{\theta}(\Delta T)r(kT) \end{aligned} \qquad (4\text{-}440)$$

By varying the value of Δ between 0 and 1, the information between the sampling instants is completely described by Eq. (4-440).

One of the interesting properties of the state transition matrix $\boldsymbol{\phi}(t)$ is that

$$\boldsymbol{\phi}^k(T) = \boldsymbol{\phi}(kT) \qquad (4\text{-}441)$$

which is proved as follows.

Using the homogeneous solution of the state equation of Eq. (4-426), we have

$$\mathbf{x}(t) = \boldsymbol{\phi}(t - t_0)\mathbf{x}(t_0) \qquad (4\text{-}442)$$

Let $t = kT$ and $t_0 = 0$; the last equation becomes

$$\mathbf{x}(kT) = \boldsymbol{\phi}(kT)\mathbf{x}(0) \qquad (4\text{-}443)$$

Also, by the recursive procedure with $t = (k + 1)T$ and $t_0 = kT, k = 0, 1,$

2, . . . , Eq. (4-442) leads to

$$\mathbf{x}(kT) = \boldsymbol{\phi}^k(T)\mathbf{x}(0) \qquad (4\text{-}444)$$

Comparison of Eqs. (4-443) and (4-444) gives the identity in Eq. (4-441).

In view of the relation of Eq. (4-441), the discrete state transition equations of Eqs. (4-437) and (4-438) are written

$$\mathbf{x}(kT) = \boldsymbol{\phi}(kT)\mathbf{x}(0) + \sum_{i=0}^{k-1} \boldsymbol{\phi}[(k - i - 1)T]\boldsymbol{\theta}(T)\mathbf{r}(iT) \qquad (4\text{-}445)$$

$$\mathbf{x}[(k + N)T] = \boldsymbol{\phi}(NT)\mathbf{x}(kT) + \sum_{i=0}^{N-1} \boldsymbol{\phi}[(N - i - 1)T]\boldsymbol{\theta}(T)\mathbf{r}[(k + i)T] \qquad (4\text{-}446)$$

respectively. These two equations can be modified to represent systems with multiple inputs simply by changing the input r into a vector $\mathbf{r}$.

When a linear system has only discrete data throughout the system, its dynamics can be described by a set of discrete state equations

$$\mathbf{x}[(k + 1)T] = \mathbf{A}\mathbf{x}(kT) + \mathbf{B}\mathbf{r}(kT) \qquad (4\text{-}447)$$

and output equations

$$\mathbf{c}(kT) = \mathbf{D}\mathbf{x}(kT) + \mathbf{E}\mathbf{r}(kT) \qquad (4\text{-}448)$$

where $\mathbf{A}$, $\mathbf{B}$, $\mathbf{D}$, and $\mathbf{E}$ are coefficient matrices of the appropriate dimensions. Notice that Eq. (4-447) is basically of the same form as Eq. (4-431). The only difference in the two situations is the starting point of the system representation. In the case of Eq. (4-431), the starting point is the continuous-data state equations of Eq. (4-426); $\boldsymbol{\phi}(T)$ and $\boldsymbol{\theta}(T)$ are determined from the $\mathbf{A}$ and $\mathbf{B}$ matrices of Eq. (4-426). In the case of Eq. (4-447), the equation itself represents an outright description of the discrete-data system, which has only discrete signals.

The solution of Eq. (4-447) follows directly from that of Eq. (4-431). Therefore, the discrete state transition equation of Eq. (4-447) is written

$$\mathbf{x}(kT) = \mathbf{A}^k\mathbf{x}(0) + \sum_{i=0}^{k-1} \mathbf{A}^{k-i-1}\mathbf{B}\mathbf{r}(iT) \qquad (4\text{-}449)$$

where

$$\mathbf{A}^k = \underbrace{\mathbf{A}\mathbf{A}\mathbf{A} \ldots \mathbf{A}}_{k} \qquad (4\text{-}450)$$

4.20 z-Transform Solution of Discrete State Equations

The discrete state equation in vector-matrix form,

$$\mathbf{x}[(k + 1)T] = \mathbf{A}\mathbf{x}(kT) + \mathbf{B}\mathbf{r}(kT) \qquad (4\text{-}451)$$

can be solved by means of the z-transform method. Taking the z-transform on both sides of Eq. (4-451) yields

$$z\mathbf{X}(z) - z\mathbf{x}(0+) = \mathbf{A}\mathbf{X}(z) + \mathbf{B}\mathbf{R}(z) \qquad (4\text{-}452)$$

Solving for $\mathbf{X}(z)$ from the last equation gives

$$\mathbf{X}(z) = (z\mathbf{I} - \mathbf{A})^{-1}z\mathbf{x}(0+) + (z\mathbf{I} - \mathbf{A})^{-1}\mathbf{B}\mathbf{R}(z) \qquad (4\text{-}453)$$

The inverse z-transform of the last equation is

$$\mathbf{x}(kT) = \mathfrak{z}^{-1}[(z\mathbf{I} - \mathbf{A})^{-1}z]\mathbf{x}(0) + \mathfrak{z}^{-1}[(z\mathbf{I} - \mathbf{A})^{-1}\mathbf{B}\mathbf{R}(z)] \qquad (4\text{-}454)$$

In order to carry out the inverse z-transform operation of the last equation, we write the z-transform of $\mathbf{A}^k$ as

$$\mathfrak{z}(\mathbf{A}^k) = \sum_{k=0}^{\infty} \mathbf{A}^k z^{-k} = \mathbf{I} + \mathbf{A}z^{-1} + \mathbf{A}^2 z^{-2} + \cdots \qquad (4\text{-}455)$$

Premultiplying both sides of the last equation by $\mathbf{A}z^{-1}$ and subtracting the result from the last equation, we get

$$(\mathbf{I} - \mathbf{A}z^{-1})\mathfrak{z}(\mathbf{A}^k) = \mathbf{I} \qquad (4\text{-}456)$$

Therefore, solving for $\mathfrak{z}(\mathbf{A}^k)$ from the last equation yields

$$\mathfrak{z}(\mathbf{A}^k) = (\mathbf{I} - \mathbf{A}z^{-1})^{-1} = (z\mathbf{I} - \mathbf{A})^{-1}z \qquad (4\text{-}457)$$

or

$$\mathbf{A}^k = \mathfrak{z}^{-1}[(z\mathbf{I} - \mathbf{A})^{-1}z] \qquad (4\text{-}458)$$

Equation (4-458) also represents a way of finding $\mathbf{A}^k$ by using the z-transform method. Similarly, we can prove that

$$\mathfrak{z}^{-1}[(z\mathbf{I} - \mathbf{A})^{-1}\mathbf{B}\mathbf{R}(z)] = \sum_{i=0}^{k-1} \mathbf{A}^{k-i-1}\mathbf{B}\mathbf{r}(iT) \qquad (4\text{-}459)$$

Now we substitute Eqs. (4-458) and (4-459), into Eq. (4-454) and we have the solution for $\mathbf{x}(kT)$ as

$$\mathbf{x}(kT) = \mathbf{A}^k\mathbf{x}(0) + \sum_{i=0}^{k-1} \mathbf{A}^{k-i-1}\mathbf{B}\mathbf{r}(iT) \qquad (4\text{-}460)$$

which is identical to the expression in Eq. (4-449).

Once a discrete-data system is represented by the dynamic equations of Eqs. (4-447) and (4-448), the transfer function relation of the system can be expressed in terms of the coefficient matrices.

Setting the initial state $\mathbf{x}(0+)$ to zero, Eq. (4-453) gives

$$\mathbf{X}(z) = (z\mathbf{I} - \mathbf{A})^{-1}\mathbf{B}\mathbf{R}(z) \qquad (4\text{-}461)$$

When this equation is substituted into the z-transformed version of Eq. (4-448), we have

$$\mathbf{C}(z) = [\mathbf{D}(z\mathbf{I} - \mathbf{A})^{-1}\mathbf{B} + \mathbf{E}]\mathbf{R}(z) \qquad (4\text{-}462)$$

Thus the transfer function matrix of the system is

$$\mathbf{G}(z) = \mathbf{D}(z\mathbf{I} - \mathbf{A})^{-1}\mathbf{B} + \mathbf{E} \qquad (4\text{-}463)$$

This equation can be written

$$\mathbf{G}(z) = \frac{\mathbf{D}\,[\text{adj}\,(z\mathbf{I} - \mathbf{A})]\mathbf{B} + |z\mathbf{I} - \mathbf{A}|\,\mathbf{E}}{|z\mathbf{I} - \mathbf{A}|} \qquad (4\text{-}464)$$

The characteristic equation of the system is defined as

$$|z\mathbf{I} - \mathbf{A}| = 0 \qquad (4\text{-}465)$$

In general, a linear time-invariant discrete-data system with one input and one output can be described by the following linear difference equation with constant coefficients:

$$c[(k + n)T] + a_1c[(k + n - 1)T] + a_2c[(k + n - 2)T] + \ldots$$
$$+ a_{n-1}c[(k + 1)T] + a_nc(kT)$$
$$= b_0r[(k + m)T] + b_1r[(k + m - 1)T] + \ldots$$
$$+ b_{m-1}r[(k + 1)T] + b_mr(kT) \qquad n \geq m \tag{4-466}$$

Taking the z-transform on both sides of this equation and rearranging terms, the transfer function of the system is written

$$\frac{C(z)}{R(z)} = \frac{b_0z^m + b_1z^{m-1} + \ldots + b_{m-1}z + b_m}{z^n + a_1z^{n-1} + \ldots + a_{n-1}z + a_n} \tag{4-467}$$

The characteristic equation is defined as

$$z^n + a_1z^{n-1} + \ldots + a_{n-1}z + a_n = 0 \tag{4-468}$$

EXAMPLE 4-33   Consider that a discrete-data system is described by the difference equation

$$c(k + 2) + 5c(k + 1) + 3c(k) = r(k + 1) + 2r(k) \tag{4-469}$$

Taking the z-transform on both sides of the last equation and assuming zero initial conditions yields

$$z^2C(z) + 5zC(z) + 3C(z) = zR(z) + 2R(z) \tag{4-470}$$

From the last equation the transfer function of the system is easily written

$$\frac{C(z)}{R(z)} = \frac{z + 2}{z^2 + 5z + 3} \tag{4-471}$$

The characteristic equation is obtained by setting the denominator polynomial of the transfer function to zero,

$$z^2 + 5z + 3 = 0 \tag{4-472}$$

The state variables of the system are arbitrarily defined as

$$x_1(k) = c(k) \tag{4-473}$$
$$x_2(k) = x_1(k + 1) - r(k) \tag{4-474}$$

Substitution of the last two relations into the original difference equation of Eq. (4-469) gives the two state equations of the system as

$$x_1(k + 1) = x_2(k) + r(k) \tag{4-475}$$
$$x_2(k + 1) = -3x_1(k) - 5x_2(k) - 3r(k) \tag{4-476}$$

from which we have the **A** matrix of the system,

$$\mathbf{A} = \begin{bmatrix} 0 & 1 \\ -3 & -5 \end{bmatrix} \tag{4-477}$$

The same characteristic equation as in Eq. (4-472) is obtained by using $|z\mathbf{I} - \mathbf{A}| = 0$.

4.21 State Diagrams for Discrete-Data Systems

When a discrete-data system is described by difference equations or discrete state equations, a discrete state diagram may be constructed for the system. Similar to the relations between the analog computer diagram and the state diagram for a continuous-data system, the elements of a discrete state diagram resemble the computing elements of a digital computer. Some of the operations of a digital computer are multiplication by a constant, addition of several machine variables, time delay, or shifting. The mathematical descriptions of these basic digital computations and their corresponding z-transform expressions are as follows:

1. *Multiplication by a constant:*

$$x_2(kT) = ax_1(kT) \tag{4-478}$$

$$X_2(z) = aX_1(z) \tag{4-479}$$

2. *Summing:*

$$x_2(kT) = x_0(kT) + x_1(kT) \tag{4-480}$$

$$X_2(z) = X_0(z) + X_1(z) \tag{4-481}$$

3. *Shifting or time delay:*

$$x_2(kT) = x_1[(k + 1)T] \tag{4-482}$$

$$X_2(z) = zX_1(z) - zx_1(0+) \tag{4-483}$$

or

$$X_1(z) = z^{-1}X_2(z) + x_1(0+) \tag{4-484}$$

The state diagram representations of these operations are illustrated in Fig. 4-29. The initial time $t = 0+$ in Eq. (4-484) can be generalized to $t = t_0^+$. Then Eq. (4-484) is written

$$X_1(z) = z^{-1}X_2(z) + x_1(t_0^+) \tag{4-485}$$

which represents the discrete-time state transition for time greater than or equal to t_0^+.

EXAMPLE 4-34   Consider again the difference equation in Eq. (4-469), which is

$$c(k + 2) + 5c(k + 1) + 3c(k) = r(k + 1) + 2r(k) \tag{4-486}$$

One way of constructing the discrete state diagram for the system is to use the state equations. In this case the state equations are available in Eqs. (4-475) and (4-476) and these are repeated here:

$$x_1(k + 1) = x_2(k) + r(k) \tag{4-487}$$

$$x_2(k + 1) = -3x_1(k) - 5x_2(k) - 3r(k) \tag{4-488}$$

Using essentially the same principle as for the state diagrams for continuous-data systems, the state diagram for Eqs. (4-487) and (4-488) is constructed in Fig. 4-30. The time delay unit z^{-1} is used to relate $x_1(k + 1)$ to $x_1(k)$. The state variables will always appear as outputs of the delay units on the state diagram.

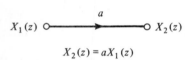

$$X_2(z) = aX_1(z)$$

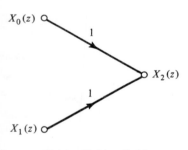

$$X_2(z) = X_0(z) + X_1(z)$$

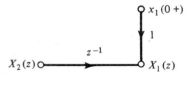

$$X_1(z) = z^{-1}X_2(z) + x_1(0+)$$

Fig. 4-29. Basic elements of a discrete state diagram.

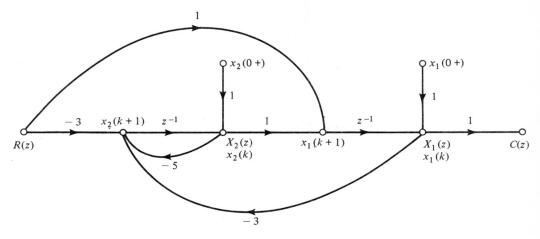

Fig. 4-30. Discrete state diagram of the system described by the difference equation of Eq. (4-486) or by the state equations of Eqs. (4-487) and (4-488).

As an alternative, the state diagram can also be drawn directly from the difference equation by means of the decomposition schemes. The decomposition of a discrete transfer function will be discussed in the following section, after we have demonstrated some of the practical applications of the discrete state diagram.

The state transition equation of the system can be obtained directly from the state diagram using the gain formula. Referring to $X_1(z)$ and $X_2(z)$ as the output nodes and to $x_1(0+)$, $x_2(0+)$, and $R(z)$ as input nodes in Fig. 4-30, the state transition equations are written in the following vector-matrix form:

$$\begin{bmatrix} X_1(z) \\ X_2(z) \end{bmatrix} = \frac{1}{\Delta} \begin{bmatrix} 1 + 5z^{-1} & z^{-1} \\ -3z^{-1} & 1 \end{bmatrix} \begin{bmatrix} x_1(0+) \\ x_2(0+) \end{bmatrix} + \frac{1}{\Delta} \begin{bmatrix} z^{-1}(1 + 5z^{-1}) - 3z^{-2} \\ -3z^{-1} - 3z^{-2} \end{bmatrix} R(z) \quad (4\text{-}489)$$

where

$$\Delta = 1 + 5z^{-1} + 3z^{-2} \quad (4\text{-}490)$$

The same transfer function between $R(z)$ and $C(z)$ as in Eq. (4-471) can be obtained directly from the state diagram by applying the gain formula between these two nodes.

Decomposition of Discrete Transfer Functions

The three schemes of decomposition discussed earlier for continuous-data systems can be applied to transfer functions of discrete-data systems without the need of modification. As an illustrative example, the following transfer function is decomposed by the three methods, and the corresponding state diagrams are shown in Fig. 4-31:

$$\frac{C(z)}{R(z)} = \frac{z + 2}{z^2 + 5z + 3} \quad (4\text{-}491)$$

Equation (4-491) is used for direct decomposition after the numerator and the denominator are both multiplied by z^{-2}. For cascade decomposition, the trans-

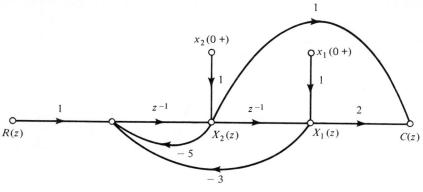

(a) Direct decomposition

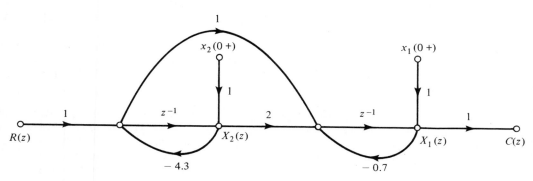

(b) Cascade decomposition

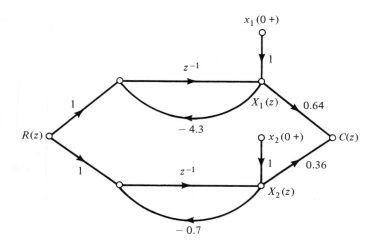

(c) Parallel decomposition

Fig. 4-31. State diagrams of the transfer function $C(z)/R(z) = (z + 2)/(z^2 + 5z + 3)$ by the three methods of decomposition. (a) Direct decomposition. (b) Cascade decomposition. (c) Parallel decomposition.

fer function is first written in factored form as

$$\frac{C(z)}{R(z)} = \frac{z+2}{(z+4.3)(z+0.7)} \tag{4-492}$$

For the parallel decomposition, the transfer function is first fractioned by partial fraction into the following form:

$$\frac{C(z)}{R(z)} = \frac{0.64}{z+4.3} + \frac{0.36}{z+0.7} \tag{4-493}$$

4.22 State Diagrams for Sampled-Data Systems

When a discrete-data system has continuous-data as well as discrete-data elements, with the two types of elements separated by sample-and-hold devices, a special treatment of the state diagram is necessary if a description of the continuous-data states is desired for all times.

Let us first establish the state diagram of the zero-order hold. Consider that the input of the zero-order hold is denoted by $e^*(t)$ which is a train of impulses, and the output by $h(t)$. Since the zero-order hold simply holds the magnitude of the input impulse at the sampling instant until the next input comes along, the signal $h(t)$ is a sequence of steps. The input–output relation in the Laplace domain is written

$$H(s) = \frac{1 - e^{-Ts}}{s} E^*(s) \tag{4-494}$$

In the time domain, the relation is simply

$$h(t) = e(kT+) \tag{4-495}$$

for $kT \le t < (k+1)T$.

In the state diagram notation, we need the relation between $H(s)$ and $e(kT+)$. For this purpose we take the Laplace transform on both sides of Eq. (4-495) to give

$$H(s) = \frac{e(kT+)}{s} \tag{4-496}$$

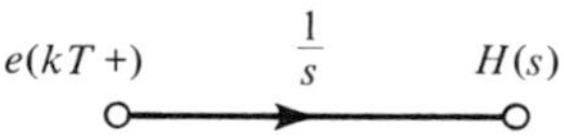

$$e(kT+) \quad\xrightarrow{\quad\frac{1}{s}\quad}\quad H(s)$$

Fig. 4-32. State diagram representation of the zero-order hold.

for $kT \le t < (k+1)T$. The state diagram representation of the zero-order hold is shown in Fig. 4-32. As an illustrative example on how the state diagram of a sampled-data system is constructed, let us consider the system shown in Fig. 4-33. We shall demonstrate the various available ways of modeling the input–output relations of the system.

First, the Laplace transform of the output of the system is written

$$C(s) = \frac{1 - e^{-Ts}}{s} \frac{1}{s+1} E^*(s) \tag{4-497}$$

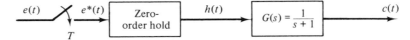

Fig. 4-33. Sampled-data system.

Taking the z-transform on both sides of the last equation yields

$$C(z) = \frac{1 - e^{-T}}{z - e^{-T}} E(z) \qquad (4\text{-}498)$$

Given information on the input $e(t)$ or $e^*(t)$, Eq. (4-498) gives the output response at the sampling instants.

A state diagram can be drawn from Eq. (4-498) using the decomposition technique. Figure 4-34 illustrates the discrete state diagram of the system through

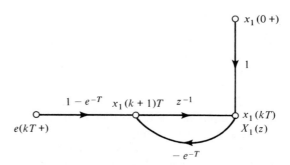

Fig. 4-34. Discrete state diagram of the system in Fig. 4-33.

decomposition. The discrete dynamic equations of the system are written directly from this state diagram:

$$x_1[(k + 1)T] = e^{-T}x_1(kT) + (1 - e^{-T})e(kT) \qquad (4\text{-}499)$$

$$c(kT) = x_1(kT) \qquad (4\text{-}500)$$

Therefore, the output response of the system can also be obtained by solving the difference equation of Eq. (4-499).

If the response of the output $c(t)$ is desired for all t, we may construct the state diagram shown in Fig. 4-35. This state diagram is obtained by cascading the state diagram representations of the zero-order hold and the process $G(s)$.

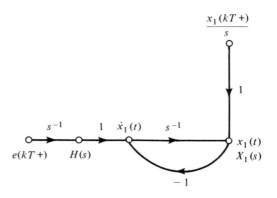

Fig. 4-35. State diagram for the system of Fig. 4-33 for the time interval $kT \leq t \leq (k + 1)T$.

To determine $c(t)$, which is also $x_1(t)$, we must first obtain $X_1(s)$ by applying the gain formula to the state diagram of Fig. 4-35. We have

$$X_1(s) = \frac{s^{-2}}{1 + s^{-1}}e(kT+) + \frac{s^{-1}}{1 + s^{-1}}x_1(kT) \tag{4-501}$$

for $kT \le t \le (k + 1)T$. Taking the inverse Laplace transform of the last equation gives

$$x_1(t) = [1 - e^{-(t-kT)}]e(kT+) + e^{-(t-kT)}x_1(kT) \tag{4-502}$$

$kT \le t \le (k + 1)T$. It is interesting to note that in Eq. (4-502) t is valid for one sampling period, whereas the result in Eq. (4-499) gives information on $x_1(t)$ only at the sampling instants. It is easy to see that if we let $t = (k + 1)T$ in Eq. (4-502), the latter becomes Eq. (4-499).

4.23 State Equations of Linear Time-Varying Systems

When a linear system has time-varying elements, it can be represented by the following dynamic equations:

$$\frac{d\mathbf{x}(t)}{dt} = \mathbf{A}(t)\mathbf{x}(t) + \mathbf{B}(t)\mathbf{r}(t) \tag{4-503}$$

$$\mathbf{c}(t) = \mathbf{D}(t)\mathbf{x}(t) + \mathbf{E}(t)\mathbf{r}(t) \tag{4-504}$$

where

$\mathbf{x}(t) = n \times 1$ state vector

$\mathbf{r}(t) = p \times 1$ input vector

$\mathbf{c}(t) = q \times 1$ output vector

$\mathbf{A}(t)$, $\mathbf{B}(t)$, $\mathbf{D}(t)$, and $\mathbf{E}(t)$ are coefficient matrices of appropriate dimensions. The elements of these coefficient matrices are functions of t.

Unlike in the time-invariant case, time-varying differential equations generally do not have closed-form solutions. Let us investigate the properties of a time-varying system by considering a scalar homogeneous state equation,

$$\frac{dx(t)}{dt} = a(t)x(t) \tag{4-505}$$

This equation can be solved by first separating the variables,

$$\frac{dx(t)}{x(t)} = a(t)\,dt \tag{5-506}$$

and then integrating both sides to get

$$\ln x(t) - \ln x(t_0) = \int_{t_0}^{t} a(\tau)\,d\tau \tag{4-507}$$

Therefore,

$$x(t) = \exp\left[\int_{t_0}^{t} a(\tau)\,d\tau\right]x(t_0) \tag{4-508}$$

where t_0 denotes the initial time.

Just as in the time-invariant situation, we can define a state transition matrix for the time-varying state equation. For the scalar case under consideration, the state transition matrix is

$$\phi(t, t_0) = \exp \int_{t_0}^{t} a(\tau) \, d\tau \tag{4-509}$$

Notice that for the time-varying case, the state transition matrix depends upon t and t_0, not simply $t - t_0$.

For the vector-matrix state equation

$$\dot{\mathbf{x}}(t) = \mathbf{A}(t)\mathbf{x}(t) \tag{4-510}$$

it is simple to show that the solution can be written

$$\mathbf{x}(t) = \phi(t, t_0)\mathbf{x}(t_0) \tag{4-511}$$

where $\phi(t, t_0)$ is the state transition matrix that satisfies Eq. (4-510). However, the problem is how to find $\phi(t, t_0)$ in general. The question is: Is $\phi(t, t_0)$ related to the $\mathbf{A}(t)$ matrix through the following relationship?

$$\phi(t, t_0) = \exp\left[\int_{t_0}^{t} \mathbf{A}(\tau) \, d\tau\right] \tag{4-512}$$

To answer the posed question, let us expand the right side of Eq. (4-512) into a power series,

$$\exp\left[\int_{t_0}^{t} \mathbf{A}(\tau) \, d\tau)\right] = \mathbf{I} + \int_{t_0}^{t} \mathbf{A}(\tau) \, d\tau + \frac{1}{2!} \int_{t_0}^{t} \mathbf{A}(\tau) \, d\tau \int_{t_0}^{t} \mathbf{A}(\sigma) \, d\sigma + \dots \tag{4-513}$$

Taking the derivative on both sides of the last equation with respect to time, we have

$$\frac{d}{dt} \exp\left[\int_{t_0}^{t} \mathbf{A}(\tau) \, d\tau\right] = \mathbf{A}(t) + \frac{1}{2}\mathbf{A}(t) \int_{t_0}^{t} \mathbf{A}(\sigma) \, d\sigma + \frac{1}{2} \int_{t_0}^{t} \mathbf{A}(\tau) \, d\tau \, \mathbf{A}(t) + \dots \tag{4-514}$$

Multiplying both sides of Eq. (4-513) by $\mathbf{A}(t)$, we have

$$\mathbf{A}(t) \exp\left[\int_{t_0}^{t} \mathbf{A}(\tau) \, d\tau\right] = \mathbf{A}(t) + \mathbf{A}(t) \int_{t_0}^{t} \mathbf{A}(\tau) \, d\tau + \dots \tag{4-515}$$

By comparison of Eqs. (4-514) and (4-515), we see that

$$\frac{d}{dt} \exp\left[\int_{t_0}^{t} \mathbf{A}(\tau) \, d\tau\right] = \mathbf{A}(t) \exp\left[\int_{t_0}^{t} \mathbf{A}(\tau) \, d\tau\right] \tag{4-516}$$

or

$$\frac{d}{dt}\phi(t, t_0) = \mathbf{A}(t)\phi(t, t_0) \tag{4-517}$$

if and only if

$$\mathbf{A}(t) \int_{t_0}^{t} \mathbf{A}(\tau) \, d\tau = \int_{t_0}^{t} \mathbf{A}(\tau) \, d\tau \, \mathbf{A}(t) \tag{4-518}$$

that is,

$$\mathbf{A}(t) \quad \text{and} \quad \int_{t_0}^{t} \mathbf{A}(\tau) \, d\tau$$

commute.

The requirement that $\mathbf{A}(t)$ and its integral commute is evidently a very stringent condition. Therefore, in general, Eq. (4-512) will not be valid.

Most of the properties of the time-invariant state transition matrix, $\boldsymbol{\phi}(t - t_0)$, can be extended to the time-varying case. These are listed as follows:

1. $\boldsymbol{\phi}(t, t_0) = \mathbf{I}$.
2. $\boldsymbol{\phi}^{-1}(t, t_0) = \boldsymbol{\phi}(t_0, t)$.
3. $\boldsymbol{\phi}(t_2, t_1)\,\boldsymbol{\phi}(t_1, t_0) = \boldsymbol{\phi}(t_2, t_0)$ for any t_0, t_1, t_2.

Solution of the Nonhomogeneous Time-Varying State Equation

Disregarding the problem of finding the state transition matrix $\boldsymbol{\phi}(t, t_0)$ for the moment, we shall solve for the solution of the nonhomogeneous state equation of Eq. (4-503).

Let the solution be

$$\mathbf{x}(t) = \boldsymbol{\phi}(t, t_0)\boldsymbol{\eta}(t) \tag{4-519}$$

where $\boldsymbol{\eta}(t)$ is an $n \times 1$ vector, and $\boldsymbol{\phi}(t, t_0)$ is the state transition matrix that satisfies Eq. (4-517). Equation (4-519) must satisfy Eq. (4-503). Substitution of Eq. (4-519) into Eq. (4-503) yields

$$\dot{\boldsymbol{\phi}}(t, t_0)\boldsymbol{\eta}(t) + \boldsymbol{\phi}(t, t_0)\dot{\boldsymbol{\eta}}(t) = \mathbf{A}(t)\boldsymbol{\phi}(t, t_0)\boldsymbol{\eta}(t) + \mathbf{B}(t)\mathbf{u}(t) \tag{4-520}$$

Substituting Eq. (4-517) into Eq. (4-520) and simplifying, we get

$$\boldsymbol{\phi}(t, t_0)\dot{\boldsymbol{\eta}}(t) = \mathbf{B}(t)\mathbf{u}(t) \tag{4-521}$$

Thus

$$\dot{\boldsymbol{\eta}}(t) = \boldsymbol{\phi}^{-1}(t, t_0)\,\mathbf{B}(t)\mathbf{u}(t) \tag{4-522}$$

and

$$\boldsymbol{\eta}(t) = \int_{t_0}^{t} \boldsymbol{\phi}^{-1}(\tau, t_0)\mathbf{B}(\tau)\mathbf{u}(\tau)\,d\tau + \boldsymbol{\eta}(t_0) \tag{4-523}$$

The vector $\boldsymbol{\eta}(t_0)$ is obtained from Eq. (4-519) by setting $t = t_0$. Thus substituting Eq. (4-523) into Eq. (4-519), we have

$$\mathbf{x}(t) = \boldsymbol{\phi}(t, t_0)\mathbf{x}(t_0) + \boldsymbol{\phi}(t, t_0)\int_{t_0}^{t} \boldsymbol{\phi}^{-1}(\tau, t_0)\mathbf{B}(\tau)\mathbf{u}(\tau)\,d\tau \tag{4-524}$$

Since

$$\boldsymbol{\phi}(t, t_0)\boldsymbol{\phi}^{-1}(\tau, t_0) = \boldsymbol{\phi}(t, t_0)\boldsymbol{\phi}(t_0, \tau) = \boldsymbol{\phi}(t, \tau) \tag{4-525}$$

Eq. (4-524) becomes

$$\mathbf{x}(t) = \boldsymbol{\phi}(t, t_0)\mathbf{x}(t_0) + \int_{t_0}^{t} \boldsymbol{\phi}(t, \tau)\mathbf{B}(\tau)\mathbf{u}(\tau)\,d\tau \tag{4-526}$$

which is the state transition equation of Eq. (4-503).

Discrete Approximation of the Linear Time-Varying System

In practice, not too many time-varying systems can be solved by using Eq. (4-526), since $\boldsymbol{\phi}(t, t_0)$ is not readily available. It is possible to discretize the system with a time increment during which the time-varying parameters do not vary appreciably. Then the problem becomes that of solving a set of linear time-varying discrete state equations. One method of discretizing the system is to approximate the derivative of $\mathbf{x}(t)$ by

$$\dot{\mathbf{x}}(t) \simeq \frac{1}{T}\{\mathbf{x}[(k+1)T] - \mathbf{x}(kT)\} \qquad kT \leq t \leq (k+1)T \qquad (4\text{-}527)$$

where T is a small time interval. The state equation of Eq. (4-503) is approximated by the time-varying difference equation

$$\mathbf{x}[(k+1)T] = \mathbf{A}^*(kT)\mathbf{x}(kT) + \mathbf{B}^*(kT)\mathbf{r}(kT) \qquad (4\text{-}528)$$

over the time interval, $kT \leq t \leq (k+1)T$, where

$$\mathbf{A}^*(kT) = T\mathbf{A}(kT) + \mathbf{I}$$
$$\mathbf{B}^*(kT) = T\mathbf{B}(kT)$$

Equation (4-528) can be solved recursively in much the same way as in the time-invariant case, Eqs. (4-433) through (4-437).

REFERENCES

State Variables and State Equations

1. L. A. ZADEH, "An Introduction to State Space Techniques," Workshop on Techniques for Control Systems, *Proceedings, Joint Automatic Control Conference*, Boulder, Colo., 1962.

2. B. C. KUO, *Linear Networks and Systems*, McGraw-Hill Book Company, New York, 1967.

3. D. W. WIBERG, *Theory and Problems of State Space and Linear Systems* (Schaum's Outline Series), McGraw-Hill Book Company, New York, 1971.

State Transition Matrix

4. R. B. KIRCHNER, "An Explicit Formula for e^{At}," *Amer. Math. Monthly*, Vol. 74, pp. 1200, 1204, 1967.

5. W. EVERLING, "On the Evaluation of e^{At} by Power Series," *Proc. IEEE*, Vol. 55, p. 413, Mar. 1967.

6. T. A. BICKART, "Matrix Exponential: Approximation by Truncated Power Series," *Proc. IEEE*, Vol. 56, pp. 872–873, May 1968.

7. T. M. APOSTOL, "Some Explicit Formulas for the Exponential Matrix e^{At}," *Amer. Math. Monthly*, Vol. 76, pp. 289–292, 1969.

8. M. Vidyasagar, "A Novel Method of Evaluating e^{At} in Closed Form," *IEEE Trans. Automatic Control*, Vol. AC-15, pp. 600–601, Oct. 1970.

9. C. G. CULLEN, "Remarks on Computing e^{At}," *IEEE Trans. Automatic Control*, Vol. AC-16, pp. 94–95, Feb. 1971.

10. J. C. JOHNSON and C. L. PHILLIPS, "An Algorithm for the Computation of the Integral of the State Transition Matrix," *IEEE Trans. Automatic Control*, Vol. AC-16, pp. 204–205, Apr. 1971.

11. M. HEALEY, "Study of Methods of Computing Transition Matrices," *Proc. IEE*, Vol. 120, No. 8, pp. 905–912, Aug. 1973.

Transformations

12. C. D. JOHNSON and W. M. WONHAM, "A Note on the Transformation to Canonical (Phase Variable) Form," *IEEE Trans. Automatic Control*, Vol. AC-9, pp. 312–313, July 1964.

13. I. H. MUFTI, "On the Reduction of a System to Canonical (Phase-Variable) Form," *IEEE Trans. Automatic Control*, Vol. AC-10, pp. 206–207, Apr. 1965.

14. M. R. CHIDAMBARA, "The Transformation to (Phase-Variable) Canonical Form," *IEEE Trans. Automatic Control*, Vol. AC-10, pp. 492–495, Oct. 1965.

15. L. M. SILVERMAN, "Transformation of Time-Variable Systems to Canonical (Phase-Variable) Form," *IEEE Trans. Automatic Control*, Vol. AC-11, pp. 300–303, Apr. 1966.

16. W. G. TUEL, JR., "On the Transformation to (Phase-Variable) Canonical Form," *IEEE Trans. Automatic Control*, Vol. AC-11, p. 607, July 1966.

17. D. G. LUENBERGER, "Canonical Forms for Linear Multivariable Systems," *IEEE Trans. Automatic Control*, Vol. AC-12, pp. 290–293, June 1967.

18. S. J. ASSEO, "Phase-Variable Canonical Transformation of Multicontroller Systems," *IEEE Trans. Automatic Control*, Vol. AC-13, pp. 129–131, Feb. 1968.

19. B. RAMASWAMI and RAMAR, "Transformation to the Phase-Variable Canonical Form," *IEEE Trans. Automatic Control*, Vol. AC-13, pp. 746–747, Dec. 1968.

20. W. B. RUBIN, "A Simple Method for Finding the Jordan Form of a Matrix," *IEEE Trans. Automatic Control*, Vol. AC-17, pp. 145–146, Feb. 1972.

21. K. OGATA, *State Space Analysis of Control Systems*, Prentice-Hall, Inc., Englewood Cliffs, N.J., 1967.

State Diagram

22. B. C. KUO, "State Transition Flow Graphs of Continuous and Sampled Dynamic Systems," *WESCON Convention Records*, 18.1, Aug. 1962.

Controllability and Observability

23. Y. C. HO, "What Constitutes a Controllable System?" *IRE Trans. Automatic Control*, Vol. AC-7, p. 76, Apr. 1962.

24. R. E. KALMAN, Y. C. HO, and K. S. NARENDRA, "Controllability of Linear Dynamical Systems," *Contribution to Differential Equations*, Vol. 1, No. 2, pp. 189–213, 1962.

25. E. G. GILBERT, "Controllability and Observability in Multivariable Control Systems," *J. SIAM Control*, Vol. 1, pp. 128–151, 1963.

26. L. A. ZADEH and C. A. DESOER, *Linear System Theory*, McGraw-Hill Book Company, New York, 1963.

27. R. E. KALMAN, "Mathematical Description of Linear Dynamical Systems," *J. Soc. Ind. Appl. Math.*, Vol. II, No. 1, Ser. A, pp. 151–192, 1963.

28. E. KREINDLER and P. SARACHIK, "On the Concept of Controllability and Observability of Linear Systems," *IEEE Trans. Automatic Control*, Vol. AC-9, pp. 129–136, Apr. 1964.

29. A. R. STUBBERUD, "A Controllability Criterion for a Class of Linear Systems," *IEEE Trans. Application and Industry*, Vol. 83, pp. 411–413, Nov. 1964.

30. R. W. BROCKETT, "Poles, Zeros, and Feedback: State Space Interpretation," *IEEE Trans. Automatic Control*, Vol. AC-10, pp. 129–135, Apr. 1965.

31. R. D. BONNELL, "An Observability Criterion for a Class of Linear Systems," *IEEE Trans. Automatic Control*, Vol. AC-11, p. 135, Jan. 1966.

CSMP (Continuous System Modeling Program)

32. *System/360 Continuous System Modeling Program (360A-CX-16X) User's Manual*, Technical Publications Dept., International Business Machines Corporation, White Plains, N.Y.

PROBLEMS

4.1. Write state equations for the electric networks shown in Fig. P4-1.

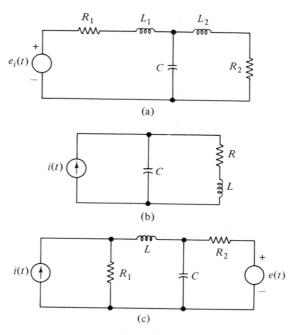

Figure P4-1.

4.2. The following differential equations represent linear time-invariant systems. Write the dynamic equations (state equations and output equation) in vector-matrix form.

(a) $\dfrac{d^2c(t)}{dt^2} + 3\dfrac{dc(t)}{dt} + c(t) = r(t)$

(b) $\dfrac{d^3c(t)}{dt^3} + 6\dfrac{dc(t)}{dt} + 5c(t) = r(t)$

(c) $3\dfrac{d^2c(t)}{dt^2} + 3\dfrac{dc(t)}{dt} + c(t) = \dfrac{dr(t)}{dt}$

(d) $\dfrac{d^2c(t)}{dt^2} + 2\dfrac{dc(t)}{dt} + c(t) + \displaystyle\int_0^t c(\tau)d\tau = r(t)$

(e) $\dfrac{d^2c(t)}{dt^2} + 6\dfrac{dc(t)}{dt} + 5c(t) = 2\dfrac{dr(t)}{dt} + r(t)$

4.3. Using Eq. (4-42), show that

$$\boldsymbol{\phi}(t) = \mathbf{I} + \mathbf{A}t + \frac{1}{2!}\mathbf{A}^2 t^2 + \frac{1}{3!}\mathbf{A}^3 t^3 + \cdots$$

4.4. The state equations of a linear time-invariant system are represented by

$$\dot{\mathbf{x}}(t) = \mathbf{A}\mathbf{x}(t) + \mathbf{B}u(t)$$

Find the state transition matrix $\boldsymbol{\phi}(t)$ for the following cases:

(a)
$$\mathbf{A} = \begin{bmatrix} 0 & 0 \\ -1 & -2 \end{bmatrix} \qquad \mathbf{B} = \begin{bmatrix} 1 \\ 1 \end{bmatrix}$$

(b)
$$\mathbf{A} = \begin{bmatrix} 0 & 1 \\ -2 & -3 \end{bmatrix} \qquad \mathbf{B} = \begin{bmatrix} 0 \\ 1 \end{bmatrix}$$

(c)
$$\mathbf{A} = \begin{bmatrix} -2 & 0 \\ 0 & -2 \end{bmatrix} \qquad \mathbf{B} = \begin{bmatrix} 10 \\ 1 \end{bmatrix}$$

(d)
$$\mathbf{A} = \begin{bmatrix} -1 & 1 & 0 \\ 0 & -1 & 1 \\ 0 & 0 & -1 \end{bmatrix} \qquad \mathbf{B} = \begin{bmatrix} 0 \\ 0 \\ 1 \end{bmatrix}$$

4.5. Find the state transition equations for the systems described in Problem 4.4 for $t \geq 0$. It is assumed that $\mathbf{x}(0+)$ is given and $u(t)$ is a unit step function.

4.6. Given the state equation

$$\dot{\mathbf{x}}(t) = \mathbf{A}\mathbf{x}(t) + \mathbf{B}u(t)$$

where

$$\mathbf{A} = \begin{bmatrix} 0 & 0 & 1 \\ 1 & 0 & -1 \\ 2 & 1 & 0 \end{bmatrix} \qquad \mathbf{B} = \begin{bmatrix} 0 \\ 0 \\ 1 \end{bmatrix}$$

find the transformation such that the state equation becomes

$$\dot{\mathbf{y}}(t) = \mathbf{A}_1 \mathbf{x}(t) + \mathbf{B}_1 u(t)$$

where $\mathbf{A}_1$ and $\mathbf{B}_1$ are in the phase-variable canonical form.

4.7. For the state equation given in Problem 4.6, if

$$\mathbf{A} = \begin{bmatrix} 1 & 0 & 0 \\ 0 & 1 & 0 \\ -2 & -3 & -1 \end{bmatrix} \qquad \mathbf{B} = \begin{bmatrix} 0 \\ 0 \\ 1 \end{bmatrix}$$

can the state equation be transformed into the phase-variable form? Explain.

4.8. Given the state equations of a linear time-invariant system as

$$\dot{\mathbf{x}}(t) = \mathbf{A}\mathbf{x}(t) + \mathbf{B}u(t)$$

where

$$\mathbf{A} = \begin{bmatrix} 0 & 1 & 0 \\ 0 & 0 & 1 \\ 0 & -2 & -3 \end{bmatrix} \qquad \mathbf{B} = \begin{bmatrix} 0 \\ 0 \\ 1 \end{bmatrix}$$

determine the transfer function relation between $X(s)$ and $U(s)$. Find the eigenvalues of $\mathbf{A}$.

4.9. For a linear time-invariant system whose state equations have coefficient matrices given by Eqs. (4-111) and (4-112) (phase-variable canonical form), show that

$$\text{adj } (s\mathbf{I} - \mathbf{A})\mathbf{B} = \begin{bmatrix} 1 \\ s \\ s^2 \\ \cdot \\ \cdot \\ \cdot \\ s^{n-1} \end{bmatrix}$$

and the characteristic equation is

$$s^n + a_1 s^{n-1} + a_2 s^{n-2} + \ldots + a_{n-1} s + a_n = 0.$$

4.10. A closed-loop control system is described by

$$\dot{\mathbf{x}}(t) = \mathbf{A}\mathbf{x}(t) + \mathbf{B}\mathbf{u}(t)$$

$$\mathbf{u}(t) = -\mathbf{G}\mathbf{x}(t)$$

where $\mathbf{x}(t) = n$-vector, $\mathbf{u}(t) = r$-vector, $\mathbf{A}$ is $n \times n$, $\mathbf{B}$ is $n \times r$, and $\mathbf{G}$ is the $r \times n$ feedback matrix. Show that the roots of the characteristic equation are eigenvalues of $\mathbf{A} - \mathbf{BG}$. Let

$$\mathbf{A} = \begin{bmatrix} 0 & 1 & 0 & 0 \\ 0 & 0 & 1 & 0 \\ 0 & 0 & 0 & 1 \\ 0 & -2 & -5 & -10 \end{bmatrix} \qquad \mathbf{B} = \begin{bmatrix} 0 \\ 0 \\ 0 \\ 1 \end{bmatrix}$$

$$\mathbf{G} = [g_1 \quad g_2 \quad g_3 \quad g_4]$$

Find the characteristic equation of the closed-loop system. Determine the elements of $\mathbf{G}$ so that the eigenvalues of $\mathbf{A} - \mathbf{BG}$ are at $-1, -2, -1 - j1$, and $-1 + j1$. Can all the eigenvalues of $\mathbf{A} - \mathbf{BG}$ be arbitrarily assigned for this problem?

4.11. A linear time-invariant system is described by the following differential equation:

$$\frac{d^2 c(t)}{dt^2} + 2\frac{dc(t)}{dt} + c(t) = r(t)$$

(a) Find the state transition matrix $\boldsymbol{\phi}(t)$.
(b) Let $c(0) = 1$, $\dot{c}(0) = 0$, and $r(t) = u_s(t)$, the unit step function; find the state transition equations for the system.
(c) Determine the characteristic equation of the system and the eigenvalues.

4.12. A linear multivariable system is described by the following set of differential equations:

$$\frac{d^2c_1(t)}{dt^2} + \frac{dc_1(t)}{dt} + 2c_1(t) - 2c_2(t) = r_1(t)$$

$$\frac{d^2c_2(t)}{dt^2} - c_1(t) + c_2(t) = r_2(t)$$

(a) Write the state equations of the system in vector-matrix form. Write the output equation in vector-matrix form.

(b) Find the transfer function between the outputs and the inputs of the system.

4.13. Given the state transition equation $\dot{\mathbf{x}}(t) = \mathbf{A}\mathbf{x}(t)$, where

$$\mathbf{A} = \begin{bmatrix} \sigma & -\omega \\ \omega & \sigma \end{bmatrix}$$

σ and ω are real numbers.

(a) Find the state transition matrix $\boldsymbol{\phi}(t)$.

(b) Find the eigenvalues of $\mathbf{A}$.

(c) Find the eigenvectors of $\mathbf{A}$.

4.14. Given the state equations of a linear system as

$$\dot{\mathbf{x}}(t) = \mathbf{A}\mathbf{x}(t) + \mathbf{B}u(t)$$

where

$$\mathbf{A} = \begin{bmatrix} 0 & 1 & 0 \\ 0 & 0 & 1 \\ -6 & -11 & -6 \end{bmatrix} \quad \mathbf{B} = \begin{bmatrix} 0 \\ 0 \\ 1 \end{bmatrix}$$

The eigenvalues of $\mathbf{A}$ are $\lambda_1 = -1$, $\lambda_2 = -2$, $\lambda_3 = -3$. Find a transformation $\mathbf{x}(t) = \mathbf{P}\mathbf{y}(t)$ that will transform $\mathbf{A}$ into a diagonal matrix $\boldsymbol{\Lambda} = \text{diag}\,[\lambda_1\ \lambda_2\ \lambda_3]$.

4.15. Given a linear system with the state equations described by

$$\dot{\mathbf{x}}(t) = \mathbf{A}\mathbf{x}(t) + \mathbf{B}u(t)$$

where

$$\mathbf{A} = \begin{bmatrix} 0 & 1 & 0 \\ 0 & 0 & 1 \\ -25 & -35 & -11 \end{bmatrix} \quad \mathbf{B} = \begin{bmatrix} 0 \\ 0 \\ 1 \end{bmatrix}$$

The eigenvalues are $\lambda_1 = -1$, $\lambda_2 = -5$, $\lambda_3 = -5$. Find the transformation $\mathbf{x}(t) = \mathbf{P}\mathbf{y}(t)$ so that $\mathbf{A}$ is transformed into the Jordan canonical form. The transformed state equations are

$$\dot{\mathbf{y}}(t) = \boldsymbol{\Lambda}\mathbf{y}(t) + \boldsymbol{\Gamma}u(t)$$

Find $\boldsymbol{\Lambda}$ and $\boldsymbol{\Gamma}$.

4.16 Draw state diagrams for the following systems:

(a) $\dot{\mathbf{x}}(t) = \mathbf{A}\mathbf{x}(t) + \mathbf{B}u(t)$

$$\mathbf{A} = \begin{bmatrix} -3 & 2 & 0 \\ -1 & -1 & 1 \\ -5 & -2 & -1 \end{bmatrix} \quad \mathbf{B} = \begin{bmatrix} 0 \\ 0 \\ 1 \end{bmatrix}$$

(b) $\dot{\mathbf{x}}(t) = \mathbf{A}\mathbf{x}(t) + \mathbf{B}u(t)$. Same $\mathbf{A}$ as in part (a) but with

$$\mathbf{B} = \begin{bmatrix} 0 & 1 \\ 1 & 0 \\ 1 & 0 \end{bmatrix}$$

4.17. The block diagram of a feedback control system is shown in Fig. P4-17.

(a) Write the dynamic equations of the system in vector-matrix form.

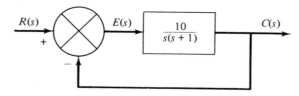

Figure P4-17.

(b) Draw a state diagram for the system.

(c) Find the state transition equations for the system. Express the equations in matrix form. The initial states are represented by $\mathbf{x}(t_0^+)$, and the input $r(t)$ is a unit step function, $u_s(t - t_0)$, which is applied at $t = t_0$.

4.18. Draw state diagrams for the following transfer functions by means of direct decomposition:

(a) $G(s) = \dfrac{10}{s^3 + 5s^2 + 4s + 10}$

(b) $G(s) = \dfrac{6(s + 1)}{s(s + 1)(s + 3)}$

Write the state equations from the state diagrams and express them in the phase-variable canonical form.

4.19. Draw state diagrams for the following systems by means of parallel decomposition:

(a) $G(s) = \dfrac{6(s + 1)}{s(s + 2)(s + 3)}$

(b) $\dfrac{d^2c(t)}{dt^2} + 6\dfrac{dc(t)}{dt} + 5c(t) = 2\dfrac{dr(t)}{dt} + r(t)$

Write the state equations from the state diagrams and show that the states are decoupled from each other.

4.20. Draw state diagrams for the systems in Problem 4.19 by means of cascade decomposition.

4.21. Given the transfer function of a linear system,

$$G(s) = \frac{10(s + 1)}{(s + 2)^2(s + 5)}$$

Draw state diagrams for the system using three different methods of decomposition. The state diagrams should contain a minimum number of integrators.

4.22. The state diagram of a linear system is shown in Fig. P4-22.

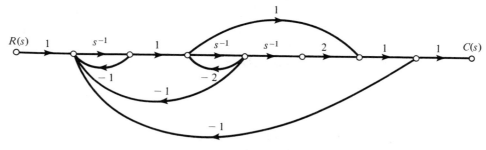

Figure P4-22.

(a) Assign the state variables and write the dynamic equations of the system.

(b) Determine the transfer function $C(s)/R(s)$.

4.23. Draw state diagrams for the electric network shown in Fig. P4-1.

4.24. The state diagram of a linear system is shown in Fig. P4-24.

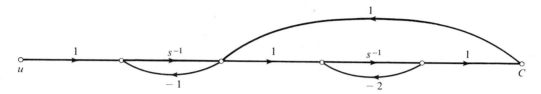

Figure P4-24.

(a) Assign state variables on the state diagram; create additional nodes if necessary, as long as the system is not altered.

(b) Write the dynamic equations for the system.

4.25. Given the state equation

$$\dot{\mathbf{x}}(t) = \mathbf{A}\mathbf{x}(t)$$

where

$$\mathbf{A} = \begin{bmatrix} -2 & 1 & 0 \\ 0 & -2 & 1 \\ 0 & 0 & -2 \end{bmatrix}$$

(a) Find the eigenvalues of $\mathbf{A}$.

(b) Determine the state transition matrix $\boldsymbol{\phi}(t)$.

4.26. Given the state equation

$$\dot{\mathbf{x}}(t) = \mathbf{A}\mathbf{x}(t) + \mathbf{B}u(t)$$

where

$$\mathbf{A} = \begin{bmatrix} 0 & 1 & 0 \\ 0 & 0 & 1 \\ -2 & -4 & -3 \end{bmatrix} \qquad \mathbf{B} = \begin{bmatrix} 0 \\ 0 \\ 1 \end{bmatrix}$$

The eigenvalues of $\mathbf{A}$ are $\lambda_1 = -1$, $\lambda_2 = -1 - j1$, $\lambda_3 = -1 + j1$. Find the transformation $\mathbf{x}(t) = \mathbf{P}y(t)$ which transforms $\mathbf{A}$ into the modal form

$$\boldsymbol{\Lambda} = \begin{bmatrix} -1 & 0 & 0 \\ 0 & -1 & 1 \\ 0 & -1 & -1 \end{bmatrix} = \mathbf{P}^{-1}\mathbf{A}\mathbf{P}$$

4.27. Given the linear system

$$\dot{\mathbf{x}}(t) = \mathbf{A}\mathbf{x}(t) + \mathbf{B}u(t)$$

where $u(t)$ is generated by state feedback,

$$u(t) = -\mathbf{G}\mathbf{x}(t)$$

The state transition matrix for the closed-loop system is

$$\boldsymbol{\phi}(t) = e^{(\mathbf{A}-\mathbf{BG})t} = \mathcal{L}^{-1}[(s\mathbf{I} - \mathbf{A} + \mathbf{BG})^{-1}]$$

Is the following relation valid?

$$e^{(\mathbf{A}-\mathbf{BG})t} = e^{\mathbf{A}t}e^{-\mathbf{BG}t}$$

where

$$e^{\mathbf{A}t} = \mathcal{L}^{-1}[(s\mathbf{I} - \mathbf{A})^{-1}]$$

$$e^{-\mathbf{B}\mathbf{G}t} = \mathcal{L}^{-1}[(s\mathbf{I} + \mathbf{B}\mathbf{G})^{-1}]$$

Explain your conclusions.

4.28. Determine the state controllability of the system shown in Fig. P4-28.

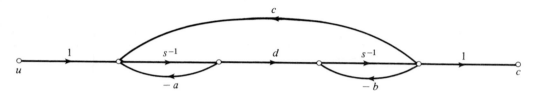

Figure P4-28.

(a) $a = 1$, $b = 2$, $c = 2$, and $d = 1$.

(b) Are there any nonzero values for a, b, c, and d such that the system is not completely state controllable?

4.29. Figure P4-29 shows the block diagram of a feedback control system. Determine the state controllability and observability of the system by the following methods, whenever applicable:

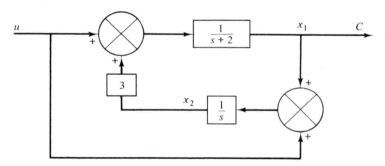

Figure P4-29.

(a) Conditions on the **A, B, D,** and **E** matrices.

(b) Transfer function.

(c) Coupling of states.

4.30. The transfer function of a linear system is given by

$$\frac{C(s)}{R(s)} = \frac{s + a}{s^3 + 6s^2 + 11s + 6}$$

(a) Determine the value of a so that the system is either uncontrollable or unobservable.

(b) Define the state variables so that one of them is uncontrollable.

(c) Define the state variables so that one of the states is unobservable.

4.31. Consider the system described by the state equation

$$\dot{\mathbf{x}}(t) = \mathbf{A}\mathbf{x}(t) + \mathbf{B}u(t)$$

where

$$\mathbf{A} = \begin{bmatrix} 0 & 1 \\ -1 & a \end{bmatrix} \qquad \mathbf{B} = \begin{bmatrix} 1 \\ b \end{bmatrix}$$

Find the region in the a-versus-b plane such that the system is completely controllable.

4.32. Draw the state diagram of a second-order system that is neither controllable nor observable.

4.33. Determine the conditions on b_1, b_2, d_1, and d_2 so that the following system is completely state controllable, output controllable, and observable:

$$\dot{\mathbf{x}}(t) = \mathbf{A}\mathbf{x}(t) + \mathbf{B}u(t)$$

$$\mathbf{A} = \begin{bmatrix} 1 & 1 \\ 0 & 1 \end{bmatrix} \qquad \mathbf{B} = \begin{bmatrix} b_1 \\ b_2 \end{bmatrix}$$

$$c(t) = \mathbf{D}\mathbf{x}(t)$$

$$\mathbf{D} = \begin{bmatrix} d_1 & d_2 \end{bmatrix}$$

4.34. The block diagram of a simplified control system for the Large Space Telescope (LST) is shown in Fig. P4-34. For simulation and control purposes, it would be desirable to represent the system by state equations and a state diagram.

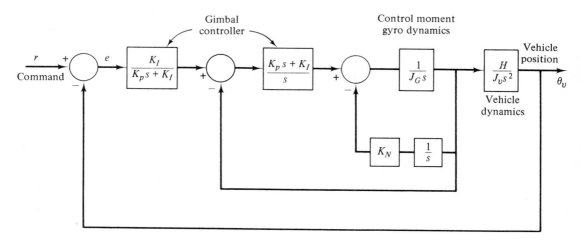

Figure P4-34.

(a) Draw a state diagram for the system and write the state equations in vector-matrix form.

(b) Find the characteristic equation of the system.

(c) A modern control design scheme, called the state feedback, utilizes the concept of feeding back every state variable through a constant gain. In this case the control law is described by

$$e = r - g_1 x_1 - g_2 x_2 - g_3 x_3 - g_4 x_4$$

Find the values of g_1, g_2, g_3, and g_4 such that the eigenvalues of the overall system are at $s = -100, -200, -1 + j1, -1 - j1$. The system parameters are given as $H = 600$, $K_I = 9700$, $J_G = 2$, $J_v = 10^5$, $K_p = 216$, and $K_N = 300$. All units are consistent.

4.35. The difference equation of a linear discrete-data system is given by

$$c[(k + 2)T] + 0.5c[(k + 1)T] + 0.1c(kT) = 1$$

(a) Write the state equations for the system.

(b) The initial conditions are given as $c(0) = 1$ and $c(T) = 0$. Find $c(kT)$ for $k = 2, 3, \ldots, 10$ by means of recursion. Can you project the final value of $c(kT)$ from the recursive results?

4.36. Given the discrete-data state equations,

$$x_1(k + 1) = 0.1x_2(k)$$

$$x_2(k + 1) = -x_1(k) + 0.7x_2(k) + r(k)$$

find the state transition matrix $\boldsymbol{\phi}(k)$.

4.37. A discrete-data system is characterized by the transfer function

$$\frac{C(z)}{R(z)} = \frac{Kz}{(z - 1)(z^2 - z + 3)}$$

(a) Draw a state diagram for the system.

(b) Write the dynamic equation for the system in vector-matrix form.

4.38. The block diagram of a discrete-data control system is shown in Fig. P4-38.

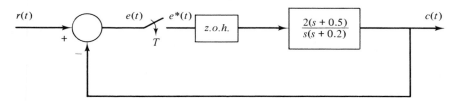

Figure P4-38.

(a) Draw a state diagram for the system.

(b) Write the state equations in vector-matrix form,

$$\mathbf{x}[(k + 1)T] = \boldsymbol{\phi}(T)\mathbf{x}(kT) + \boldsymbol{\theta}(T)r(kT)$$

(c) Find $\boldsymbol{\phi}(T)$ when $T = 0.1$ sec.

5

Mathematical Modeling

of Physical Systems

5.1 Introduction

One of the most important tasks in the analysis and design of control systems is the mathematical modeling of the systems. In preceding chapters we have introduced a number of well-known methods of modeling linear systems. The two most common methods are the transfer function approach and the state-variable approach. However, in reality most physical systems have nonlinear characteristics to some extent. A physical system may be portrayed by a linear mathematical model only if the true characteristics and the range of operation of the system justify the assumption of linearity.

Although the analysis and design of linear control systems have been well developed, their counterparts for nonlinear systems are usually quite complex. Therefore, the control systems engineer often has the task of determining not only how to accurately describe a system mathematically, but, more important, how to make proper assumptions and approximations, whenever necessary, so that the system may be adequately characterized by a linear mathematical model.

It is important to point out that the modern control engineer should place special emphasis on the mathematical modeling of the system so that the analysis and design problems can be adaptable for computer solutions. Therefore, the main objectives of this chapter are

1. To demonstrate the mathematical modeling of control systems and components.
2. To demonstrate how the modeling will lead to computer solutions.

The modeling of many system components and control systems will be illustrated in this chapter. However, the emphasis is placed on the approach to the problem, and no attempt is being made to cover all possible types of systems encountered in practice.

5.2 Equations of Electrical Networks

The classical way of writing network equations of an electrical network is the loop method and the node method, which are formulated from the two laws of Kirchhoff. However, although the loop and node equations are easy to write, they are not natural for computer solutions. A more modern method of writing network equations is the state-variable method. We shall treat the subject briefly in this section. More detailed discussions on the state equations of electrical networks may be found in texts on network theory.[1,2]

Let us use the *RLC* network of Fig. 5-1 to illustrate the basic principle of

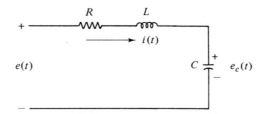

Fig. 5-1. *RLC* network.

writing state equations for electric networks. It is relatively simple to write the loop equation of this network:

$$e(t) = L\frac{d^2q(t)}{dt^2} + R\frac{dq(t)}{dt} + \frac{1}{C}q(t) \tag{5-1}$$

where $q(t)$ is the electric charge and is related to the current $i(t)$ by

$$q(t) = \int_0^t i(\tau)\,d\tau \tag{5-2}$$

It is shown in Chapter 4 that the second-order differential equation in Eq. (5-1) can be replaced by two first-order differential equations called the *state equations*. In this case it is convenient to define the state variables as

$$x_1(t) = \frac{q(t)}{C} = e_c(t) \tag{5-3}$$

where $e_c(t)$ is the voltage across the capacitor and

$$x_2(t) = \frac{dq(t)}{dt} = i(t) = C\frac{dx_1(t)}{dt} \tag{5-4}$$

Substituting Eqs. (5-3) and (5-4) into Eq. (5-1) yields

$$e(t) = L\frac{dx_2(t)}{dt} + Rx_2(t) + x_1(t) \tag{5-5}$$

Thus, from Eqs. (5-4) and (5-5) the state equations of the network are

$$\frac{dx_1(t)}{dt} = \frac{1}{C}x_2(t) \tag{5-6}$$

$$\frac{dx_2(t)}{dt} = -\frac{1}{L}x_1(t) - \frac{R}{L}x_2(t) + \frac{1}{L}e(t) \tag{5-7}$$

A more direct way of arriving at the state equations is to assign the current in the inductor L, $i(t)$, and the voltage across the capacitor C, $e_c(t)$, as the state variables. Then the state equations are written by equating the current in C and the voltage across L in terms of the state variables and the input source. This way the state equations are written by inspection from the network. Therefore,

Current in C: $$C\frac{de_c(t)}{dt} = i(t) \tag{5-8}$$

Voltage across L: $$L\frac{di(t)}{dt} = -e_c(t) - Ri(t) + e(t) \tag{5-9}$$

Since $x_1(t) = e_c(t)$ and $x_2(t) = i(t)$, it is apparent that these state equations are identical to those of Eqs. (5-6) and (5-7).

In general, it is appropriate to assign the voltages across the capacitors and currents in the inductors as state variables in an electric network, although there are exceptions.[1,2]

One must recognize that the basic laws used in writing state equations for electric networks are still the Kirchhoff's laws. Although the state equations in Eqs. (5-8) and (5-9) are arrived at by inspection, in general, the inspection method does not always work, especially for complicated networks. However, a general method using the theory of linear graphs of network analysis is available.[1]

EXAMPLE 5-1   As another example of writing the state equations of an electric network, consider the network shown in Fig. 5-2. According to the foregoing discussion, the voltage across the capacitor e_c and the currents in the inductors i_1 and i_2 are assigned as state variables, as shown in Fig. 5-2.

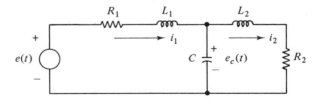

Fig. 5-2. Network in Example 5-1.

The state equations of the network are obtained by writing the voltages across the inductors and the currents in the capacitor in terms of the three state variables. The state equations are

$$L_1\frac{di_1(t)}{dt} = -R_1 i_1(t) - e_c(t) + e(t) \tag{5-10}$$

$$L_2\frac{di_2(t)}{dt} = -R_2 i_2(t) + e_c(t) \tag{5-11}$$

$$C \frac{de_c(t)}{dt} = i_1(t) - i_2(t) \qquad (5\text{-}12)$$

Rearranging the constant coefficients, the state equations are written in the following canonical form:

$$
\begin{bmatrix} \dfrac{di_1(t)}{dt} \\[2mm] \dfrac{di_2(t)}{dt} \\[2mm] \dfrac{de_c(t)}{dt} \end{bmatrix}
=
\begin{bmatrix} -\dfrac{R_1}{L_1} & 0 & -\dfrac{1}{L_1} \\[2mm] 0 & -\dfrac{R_2}{L_2} & \dfrac{1}{L_2} \\[2mm] \dfrac{1}{C} & -\dfrac{1}{C} & 0 \end{bmatrix}
\begin{bmatrix} i_1(t) \\[2mm] i_2(t) \\[2mm] e_c(t) \end{bmatrix}
+
\begin{bmatrix} 1 \\[2mm] 0 \\[2mm] 0 \end{bmatrix} \dfrac{1}{L_1} \, e(t) \qquad (5\text{-}13)
$$

5.3 Modeling of Mechanical System Elements[3]

Most feedback control systems contain mechanical as well as electrical components. From a mathematical viewpoint, the descriptions of electrical and mechanical elements are analogous. In fact, we can show that given an electrical device, there is usually an analogous mechanical counterpart, and vice versa. The analogy, of course, is a mathematical one; that is, two systems are analogous to each other if they are described mathematically by similar equations.

The motion of mechanical elements can be described in various dimensions as translational, rotational, or a combination of both. The equations governing the motions of mechanical systems are often directly or indirectly formulated from Newton's law of motion.

Translational Motion

The motion of translation is defined as a motion that takes place along a straight line. The variables that are used to describe translational motion are acceleration, velocity, and displacement.

Newton's law of motion states that *the algebraic sum of forces acting on a rigid body in a given direction is equal to the product of the mass of the body and its acceleration in the same direction.* The law can be expressed as

$$\Sigma \text{ forces} = Ma \qquad (5\text{-}14)$$

where M denotes the mass and a is the acceleration in the direction considered.

For translational motion, the following elements are usually involved:

1. Mass: Mass is considered as an indication of the property of an element which stores the kinetic energy of translational motion. It is analogous to inductance of electrical networks. If W denotes the weight of a body, then M is given by

$$M = \frac{W}{g} \qquad (5\text{-}15)$$

where g is the acceleration of the body due to gravity of the acceleration of free fall. Three consistent sets of units for the elements in Eqs. (5-14) and (5-15) are as follows:

Units	Mass M	Weight W	Acceleration	Force
MKS	newtons/m/sec²	newton	m/sec²	newton
CGS	dynes/cm/sec²	dyne	cm/sec²	dyne
British	lb/ft/sec² (slug)	lb	ft/sec²	lb

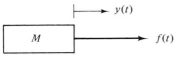

Fig. 5-3. Force–mass system.

Figure 5-3 illustrates the situation where a force is acting on a body with mass M. The force equation is written

$$f(t) = Ma(t) = M \frac{d^2y(t)}{dt^2} = M \frac{dv(t)}{dt} \qquad (5\text{-}16)$$

where $y(t)$ represents displacement, $v(t)$ the velocity, and $a(t)$ is the acceleration, all referenced in the direction of the applied force.

2. Linear spring: A linear spring in practice may be an actual spring or the compliance of a cable or a belt. In general, a spring is considered to be an element that stores potential energy. It is analogous to a capacitor in electric networks. In practice, all springs are nonlinear to some extent. However, if the deformation of a spring is small, its behavior may be approximated by a linear relationship,

$$f(t) = Ky(t) \qquad (5\text{-}17)$$

where K is the spring constant, or simply stiffness. The three unit systems for the spring constant are:

Units	K
MKS	newtons/m
CGS	dynes/cm
British	lb/ft

Equation (5-17) implies that the force acting on the spring is directly proportional to the displacement (deformation) of the spring. The model representing a linear spring element is shown in Fig. 5-4.

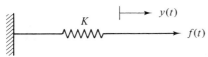

Fig. 5-4. Force–spring system.

If the spring is preloaded with a preload tension of T, then Eq. (5-17) should be modified to

$$f(t) - T = Ky(t) \qquad (5\text{-}18)$$

Friction for translational motion. Whenever there is motion or tendency of motion between two elements, frictional forces exist. The frictional forces encountered in physical systems are usually of a nonlinear nature. The characteristics of the frictional forces between two contacting surfaces often depend on such factors as the composition of the surfaces, the pressure between the surfaces, their relative velocity, and others, so that an exact mathematical description of the frictional force is difficult. However, for practical purposes, frictional forces can be divided into three basic catagories: viscous friction, static friction, and Coulomb friction. These are discussed separately in detail in the following.

1. *Viscous friction.* Viscous friction represents a retarding force that is a linear relationship between the applied force and velocity. The schematic diagram element for friction is often represented by a dashpot such as that shown in Fig. 5-5. The mathematical expression of viscous friction is

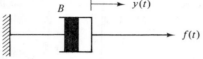

$$f(t) = B \frac{dy(t)}{dt} \qquad (5\text{-}19)$$

Fig. 5-5. Dashpot for viscous friction.

where B is the viscous frictional coefficient. The dimensions of B in the three unit systems are as follows:

Units	B
MKS	newton/m/sec
CGS	dyne/cm/sec
British	lb/ft/sec

Figure 5-6(a) shows the functional relation between the viscous frictional force and velocity.

2. *Static friction.* Static friction represents a retarding force that tends to prevent motion from beginning. The static frictional force can be represented by the following expression:

$$f(t) = \pm(F_s)_{\dot{y}=0} \qquad (5\text{-}20)$$

where $(F_s)_{\dot{y}=0}$ is defined as the static frictional force that exists only when the body is stationary but has a tendency of moving. The sign of the friction depends on the direction of motion or the initial direction of velocity. The force–velocity relation of static friction is illustrated in Fig. 5-6(b). Notice that once motion begins, the static frictional force vanishes, and other frictions take over.

3. *Coulomb friction.* Coulomb friction is a retarding force that has a

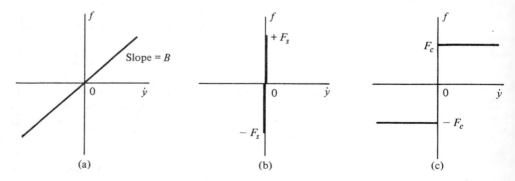

Fig. 5-6. Functional relationships of linear and nonlinear frictional forces. (a) Viscous friction. (b) Static friction. (c) Coulomb friction.

constant amplitude with respect to the change in velocity, but the sign of the frictional force changes with the reversal of the direction of velocity. The mathematical relation for the Coulomb friction is given by

$$f(t) = F_c\left(\frac{dy}{dt}\Big/\left|\frac{dy}{dt}\right|\right) \tag{5-21}$$

where F_c is the Coulomb friction coefficient. The functional description of the friction to velocity relation is shown in Fig. 5-6(c).

Rotational Motion

The rotational motion of a body may be defined as motion about a fixed axis. The variables generally used to describe the motion of rotation are torque; angular acceleration, α; angular velocity, ω; and angular displacement, θ. The following elements are usually involved with the rotational motion.

Inertia. Inertia, J, is considered as an indication of the property of an element which stores the kinetic energy of rotational motion. The inertia of a given element depends on the geometric composition about the axis of rotation and its density.

For instance, the inertia of a circular disk or a circular shaft about its geometric axis is given by

$$J = \tfrac{1}{2}Mr^2 \tag{5-22}$$

where M is the mass of the disk or shaft and r is its radius.

EXAMPLE 5-2   Given a disk that is 1 in. in diameter, 0.25 in. thick, and weighing 5 oz, its inertia is

$$J = \frac{1}{2}\frac{Wr^2}{g} = \frac{1}{2}\frac{(5 \text{ oz})(1 \text{ in})^2}{386 \text{ in/sec}^2} \tag{5-23}$$
$$= 0.00647 \text{ oz-in-sec}^2$$

Usually the density of the material is given in weight per unit volume. Then, for a circular disk or shaft it can be shown that the inertia is proportional to the fourth power of the radius and the first power of the thickness or length. Therefore, if the weight W is expressed as

$$W = \rho(\pi r^2 h) \tag{5-24}$$

where ρ is the density in weight per unit volume, r the radius, and h the thickness or length, then Eq. (5-22) is written

$$J = \frac{1}{2}\frac{\rho\pi h r^4}{g} = 0.00406\rho\ hr^4 \tag{5-25}$$

where h and r are in inches.

For steel, ρ is 4.53 oz/in^3; Eq. (5-25) becomes

$$J = 0.0184\ hr^4 \tag{5-26}$$

For aluminum, ρ is 1.56 oz/in^3; Eq. (5-25) becomes

$$J = 0.00636\ hr^4 \tag{5-27}$$

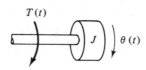

Fig. 5-7. Torque–inertia system.

When a torque is applied to a body with inertia J, as shown in Fig. 5-7, the torque equation is written

$$T(t) = J\alpha(t) = J\frac{d\omega(t)}{dt} = J\frac{d^2\theta(t)}{dt^2} \qquad (5\text{-}28)$$

The three generally used sets of units for the quantities in Eq. (5-28) are tabulated as follows:

Units	Inertia	Torque	Angular Displacement
MKS	kg-m^2	newton-m	radian
CGS	g-cm^2	dyne-cm	radian
English	slug-ft^2	lb-ft	
	or lb-ft-sec^2		
	oz-in-sec^2	oz-in	radian

The following conversion factors are often found useful:

Angular displacement.

$$1 \text{ rad} = \frac{180}{\pi} = 57.3°$$

Angular velocity.

$$1 \text{ rpm} = \frac{2\pi}{60} = 0.1047 \text{ rad/sec}$$

$$1 \text{ rpm} = 6 \text{ deg/sec}$$

Torque.

$$1 \text{ g-cm} = 0.0139 \text{ oz-in}$$

$$1 \text{ lb-ft} = 192 \text{ oz-in}$$

$$1 \text{ oz-in} = 0.00521 \text{ lb-ft}$$

Inertia.

$$1 \text{ g-cm}^2 = 1.417 \times 10^{-5} \text{ oz-in-sec}^2$$

$$1 \text{ lb-ft-sec}^2 = 192 \text{ oz-in-sec}^2$$

$$1 \text{ oz-in-sec}^2 = 386 \text{ oz-in}^2$$

$$1 \text{ g-cm-sec}^2 = 980 \text{ g-cm}^2$$

$$1 \text{ lb-ft-sec}^2 = 32.2 \text{ lb-ft}^2$$

Torsional spring. As with the linear spring for translational motion, a torsional spring constant K, in torque per unit angular displacement, can be devised to represent the compliance of a rod or a shaft when it is subject to an applied torque. Figure 5-8 illustrates a simple torque–spring system that can be represented by the following equation:

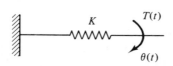

Fig. 5-8. Torque–torsional spring system.

$$T(t) = K\theta(t) \qquad (5\text{-}29)$$

The dimension for K is given in the following units:

Units	K
MKS	newton-m/rad
CGS	dyne-cm/rad
British	oz-in/rad

If the torsional spring is preloaded by a preload torque of TP, Eq. (5-29) is modified to

$$T(t) - TP = K\theta(t) \tag{5-30}$$

Friction for rotational motion. The three types of friction described for translational motion can be carried over to the motion of rotation. Therefore, Eqs. (5-19), (5-20), and (5-21) can be replaced, respectively, by their counterparts:

$$T(t) = B\frac{d\theta(t)}{dt} \tag{5-31}$$

$$T(t) = \pm(F_s)_{\theta=0} \tag{5-32}$$

$$T(t) = F_c\left(\frac{d\theta}{dt}\bigg/\left|\frac{d\theta}{dt}\right|\right) \tag{5-33}$$

where B is the viscous frictional coefficient in torque per unit angular velocity, $(F_s)_{\theta=0}$ is the static friction, and F_c is the Coulomb friction coefficient.

Relation Between Translational and Rotational Motions

In motion control problems it is often necessary to convert rotational motion into a translational one. For instance, a load may be controlled to move along a straight line through a rotary motor and screw assembly, such as that shown in Fig. 5-9. Figure 5-10 shows a similar situation in which a rack and

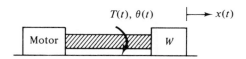

Fig. 5-9. Rotary-to-linear motion-control system.

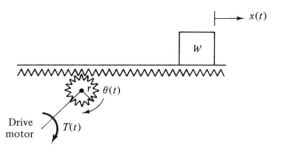

Fig. 5-10. Rotary-to-linear motion-control system.

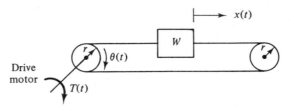

Fig. 5-11. Rotary-to-linear motion-control system.

pinion is used as the mechanical linkage. Another common system in motion control is the control of a mass through a pulley by a rotary prime mover, such as that shown in Fig. 5-11. The systems shown in Figs. 5-9, 5-10, and 5-11 can all be represented by a simple system with an equivalent inertia connected directly to the drive motor. For instance, the mass in Fig. 5-11 can be regarded as a point mass which moves about the pulley, which has a radius r. Disregarding the inertia of the pulley, the equivalent inertia that the motor sees is

$$J = Mr^2 = \frac{W}{g}r^2 \tag{5-34}$$

If the radius of the pinion in Fig. 5-10 is r, the equivalent inertia which the motor sees is also given by Eq. (5-34).

Now consider the system of Fig. 5-9. The lead of the screw, L, is defined as the linear distance which the mass travels per revolution of the screw. In principle, the two systems in Fig. 5-10 and 5-11 are equivalent. In Fig. 5-10, the distance traveled by the mass per revolution of the pinion is $2\pi r$. Therefore, using Eq. (5-34) as the equivalent inertia for the system of Fig. 5-9,

$$J = \frac{W}{g}\left(\frac{L}{2\pi}\right)^2 \tag{5-35}$$

where in the British system

$$J = \text{inertia (oz-in-sec}^2)$$
$$W = \text{weight (oz)}$$
$$L = \text{screw lead (in)}$$
$$g = \text{gravitational force (386.4 in/sec}^2)$$

Mechanical Energy and Power

Energy and power play an important role in the design of electromechanical systems. Stored energy in the form of kinetic and potential energy controls the dynamics of the system, whereas dissipative energy usually is spent in the form of heat, which must be closely controlled.

The mass or inertia of a body indicates its ability to store kinetic energy. The kinetic energy of a moving mass with a velocity v is

$$W_k = \tfrac{1}{2}Mv^2 \tag{5-36}$$

The following consistent sets of units are given for the kinetic energy relation:

Units	Energy	Mass	Velocity
MKS	joule or newton-m	newton/m/sec²	m/sec
CGS	dyne-cm	dyne-cm-sec²	cm/sec
British	ft-lb	lb/ft/sec² (slug)	ft/sec

For a rotational system, the kinetic energy relation is written

$$W_k = \tfrac{1}{2} J \omega^2 \tag{5-37}$$

where J is the moment of intertia and ω the angular velocity. The following units are given for the rotational kinetic energy:

Units	Energy	Inertia	Angular Velocity
MKS	joule or newton-m	kg-m²	rad/sec
CGS	dyne-cm	gm-cm²	rad/sec
British	oz-in	oz-in-sec²	rad/sec

Potential energy stored in a mechanical element represents the amount of work required to change the configuration. For a linear spring that is deformed by y in length, the potential energy stored in the spring is

$$W_p = \tfrac{1}{2} K y^2 \tag{5-38}$$

where K is the spring constant. For a torsional spring, the potential energy stored is given by

$$W_p = \tfrac{1}{2} K \theta^2 \tag{5-39}$$

When dealing with a frictional element, the form of energy differs from the previous two cases in that the energy represents a loss or dissipation by the system in overcoming the frictional force. Power is the time rate of doing work. Therefore, the power dissipated in a frictional element is the product of force and velocity; that is,

$$P = fv \tag{5-40}$$

Since $f = Bv$, where B is the frictional coefficient, Eq. (5-40) becomes

$$P = Bv^2 \tag{5-41}$$

The MKS unit for power is in newton-m/sec or watt; for the CGS system it is dyne-cm/sec. In the British unit system, power is represented in ft-lb/sec or horsepower (hp). Furthermore,

$$\begin{aligned} 1 \text{ hp} &= 746 \text{ watt} \\ &= 550 \text{ ft-lb/sec} \end{aligned} \tag{5-42}$$

Since power is the rate at which energy is being dissipated, the energy dissipated in a frictional element is

$$W_d = B \int v^2 \, dt \tag{5-43}$$

Gear Trains, Levers, and Timing Belts

A gear train, a lever, or a timing belt over pulleys is a mechanical device that transmits energy from one part of a system to another in such a way that force, torque, speed, and displacement are altered. These devices may also be regarded as matching devices used to attain maximum power transfer. Two gears are shown coupled together in Fig. 5-12. The inertia and friction of the gears are neglected in the ideal case considered.

The relationships between the torques T_1 and T_2, angular displacements θ_1 and θ_2, and the teeth numbers N_1 and N_2, of the gear train are derived from the following facts:

1. The number of teeth on the surface of the gears is proportional to the radii r_1 and r_2 of the gears; that is,

$$r_1 N_2 = r_2 N_1 \qquad (5\text{-}44)$$

2. The distance traveled along the surface of each gear is the same. Therefore,

$$\theta_1 r_1 = \theta_2 r_2 \qquad (5\text{-}45)$$

3. The work done by one gear is equal to that of the other since there is assumed to be no loss. Thus

$$T_1 \theta_1 = T_2 \theta_2 \qquad (5\text{-}46)$$

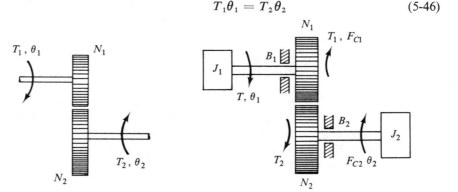

Fig. 5-12. Gear train.   **Fig. 5-13.** Gear train with friction and inertia.

If the angular velocities of the two gears, ω_1 and ω_2, are brought into the picture, Eqs. (5-44) through (5-46) lead to

$$\frac{T_1}{T_2} = \frac{\theta_2}{\theta_1} = \frac{N_1}{N_2} = \frac{\omega_2}{\omega_1} = \frac{r_1}{r_2} \qquad (5\text{-}47)$$

In practice, real gears do have inertia and friction between the coupled gear teeth which often cannot be neglected. An equivalent representation of a gear train with viscous friction, Coulomb friction, and inertia considered as lumped elements is shown in Fig. 5-13. The following variables and parameters are defined for the gear train:

$$T = \text{applied torque}$$
$$\theta_1, \theta_2 = \text{angular displacements}$$
$$T_1, T_2 = \text{torque transmitted to gears}$$
$$J_1, J_2 = \text{inertia of gears}$$
$$N_1, N_2 = \text{number of teeth}$$
$$F_{c1}, F_{c2} = \text{Coulomb friction coefficients}$$
$$B_1, B_2 = \text{viscous frictional coefficients}$$

The torque equation for gear 2 is written

$$T_2(t) = J_2 \frac{d^2\theta_2(t)}{dt^2} + B_2 \frac{d\theta_2(t)}{dt} + F_{c2} \frac{\dot{\theta}_2}{|\dot{\theta}_2|} \qquad (5\text{-}48)$$

The torque equation on the side of gear 1 is

$$T(t) = J_1 \frac{d^2\theta_1(t)}{dt^2} + B_1 \frac{d\theta_1(t)}{dt} + F_{c1} \frac{\dot{\theta}_1}{|\dot{\theta}_1|} + T_1(t) \qquad (5\text{-}49)$$

By the use of Eq. (5-47), Eq. (5-48) is converted to

$$T_1(t) = \frac{N_1}{N_2} T_2(t) = \left(\frac{N_1}{N_2}\right)^2 J_2 \frac{d^2\theta_1(t)}{dt^2} + \left(\frac{N_1}{N_2}\right)^2 B_2 \frac{d\theta_1(t)}{dt} + \frac{N_1}{N_2} F_{c2} \frac{\dot{\theta}_2}{|\dot{\theta}_2|} \qquad (5\text{-}50)$$

Equation (5-50) indicates that it is possible to reflect inertia, friction, (and compliance) torque, speed, and displacement from one side of a gear train to the other.

Therefore, the following quantities are obtained when reflecting from gear 2 to gear 1:

Inertia: $\left(\dfrac{N_1}{N_2}\right)^2 J_2$

Viscous frictional coefficient: $\left(\dfrac{N_1}{N_2}\right)^2 B_2$

Torque: $\dfrac{N_1}{N_2} T_2$

Angular displacement: $\dfrac{N_1}{N_2} \theta_2$

Angular velocity: $\dfrac{N_1}{N_2} \omega_2$

Coulomb frictional torque: $\dfrac{N_1}{N_2} F_{c2} \dfrac{\omega_2}{|\omega_2|}$

If torsional spring effect were present, the spring constant is also multiplied by $(N_1/N_2)^2$ in reflecting from gear 2 to gear 1. Now, substituting Eq. (5-50) into Eq. (5-49), we get

$$T(t) = J_{1e} \frac{d^2\theta_1(t)}{dt^2} + B_{1e} \frac{d\theta_1(t)}{dt} + T_F \qquad (5\text{-}51)$$

where

$$J_{1e} = J_1 + \left(\frac{N_1}{N_2}\right)^2 J_2 \qquad (5\text{-}52)$$

$$B_{1e} = B_1 + \left(\frac{N_1}{N_2}\right)^2 B_2 \qquad (5\text{-}53)$$

$$T_F = F_{c1} \frac{\dot{\theta}_1}{|\dot{\theta}_1|} + \frac{N_1}{N_2} F_{c2} \frac{\dot{\theta}_2}{|\dot{\theta}_2|} \qquad (5\text{-}54)$$

EXAMPLE 5-3   Given a load that has inertia of 0.05 oz-in-sec² and a Coulomb friction torque of 2 oz-in, find the inertia and frictional torque reflected
through a 1: 5 gear train ($N_1/N_2 = \frac{1}{5}$ with N_2 on the load side). The
reflected inertia on the side of N_1 is $(\frac{1}{5})^2 \times 0.05 = 0.002$ oz-in-sec². The reflected
Coulomb friction is $(\frac{1}{5})2 = 0.4$ oz-in.

Timing belts and chain drives serve the same purposes as the gear train
except that they allow the transfer of energy over a longer distance without using
an excessive number of gears. Figure 5-14 shows the diagram of a belt or chain
drive between two pulleys. Assuming that there is no slippage between the belt
and the pulleys, it is easy to see that Eq. (5-47) still applies to this case. In fact,
the reflection or transmittance of torque, inertia, friction, etc., is similar to that of
a gear train.

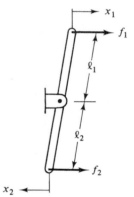

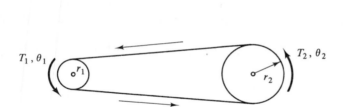

Fig. 5-14. Belt or chain drive.   Fig. 5-15. Lever system.

The lever system shown in Fig. 5-15 transmits translational motion and
force in the same way that gear trains transmit rotational motion. The relation
between the forces and distances is

$$\frac{f_1}{f_2} = \frac{l_2}{l_1} = \frac{x_2}{x_1} \qquad (5\text{-}55)$$

Backlash and Dead Zone

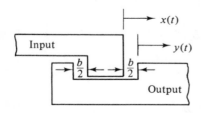

Fig. 5-16. Physical model of backlash
between two mechanical elements.

Backlash and dead zone usually play an important
role in gear trains and similar mechanical linkages. In a
great majority of situations, backlash may give rise to
undesirable oscillations and instability in control systems.
In addition, it has a tendency to wear down the mechanical elements. Regardless of the actual mechanical elements, a physical model of backlash or dead zone between
an input and an output member is shown in Fig. 5-16.
The model can be used for a rotational system as well as

for a translational system. The amount of backlash is $b/2$ on either side of the reference position.

In general, the dynamics of the mechanical linkage with backlash depend upon the relative inertia-to-friction ratio of the output member. If the inertia of the output member is very small compared with that of the input member, the motion is controlled predominantly by friction. This means that the output member will not coast whenever there is no contact between the two members. When the output is driven by the input, the two members will travel together until the input member reverses its direction; then the output member will stand still until the backlash is taken up on the other side, at which time it is assumed that the output member instantaneously takes on the velocity of the input member. The transfer characteristic between the input and the output displacements of a backlash element with negligible output inertia is shown in Figure 5-17. To illustrate the relative motion between the input and the output

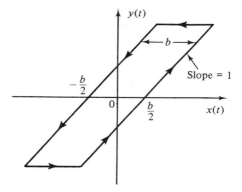

Fig. 5-17. Input–output characteristic of backlash with negligible output inertia.

members, let us assume that the input displacement is driven sinusoidally with respect to time. The displacements and velocities of the input and output members are illustrated as shown in Fig. 5-18. Note that the reference position of the two members is taken to be that of Fig. 5-16, that is, with the input member starting at the center of the total backlash. For Fig. 5-18, it is assumed that when motion begins, the input member is in contact with the output member on the right, so that $x(0) = 0$ and $y(0) = -b/2$. At the other extreme, if the friction on the output member is so small that it may be neglected, the inertia of the output member remains in contact with the input member as long as the acceleration is in the direction to keep the two members together. When the acceleration of the input member becomes zero, the output member does not stop immediately but leaves the input member and coasts at a constant velocity that is equal to the maximum velocity attained by the input member. When the output member has traversed a distance, relative to the input member, equal to the full width of the backlash, it will be restrained by the opposite side of the input mem-

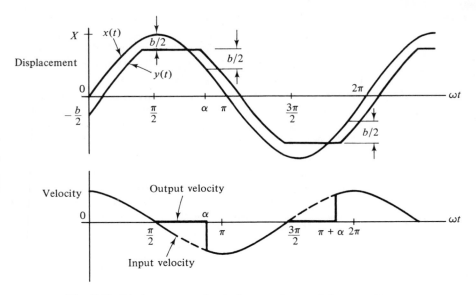

Fig. 5-18. Displacement and velocity waveforms of input and output members of a backlash element with a sinusoidal input displacement.

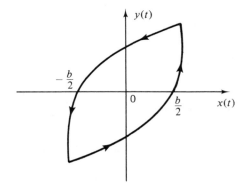

Fig. 5-19. Input–output displacement characteristic of a backlash element without friction.

ber. At that time the output member will again assume the velocity of the input member. The transfer characteristic between the input and the output displacement of a backlash element with negligible output friction is shown in Fig. 5-19. The displacement, velocity, and acceleration waveforms of the input and output members, when the input displacement is driven sinusoidally, is shown in Fig. 5-20.

In practice, of course, the output member of a mechanical linkage with backlash usually has friction as well as inertia. Then the output waveforms in response to a sinusoidally driven input displacement should lie between those of Figs. 5-18 and 5-20.

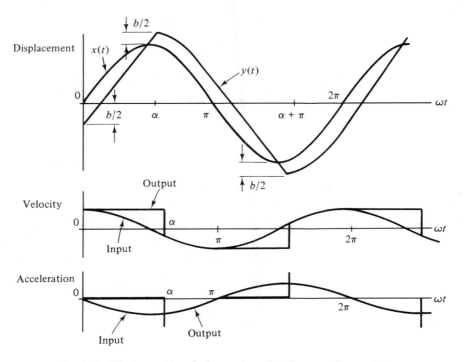

Fig. 5-20. Displacement, velocity, and acceleration waveforms of input and output members of a backlash element when the input displacement is driven sinusoidally.

5.4 Equations of Mechanical Systems[3]

The equations of a linear mechanical system are written by first constructing a model of the system containing interconnected linear elements, and then the system equations are written by applying Newton's law of motion to the free-body diagram.

EXAMPLE 5-4   Let us consider the mechanical system shown in Fig. 5-21(a). The free-body diagram of the system is shown in Fig. 5-21(b). The force

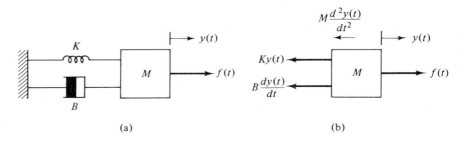

Fig. 5-21. (a) Mass–spring–friction system. (b) Free-body diagram.

equation of the system is written

$$f(t) = M\frac{d^2y(t)}{dt^2} + B\frac{dy(t)}{dt} + Ky(t) \tag{5-56}$$

This second-order differential equation can be decomposed into two first-order state equations, using the method discussed in Chapter 4. Let us assign $x_1 = y$ and $x_2 = dy/dt$ as the state variables. Then Eq. (5-56) is written

$$\frac{dx_1(t)}{dt} = x_2(t) \tag{5-57}$$

$$\frac{dx_2(t)}{dt} = -\frac{K}{M}x_1(t) - \frac{B}{M}x_2(t) + \frac{1}{M}f(t) \tag{5-58}$$

It is not difficult to see that this mechanical system is analogous to a series RLC electric network. With this analogy it is simple to formulate the state equations directly from the mechanical system using a different set of state variables. If we consider that mass is analogous to inductance, and the spring constant K is analogous to the inverse of capacitance, $1/C$, it is logical to assign $v(t)$, the velocity, and $f_k(t)$, the force acting on the spring, as state variables, since the former is analogous to the current in an inductor and the latter is analogous to the voltage across a capacitor.

Then the state equations of the system are

Force on mass: $$M\frac{dv(t)}{dt} = -Bv(t) - f_k(t) + f(t) \tag{5-59}$$

Velocity of spring: $$\frac{1}{K}\frac{df_k(t)}{dt} = v(t) \tag{5-60}$$

Notice that the first state equation is similar to writing the equation on the voltage across an inductor; the second is like that of the current through a capacitor.

This simple example further illustrates the points made in Chapter 4 regarding the fact that the state equations and state variables of a dynamic system are not unique.

EXAMPLE 5-5   As a second example of writing equations for mechanical systems, consider the system shown in Fig. 5-22(a). Since the spring is deformed when it is subjected to the force $f(t)$, two displacements, y_1 and y_2, must be assigned to the end points of the spring. The free-body diagrams of the system are given in Fig. 5-22(b). From these free-body diagrams the force equations of the system are written

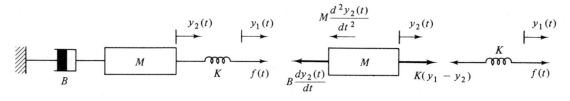

(a)   (b)

Fig. 5-22. Mechanical system for Example 5-5. (a) Mass–spring–friction system. (b) Free-body diagrams.

$$f(t) = K[y_1(t) - y_2(t)] \tag{5-61}$$

$$K[y_1(t) - y_2(t)] = M\frac{d^2y_2(t)}{dt^2} + B\frac{dy_2(t)}{dt} \tag{5-62}$$

Now let us write the state equations of the system. Since the differential equation of the system is already available in Eq. (5-62), the most direct way is to decompose this equation into two first-order differential equations.

Therefore, letting $x_1(t) = y_2(t)$ and $x_2(t) = dy_2(t)/dt$, Eqs. (5-61) and (5-62) give

$$\frac{dx_1(t)}{dt} = x_2(t) \tag{5-63}$$

$$\frac{dx_2(t)}{dt} = -\frac{B}{M}x_2(t) + \frac{1}{M}f(t) \tag{5-64}$$

As an alternative, we can assign the velocity $v(t)$ of the body with mass M as one state variable, and the force $f_k(t)$ on the spring as the other state variable, so we have

$$\frac{dv(t)}{dt} = -\frac{B}{M}v(t) + \frac{1}{M}f_k(t) \tag{5-65}$$

and

$$f_k(t) = f(t) \tag{5-66}$$

One may wonder at this point if the two equations in Eqs. (5-65) and (5-66) are correct as state equations, since it seems that only Eq. (5-65) is a state equation, but we do have two state variables in $v(t)$ and $f_k(t)$. Why do we need only one state equation here, whereas Eqs. (5-63) and (5-64) clearly are two independent state equations? The situation is better explained (at least for electrical engineers) by referring to the analogous electric network of the system, shown in Fig. 5-23. It is clear that although the network has two reactive elements in L and C and thus there should be two state variables, the capacitance in this case is a "redundant" element, since $e_c(t)$ is equal to the applied voltage $e(t)$. However, the equations in Eqs. (5-65) and (5-66) can provide only the solution to the velocity of M once $f(t)$ is specified. If we need to find the displacement $y_1(t)$ at the point where $f(t)$ is applied, we have to use the relation

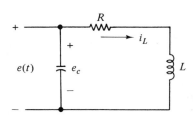

Fig. 5-23. Electric network analogous to the mechanical system of Fig. 5-22.

$$y_1(t) = \frac{f_k(t)}{K} + y_2(t) = \frac{f(t)}{K} + \int_{0+}^{t} v(\tau)\,d\tau + y_2(0+) \tag{5-67}$$

where $y_2(0+)$ is the initial displacement of the body with mass M. On the other hand, we can solve for $y_2(t)$ from the two state equations of Eqs. (5-63) and (5-64), and then $y_1(t)$ is determined from Eq. (5-61).

EXAMPLE 5-6   In this example the equations for the mechanical system in Fig. 5-24(a) are to be written. Then we are to draw state diagrams and derive transfer functions for the system.

The free-body diagrams for the two masses are shown in Fig. 5-24(b), with the reference directions of the displacements y_1 and y_2 as indicated. The Newton's force

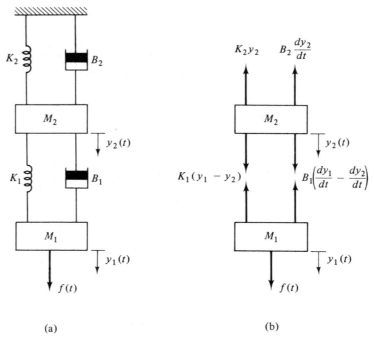

(a)   (b)

Fig. 5-24. Mechanical system for Example 5-6.

equations for the system are written directly from the free-body diagram:

$$f(t) = M_1 \frac{d^2y_1(t)}{dt^2} + B_1\left[\frac{dy_1(t)}{dt} - \frac{dy_2(t)}{dt}\right] + K_1[y_1(t) - y_2(t)] \qquad (5\text{-}68)$$

$$0 = -B_1\left[\frac{dy_1(t)}{dt} - \frac{dy_2(t)}{dt}\right] - K_1[y_1(t) - y_2(t)]$$
$$+ M_2 \frac{d^2y_2(t)}{dt^2} + B_2 \frac{dy_2(t)}{dt} + K_2 y_2(t) \qquad (5\text{-}69)$$

We may now decompose these two second-order simultaneous differential equations into four state equations by defining the following state variables:

$$x_1 = y_1 \qquad (5\text{-}70)$$

$$x_2 = \frac{dy_1}{dt} = \frac{dx_1}{dt} \qquad (5\text{-}71)$$

$$x_3 = y_2 \qquad (5\text{-}72)$$

$$x_4 = \frac{dy_2}{dt} = \frac{dx_3}{dt} \qquad (5\text{-}73)$$

Equations (5-71) and (5-73) form two state equations naturally; the other two are obtained by substituting Eqs. (5-70) through (5-73) into Eqs. (5-68) and (5-69), and rearranging we have

$$\frac{dx_1}{dt} = x_2 \qquad (5\text{-}74)$$

$$\frac{dx_2}{dt} = -\frac{K_1}{M_1}(x_1 - x_3) - \frac{B_1}{M_1}(x_2 - x_4) + \frac{1}{M_1}f(t) \tag{5-75}$$

$$\frac{dx_3}{dt} = x_4 \tag{5-76}$$

$$\frac{dx_4}{dt} = \frac{K_1}{M_2}x_1 + \frac{B_1}{M_2}x_2 - \frac{K_1 + K_2}{M_2}x_3 - \frac{1}{M_2}(B_1 + B_2)x_4 \tag{5-77}$$

If we are interested in the displacements y_1 and y_2, the output equations are written

$$y_1(t) = x_1(t) \tag{5-78}$$

$$y_2(t) = x_3(t) \tag{5-79}$$

The state diagram of the system, according to the equations written above, is drawn as shown in Fig. 5-25. The transfer functions $Y_1(s)/F(s)$ and $Y_2(s)/F(s)$ are

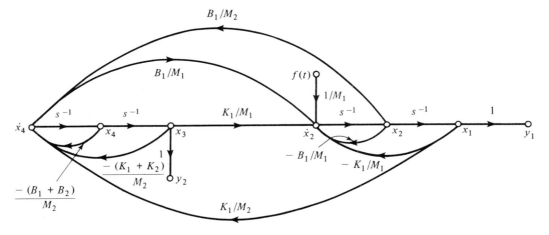

Fig. 5-25. State diagram for the mechanical system of Fig. 5-24.

obtained from the state diagram by applying Mason's gain formula. The reader should verify the following results (make sure that all the loops and nontouching loops are taken into account):

$$\frac{Y_1(s)}{F(s)} = \frac{M_2 s^2 + (B_1 + B_2)s + (K_1 + K_2)}{\Delta} \tag{5-80}$$

$$\frac{Y_2(s)}{F(s)} = \frac{B_1 s + K_1}{\Delta} \tag{5-81}$$

where

$$\Delta = M_1 M_2 s^4 + [M_1(B_1 + B_2) + B_1 M_2 - B_1^2]s^3 + [M_1(K_1 + K_2) + K_1 M_2 \\ + B_1(B_1 + B_2) - B_1 K_1]s^2 + [K_1 B_2 + B_1(K_1 + K_2)]s + K_1 K_2 \tag{5-82}$$

The state equations can also be written directly from the diagram of the mechanical system. The state variables are assigned as $v_1 = dy_1/dt$, $v_2 = dy_2/dt$, and the forces on the two springs, f_{K1} and f_{K2}. Then, if we write the forces acting on the masses and the velocities of the springs, as functions of the four state variables and the external force, the state equations are

Force on M_1: $$M_1 \frac{dv_1}{dt} = -B_1 v_1 + B_1 v_2 - f_{K1} + f \tag{5-83}$$

Force on M_2: $\qquad M_2 \dfrac{dv_2}{dt} = B_1 v_1 - (B_1 + B_2)v_2 + f_{K1} - f_{K2}$ $\qquad$ (5-84)

Velocity on K_1: $\qquad \dfrac{df_{K1}}{dt} = K_1(v_1 - v_2)$ $\qquad\qquad\qquad$ (5-85)

Velocity on K_2: $\qquad \dfrac{df_{K2}}{dt} = K_2 v_2$ $\qquad\qquad\qquad\qquad$ (5-86)

EXAMPLE 5-7 $\quad$ The rotational system shown in Fig. 5-26(a) consists of a disk mounted on a shaft that is fixed at one end. The moment of inertia of the disk about its axis is J. The edge of the disk is riding on a surface, and the viscous friction coefficient between the two surfaces is B. The inertia of the shaft is negligible, but the stiffness is K.

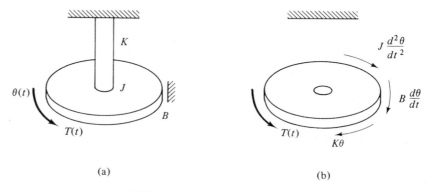

(a) $\hspace{7cm}$ (b)

Fig. 5-26. Rotational system for Example 5-7.

Assume that a torque is applied to the disk as shown; then the torque or moment equation about the axis of the shaft is written from the free-body diagram of Fig. 5-26(b):

$$T(t) = J \frac{d^2\theta(t)}{dt^2} + B \frac{d\theta(t)}{dt} + K\theta(t) \qquad (5-87)$$

Notice that this system is analogous to the translational system of Fig. 5-21. The state equations may be written by defining the state variables as $x_1(t) = \theta(t)$ and $dx_1(t)/dt = x_2(t)$. The reader may carry out the next step of writing the state equations as an exercise.

5.5 Error-Sensing Devices in Control Systems[4,5]

In feedback control systems it is often necessary to compare several signals at a certain point of a system. For instance, it is usually the case to compare the reference input with the controlled variable; the difference between the two signals is called the *error*. The error signal is then used to actuate the system. The block-diagram notation for the algebraic sum of several signals is defined in Fig. 3-5. In terms of physical components, an error-sensing device can be a simple potentiometer or combination of potentiometers, a differential gear, a transformer, a differential amplifier, a synchro, or a similar element. The mathematical modeling of some of these devices is discussed in the following.

Potentiometers. Since the output voltage of a potentiometer is proportional to the shaft displacement, when a voltage is applied across its fixed terminals, the device can be used to compare two shaft positions. In this case one shaft may be fastened to the potentiometer case and the other to the shaft of the potentiometer. When a constant voltage is applied to the fixed terminals of the potentiometer, the voltage across the variable and the reference terminals will be proportional to the difference between the two shaft positions. The arrangement shown in Fig. 5-27(b) is a one-potentiometer realization of the error-sensing devices shown in Fig. 5-27(a). A more versatile arrangement may be obtained by using two potentiometers connected in parallel as shown in Fig. 5-27(c). This

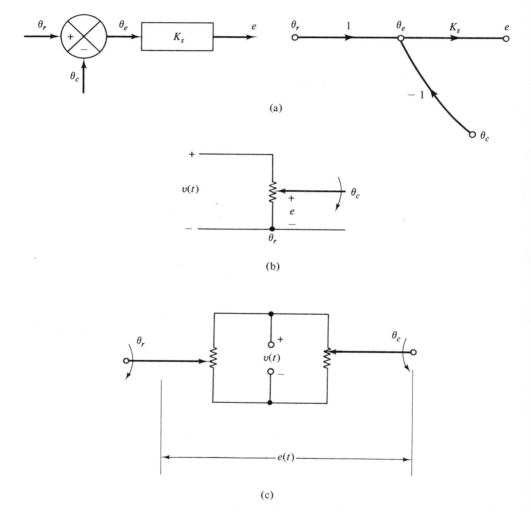

Fig. 5-27. (a) Block-diagram and signal-flow-graph symbols for an error sensor. (b) Position error sensor using one potentiometer. (c) Position error sensor using two potentiometers.

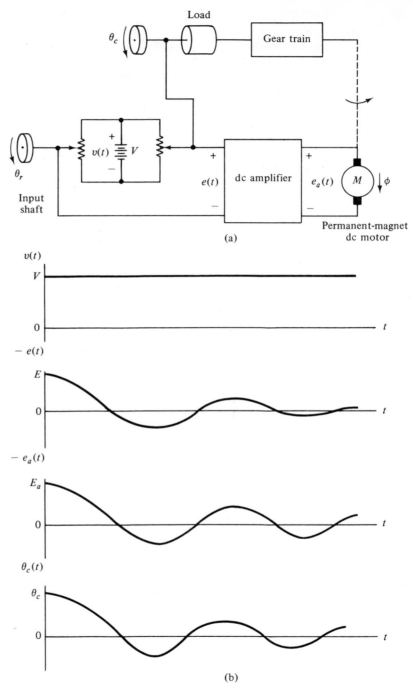

Fig. 5-28. (a) Direct current control system with potentiometers as error detectors. (b) Typical waveforms of signals in the control system of (a).

allows comparison of two remotely located shaft positions. The applied voltage $v(t)$ can be ac or dc, depending upon the types of transducers that follow the error sensor. If $v(t)$ is a dc voltage, the *polarity* of the output voltage $e(t)$ determines the relative position of the two shafts. In the case of an ac applied voltage, the *phase* of $e(t)$ acts as the indicator of the relative shaft directions. In either case the transfer relation of the two error-sensor configurations can be written

$$e(t) = K_s[\theta_r(t) - \theta_c(t)] \tag{5-88}$$

where

$e(t)$ = error voltage, volts

K_s = sensitivity of the error sensor, volts per radian

The value of K_s depends upon the applied voltage and the total displacement capacity of the potentiometers. For instance, if the magnitude of $v(t)$ is V volts and each of the potentiometers is capable of rotating 10 turns, $K_s = V/20\pi$ V/rad.

A simple example that illustrates the use of a pair of potentiometers as an error detector is shown in Fig. 5-28(a). In this case the voltage supplied to the error detector, $v(t)$, is a dc voltage. An unmodulated or dc electric signal, $e(t)$, proportional to the misalignment between the reference input shaft and the controlled shaft, appears as the output of the potentiometer error detector. In control system terminology, a *dc signal* usually refers to an unmodulated signal. On the other hand, an ac signal in control systems is modulated by a modulation process. These definitions are different from those commonly used in electrical engineering, where dc simply refers to unidirectional and ac indicates alternating.

As shown in Fig. 5-28(a), the error signal is amplified by a dc amplifier whose output drives the armature of a permanent-magnet dc motor. If the system works properly, wherever there is a misalignment between the input and the output shafts, the motor will rotate in such a direction as to reduce the error to a minimum. Typical waveforms of the signals in the system are shown in Fig. 5-28(b). Note that the electric signals are all unmodulated and the output displacements of the motor and the load are essentially of the same form as the error signal. Figure 5-29(a) illustrates a control system which could serve essentially the same purpose as that of the system of Fig. 5-28(a) except that ac signals prevail. In this case the voltage applied to the error sensor is sinusoidal. The frequency of this signal is usually much higher than the frequency of the true signal that is being transmitted through the system. Typical signals of the ac control system are shown in Fig. 5-29(b). The signal $v(t)$ is referred to as the carrier signal whose frequency is ω_c, or

$$v(t) = V \sin \omega_c t \tag{5-89}$$

Analytically, the output of the error sensor is given by

$$e(t) = K_s \theta_e(t) v(t) \tag{5-90}$$

where $\theta_e(t)$ is the difference between the input displacement and the load displacement, $\theta_e(t) = \theta_r(t) - \theta_c(t)$. For the $\theta_e(t)$ shown in Fig. 5-29(b), $e(t)$ becomes a *suppressed-carrier modulated* signal. A reversal in phase of $e(t)$ occurs whenever

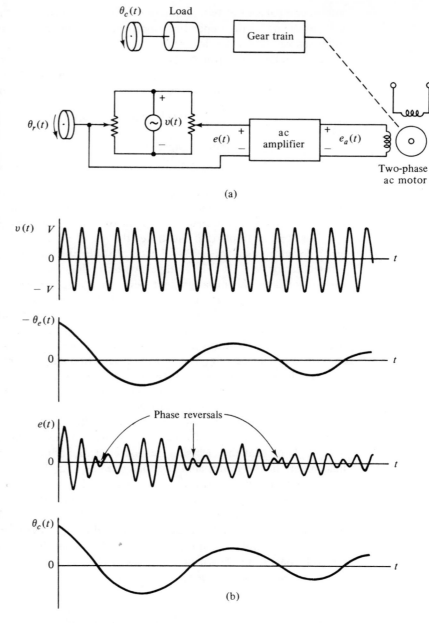

Fig. 5-29. (a) AC control system with potentiometers as error detector. (b) Typical waveforms of signals in the control system of (a).

the signal crosses the zero-magnitude axis. This reversal in phase causes the ac motor to reverse in direction according to the desired sense of correction of the error $\theta_e(t)$. The name "suppressed-carrier modulation" stems from the fact that when a signal $\theta_e(t)$ is modulated by a carrier signal $v(t)$ according to Eq. (5-90), the resultant signal $e(t)$ no longer contains the original carrier frequency ω_c. To illustrate this, let us assume that $\theta_e(t)$ is also a sinusoid given by

$$\theta_e(t) = \sin \omega_s(t) \tag{5-91}$$

where normally, $\omega_s \ll \omega_c$. By use of familiar trigonometric relations, substituting Eqs. (5-89) and (5-91) into Eq. (5-90) we get

$$e(t) = \tfrac{1}{2}K_sV[\cos(\omega_c - \omega_s)t - \cos(\omega_c + \omega_s)t] \tag{5-92}$$

Therefore, $e(t)$ no longer contains the carrier frequency ω_c or the signal frequency ω_s, but it does have the two side bands $\omega_c + \omega_s$ and $\omega_c - \omega_s$.

Interestingly enough, when the modulated signal is transmitted through the system, the motor acts as a demodulator, so that the displacement of the load will be of the same form as the dc signal before modulation. This is clearly seen from the waveforms of Fig. 5-29(b).

It should be pointed out that a control system need not contain all-dc or all-ac components. It is quite common to couple a dc component to an ac component through a modulator or an ac device to a dc device through a demodulator. For instance, the dc amplifier of the system in Fig. 5-28(a) may be replaced by an ac amplifier that is preceded by a modulator and followed by a demodulator.

Synchros. Among the various types of sensing devices for mechanical shaft errors, the most widely used is a pair of synchros. Basically, a synchro is a rotary device that operates on the same principle as a transformer and produces a correlation between an angular position and a voltage or a set of voltages. Depending upon the manufacturers, synchros are known by such trade names as Selsyn, Autosyn, Diehlsyn, and Telesyn. There are many types and different applications of synchros, but in this section only the *synchro transmitter* and the *synchro control transformer* will be discussed.

Synchro Transmitter

A synchro transmitter has a Y-connected stator winding which resembles the stator of a three-phase induction motor. The rotor is a salient-pole, dumbbell-shaped magnet with a single winding. The schematic diagram of a synchro transmitter is shown in Fig. 5-30. A single-phase ac voltage is applied to the

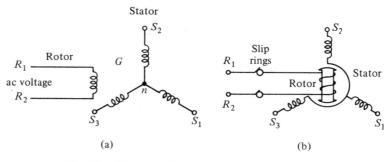

(a)   (b)

Fig. 5-30. Schematic diagrams of a synchro transmitter.

rotor through two slip rings. The symbol G is often used to designate a synchro transmitter, which is sometimes also known as a synchro generator.

Let the ac voltage applied to the rotor of a synchro transmitter be

$$e_r(t) = E_r \sin \omega_c t \tag{5-93}$$

When the rotor is in the position shown in Fig. 5-30, which is defined as the *electric zero*, the voltage induced across the stator winding between S_2 and the neutral n is maximum and is written

$$e_{S_2n}(t) = KE_r \sin \omega t \tag{5-94}$$

where K is a proportional constant. The voltages across the terminals $S_1 n$ and $S_3 n$ are

$$e_{S_1n}(t) = KE_r \cos 240° \sin \omega t = -0.5KE_r \sin \omega t \tag{5-95}$$

$$e_{S_3n}(t) = KE_r \cos 120° \sin \omega t = -0.5KE_r \sin \omega t \tag{5-96}$$

Then the terminal voltages of the stator are

$$e_{S_1S_2} = e_{S_1n} - e_{S_2n} = -1.5KE_r \sin \omega t$$
$$e_{S_2S_3} = e_{S_2n} - e_{S_3n} = 1.5KE_r \sin \omega t \tag{5-97}$$

$$e_{S_3S_1} = e_{S_3n} - e_{S_1n} = 0 \tag{5-98}$$

The above equations show that, despite the similarity between the construction of the stator of a synchro and that of a three-phase machine, there are only single-phase voltages induced in the stator.

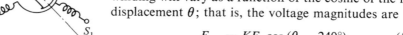

Consider now that the rotor of the synchro transmitter is allowed to rotate in a counterclockwise direction, as shown in Fig. 5-31. The voltages in each stator winding will vary as a function of the cosine of the rotor displacement θ; that is, the voltage magnitudes are

$$E_{S_1n} = KE_r \cos(\theta - 240°) \tag{5-99}$$

$$E_{S_2n} = KE_r \cos \theta \tag{5-100}$$

Fig. 5-31. Rotor position of a synchro transmitter.

$$E_{S_3n} = KE_r \cos(\theta - 120°) \tag{5-101}$$

The magnitudes of the stator terminal voltages become

$$E_{S_1S_2} = E_{S_1n} - E_{S_2n} = \sqrt{3} \, KE_r \sin(\theta + 240°) \tag{5-102}$$

$$E_{S_2S_3} = E_{S_2n} - E_{S_3n} = \sqrt{3} \, KE_r \sin(\theta + 120°) \tag{5-103}$$

$$E_{S_3S_1} = E_{S_3n} - E_{S_1n} = \sqrt{3} \, KE_r \sin \theta \tag{5-104}$$

A plot of these terminal voltages as a function of the rotor shaft position is shown in Fig. 5-32. Notice that each rotor position corresponds to one unique set of stator voltages. This leads to the use of the synchro transmitter to identify angular positions by measuring and identifying the set of voltages at the three stator terminals.

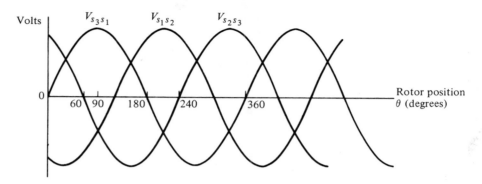

Fig. 5-32. Variation of the terminal voltages of a synchro transmitter as a function of the rotor position. θ is measured counterclockwise from the electric zero.

Synchro Control Transformer

Since the function of an error detector is to convert the difference of two shaft positions into an electrical signal, a single synchro transmitter is apparently inadequate. A typical arrangement of a synchro error detector involves the use of two synchros: a transmitter and a control transformer, as shown in Fig. 5-33.

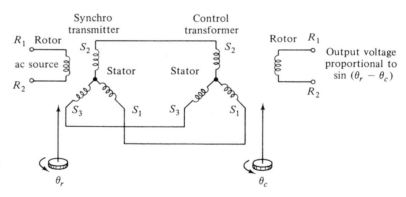

Fig. 5-33. Synchro error detector.

For small angular deviations between the two rotor positions, a proportional voltage is generated at the rotor terminals of the control transformer.

Basically, the principle of operation of a synchro control transformer is identical to that of the synchro transmitter, except that the rotor is cylindrically shaped so that the air-gap flux is uniformly distributed around the rotor. This feature is essential for a control transformer, since its rotor terminals are usually connected to an amplifier or similar electrical device, in order that the latter sees a constant impedance. The change in the rotor impedance with rotations of the shaft position should be minimized.

The symbol CT is often used to designate a synchro control transformer.

Referring to the arrangement shown in Fig. 5-33, the voltages given by Eqs. (5-102), (5-103), and (5-104) are now impressed across the corresponding stator terminals of the control transformer. Because of the similarity in the magnetic construction, the flux patterns produced in the two synchros will be the same if all losses are neglected. For example, if the rotor of the transmitter is in its electric zero position, the fluxes in the transmitter and in the control transformer are as shown in Fig. 5-34(a) and (b).

When the rotor of the control transformer is in the position shown in Fig. 5-34(b), the induced voltage at its rotor terminals is zero. The shafts of the two synchros are considered to be in alignment. When the rotor position of the control transformer is rotated 180° from the position shown, its terminal voltage is again zero. These are known as the two *null positions* of the error detector. If the control transformer rotor is at an angle α from either of the null positions,

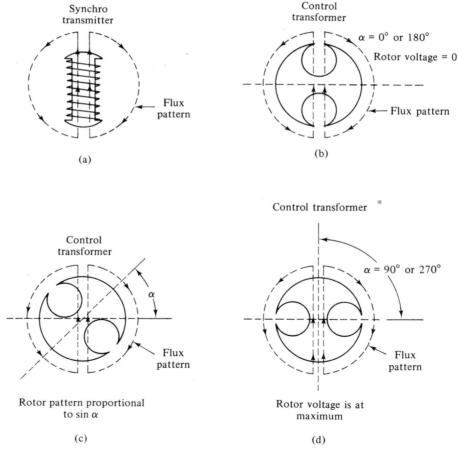

Fig. 5-34. Relations among flux patterns, rotor positions, and the rotor voltage of synchro error detector.

such as that shown in Fig. 5-34(c) and (d), the magnitude of the rotor voltage is proportional to sin α. Similarly, it can be shown that when the transmitter shaft is in any position other than that shown in Fig. 5-34(a), the flux patterns will shift accordingly, and the rotor voltage of the control transformer will be proportional to the sine of the difference of the rotor positions, α. The rotor voltage of the control transformer versus the difference in positions of the rotors of the transmitter and the control transformer is shown in Fig. 5-35.

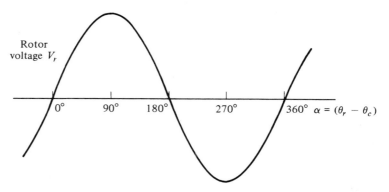

Fig. 5-35. Rotor voltage of control transformer as a function of the difference of rotor positions.

From Fig. 5-35 it is apparent that the synchro error detector is a nonlinear device. However, for small angular deviations of up to 15 degrees in the vicinity of the two null positions, the rotor voltage of the control transformer is approximately proportional to the difference between the positions of the rotors of the transmitter and the control transformer. Therefore, for small deviations, the transfer function of the synchro error detector can be approximated by a constant K_S:

$$K_S = \frac{E}{\theta_r - \theta_c} = \frac{E}{\theta_e} \tag{5-105}$$

where

$E =$ error voltage
$\theta_r =$ shaft position of synchro transmitter, degrees
$\theta_c =$ shaft position of synchro control transformer, degrees
$\theta_e =$ error in shaft positions
$K_S =$ sensitivity of the error detector, volts per degree

The schematic diagram of a positional control system employing a synchro error detector is shown in Fig. 5-36(a). The purpose of the control system is to make the controlled shaft follow the angular displacement of the reference input shaft as closely as possible. The rotor of the control transformer is mechanically connected to the controlled shaft, and the rotor of the synchro transmitter is connected to the reference input shaft. When the controlled shaft is aligned with the reference shaft, the error voltage is zero and the motor does not turn. When an angular misalignment exists, an error voltage of relative polarity

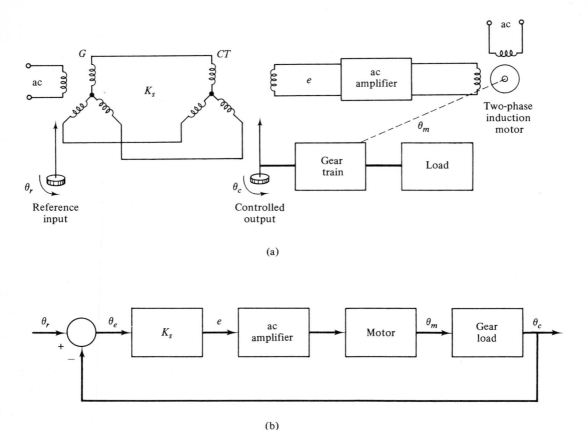

(a)

(b)

Fig. 5-36. (a) Alternating current control system employing synchro error detector. (b) Block diagram of system in (a).

appears at the amplifier input, and the output of the amplifier will drive the motor in such a direction as to reduce the error. For small deviations between the controlled and the reference shafts, the synchro error detector can be represented by the constant K_S given by Eq. (5-105). Then the linear operation of the positional control system can be represented by the block diagram of Fig. 5-36(b). From the characteristic of the error detector shown in Fig. 5-35, it is clear that K_S has opposite signs at the two null positions. However, in closed-loop systems, only one of the two null positions is a true null; the other one corresponds to an unstable operating point.

Suppose that, in the system shown in Fig. 5-36(a), the synchro positions are close to the true null and the controlled shaft lags behind the reference shaft; a positive error voltage will cause the motor to turn in the proper direction to correct this lag. But if the synchros are operating close to the false null, for the same lag between θ_r and θ_c, the error voltage is negative and the motor is driven in the direction that will increase the lag. A larger lag in the controlled shaft position will increase the magnitude of the error voltage still further and cause the motor to rotate in the same direction, until the true null position is reached.

In reality, the error signal at the rotor terminals of the synchro control transformer may be represented as a function of time. If the ac signal applied to the rotor terminals of the transmitter is denoted by $\sin \omega_c t$, where ω_c is known as the carrier frequency, the error signal is given by

$$e(t) = K_s \theta_e(t) \sin \omega_c t \qquad (5\text{-}106)$$

Therefore, as explained earlier, $e(t)$ is again a suppressed-carrier modulated signal.

5.6 Tachometers[4, 5]

Tachometers are electromechanical devices that convert mechanical energy into electrical energy. The device works essentially as a generator with the output voltage proportional to the magnitude of the angular velocity.

In control systems tachometers are used for the sensing of shaft velocity and for the improvement of system performance. For instance, in a control system with the output displacement designated as the state variable x_1 and the output velocity as the state variable x_2, the first state variable may be excessed to by means of a potentiometer while x_2 is monitored by a tachometer.

In general, tachometers may be classified into two types: ac and dc. The

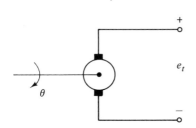

simplified schematic diagrams of these two versions are shown in Fig. 5-37. For the ac tachometer, a sinusoidal voltage of rated value is applied to the primary winding. A secondary winding is placed at a 90° angle mechanically with respect to the primary winding. When the rotor of the tachometer is stationary, the output voltage at the secondary winding is zero. When the rotor shaft is rotated, the output voltage of the tachometer is closely proportional to the rotor velocity. The polarity of the voltage is dependent upon the direction of rotation. The input–output relation of an ac tachometer can be represented by a first-order differential equation

Fig. 5-37. Schematic diagram of a tachometer.

$$e_t(t) = K_t \frac{d\theta(t)}{dt} \qquad (5\text{-}107)$$

where $e_t(t)$ is the output voltage, $\theta(t)$ the rotor displacement, and K_t is defined as the *tachometer constant*, usually represented in units of volts per rpm or volts per 1000 rpm. The transfer function of an ac tachometer is obtained by taking the Laplace transform of Eq. (5-107); thus

$$\frac{E_t(s)}{\theta(s)} = K_t s \qquad (5\text{-}108)$$

A dc tachometer serves exactly the same purpose as the ac tachometer described above. One advantage of the dc tachometer is that the magnetic field of the device may be set up by permanent magnet, and therefore no separate excitation voltage is required. In principle, the equations of Eqs. (5-107) and (5-108) are also true for a dc tachometer.

A dc tachometer can also replace an ac tachometer in a control system by use of a modulator to convert its dc output signal into ac. Similarly, a dc tachometer can be replaced by an ac one if a phase-sensitive demodulator is used to convert the ac output to dc.

5.7 DC Motors in Control Systems

Direct current motors are one of the most widely used prime movers in industry. The advantages of dc motors are that they are available in a great variety of types and sizes and that control is relatively simple. The primary disadvantage of a dc motor relates to its brushes and commutator.

For general purposes, dc motors are classified as *series-excited*, *shunt-excited*, and *separately-excited*, all of which refer to the way in which the field is excited. However, the characteristics of the first two types of motor are highly nonlinear, so for control systems applications, the separately-excited dc motor is the most popular.

The separately-excited dc motor is divided into two groups, depending on whether the control action is applied to the field terminals or to the armature terminals of the motor. These are referred to as *field controlled* and *armature controlled*, where usually the rotor of the motor is referred to as the armature (although there are exceptions).

In recent years, advanced design and manufacturing techniques have produced dc motors with permanent-magnet fields of high field intensity and rotors of very low inertia—motors with very high torque-to-inertia ratios, in other words. It is possible to have a 4-hp motor with a mechanical time constant as low as 2 milliseconds. The high torque-to-inertia ratio of dc motors has opened new applications for motors in computer peripheral equipment such as tape drives, printers, and disk packs, as well as in the machine tool industry. Of course, when a dc motor has a permanent-magnetic field it is necessarily armature controlled.

The mathematical modeling of the armature-controlled and field-controlled dc motors, including the permanent-magnet dc motor, is discussed in the following.

Field-Controlled DC Motor

The schematic diagram of a field-controlled dc motor is shown in Fig. 5-38. The following motor variables and parameters are defined:

$e_a(t)$	armature voltage
R_a	armature resistance
$\phi(t)$	air-gap flux
$e_b(t)$	back emf (electromotive force) voltage
K_b	back emf constant

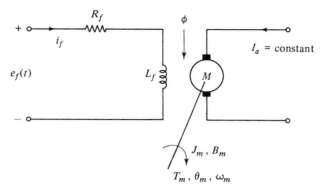

Fig. 5-38. Schematic diagram of a field-controlled dc motor.

K_i	torque constant
$i_a(t)$	armature current
$i_f(t)$	field current
$e_f(t)$	field voltage
$T_m(t)$	torque developed by motor
J_m	rotor inertia of motor
B_m	viscous frictional coefficient
$\theta_m(t)$	angular rotor displacement

To carry out a linear analysis, the following assumptions are made:

1. The armature current is held constant, $i_a = I_a$.
2. The air-gap flux is proportional to the field current; that is,

$$\phi(t) = K_f i_f(t) \tag{5-109}$$

3. The torque developed by the motor is proportional to the air-gap flux and the armature current. Thus

$$T_m(t) = K_m I_a \phi(t) \tag{5-110}$$

where K_m is a constant. Substituting Eq. (5-109) into Eq. (5-110) gives

$$T_m(t) = K_m K_f I_a i_f(t) \tag{5-111}$$

If we let

$$K_i = K_m K_f I_a \tag{5-112}$$

be the torque constant, Eq. (5-111) becomes

$$T_m(t) = K_i i_f(t) \tag{5-113}$$

Referring to the motor circuit diagram of Fig. 5-38, the state variables are assigned as $i_f(t)$, $\omega_m(t)$, and $\theta_m(t)$. The first-order differential equations relating these state variables and the other variables are written

$$L_f \frac{di_f(t)}{dt} = -R_f i_f(t) + e_f(t) \tag{5-114}$$

$$J_m \frac{d\omega_m(t)}{dt} = -B_m\omega_m(t) + T_m(t) \qquad (5\text{-}115)$$

$$\frac{d\theta_m(t)}{dt} = \omega_m(t) \qquad (5\text{-}116)$$

By proper substitution, the last three equations are written in the form of state equations:

$$
\begin{bmatrix}
\dfrac{di_f(t)}{dt} \\[2mm]
\dfrac{d\omega_m(t)}{dt} \\[2mm]
\dfrac{d\theta_m(t)}{dt}
\end{bmatrix}
=
\begin{bmatrix}
-\dfrac{R_f}{L_f} & 0 & 0 \\[2mm]
\dfrac{K_i}{J_m} & -\dfrac{B_m}{J_m} & 0 \\[2mm]
0 & 1 & 0
\end{bmatrix}
\begin{bmatrix}
i_f(t) \\[2mm]
\omega_m(t) \\[2mm]
\theta_m(t)
\end{bmatrix}
+
\begin{bmatrix}
\dfrac{1}{L_f} \\[2mm]
0 \\[2mm]
0
\end{bmatrix}
e_f(t) \qquad (5\text{-}117)
$$

The state diagram of the system is drawn as shown in Fig. 5-39. The transfer

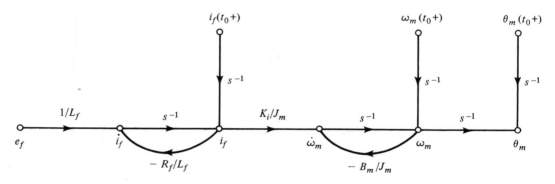

Fig. 5-39. State diagram of a field-controlled dc motor.

function between the motor displacement and the input voltage is obtained from the state diagram as

$$\frac{\theta_m(s)}{E_f(s)} = \frac{K_i}{L_f J_m s^3 + (R_f J_m + B_m L_f)s^2 + R_a B_m s} \qquad (5\text{-}118)$$

or

$$\frac{\theta_m(s)}{E_f(s)} = \frac{K_i}{R_a B_m s(1 + \tau_m s)(1 + \tau_f s)} \qquad (5\text{-}119)$$

where

$$\tau_m = \frac{J_m}{B_m} = \text{mechanical time constant of motor}$$

$$\tau_f = \frac{L_f}{R_f} = \text{field electrical time constant of motor}$$

Armature-Controlled DC Motor

The schematic diagram of an armature-controlled dc motor is shown in Fig. 5-40. In this case for linear operation it is necessary to hold the field current of the motor constant. The torque constant K_i relates the motor torque and the

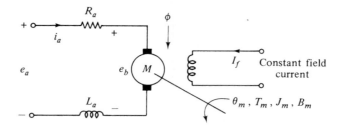

Fig. 5-40. Schematic diagram of an armature-controlled dc motor.

armature current, $i_a(t)$; thus

$$T_m(t) = K_i i_a(t) \tag{5-120}$$

where K_i is a function of the air-gap flux, which is constant in this case. Of course, in the case of a permanent-magnet motor, ϕ is constant also. The back emf voltage is proportional to the motor speed,

$$e_b(t) = K_b \frac{d\theta_m(t)}{dt} = K_b \omega_m(t) \tag{5-121}$$

With reference to Fig. 5-40, the state equations of the armature-controlled dc motor are written

$$\begin{bmatrix} \dfrac{di_a(t)}{dt} \\[2mm] \dfrac{d\omega_m(t)}{dt} \\[2mm] \dfrac{d\theta_m(t)}{dt} \end{bmatrix} = \begin{bmatrix} -\dfrac{R_a}{L_a} & -\dfrac{K_b}{L_a} & 0 \\[2mm] \dfrac{K_i}{J_m} & -\dfrac{B_m}{J_m} & 0 \\[2mm] 0 & 1 & 0 \end{bmatrix} \begin{bmatrix} i_a(t) \\[2mm] \omega_m(t) \\[2mm] \theta_m(t) \end{bmatrix} + \begin{bmatrix} \dfrac{1}{L_a} \\[2mm] 0 \\[2mm] 0 \end{bmatrix} e_a(t) \tag{5-122}$$

The state diagram of the system is drawn as shown in Fig. 5-41. The transfer function between the motor displacement and the input voltage is obtained from the state diagram as

$$\frac{\theta_m(s)}{E_a(s)} = \frac{K_i}{L_a J_m s^3 + (R_a J_m + B_m L_a)s^2 + (K_b K_i + R_a B_m)s} \tag{5-123}$$

Although a dc motor is basically an open-loop system, as in the case of the field-controlled motor, the state diagram of Fig. 5-41 shows that the armature-controlled motor has a "built-in" feedback loop caused by the back emf. Physically, the back emf voltage represents the feedback of a signal that is proportional to the negative of the speed of the motor. As seen from Eq. (5-123), the back emf constant, K_b, represents an added term to the resistance and frictional coefficient. Therefore, the back emf effect is equivalent to an "electrical friction," which tends to improve the stability of the motor.

It should be noted that in the English unit system, K_i is in lb-ft/amp or oz-in/amp and the back emf constant K_b is in V/rad/sec. With these units, K_i and K_b are related by a constant factor. We can write the mechanical power

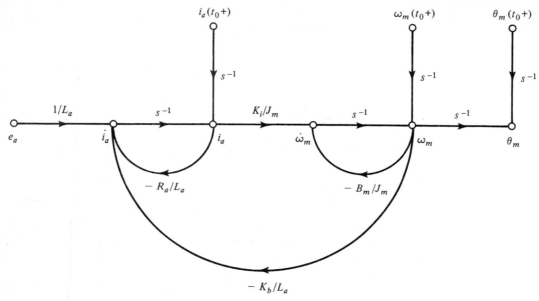

Fig. 5-41. State diagram of an armature-controlled dc motor.

developed in the motor armature as

$$P = e_b(t)i_a(t) \qquad \text{watts}$$

$$= \frac{e_b(t)i_a(t)}{746} \qquad \text{hp} \tag{5-124}$$

Substituting Eqs. (5-120) and (5-121) into Eq. (5-124), we get

$$P = \frac{K_b T_m}{746 K_i} \frac{d\theta_m(t)}{dt} \qquad \text{hp} \tag{5-125}$$

This power is equal to the mechanical power

$$P = \frac{T_m}{550} \frac{d\theta_m(t)}{dt} \qquad \text{hp} \tag{5-126}$$

Therefore, equating Eqs. (5-125) and (5-126), we have

$$K_i = 0.737 K_b \tag{5-127}$$

We can also determine the constants of the motor using torque–speed curves. A typical set of torque–speed curves for various applied voltages of a dc motor is shown in Fig. 5-42. The rated voltage is denoted by E_r. At no load, $T_m = 0$, the speed is given by the intersect on the abscissa for a given applied voltage. Then the back emf constant K_b is given by

$$K_b = \frac{E_r}{\Omega_r} \tag{5-128}$$

where in this case the rated values are used for voltage and angular velocity.

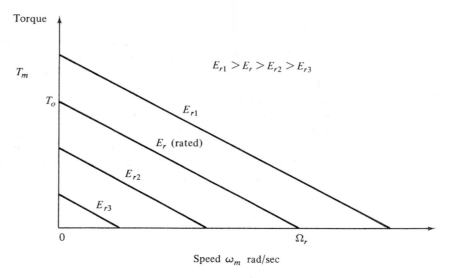

Fig. 5-42. Typical torque–speed curves of a dc motor.

When the motor is stalled, the blocked-rotor torque at the rated voltage is designated by $T_0(t)$. Let

$$k = \frac{\text{blocked-rotor torque at rated voltage}}{\text{rated voltage}} = \frac{T_0}{E_r} \qquad (5\text{-}129)$$

Also,

$$T_0(t) = K_i i_a(t) = \frac{K_i}{R_a} E_r \qquad (5\text{-}130)$$

Therefore, from the last two equations, the torque constant is determined:

$$K_i = k R_a \qquad (5\text{-}131)$$

5.8 Two-Phase Induction Motor[6,7]

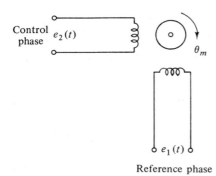

Fig. 5-43. Schematic diagram of a two-phase induction motor.

For low-power applications in control systems ac motors are preferred over dc motors because they are more rugged and have no brushes to maintain. Most ac motors used in control systems are of the two-phase induction type, which generally are rated from a fraction of a watt up to a few hundred watts. The frequency of the motor is normally at 60, 400, or 1000 Hz.

A schematic diagram of a two-phase induction motor is shown in Fig. 5-43. The motor consists of a stator with two distributed windings displaced 90 electrical degrees apart. Under normal operating conditions in control applications, a fixed voltage from a constant-voltage source is applied to one phase, the *fixed* or *reference phase*. The other phase, the *control phase*, is energized by a voltage that is 90° out of phase with respect to the voltage of the fixed phase. The control-phase voltage is usually supplied from a servo amplifier, and the voltage has a variable amplitude and polarity. The direction of rotation of the motor reverses when the control phase signal changes its sign.

Unlike that of a dc motor, the torque–speed curve of a two-phase induction motor is quite nonlinear. However, for linear analysis, it is generally considered an acceptable practice to approximate the torque–speed curves of a two-phase induction motor by straight lines, such as those shown in Fig. 5-44. These curves are assumed to be straight lines parallel to the torque–speed curve at a rated control voltage ($E_2 = E_1 =$ rated value), and they are equally spaced for equal increments of the control voltage.

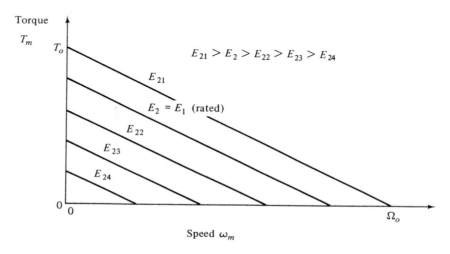

Fig. 5-44. Typical linearized torque–speed curves of a two-phase induction motor.

The state equations of the motor are determined as follows. Let k be the *blocked-rotor torque at rated voltage* per unit control voltage; that is,

$$k = \frac{\text{blocked-rotor torque at } E_2 = E_1}{\text{rated control voltage } E_1} = \frac{T_0}{E_1} \qquad (5\text{-}132)$$

Let m be a negative number which represents the slope of the linearized torque–speed curve shown in Fig. 5-44. Then

$$m = -\frac{\text{blocked-rotor torque}}{\text{no-load speed}} = -\frac{T_0}{\Omega_0} \qquad (5\text{-}133)$$

For any torque T_m, the family of straight lines in Fig. 5-44 is represented by the equation

$$T_m(t) = m\omega_m(t) + ke_2(t) \qquad (5\text{-}134)$$

where $\omega_m(t)$ is the speed of the motor and $e_2(t)$ the control voltage. Now, if we designate $\omega_m(t)$ as a state variable, one of the state equations may be obtained from

$$J_m \frac{d\omega_m(t)}{dt} = -B_m \omega_m(t) + T_m(t) \tag{5-135}$$

Substituting Eq. (5-134) into Eq. (5-135) and recognizing that $\theta_m(t)$ is the other state variable, we have the two state equations

$$\frac{d\theta_m(t)}{dt} = \omega_m(t) \tag{5-136}$$

$$\frac{d\omega_m(t)}{dt} = \frac{1}{J_m}(m - B_m)\omega_m + \frac{k}{J_m}e_2(t) \tag{5-137}$$

The state diagram of the two-phase induction motor is shown in Fig. 5-45. The

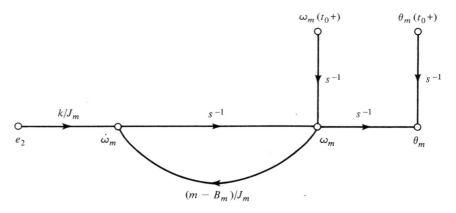

Fig. 5-45. State diagram of the two-phase induction motor.

transfer function of the motor between the control voltage and the motor displacement is obtained as

$$\frac{\theta_m(s)}{E_2(s)} = \frac{k}{(B_m - m)s[1 + J_m/(B_m - m)s]} \tag{5-138}$$

or

$$\frac{\theta_m(s)}{E_2(s)} = \frac{K_m}{s(1 + \tau_m s)} \tag{5-139}$$

where

$$K_m = \frac{k}{B_m - m} = \text{motor gain constant} \tag{5-140}$$

$$\tau_m = \frac{J_m}{B_m - m} = \text{motor time constant} \tag{5-141}$$

Since m is a negative number, the equations above show that the effect of the slope of the torque–speed curve is to add more friction to the motor, thus improving the damping or stability of the motor. Therefore, the slope of the torque–speed curve of a two-phase induction motor is analogous to the back emf effect of a dc motor. However, if m is a positive number, negative damping occurs for $m > B_m$, it can be shown that the motor becomes unstable.

5.9 Step Motors[8-10]

A step motor is an electromagnetic incremental actuator that converts digital pulse inputs to analog output shaft motion. In a rotary step motor, the output shaft of the motor rotates in equal increments in response to a train of input pulses. When properly controlled, the output steps of a step motor are equal in number to the number of input pulses. Because modern control systems often have incremental motion of one type or another, step motors have become an important actuator in recent years. For instance, incremental motion control is found in all types of computer peripheral equipment, such as printers, tape drives, capstan drives, and memory-access mechanisms, as well as in a great variety of machine tool and process control systems. Figure 5-46 illustrates the

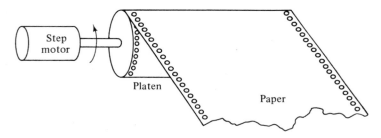

Fig. 5-46. Paper-drive mechanism using a step motor.

application of a step motor in the paper-drive mechanism of a printer. In this case the motor is coupled directly to the platen so that the paper is driven a certain increment at a time. Typical resolution of step motors available commercially ranges from several steps per revolution to as much as 800 steps per revolution, and even higher.

Step motors come in a variety of types, depending upon the principle of operation. The two most common types of step motors are the variable-reluctance type and the permanent-magnet type. The complete mathematical analysis of these motors is highly complex, since the motor characteristics are very nonlinear. Unlike dc and induction motors, linearized models of a step motor are usually unrealistic. In this section we shall describe the principle of operation and a simplified mathematical model of the variable-reluctance motor.

The variable-reluctance step motor has a wound stator and an unexcited rotor. The motor can be of the single- or the multiple-stack type. In the multiple-stack version, the stator and the rotor consist of three or more separate sets of teeth. The separate sets of teeth on the rotor, usually laminated, are mounted on the same shaft. The teeth on all portions of the rotor are perfectly aligned. Figure 5-47 shows a typical rotor-stator model of a motor that has three sepa-

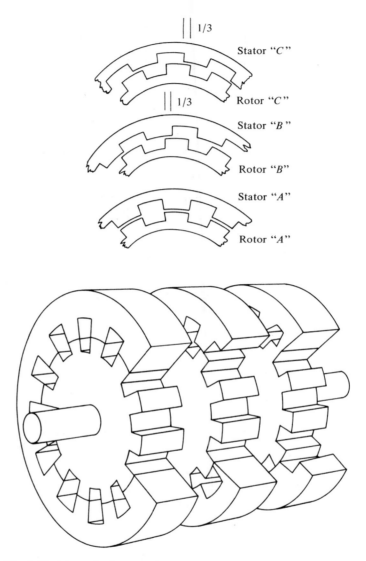

Fig. 5-47. Schematic diagram of the arrangement of rotor and stator teeth in a multiple-stack, three-phase variable-reluctance step motor. The motor is shown to have 12 teeth on each stack or phase.

rate sections on the rotor, or a three-phase motor. A variable-reluctance step motor must have at least three phases in order to have directional control. The three sets of rotor teeth are magnetically independent, and are assembled to one shaft, which is supported by bearings. Arranged around each rotor section is a stator core with windings. The windings are not shown in Fig. 5-47. Figure 5-48 is a schematic diagram of the windings on the stator. The end view of the stator of one phase, and the rotor, of a practical motor is shown in Fig. 5-49. In this

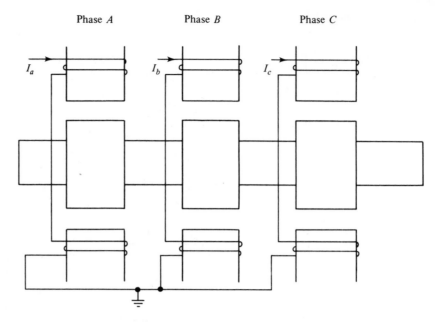

Phase *A*   Phase *B*   Phase *C*

I_a I_b I_c

Fig. 5-48. Schematic diagram of a multiple-stack three-phase variable-reluctance step motor.

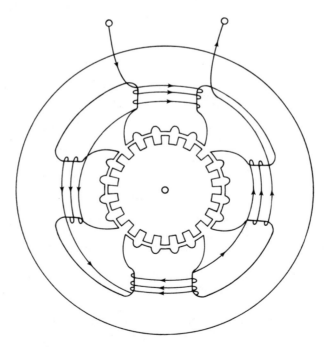

Fig. 5-49. End view of the stator of one phase of a multiple-stack variable-reluctance step motor.

case the rotor is shown at a position where its teeth are in alignment with that of the particular phase of the stator.

The rotor and stator have the same number of teeth, which means that the tooth pitch on the stator and the rotor are the same. To make the motor rotate, the stator sections of the three-phase motor are indexed one-third of a tooth pitch, in the same direction. Figure 5-50 shows this arrangement for a 10-tooth

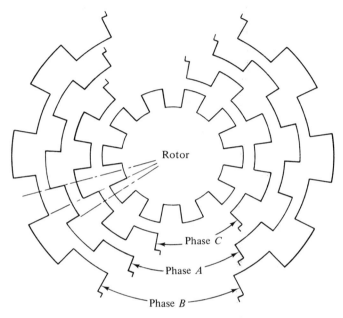

Fig. 5-50. Rotor and stator teeth arrangements of a multiple-stack three-phase variable-reluctance step motor. The rotor has 10 teeth.

rotor. Therefore, the teeth on one stator phase are displaced 12° with respect to the stator phase. Here the teeth of phase C of the stator are shown to be aligned with the corresponding rotor teeth. The teeth of phase A of the stator are displaced clockwise by 12° with respect to the teeth of phase C. The teeth of phase B of the stator are displaced 12° clockwise with respect to those of phase A, or 12° counterclockwise with respect to those of phase C. It is easy to see that a minimum of three phases is necessary to give directional control. In general, four- and five-phase motors are also common, and motors with as many as eight phases are available commercially. For an n-phase motor, the stator teeth are displaced by $1/n$ of a tooth pitch from section to section.

The operating principle of the variable-reluctance stepping motor is straightforward. Let any one phase of the windings be energized with a dc signal. The magnetomotive force setup will position the rotor such that the teeth of the rotor section under the excited phase are aligned opposite the teeth on the excited phase of the stator. This is the position of minimum reluctance, and the motor is in a stable equilibrium.

If phase C is energized in Fig. 5-50, the rotor would be (in steady state) positioned as shown. It can also be visualized from the same figure that if the dc signal is switched to phase A, the rotor will rotate by 12°, clockwise, and the rotor teeth will be aligned opposite the teeth of phase A of the stator. Continuing in the same way, the input sequence CABCAB will rotate the motor clockwise in steps of 12°.

Reversing the input sequence will reverse the direction of rotation. That is, the input sequence CBACB will rotate the motor in the counterclockwise direction in steps of 12°.

The steady-state torque curve of each phase is approximately as shown in Fig. 5-51. The 0° line represents the axis of any tooth of the energized stator phase. The nearest rotor tooth axis will always lie within 18° on either side of this line. The corresponding starting torque exerted when this phase is energized can be seen in Fig. 5-51. The arrows mark the direction of motion of the rotor.

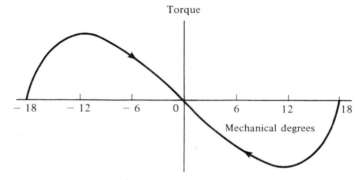

Fig. 5-51. Torque curve for one phase of a step motor.

Let positive angular displacements represent clockwise motion. Suppose also that phase C has been excited for a long time. This means that the initial condition of the rotor will be as shown in Fig. 5-50. If phase A is now energized and Fig. 5-51 represents the torque variation of phase A, the initial position of the rotor teeth will be at −12°. As soon as phase A is energized, the rotor will finally settle after some oscillations, assuming that the inertia and friction are such that there is no overshoot beyond the 18° point.

It may be noticed that the position, ±18°, also represents an equilibrium point. This is because in that position the deflecting torque is zero. It is, however, a position of unstable equilibrium since the slightest shift from this position will send the motor straight to 0°.

If on energizing one phase the rotor happens to lie exactly at the ±18° point, theoretically it will stay there. In practice, however, there will always be some mechanical imperfections in construction, and the resulting asymmetry will prevent any locking at the unstable point.

We now look upon the stepping motor from a single-step point of view and try to develop the equations that govern its performance. Several assumptions will be made initially to simplify the development. Subsequent modifications may be made if any of these assumptions are found to be invalidated. We start by writing the equation for the electrical circuit of the stator winding. Let

$$e(t) = \text{applied voltage per phase}$$
$$R = \text{winding resistance per phase}$$
$$L(\theta) = \text{winding inductance per phase}$$
$$i(t) = \text{current per phase}$$
$$\theta(t) = \text{angular displacement}$$

The voltage–current equation of one stator phase is written

$$e(t) = Ri(t) + \frac{d}{dt}[iL(\theta)]$$
$$= Ri(t) + L(\theta)\frac{di}{dt} + i\frac{d}{dt}L(\theta) \tag{5-142}$$

or

$$e(t) = Ri(t) + L(\theta)\frac{di}{dt} + i\frac{d}{d\theta}L(\theta)\frac{d\theta}{dt} \tag{5-143}$$

The term, $L(\theta)(di/dt)$ represents transformer electromotive force, or self-induced electromotive force, and the term $i[dL(\theta)/d\theta](d\theta/dt)$ represents the back emf due to the rotation of the rotor. We have assumed above that the inductance is a function of $\theta(t)$, the angular displacement, only. No dependence on the current has been assumed. This will reflect directly on our procedure for getting the torque developed by the motor.

The energy in the air gap can be written

$$W = \tfrac{1}{2}L(\theta)i^2(t) \tag{5-144}$$

From elementary electromechanical energy-conversion principles, we know that the torque in a single excited rotational system is given by

$$T = \frac{\partial}{\partial\theta}[W(i, \theta)] \tag{5-145}$$

where W is the energy of the system expressed explicitly in terms of $i(t)$ and $\theta(t)$. Therefore,

$$T = \frac{1}{2}i^2(t)\frac{d}{d\theta}[L(\theta)] \tag{5-146}$$

This torque is then applied to the rotor and the equation of motion is obtained as

$$T = J_m\frac{d^2\theta(t)}{dt^2} + B_m\frac{d\theta(t)}{dt} \tag{5-147}$$

where J_m is the rotor inertia and B_m the viscous frictional coefficient. J_m and B_m also may include the effects of any load.

To complete the torque expression of Eq. (5-146), we need to know the form of the inductance $L(\theta)$. In practice the motor inductance as a function of displacement may be approximated by a cosine wave; that is,

$$L(\theta) = L_1 + L_2 \cos r\theta \tag{5-148}$$

where L_1 and L_2 are constants and r is the number of teeth on each rotor section. Substituting Eq. (5-148) into Eq. (5-146), we get

$$T = -\tfrac{1}{2}L_2 ri^2(t) \sin r\theta = -Ki^2(t) \sin r\theta \qquad (5\text{-}149)$$

which is the sinusoidal approximation of the static torque curve of Fig. 5-51.

Now let us apply these equations to a three-phase motor. Let the equilibrium position be the situation when phase A is energized. Then the inductance and torque for phase A are given by

$$L_A = L_1 + L_2 \cos r\theta \qquad (5\text{-}150)$$

$$T_A = -Ki_A^2 \sin r\theta \qquad (5\text{-}151)$$

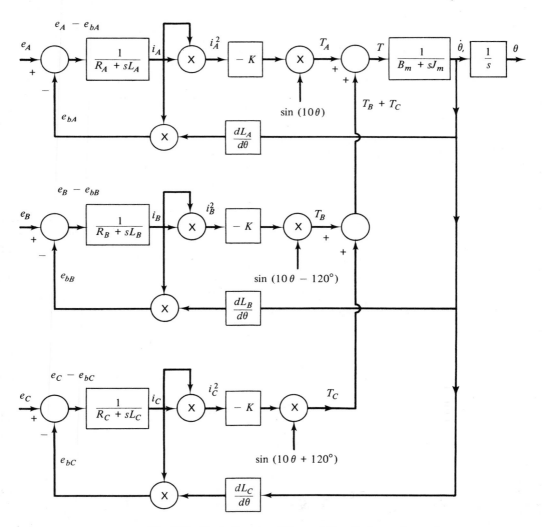

Fig. 5-52. Block diagram of the variable-reluctance step motor.

respectively. For the 10-teeth rotor considered earlier, $r = 10$, phase B has its equilibrium position 12° behind the reference point, assuming that the sequence ABC represents forward motion. Similarly, the equilibrium position of phase C is 12° ahead of the reference point. Therefore, the inductances and the torques for phases B and C are written as follows:

$$L_B = L_1 + L_2 \cos (10\theta - 120°) \tag{5-152}$$

$$L_C = L_1 + L_2 \cos (10\theta + 120°) \tag{5-153}$$

$$T_B = -Ki_B^2(t) \sin (10\theta - 120°) \tag{5-154}$$

$$T_C = -Ki_C^2(t) \sin (10\theta + 120°) \tag{5-155}$$

The electrical circuits of the three phases are isolated so that each phase has its differential equation of the form of Eq. (5-143).

The total torque developed on the rotor is the algebraic sum of torques of the three phases. Thus

$$T = T_A + T_B + T_C \tag{5-156}$$

The nonlinearity of the torque equations precludes the use of linearized models for the portrayal of a step motor. Therefore, realistic studies of a step motor using the equations presented above can be made only through computer simulation. A block-diagram representation of the motor, which may be used for analog or digital computer simulation, is shown in Fig. 5-52.

5.10 Tension-Control System

The problem of proper tension control exists in a great variety of winding and unwinding industrial processes. Such industries as paper, plastic, and wire all have processes that involve unwinding and rewinding processes. For example, in the paper industry, the paper is first wound on a roll in a form that is nonsaleable, owing to nonuniform width and breaks. This roll is rewound to trim edges, splice breaks, and slit to required widths. Proper tension during this rewinding is mandatory for several reasons: slick paper will telescope if not wound tightly enough and the width will vary inversely as the tension. Conversely, during storage, a roll wound at varying tensions has internal stresses that will cause it to explode. Similar examples could be cited, but the need for proper tension control is relatively simple to understand.

Most rewind systems contain an unwind roll, a windup roll driven by a motor, and some type of dancer and/or pinch-roller assemblies between the two. Some systems employ spring-with-damper idlers with feedback to motor drives to control tension. Some use tension-measuring devices and feedback to a motor–generator or brake on the unwind reel to hold tension at a constant value.

In this section a specific type of tension-control system for unwind processes is investigated. As shown in Fig. 5-53, the system has a dc-motor-driven windup reel. The tension of the web is controlled by control of the armature voltage $e_a(t)$ of the motor. .

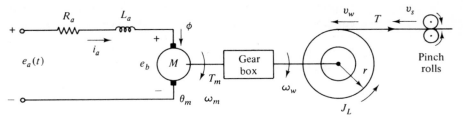

Fig. 5-53. Tension-control system for a winding process.

The mathematical modeling of the system is conducted by writing the equations of the dc motor.

Armature:

$$e_a(t) = R_a i_a(t) + L_a \frac{di_a(t)}{dt} + K_b \omega_m(t) \qquad (5\text{-}157)$$

where

$$K_b = \text{back emf constant of motor}$$
$$\omega_m(t) = \text{angular velocity of motor}$$

Torque equation:

$$T_m(t) = \frac{d}{dt}[J_{me} \omega_m(t)] + B_{me} \omega_m(t) + nrT(t) \qquad (5\text{-}158)$$

where

$$r = \text{effective radius of windup reel}$$
$$T_m(t) = \text{motor torque} = K_i i_a(t)$$
$$n = \text{gear-train ratio}$$
$$T(t) = \text{web tension}$$
$$J_{me} = J_m + n^2 J_L = \text{equivalent inertia at motor shaft}$$
$$B_{me} = B_m + n^2 B_L = \text{equivalent viscous friction coefficient}$$
$$\text{at motor shaft}$$
$$J_L = \text{effective inertia of windup reel}$$
$$B_L = \text{viscous friction coefficient of windup reel}$$

Since the web material is taken up by the windup reel as the process proceeds, the inertia J_L and the radius r of the windup reel increase as functions of time. This explains the reason the derivative of $J_{me} \omega_m$ is taken in Eq. (5-158). Furthermore, if h denotes the thickness of the web,

$$\frac{dr}{dt} = \frac{h}{2} \omega_w \qquad (5\text{-}159)$$

$$J_L = \frac{\rho \pi}{2} r^4 \qquad (5\text{-}160)$$

Thus

$$\frac{dJ_L}{dt} = 2\pi \rho r^3 \frac{dr}{dt} \qquad (5\text{-}161)$$

where

$$\rho = \text{mass density of the web material per width}$$

Assume now that the web material has a coefficient of elasticity of C and that Hooke's law is obeyed. Then

$$\frac{dT(t)}{dt} = C[v_w(t) - v_s(t)] \tag{5-162}$$

where $v_s(t)$ is the web velocity at the pinch rolls. Assuming that the pinch rolls are driven at constant speed, $v_s(t) = \text{constant} = V_s$. Also,

$$v_w(t) = r\omega_w(t) = nr\omega_m(t) \tag{5-163}$$

It is apparent now that because r and J_L are functions of time, Eq. (5-158) is a time-varying nonlinear differential equation. However, if the web is very thin, $h \simeq 0$, we may consider that over a certain time period r and J_L are constant. Then, the linearized state equations of the system are written

$$\frac{di_a(t)}{dt} = -\frac{R_a}{L_a}i_a(t) - \frac{K_b}{L_a}\omega_m(t) + \frac{1}{L_a}e_a(t) \tag{5-164}$$

$$\frac{d\omega_m(t)}{dt} = \frac{K_i}{J_{me}}i_a(t) - \frac{B_{me}}{J_{me}}\omega_m(t) - \frac{nr}{J_{me}}T(t) \tag{5-165}$$

$$\frac{dT(t)}{dt} = Cnr\omega_m(t) - CV_s \tag{5-166}$$

5.11 Edge-Guide Control System

Whenever there is a tension-control problem there is a desire for edge-guide control. Therefore, the problem of edge guiding has become very important in modern paper manufacturing, steel strip processing lines, flexographic and textile industries, and similar fields. To maintain optimum sheet quality and maximum process line speed, the moving web must be maintained at a constant lateral-edge position. In general, there are many different ways of measuring and tracking the edge of a moving web. However, to achieve stable and accurate edge guiding, a feedback control system should be used.

The schematic diagram of an edge-guide system using the pivot-roll method is shown in Fig. 5-54. The pivot roll is controlled to rotate about the pivot point in guiding the direction of the web. The source of controlling the motion of the pivot roll may be a dc motor coupled to a lead screw or rack and pinion, or a linear actuator.

Figure 5-55 shows the side view of the edge-guide system. The axes of the rollers 1 and 2 are assumed to be fixed or uncontrolled. S_1 and S_2 represent sensors that are placed at the indicated points to sense the centering of the web at the respective points.

Let

$v(t) = $ linear velocity of web
$z_R(t) = $ initial error of web position in the z direction at roll 1
$z_1(t) = $ error of web position in the z direction at the leading side of the pivot roll (see Figs. 5-54 and 5-55)

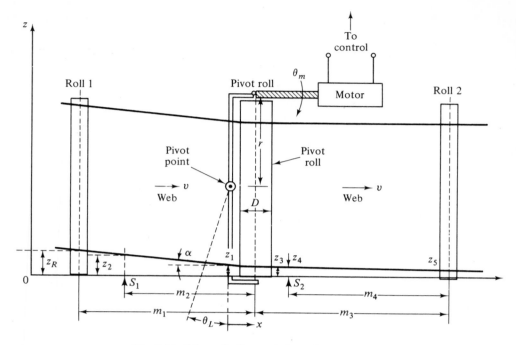

Fig. 5-54. Schematic diagram of an edge-guide control system.

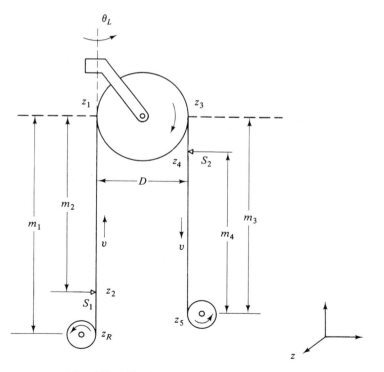

Fig. 5-55. Side view of an edge-guide system.

Assuming that there is no slippage when the web moves over the pivot roll, the following equation is written for $v(t)$ and $z_1(t)$:

$$\frac{dz_1(t)}{dt} = v(t)\tan\alpha \qquad (5\text{-}167)$$

If the angle α is small, from Fig. 5-54 it is seen that

$$\tan\alpha \simeq \frac{z_R(t) - z_1(t)}{m_1} \qquad (5\text{-}168)$$

Thus Eq. (5-167) can be written approximately as

$$\frac{dz_1(t)}{dt} = v(t)\frac{z_R(t) - z_1(t)}{m_1} \qquad (5\text{-}169)$$

If the linear velocity of the web is constant, $v(t) = v$, and Eq. (5-169) is written

$$\frac{dz_1(t)}{dt} + \frac{v}{m_1}z_1(t) = \frac{v}{m_1}z_R(t) \qquad (5\text{-}170)$$

Taking the Laplace transform on both sides of Eq. (5-170), the transfer relation between $z_1(t)$ and $z_R(t)$ is

$$\frac{Z_1(s)}{Z_R(s)} = \frac{1}{1 + \tau_1 s} \qquad (5\text{-}171)$$

where $\tau_1 = m_1/v$.

Assuming that there is no stretching in the web, from Fig. 5-55,

$$\frac{z_2(t) - z_1(t)}{m_2} = \frac{z_R(t) - z_1(t)}{m_1} \qquad (5\text{-}172)$$

or

$$z_2(t) = z_1(t) + \frac{m_2}{m_1}[z_R(t) - z_1(t)] \qquad (5\text{-}173)$$

Taking the Laplace transform on both sides of Eq. (5-173) and solving for $Z_R(s)$, we have

$$Z_R(s) = \frac{1 + \tau_1 s}{1 + (m_2/m_1)\tau_1 s}Z_2(s) \qquad (5\text{-}174)$$

Substitution of $Z_R(s)$ from Eq. (174) into Eq. (171) gives the transfer relation between $Z_1(s)$ and $Z_2(s)$,

$$\frac{Z_1(s)}{Z_2(s)} = \frac{1}{1 + (m_2/m_1)\tau_1 s} \qquad (5\text{-}175)$$

When the pivot roll is rotated by an angle θ_L from its reference position, the error $z_3(t)$ will be affected by approximately $D\sin\theta_L(t)$. With no slippage of the web over the pivot roll, the error $z_3(t)$ due to the error $z_1(t)$ is written

$$z_3(t) = z_1(t - T) - D\theta_L(t) \qquad (5\text{-}176)$$

where $T = \pi D/2v$, and for small $\theta_L(t)$, $\sin\theta_L(t)$ is approximated by $\theta_L(t)$. Take the inverse Laplace transform on both sides of Eq. (5-176) which yields

$$Z_3(s) = e^{-Ts}Z_1(s) - D\theta_L(s) \qquad (5\text{-}177)$$

Similar to the relationship between z_R and z_1, the transfer function between

$z_3(t)$ and $z_5(t)$ is

$$\frac{Z_5(s)}{Z_3(s)} = \frac{1}{1 + \tau_3 s} \tag{5-178}$$

where $\tau_3 = m_3/v$. Also, in analogy to Eq. (5-174), the transfer relation between z_3 and z_4 is

$$\frac{Z_4(s)}{Z_3(s)} = \frac{1 + (m_4/m_3)\tau_3 s}{1 + \tau_3 s} \tag{5-179}$$

Now consider that the drive motor of the system is an armature-controlled dc motor with negligible inductance. The equation of the motor armature is

$$I_a(s) = \frac{E_a(s) - sK_b\theta_m(s)}{R_a} \tag{5-180}$$

where $I_a(s)$ is the armature current, K_b the back emf constant, R_a the armature resistance, and $\theta_m(s)$ the motor displacement. The torque is given by

$$T_m(s) = K_i I_a(s) \tag{5-181}$$

where K_i is the torque constant.

The torque equation of the motor and load is

$$T_m(s) = (J_m s^2 + B_m s)\theta_m(s) + LF(s) \tag{5-182}$$

where

$$J_m = \text{inertia of motor}$$
$$B_m = \text{viscous frictional coefficient of motor}$$
$$L = \text{lead of screw}$$
$$F(s) = \text{transform of force in the } x \text{ direction}$$

and

$$F(s) = \frac{1}{2\pi r}(J_L s^2 + B_L s + K_L)\theta_L(s) \tag{5-183}$$

where

$$J_L = \text{inertia of pivot roll about pivot point}$$
$$B_L = \text{viscous frictional coefficient at pivot point}$$
$$K_L = \text{spring constant at pivot roll due to tension of web}$$

Combining Eqs. (5-182) and (5-183), we have

$$\frac{\theta_m(s)}{T_m(s)} = \frac{1}{J_{me} s^2 + B_{me} s + K_{me}} \tag{5-184}$$

where

$$J_{me} = J_m + \left(\frac{L}{2\pi r}\right)^2 J_L \tag{5-185}$$

$$B_{me} = B_m + \left(\frac{L}{2\pi r}\right)^2 B_L \tag{5-186}$$

$$K_{me} = \left(\frac{L}{2\pi r}\right)^2 K_L \tag{5-187}$$

Also,

$$X(s) = r\theta_L(s) \tag{5-188}$$

$$\theta_L(s) = \frac{L}{2\pi r}\theta_m(s) \tag{5-189}$$

A block diagram of the overall system is drawn as shown in Fig. 5-56 using

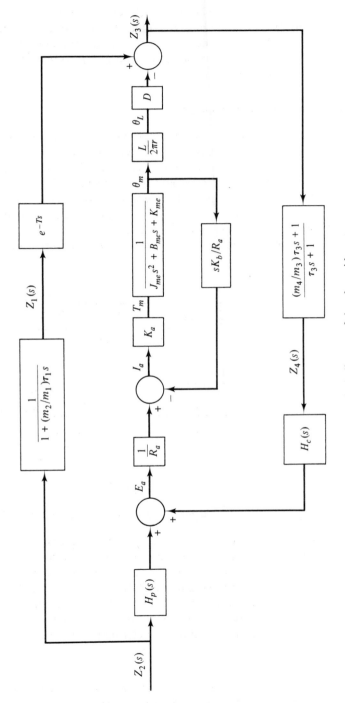

Fig. 5-56. Block diagram of the edge-guide system.

Eqs. (5-175), (5-177), (5-179), (5-180), (5-181), (5-184), and (5-189). The blocks with transfer functions $H_p(s)$ and $H_c(s)$ represent possible locations of controllers of the edge-guide system. The design problem may involve the determination of the transfer functions $H_p(s)$ and $H_c(s)$, so that for a given error z_2 the error z_3 is minimized.

5.12 Systems with Transportation Lags

Thus far a great majority of the systems considered have transfer functions that are quotients of polynomials. However, in the edge-guide control system of Section 5.11, the relation between the variables $z_1(t)$ and $z_3(t)$ is that of a pure time delay. Then $Z_1(s)$ and $Z_3(s)$ are related through an exponential transfer function e^{-Ts}. In general, pure time delays may be encountered in various types of systems, especially systems with hydraulic, pneumatic, or mechanical transmissions. In these systems the output will not begin to respond to an input until after a given time interval. Figure 5-57 illustrates examples in which transporta-

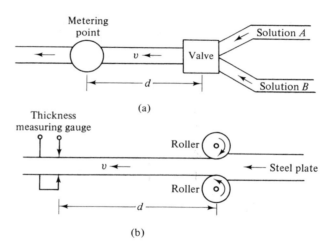

Fig. 5-57. Physical systems with transportation lags.

tion lags are observed. Figure 5-57(a) outlines an arrangement in which two different fluids are to be mixed in appropriate proportions. To assure that a homogeneous solution is measured, the monitoring point is located some distance from the mixing point. A transportation lag therefore exists between the mixing point and the place where the change in concentration is detected. If the rate of flow of the mixed solution is v inches per second and d is the distance between the mixing and the metering points, the time lag is given by

$$T = \frac{d}{v} \qquad \text{sec} \qquad (5\text{-}190)$$

If it is assumed that the concentration at the mixing point is $c(t)$ and that it is reproduced without change T seconds later at the monitoring point, the measured quantity is

$$b(t) = c(t - T) \qquad (5\text{-}191)$$

The Laplace transform of the last equation is

$$B(s) = e^{-Ts}C(s) \qquad (5\text{-}192)$$

Thus the transfer function between $b(t)$ and $c(t)$ is

$$\frac{B(s)}{C(s)} = e^{-Ts} \qquad (5\text{-}193)$$

The arrangement shown in Fig. 5-57(b) may be thought of as a thickness control of the rolling of steel plates. As in the case above, the transfer function between the thickness at the rollers and the measuring point is given by Eq. (5-193).

Other examples of transportation lags are found in human beings as control systems where action and reaction are always accompanied by pure time delays. The operation of the sample-and-hold device of a sampled-data system closely resembles a pure time delay; it is sometimes approximated by a simple time-lag term, e^{-Ts}.

In terms of state variables, a system with pure time delay can no longer be described by the matrix-state equation

$$\frac{d\mathbf{x}(t)}{dt} = \mathbf{A}\mathbf{x}(t) + \mathbf{B}\mathbf{u}(t) \qquad (5\text{-}194)$$

A general state description of a system containing time lags is given by the following matrix differential-difference equation:

$$\frac{d\mathbf{x}(t)}{dt} = \sum_{i=1}^{p} \mathbf{A}_i\mathbf{x}(t - T_i) + \sum_{j=1}^{q} \mathbf{B}_j\mathbf{u}(t - T_j) \qquad (5\text{-}195)$$

where T_i and T_j are fixed time delays. In this case Eq. (5-195) represents a general situation where time delays may exist on both the inputs as well as the states.

5.13 Sun-Seeker System

In this section we shall model a sun-seeker control system whose purpose is to control the attitude of a space vehicle so that it will track the sun with high accuracy. In the system described, tracking of the sun in one plane is accomplished. A schematic diagram of the system is shown in Fig. 5-58. The principal elements of the error discriminator are two small rectangular silicon photovoltaic cells mounted behind a rectangular slit in an enclosure. The cells are mounted in such a way that when the sensor is pointed at the sun, the beam of light from the slit overlaps both cells. The silicon cells are used as current sources and connected in opposite polarity to the input of an operational amplifier. Any difference in the short-circuit current of the two cells is sensed and amplified by the operational amplifier. Since the current of each cell is proportional to the illumination on the cell, an error signal will be present at the output of the amplifier when the light from the slit is not precisely centered on the cells. This error

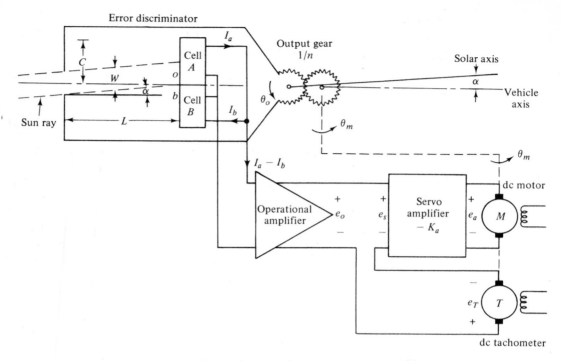

Fig. 5-58. Schematic diagram of a sun-seeker system.

voltage, when fed to the servoamplifier, will cause the servo to drive the system back into alignment. A description of each part of the system is given as follows.

Coordinate System

The center of the coordinate system is considered to be at the output gear of the system. The reference axis is taken to be the fixed frame of the dc motor, and all rotations are measured with respect to this axis. The solar axis or the line from the output gear to the sun makes an angle θ_r with respect to the reference axis, and θ_0 denotes the vehicle axis with respect to the reference axis. The objective of the control system is to maintain the error between θ_r and θ_0, α, near zero:

$$\alpha = \theta_r - \theta_0 \tag{5-196}$$

Figure 5-59 shows the coordinate system described.

Error Discriminator

When the vehicle is aligned perfectly with the sun, $\alpha = 0$, and $I_a = I_b = I$, or $I_a - I_b = 0$. From the geometry of the sun's rays and the photovoltaic cells shown in Fig. 5-58, we have

$$oa = \frac{W}{2} + L \tan \alpha \tag{5-197}$$

$$ob = \frac{W}{2} - L \tan \alpha \tag{5-198}$$

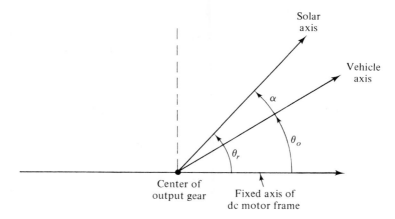

Fig. 5-59. Coordinate system of the sun-seeker system.

where *oa* denotes the width of the sun's ray that shines on cell *A*, and *ob* is the same on cell *B*, for a given α. Since the current I_a is proportional to *oa*, and I_b to *ob*, we have

$$I_a = I + \frac{2LI}{W} \tan \alpha \qquad (5\text{-}199)$$

and

$$I_b = I - \frac{2LI}{W} \tan \alpha \qquad (5\text{-}200)$$

for $0 \leq \tan \alpha \leq W/2L$. For $W/2L \leq \tan \alpha \leq (C - W/2)/L$, the sun's ray is completely on cell *A*, and $I_a = 2I$, $I_b = 0$. For $(C - W/2)/L \leq \tan \alpha \leq (C + W/2)/L$, I_a decreases linearly from $2I$ to zero. $I_a = I_b = 0$ for $\tan \alpha \geq (C + W/2)/L$. Therefore, the error discriminator may be represented by the nonlinear characteristic of Fig. 5-60, where for small angle α, $\tan \alpha$ has been approximated by α on the abscissa.

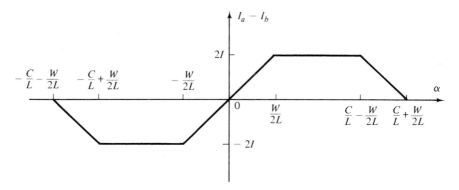

Fig. 5-60. Nonlinear characteristic of the error discriminator. The abscissa is $\tan \alpha$ but is approximated by α for small values of α.

Operational Amplifier

The schematic diagram of the operational amplifier is shown in Fig. 5-61. Summing the currents at point G and assuming that the amplifier does not draw any current, we have

$$I_a - I_b - \frac{e_g}{R} + \frac{e_0 - e_g}{R_F} = 0 \qquad (5\text{-}201)$$

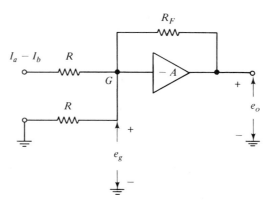

Fig. 5-61. Operational amplifier.

Since $e_0 = -Ae_g$, $e_g = -e_0/A$, Eq. (5-201) becomes

$$I_a - I_b + \left(\frac{1}{AR} + \frac{1}{R_F} + \frac{1}{AR_F}\right)e_0 = 0 \qquad (5\text{-}202)$$

If A approaches infinity, as in operational amplifiers, A may reach 10^6; then Eq. (5-202) is written

$$e_0 = -R_F(I_a - I_b) \qquad (5\text{-}203)$$

This equation gives the transfer relationship between $I_a - I_b$ and e_0.

Servoamplifier

The gain of the servoamplifier is $-K_a$. With reference to Fig. 5-58, the output of the servo amplifier is expressed as

$$\begin{aligned} e_a &= -K_a(e_0 + e_T) \\ &= -K_a e_s \end{aligned} \qquad (5\text{-}204)$$

Tachometer

The output voltage of the tachometer e_T is related to the angular velocity of the motor through the tachometer constant K_T; that is,

$$e_T = K_T \omega_m \qquad (5\text{-}205)$$

The angular position of the output gear is related to the motor position through the gear ratio $1/n$. Thus

$$\theta_0 = \frac{1}{n}\theta_m \qquad (5\text{-}206)$$

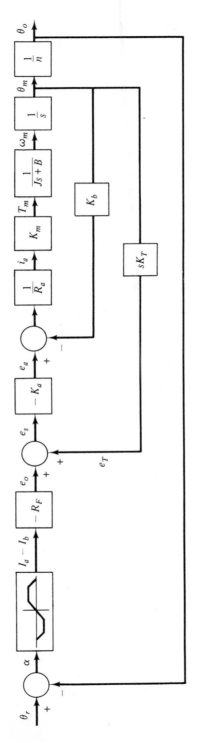

Fig. 5-62. Block diagram of the sun-seeker system.

Armature-Controlled DC Motor

The armature-controlled dc motor has been modeled earlier. The equations are

$$e_a = R_a i_a + e_b \tag{5-207}$$

$$e_b = K_b \omega_m \tag{5-208}$$

$$T_m = K_m i_a \tag{5-209}$$

$$T_m = J \frac{d\omega_m}{dt} + B\omega_m \tag{5-210}$$

where J and B are the inertia and viscous friction coefficient seen at the motor shaft. The inductance in the motor armature is neglected in Eq. (5-207).

A block diagram that characterizes all the functional relations of the system is shown in Fig. 5-62.

REFERENCES

State-Variable Analysis of Electric Networks

1. B. C. Kuo, *Linear Circuits and Systems*, McGraw-Hill Book Company, New York, 1967.

2. R. Rohrer, *Circuit Analysis: An Introduction to the State Variable Approach*, McGraw-Hill Book Company, New York, 1970.

Mechanical Systems

3. R. Cannon, *Dynamics of Physical Systems*, McGraw-Hill Book Company, New York, 1967.

Control System Components

4. W. R. Ahrendt, *Servomechanism Practice*, McGraw-Hill Book Company, New York, 1954.

5. J. E. Gibson and F. B. Tuteur, *Control System Components*, McGraw-Hill Book Company, New York, 1958.

Two-Phase Induction Motor

6. W. A. Stein and G. J. Thaler, "Effect of Nonlinearity in a 2-Phase Servomotor," *AIEE Trans.*, Vol. 73, Part II, pp. 518–521, 1954.

7. B. C. Kuo, "Studying the Two-Phase Servomotor," *Instrument Soc. Amer. J.*, Vol. 7, No. 4, pp. 64–65, Apr. 1960.

Step Motors

8. B. C. Kuo (ed.), *Proceedings, Symposium on Incremental Motion Control Systems and Devices, Part I, Step Motors and Controls*, University of Illinois, Urbana, Ill., 1972.

9. B. C. Kuo (ed.), *Proceedings, Second Annual Symposium on Incremental Motion Control Systems and Devices*, University of Illinois, Urbana, Ill., 1973.

10. B. C. Kuo, *Theory and Applications of Step Motors*, West Publishing Company, St. Paul, Minn., 1974.

PROBLEMS

5.1. Write the force or torque equations for the mechanical systems shown in Fig. P5-1. Write the state equations from the force or torque equations.

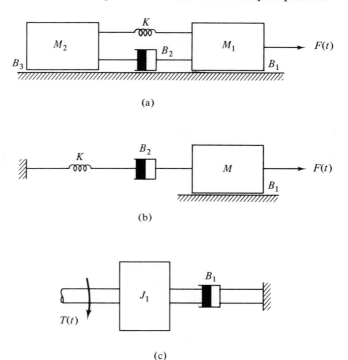

(a)

(b)

(c)

Figure P5-1.

5.2. Write a set of state equations for the mechanical system shown in Fig. P5-2. On the first try, one will probably end up with four state equations with the state variables defined as θ_2, ω_2, θ_1, and ω_1. However, it is apparent that there are only three energy-storage elements in J_1, K, and J_2, so the system has a minimum order of three.

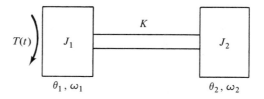

Figure P5-2.

(a) Write the state equations in vector-matrix form with the state variables defined as above.
(b) Redefine the state variables so that there are only three state equations.
(c) Draw state diagrams for both cases.
(d) Derive the transfer function $\omega_2(s)/T(s)$ for each case, and compare the results.
(e) Determine the controllability of the system. Does the fact that the system can be modeled by three state equations mean that the four-state model is uncontrollable? Explain.

5.3. For the system shown in Fig. P5-3, determine the transfer function $E_0(s)/T_m(s)$. The potentiometer rotates through 10 turns, and the voltage applied across the potentiometer terminals is E volts.

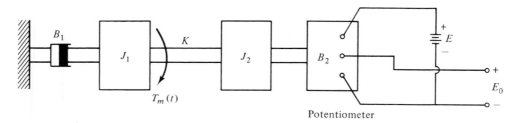

B_2 = viscous friction coefficient of potentiometer contact

Figure P5-3.

5.4. Write the torque equations of the gear-train system shown in Fig. P5-4. The moments of inertia of the gears and shafts are J_1, J_2, and J_3. $T(t)$ is the applied torque. N denotes the number of gear teeth. Assume rigid shafts.

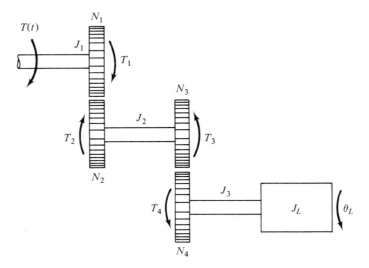

Figure P5-4.

5.5. Figure P5-5 shows the diagram of an epicyclic gear train.

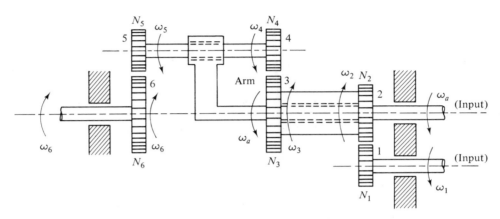

Figure P5-5.

(a) Using the reference directions of the angular velocity variables as indicated, write algebraic equations that relate these variables.

(b) Draw a signal flow graph to relate among the inputs ω_a and ω_1 and the output ω_6.

(c) Find the transfer function relation between ω_6 and ω_1 and ω_a.

5.6. The block diagram of the automatic braking control of a high-speed train is shown in Fig. P5-6a, where

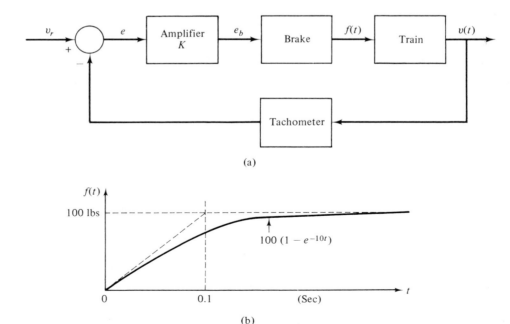

(a)

(b)

Figure P5-6.

V_r = voltage representing desired speed
v = velocity of train
K = amplifier gain = 100
M = mass of train = 5×10^4 lb/ft/sec^2
K_t = tachometer constant = 0.15 volt/ft/sec
$e_t = K_t v$

The force characteristics of the brake are shown in Fig. P5-6b when $e_b = 1$ volt. (Neglect all frictional forces.)

(a) Draw a block diagram of the system and include the transfer function of each block.

(b) Determine the closed-loop transfer function between V_r and velocity v of the train.

(c) If the steady-state velocity of the train is to be maintained at 20 ft/sec, what should be the value of V_r?

5.7. Figure P5-7 illustrates a winding process of newsprint. The system parameters and variables are defined as follows:

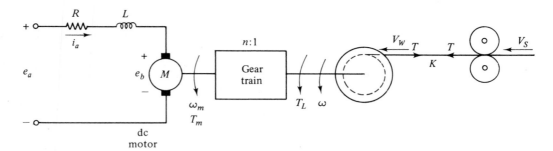

Figure P5-7.

e_a = applied voltage
R = armature resistance of dc motor
L = armature inductance of dc motor
i_a = armature current
K_b = back emf of dc motor
T_m = motor torque = $K_m i_a$
J_m = motor inertia
B_m = motor friction coefficient
J_L = inertia of windup reel
ω_m = angular velocity of dc motor
ω = angular velocity of windup reel
T_L = torque at the windup reel
r = effective radius of windup reel
V_w = linear velocity of web at windup reel
T = tension
V_s = linear velocity of web at input pinch rolls

Assume that the linear velocity at the input pinch rolls, V_s, is constant. The elasticity of the web material is assumed to satisfy Hooke's law; that is, the distortion of the material is directly proportional to the force applied, and the proportional constant is K (force/displacement).

(a) Write the nonlinear state equations for the system using i_a, ω_m, and T as state variables.

(b) Assuming that r is constant, draw a state diagram of the system with e_a and V_s as inputs.

5.8. Write state equations and output equation for the edge-guide control system whose block diagram is shown in Fig. 5-56.

5.9. The schematic diagram of a steel rolling process is shown in Fig. P5-9.

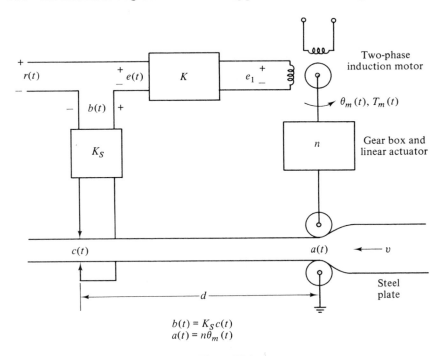

$$b(t) = K_S c(t)$$
$$a(t) = n\theta_m(t)$$

Figure P5-9.

(a) Describe the system by a set of differential-difference equation of the form of Eq. (5-195).

(b) Derive the transfer function between $c(t)$ and $r(t)$.

5.10. Figure P5-10a shows an industrial process in which a dc motor drives a capstan and tape assembly. The objective is to drive the tape at a certain constant speed. Another tape driven by a separate source is made to be in contact with the primary tape by the action of a pinch roll over certain periods of time. When the two tapes are in contact, we may consider that a constant frictional torque of T_F is seen at the load. The following system parameters are defined:

e_t = applied motor voltage, volts
i_a = armature current, amperes
e_b = back emf voltage = $K_b\omega_m$, volts
K_b = back emf constant = 0.052 volt/rad/sec
K_m = torque constant = 10 oz-in./amp
T_m = torque, oz-in.

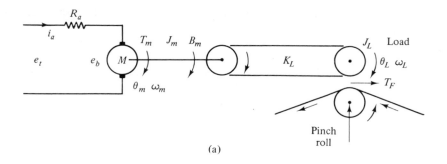

(a)

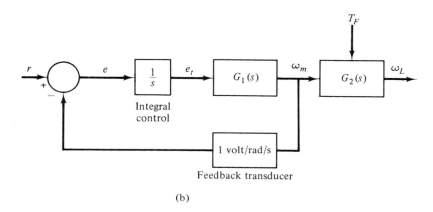

(b)

Figure P5-10.

θ_m = motor displacement, rad
ω_m = motor speed, rad/sec
R_a = motor resistance = 1 Ω
J_m = motor inertia = 0.1 oz-in./rad/sec² (includes capstan inertia)
B_m = motor viscous friction = 0.1 oz-in./rad/sec
K_L = spring constant of tape = 100 oz-in./rad (converted to rotational)
J_L = load inertia = 0.1 oz-in./rad/sec²

(a) Write the equations of the system in vector-matrix state equation form.
(b) Draw a state diagram for the system.
(c) Derive the transfer function for $\omega_L(s)$ with $E_t(s)$ and $T_F(s)$ as inputs.
(d) If a constant voltage $e_t(t) = 10$ V is applied to the motor, find the steady-state speed of the motor in rpm when the pinch roll is not activated. What is the steady-state speed of the load?
(e) When the pinch roll is activated, making the two tapes in contact, the constant friction torque T_F is 1 oz-in. Find the change in the steady-state speed ω_L when $e_t = 10$ V.
(f) To overcome the effect of the frictional torque T_F it is suggested that a closed-loop system should be formed as shown by the block diagram in

Fig. P5-10(b). In this case the motor speed is fed back and compared with the reference input. The closed-loop system should give accurate speed control, and the integral control should give better regulation to the frictional torque. Draw a state diagram for the closed-loop system.

(g) Determine the steady-state speed of the load when the input is 1 V. First consider that the pinch roll is not activated, and then is activated.

5.11. This problem deals with the attitude control of a guided missile. When traveling through the atmosphere, a missile encounters aerodynamic forces that usually tend to cause instability in the attitude of the missile. The basic concern from the flight control standpoint is the lateral force of the air, which tends to rotate the missile about its center of gravity. If the missile centerline is not aligned with the direction in which the center of gravity C is traveling, as shown in Fig. P5-11

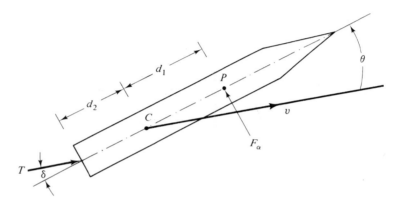

Figure P5-11.

with the angle θ (θ is also called the angle of attack), a side force is produced by the resistance of the air through which the missile is traveling. The total force F_α may be considered to be centered at the center of pressure P. As shown in Fig. P5-11, this side force has a tendency to cause the missile to tumble, especially if the point P is in front of the center of gravity C. Let the angular acceleration of the missile about the point C, due to the side force, be denoted by α_F. Normally, α_F is directly proportional to the angle of attack θ and is given by

$$\alpha_F = a\theta$$

where a is a constant described by

$$a = \frac{K_F d_1}{J}$$

K_F is a constant that depends on such parameters as dynamic pressure, velocity of the missile, air density, and so on, and

$$J = \text{missile moment of inertia about } C$$

$$d_1 = \text{distance between } C \text{ and } P$$

The main object of the flight control system is to provide the stabilizing action to counter the effect of the side force. One of the standard control means is to use gas injection at the tail of the missile to deflect the direction of the rocket engine thrust T, as shown in Fig. P5-11.

(a) Write a torque differential equation to relate among T, δ, θ, and the system parameters. Assume that δ is very small.

(b) Assume that T is constant and find the transfer function $\theta(s)/\delta(s)$ for small δ.

(c) Repeat (a) and (b) with the points C and P interchanged.

5.12. (a) Draw a state diagram for the tension-control system of Fig. 5-53, using the state equations of Eqs. (5-164), (5-165), and (5-166).

(b) Write the relation among $E_a(s)$, V_s, and $T(s)$, with $E_a(s)$ and V_s as inputs and $T(s)$ as the output.

5.13. The following equations describe the motion of an electric train in a traction system:

$$\dot{x}(t) = v(t)$$

$$\dot{v}(t) = -k(v) - g(x) + T(t)$$

where

$x(t) = $ linear displacement of train
$v(t) = $ linear velocity of train
$k(v) = $ train resistance force [odd function of v, with the properties $k(0) = 0$ and $dk(v)/dv > 0$]
$g(x) = $ force due to gravity for a nonlevel track or due to curvature of track
$T(t) = $ tractive force

The electric motor that provides the traction force is described by the following relations:

$$e(t) = K_b \phi(t) v(t) + R i_a(t)$$

$$T(t) = K_m \phi(t) i_a(t)$$

where

$R = $ armature resistance
$\phi(t) = $ magnetic flux $= K_f i_f(t)$
$e(t) = $ applied voltage
$K_m, K_b = $ proportional constants
$i_a(t) = $ armature current
$i_f(t) = $ field current

(a) Consider that the motor is a *dc* series motor so that $i_a(t) = i(t)$; $g(x) = 0$, $k(v) = Bv(t)$, and $R = 0$. The voltage $e(t)$ is the input. Show that the system is described by the following set of nonlinear state equations:

$$\dot{x}(t) = v(t)$$

$$\dot{v}(t) = -Bv(t) + \frac{K_m}{K_b^2 K_f v^2(t)} e^2(t)$$

(b) Consider that $i_a(t) = i_f(t)$ is the input and derive the state equations of the system. $g(x) = 0$, $k(v) = Bv(t)$.

(c) Consider that $\phi(t)$ is the input, $g(x) = 0$, $k(v) = Bv(t)$, and derive the state equations of the system.

5.14. Figure P5-14 shows a gear-coupled mechanical system.

(a) Find the optimum gear ratio n such that the load acceleration, α_L, is maximized.

(b) Repeat part (a) when the load drag torque T_L is zero.

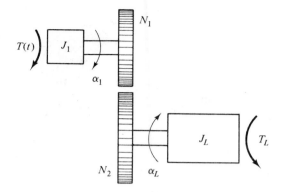

Figure P5-14.

5.15. (a) Write the torque equations of the system in Fig. P5-15 in the form

$$\ddot{\theta} + J^{-1}K\theta = 0$$

where θ is a 3×1 vector that contains all the displacement variables, θ_1, θ_2, and θ_3. J is the inertia matrix and K contains all the spring constants. Determine J and K.

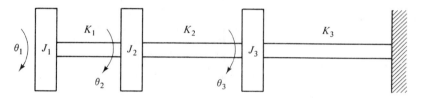

Figure P5-15.

(b) Show that the torque equations can be expressed as a set of state equations of the form

$$\dot{x} = Ax$$

where

$$A = \left[\begin{array}{c|c} 0 & I \\ \hline -J^{-1}K & 0 \end{array} \right]$$

(c) Consider the following set of parameters with consistent units: $K_1 = 1000$, $K_2 = 3000$, $J_1 = 1$, $J_2 = 5$, $J_3 = 2$, and $K_3 = 1000$. Find the matrix A.

5.16. Figure P5-16 shows the layout of the control of the unwind process of a cable reel with the object of maintaining constant linear cable velocity. Control is established by measuring the cable velocity, comparing it with a reference signal, and using the error to generate a control signal. A tachometer is used to sense the cable velocity. To maintain a constant linear cable velocity, the angular reel velocity $\dot{\theta}_R$ must increase as the cable unwinds; that is, as the effective radius of the reel decreases. Let

$$D = \text{cable diameter} = 0.1 \text{ ft}$$
$$W = \text{width of reel} = 2 \text{ ft}$$
$$R_0 = \text{effective radius of reel (empty reel)} = 2 \text{ ft}$$

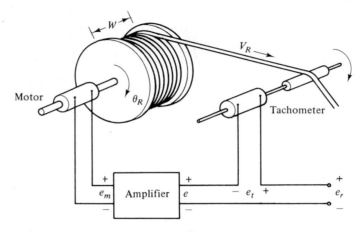

Figure P5-16.

$$R_f = \text{effective radius of reel (full reel)} = 4 \text{ ft}$$
$$R = \text{effective radius of reel}$$
$$J_R = \text{moment of inertia of reel} = 18R^4 - 200 \text{ ft-lb-sec}^2$$
$$v_R = \text{linear speed of cable, ft/sec}$$
$$e_t = \text{output voltage of tachometer, volts}$$
$$T_m(t) = \text{motor torque, ft-lb}$$
$$e_m(t) = \text{motor input voltage, volts}$$
$$K = \text{amplifier gain}$$

Motor inertia and friction are negligible. The tachometer transfer function is

$$\frac{E_t(s)}{V_R(s)} = \frac{1}{1 + 0.5s}$$

and the motor transfer function is

$$\frac{T_m(s)}{E_m(s)} = \frac{50}{s + 1}$$

(a) Write an expression to describe the change of the radius of the reel R as a function of θ_R.

(b) Between layers of the cable, R and J_R are assumed to be constant, and the system is considered linear. Draw a block diagram for the system and indicate all the transfer functions. The input is e_r and the output is v_R.

(c) Derive the closed-loop transfer function $V_R(s)/E_r(s)$.

6

Time-Domain Analysis

of Control Systems

6.1 Introduction

Since time is used as an independent variable in most control systems, it is usually of interest to evaluate the time response of the systems. In the analysis problem, a reference input signal is applied to a system, and the performance of the system is evaluated by studying the response in the time domain. For instance, if the objective of the control system is to have the output variable follow the input signal as closely as possible, it is necessary to compare the input and the output as functions of time.

The time response of a control system is usually divided into two parts: the *transient response* and the *steady-state response*. If $c(t)$ denotes a time response, then, in general, it may be written

$$c(t) = c_t(t) + c_{ss}(t) \tag{6-1}$$

where

$$c_t(t) = \text{transient response}$$

$$c_{ss}(t) = \text{steady-state response}$$

The definition of the steady state has not been entirely standardized. In circuit analysis it is sometimes useful to define a steady-state variable as being a constant with respect to time. In control systems applications, however, when a response has reached its steady state it can still vary with time. In control systems the steady-state response is simply the fixed response when time reaches infinity. Therefore, a sine wave is considered as a steady-state response because its behavior is fixed for any time interval, as when time approaches infinity. Similarly, if a response is described by $c(t) = t$, it may be defined as a steady-state response.

Transient response is defined as the part of the response that goes to zero as time becomes large. Therefore, $c_t(t)$ has the property of

$$\lim_{t \to \infty} c_t(t) = 0 \qquad (6\text{-}2)$$

It can also be stated that the steady-state response is that part of the response which remains after the transient has died out.

All control systems exhibit transient phenomenon to some extent before a steady state is reached. Since inertia, mass, and inductance cannot be avoided in physical systems, the responses cannot follow sudden changes in the input instantaneously, and transients are usually observed.

The transient response of a control system is of importance, since it is part of the dynamic behavior of the system; and the deviation between the response and the input or the desired response, before the steady state is reached, must be closely watched. The steady-state response, when compared with the input, gives an indication of the final accuracy of the system. If the steady-state response of the output does not agree with the steady state of the input exactly, the system is said to have a *steady-state error*.

6.2 Typical Test Signals for Time Response of Control Systems

Unlike many electrical circuits and communication systems, the input excitations to many practical control systems are not known ahead of time. In many cases, the actual inputs of a control system may vary in random fashions with respect to time. For instance, in a radar tracking system, the position and speed of the target to be tracked may vary in an unpredictable manner, so that they cannot be expressed deterministically by a mathematical expression. This poses a problem for the designer, since it is difficult to design the control system so that it will perform satisfactorily to any input signal. For the purposes of analysis and design, it is necessary to assume some basic types of input functions so that the performance of a system can be evaluated with respect to these test signals. By selecting these basic test signals properly, not only the mathematical treatment of the problem is systematized, but the responses due to these inputs allow the prediction of the system's performance to other more complex inputs. In a design problem, performance criteria may be specified with respect to these test signals so that a system may be designed to meet the criteria.

When the response of a linear time-invariant system is analyzed in the frequency domain, a sinusoidal input with variable frequency is used. When the input frequency is swept from zero to beyond the significant range of the system characteristics, curves in terms of the amplitude ratio and phase between input and output are drawn as functions of frequency. It is possible to predict the time-domain behavior of the system from its frequency-domain characteristics.

To facilitate the time-domain analysis, the following deterministic test signals are often used.

Step input function. The step input function represents an instantaneous change in the reference input variable. For example, if the input is the angular position of a mechanical shaft, the step input represents the sudden rotation of the shaft. The mathematical representation of a step function is

$$r(t) = \begin{cases} R & t > 0 \\ 0 & t = 0 \end{cases} \qquad (6\text{-}3)$$

where R is a constant. Or

$$r(t) = Ru_s(t) \qquad (6\text{-}4)$$

where $u_s(t)$ is the unit step function. The step function is not defined at $t = 0$. The step function as a function of time is shown in Fig. 6-1(a).

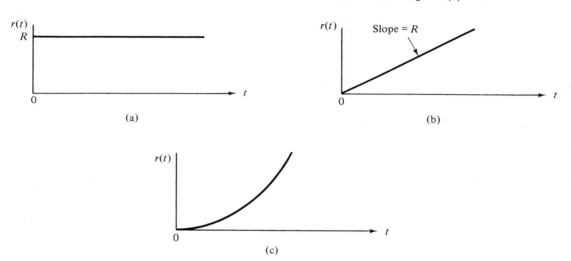

Fig. 6-1. Basic time-domain test signals for control systems. (a) Step function input, $r(t) = Ru_s(t)$. (b) Ramp function input, $r(t) = Rtu_s(t)$. (c) Parabolic function input, $r(t) = Rt^2u_s(t)$.

Ramp input function. In the case of the ramp function, the signal is considered to have a constant change in value with respect to time. Mathematically, a ramp function is represented by

$$r(t) = \begin{cases} Rt & t \geq 0 \\ 0 & t < 0 \end{cases} \qquad (6\text{-}5)$$

or simply

$$r(t) = Rtu_s(t) \qquad (6\text{-}6)$$

The ramp function is shown in Fig. 6-1(b). If the input variable is of the form of the angular displacement of a shaft, the ramp input represents the constant-speed rotation of the shaft.

Parabolic input function. The mathematical representation of a parabolic input function is

$$r(t) = \begin{cases} Rt^2 & t \geq 0 \\ 0 & t < 0 \end{cases} \qquad (6\text{-}7)$$

or simply

$$r(t) = Rt^2 u_s(t) \tag{6-8}$$

The graphical representation of the parabolic function is shown in Fig. 6-1(c).

These test signals all have the common feature that they are simple to describe mathematically, and from the step function to the parabolic function they become progressively faster with respect to time. The step function is very useful as a test signal since its initial instantaneous jump in amplitude reveals a great deal about a system's quickness to respond. Also, since the step function has, in principle, a wide band of frequencies in its spectrum, as a result of the jump discontinuity, as a test signal it is equivalent to the application of numerous sinusoidal signals with a wide range of frequencies.

The ramp function has the ability to test how the system would respond to a signal that changes linearly with time. A parabolic function is one degree faster than a ramp function. In practice, we seldom find it necessary to use a test signal faster than a parabolic function. This is because, as we shall show later, to track or follow a high-order input, the system is necessarily of high order, which may mean that stability problems will be encountered.

6.3 Time-Domain Performance of Control Systems—Steady-State Response

In this section we shall discuss the typical criteria used for the measurement of the time-domain performance of a control system. The time response of a control system may be characterized by the transient response and the steady-state response or, as an alternative, by a performance index that gives a qualitative measure on the time response as a whole. These criteria will be discussed in the following.

Steady-State Error

It was mentioned earlier that the steady-state error is a measure of system accuracy when a specific type of input is applied to a control system. In a physical system, because of friction and the nature of the particular system, the steady state of the output response seldom agrees exactly with the reference input. Therefore, steady-state errors in control systems are almost unavoidable, and in a design problem one of the objectives is to keep the error to a minimum or below a certain tolerable value. For instance, in a positional control system, it is desirable to have the final position of the output be in exact correspondence with the reference position. In a velocity-control system, the objective is to have the output velocity be as close as possible to the reference value.

If the reference input $r(t)$ and the controlled output $c(t)$ are dimensionally the same, for example, a voltage controlling a voltage, a position controlling a position, and so on, and are at the same level or of the same order of magnitude, the error signal is simply

$$e(t) = r(t) - c(t) \tag{6-9}$$

However, sometimes it may be impossible or inconvenient to provide a reference input that is at the same level or even of the same dimension as the controlled variable. For instance, it may be necessary to use a low-voltage source for the control of the output of a high-voltage power source; for a velocity-control system it is more practical to use a voltage source or position input to control the velocity of the output shaft. Under these conditions, the error signal cannot be defined simply as the difference between the reference input and the controlled output, and Eq. (6-9) becomes meaningless. The input and the output signals must be of the same dimension and at the same level before subtraction. Therefore, a nonunity element, $H(s)$, is usually incorporated in the feedback path, as shown in Fig. 6-2. The error of this nonunity-feedback

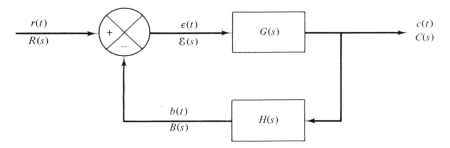

Fig. 6-2. Nonunity feedback control system.

control system is defined as

$$\epsilon(t) = r(t) - b(t) \tag{6-10}$$

or

$$\mathcal{E}(s) = R(s) - B(s) = R(s) - H(s)C(s) \tag{6-11}$$

For example, if a 10-v reference is used to regulate a 100-v voltage supply, H is a constant and is equal to 0.1. When the output voltage is exactly 100 v, the error signal is

$$\epsilon(t) = 10 - 0.1 \cdot 100 = 0 \tag{6-12}$$

As another example let us consider that the system shown in Fig. 6-2 is a velocity-control system in that the input $r(t)$ is used as a reference to control the output velocity of the system. Let $c(t)$ denote the output displacement. Then, we need a device such as a tachometer in the feedback path, so that $H(s) = K_t s$. Thus the error in velocity is defined as

$$\epsilon(t) = r(t) - b(t)$$
$$= r(t) - K_t \frac{dc(t)}{dt} \tag{6-13}$$

The error becomes zero when the output velocity $dc(t)/dt$ is equal to $r(t)/K_t$. The steady-state error of a feedback control system is defined as the error when time reaches infinity; that is,

$$\text{steady-state error} = e_{ss} = \lim_{t \to \infty} \epsilon(t) \tag{6-14}$$

With reference to Fig. 6-2, the Laplace-transformed error function is

$$\mathcal{E}(s) = \frac{R(s)}{1 + G(s)H(s)} \tag{6-15}$$

By use of the final-value theorem, the steady-state error of the system is

$$e_{ss} = \lim_{t \to \infty} e(t) = \lim_{s \to 0} s\mathcal{E}(s) \tag{6-16}$$

where $s\mathcal{E}(s)$ is to have no poles that lie on the imaginary axis and in the right half of the s-plane. Substituting Eq. (6-15) into Eq. (6-16), we have

$$e_{ss} = \lim_{s \to 0} \frac{sR(s)}{1 + G(s)H(s)} \tag{6-17}$$

which shows that the steady-state error depends on the reference input $R(s)$ and the loop transfer function $G(s)H(s)$.

Let us first establish the *type* of control system by referring to the form of $G(s)H(s)$. In general, $G(s)H(s)$ may be written

$$G(s)H(s) = \frac{K(1 + T_1 s)(1 + T_2 s) \ldots (1 + T_m s)}{s^j(1 + T_a s)(1 + T_b s) \ldots (1 + T_n s)} \tag{6-18}$$

where K and all the Ts are constants. The *type* of feedback control system refers to the *order* of the pole of $G(s)H(s)$ at $s = 0$. Therefore, the system that is described by the $G(s)H(s)$ of Eq. (6-18) is of type j, where $j = 0, 1, 2, \ldots$. The values of m, n, and the Ts are not important to the system type and do not affect the value of the steady-state error. For instance, a feedback control system with

$$G(s)H(s) = \frac{K(1 + 0.5s)}{s(1 + s)(1 + 2s)} \tag{6-19}$$

is of type 1, since $j = 1$.

Now let us consider the effects of the types of inputs on the steady-state error. We shall consider only the step, ramp, and parabolic inputs.

Steady-state error due to a step input. If the reference input to the control system of Fig. 6-2 is a step input of magnitude R, the Laplace transform of the input is R/s. Equation (6-17) becomes

$$e_{ss} = \lim_{s \to 0} \frac{sR(s)}{1 + G(s)H(s)} = \lim_{s \to 0} \frac{R}{1 + G(s)H(s)} = \frac{R}{1 + \lim_{s \to 0} G(s)H(s)} \tag{6-20}$$

For convenience we define

$$K_p = \lim_{s \to 0} G(s)H(s) \tag{6-21}$$

where K_p is the *positional error constant*. Then Eq. (6-20) is written

$$e_{ss} = \frac{R}{1 + K_p} \tag{6-22}$$

We see that for e_{ss} to be zero, when the input is a step function, K_p must be infinite. If $G(s)H(s)$ is described by Eq. (6-18), we see that for K_p to be infinite, j must be at least equal to unity; that is, $G(s)H(s)$ must have at least one pure integration. Therefore, we can summarize the steady-state error due to a step input as follows:

type 0 system: $e_{ss} = \dfrac{R}{1 + K_p} = \text{constant}$

type 1 (or higher) system: $e_{ss} = 0$

Steady-state error due to a ramp input. If the input to the control system of Fig. 6-2 is

$$r(t) = Rtu_s(t) \tag{6-23}$$

where R is a constant, the Laplace transform of $r(t)$ is

$$R(s) = \frac{R}{s^2} \tag{6-24}$$

Substituting Eq. (6-24) into Eq. (6-17), we have

$$e_{ss} = \lim_{s \to 0} \frac{R}{s + sG(s)H(s)} = \frac{R}{\lim\limits_{s \to 0} sG(s)H(s)} \tag{6-25}$$

If we define

$$K_v = \lim_{s \to 0} sG(s)H(s) = \text{velocity error constant} \tag{6-26}$$

Eq. (6-25) reads

$$e_{ss} = \frac{R}{K_v} \tag{6-27}$$

which is the steady-state error when the input is a ramp function. A typical e_{ss} due to a ramp input is shown in Fig. 6-3.

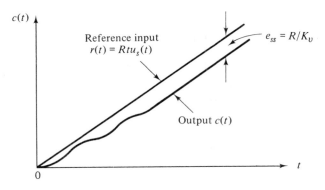

Fig. 6-3. Typical steady-state error due to a ramp input.

Equation (6-27) shows that for e_{ss} to be zero when the input is a ramp function, K_v must be infinite. Using Eq. (6-18) and (6-26),

$$K_v = \lim_{s \to 0} sG(s)H(s) = \lim_{s \to 0} \frac{K}{s^{j-1}} \qquad j = 0, 1, 2, \ldots \tag{6-28}$$

Therefore, in order for K_v to be infinite, j must be at least equal to two, or the system must be of type 2 or higher. The following conclusions may be stated with regard to the steady-state error of a system with ramp input:

type 0 system: $\qquad e_{ss} = \infty$

type 1 system: $\qquad e_{ss} = \dfrac{R}{K_v} = \text{constant}$

type 2 (or higher) system: $\quad e_{ss} = 0$

Steady-state error due to a parabolic input. If the input is described by

$$r(t) = \frac{Rt^2}{2} u_s(t) \tag{6-29}$$

the Laplace transform of $r(t)$ is

$$R(s) = \frac{R}{s^3} \tag{6-30}$$

The steady-state error of the system of Fig. 6-2 is

$$e_{ss} = \frac{R}{\lim_{s \to 0} s^2 G(s)H(s)} \tag{6-31}$$

Defining the *acceleration error constant* as

$$K_a = \lim_{s \to 0} s^2 G(s)H(s) \tag{6-32}$$

the steady-state error is

$$e_{ss} = \frac{R}{K_a} \tag{6-33}$$

The following conclusions can be made with regard to the steady-state error of a system with parabolic input:

type 0 system: $\qquad e_{ss} = \infty$

type 1 system: $\qquad e_{ss} = \infty$

type 2 system: $\qquad e_{ss} = \dfrac{R}{K_a} = \text{constant}$

type 3 (or higher) system: $\quad e_{ss} = 0$

As a summary of the error analysis, the relations among the error constants, the types of the system, and the input types are tabulated in Table 6-1. The transfer function of Eq. (6-18) is used as a reference.

Table 6-1   Summary of the Steady-State Errors Due to Step, Ramp, and Parabolic Inputs

Type of System j	K_p	K_v	K_a	Step Input, $e_{ss} = \dfrac{R}{1 + K_p}$	Ramp Input, $e_{ss} = \dfrac{R}{K_v}$	Parabolic Input, $e_{ss} = \dfrac{R}{K_a}$
0	K	0	0	$e_{ss} = \dfrac{R}{1 + K}$	$e_{ss} = \infty$	$e_{ss} = \infty$
1	∞	K	0	$e_{ss} = 0$	$e_{ss} = \dfrac{R}{K}$	$e_{ss} = \infty$
2	∞	∞	K	$e_{ss} = 0$	$e_{ss} = 0$	$e_{ss} = \dfrac{R}{K}$
3	∞	∞	∞	$e_{ss} = 0$	$e_{ss} = 0$	$e_{ss} = 0$

It should be noted that the position, velocity, and acceleration error constants are significant in the error analysis only when the input signal is a step function, a ramp function, and a parabolic function, respectively.

It should be noted further that the steady-state error analysis in this section is conducted by applying the final-value theorem to the error function, which is defined as the difference between the actual output and the desired output signal. In certain cases the error signal may be defined as the difference between the output and the reference input, whether or not the feedback element is unity. For instance, one may define the error signal for the system of Fig. 6-2 as

$$e(t) = r(t) - c(t) \tag{6-34}$$

Then

$$E(s) = \frac{1 + G(s)[H(s) - 1]}{1 + G(s)H(s)} R(s) \tag{6-35}$$

and

$$e_{ss} = \lim_{s \to 0} s \frac{1 + G(s)[H(s) - 1]}{1 + G(s)H(s)} R(s) \tag{6-36}$$

It should be kept in mind that since the steady-state error analysis discussed here relies on the use of the final-value theorem, it is important to first check to see if $sE(s)$ has any poles on the $j\omega$ axis or in the right half of the s-plane.

One of the disadvantages of the error constants is, of course, that they do not give information on the steady-state error when inputs are other than the three basic types mentioned. Another difficulty is that when the steady-state error is a function of time, the error constants give only an answer of infinity, and do not provide any information on how the error varies with time. We shall present the error series in the following section, which gives a more general representation of the steady-state error.

Error Series

In this section, the error-constant concept is generalized to include inputs of almost any arbitrary function of time. We start with the transformed error function of Eq. (6-15),

$$\mathcal{E}(s) = \frac{R(s)}{1 + G(s)H(s)} \tag{6-37}$$

or of Eq. (6-35), as the case may be.

Using the principle of the convolution integral as discussed in Section 3.3, the error signal $\epsilon(t)$ may be written

$$\epsilon(t) = \int_{-\infty}^{t} w_e(\tau) r(t - \tau) \, d\tau \tag{6-38}$$

where $w_e(\tau)$ is the inverse Laplace transform of

$$W_e(s) = \frac{1}{1 + G(s)H(s)} \tag{6-39}$$

which is known as the *error transfer function*.

If the first n derivatives of $r(t)$ exist for all values of t, the function $r(t - \tau)$

can be expanded into a Taylor series; that is,

$$r(t - \tau) = r(t) - \tau \dot{r}(t) + \frac{\tau^2}{2!}\ddot{r}(t) - \frac{\tau^3}{3!}\dddot{r}(t) + \ldots \qquad (6\text{-}40)$$

where $\dot{r}(t)$ represents the first derivative of $r(t)$ with respect to time.

Since $r(t)$ is considered to be zero for negative time, the limit of the convolution integral of Eq. (6-38) may be taken from 0 to t. Substituting Eq. (6-40) into Eq. (6-38), we have

$$\epsilon(t) = \int_0^t w_e(\tau)\left[r(t) - \tau \dot{r}(t) + \frac{\tau^2}{2!}\ddot{r}(t) - \frac{\tau^3}{3!}\dddot{r}(t) + \ldots\right] d\tau$$

$$= r(t)\int_0^t w_e(\tau)\, d\tau - \dot{r}(t)\int_0^t \tau w_e(\tau)\, d\tau + \ddot{r}(t)\int_0^t \frac{\tau^2}{2!}w_e(\tau)\, d\tau - \ldots \qquad (6\text{-}41)$$

As before, the steady-state error is obtained by taking the limit of $\epsilon(t)$ as t approaches infinity; thus

$$e_{ss} = \lim_{t \to \infty} \epsilon(t) = \lim_{t \to \infty} \epsilon_s(t) \qquad (6\text{-}42)$$

where $\epsilon_s(t)$ denotes the steady-state part of $\epsilon(t)$ and is given by

$$\epsilon_s(t) = r_s(t)\int_0^\infty w_e(\tau)\, d\tau - \dot{r}_s(t)\int_0^\infty \tau w_e(\tau)\, d\tau + \ddot{r}_s(t)\int_0^\infty \frac{\tau^2}{2!}w_e(\tau)\, d\tau$$

$$- \dddot{r}_s(t)\int_0^\infty \frac{\tau^3}{3!}w_e(\tau)\, d\tau + \ldots \qquad (6\text{-}43)$$

and $r_s(t)$ denotes the steady-state part of $r(t)$.

Let us define

$$C_0 = \int_0^\infty w_e(\tau)\, d\tau$$

$$C_1 = -\int_0^\infty \tau w_e(\tau)\, d\tau$$

$$C_2 = \int_0^\infty \tau^2 w_e(\tau)\, d\tau \qquad (6\text{-}44)$$

$$\vdots$$

$$C_n = (-1)^n \int_0^\infty \tau^n w_e(\tau)\, d\tau$$

Equation (6-42) is written

$$e_s(t) = C_0 r_s(t) + C_1 \dot{r}_s(t) + \frac{C_2}{2!}\ddot{r}_s(t) + \ldots + \frac{C_n}{n!}r_s^{(n)}(t) + \ldots \qquad (6\text{-}45)$$

which is called the *error series*, and the coefficients, $C_0, C_1, C_2, \ldots, C_n$ are defined as the *generalized error coefficients*, or simply as the *error coefficients*.

The error coefficients may be readily evaluated directly from the error transfer function, $W_e(s)$. Since $W_e(s)$ and $w_e(\tau)$ are related through the Laplace transform, we have

$$W_e(s) = \int_0^\infty w_e(\tau)e^{-\tau s}\, d\tau \qquad (6\text{-}46)$$

Taking the limit on both sides of Eq. (6-46) as s approaches zero, we have

$$\lim_{s \to 0} W_e(s) = \lim_{s \to 0} \int_0^\infty w_e(\tau) e^{-\tau s} \, d\tau$$
$$= C_0 \tag{6-47}$$

The derivative of $W_e(s)$ of Eq. (6-46) with respect to s gives

$$\frac{dW_e(s)}{ds} = -\int_0^\infty \tau w_e(\tau) e^{-\tau s} \, d\tau$$
$$= C_1 e^{-\tau s} \tag{6-48}$$

from which we get

$$C_1 = \lim_{s \to 0} \frac{dW_e(s)}{ds} \tag{6-49}$$

The rest of the error coefficients are obtained in a similar fashion by taking successive differentiation of Eq. (6-46) with respect to s. Therefore,

$$C_2 = \lim_{s \to 0} \frac{d^2 W_e(s)}{ds^2} \tag{6-50}$$

$$C_3 = \lim_{s \to 0} \frac{d^3 W_e(s)}{ds^3} \tag{6-51}$$

.
.
.

$$C_n = \lim_{s \to 0} \frac{d^n W_e(s)}{ds^n} \tag{6-52}$$

The following examples illustrate the general application of the error series and its advantages over the error constants.

EXAMPLE 6-1   In this illustrative example the steady-state error of a feedback control system will be evaluated by use of the error series and the error coefficients. Consider a unity feedback control system with the open-loop transfer function given as

$$G(s) = \frac{K}{s + 1} \tag{6-53}$$

Since the system is of type 0, the error constants are $K_p = K$, $K_v = 0$, and $K_a = 0$. Thus the steady-state errors of the system due to the three basic types of inputs are as follows:

unit step input, $u_s(t)$: $e_{ss} = \dfrac{1}{1 + K}$

unit ramp input, $tu_s(t)$: $e_{ss} = \infty$

unit parabolic input, $t^2 u_s(t)$: $e_{ss} = \infty$

Notice that when the input is either a ramp or a parabolic function, the steady-state error is infinite in magnitude, since it apparently increases with time. It is apparent that the error constants fail to indicate the exact manner in which the steady-state function increases with time. Therefore, ordinarily, if the steady-state response of this system due to a ramp or parabolic input is desired, the differential equation of the system must be solved. We now show that the steady-state response of the system can actually be determined from the error series.

Using Eq. (6-39), we have for this system

$$W_e(s) = \frac{1}{1 + G(s)} = \frac{s + 1}{s + K + 1} \tag{6-54}$$

The error coefficients are evaluated as

$$C_0 = \lim_{s \to 0} W_e(s) = \frac{1}{K + 1} \tag{6-55}$$

$$C_1 = \lim_{s \to 0} \frac{dW_e(s)}{ds} = \frac{K}{(1 + K)^2} \tag{6-56}$$

$$C_2 = \lim_{s \to 0} \frac{d^2 W_e(s)}{ds^2} = \frac{-2K}{(1 + K)^3} \tag{6-57}$$

Although higher-order coefficients can be obtained, they will become less significant as their values will be increasingly smaller. The error series is written

$$e_s(t) = \frac{1}{1 + K} r_s(t) + \frac{K}{(1 + K)^2} \dot{r}_s(t) + \frac{-K}{(1 + K)^3} \ddot{r}_s(t) + \cdots \tag{6-58}$$

Now let us consider the three basic types of inputs.

1. When the input signal is a unit step function, $r_s(t) = u_s(t)$, and all derivatives of $r_s(t)$ are zero. The error series gives

$$e_s(t) = \frac{1}{1 + K} \tag{6-59}$$

which agrees with the result given by the error-constant method.

2. When the input signal is a unit ramp function, $r_s(t) = tu_s(t)$, $\dot{r}_s(t) = u_s(t)$, and all higher-order derivatives of $\dot{r}_s(t)$ are zero. Therefore, the error series is

$$e_s(t) = \left[\frac{1}{1 + K} t + \frac{K}{(1 + K)^2} \right] u_s(t) \tag{6-60}$$

which indicates that the steady-state error increases linearly with time. The error-constant method simply yields the result that the steady-state error is infinite but fails to give details of the time dependence.

3. For a parabolic input, $r_s(t) = (t^2/2)u_s(t)$, $\dot{r}_s(t) = tu_s(t)$, $\ddot{r}_s(t) = u_s(t)$, and all higher derivatives are zero. The error series becomes

$$e_s(t) = \left[\frac{1}{1 + K} \frac{t^2}{2} + \frac{K}{(1 + K)^2} t - \frac{K}{(1 + K)^3} \right] u_s(t) \tag{6-61}$$

In this case the error increases in magnitude as the second power of t.

4. Consider that the input signal is represented by a polynomial of t and an exponential term,

$$r(t) = \left[a_0 + a_1 t + \frac{a_2 t^2}{2} + e^{-a_3 t} \right] u_s(t) \tag{6-62}$$

where a_0, a_1, a_2, and a_3 are constants. Then,

$$r_s(t) = \left[a_0 + a_1 t + \frac{a_2 t^2}{2} \right] u_s(t) \tag{6-63}$$

$$\dot{r}_s(t) = (a_1 + a_2 t) u_s(t) \tag{6-64}$$

$$\ddot{r}_s(t) = a_2 u_s(t) \tag{6-65}$$

In this case the error series becomes

$$e_s(t) = \frac{1}{1 + K} r_s(t) + \frac{K}{(1 + K)^2} \dot{r}_s(t) - \frac{K}{(1 + K)^3} \ddot{r}_s(t) \tag{6-66}$$

EXAMPLE 6-2   In this example we shall consider a situation in which the error constant is totally inadequate in providing a solution to the steady-state error. Let us consider that the input to the system described in Example 6-1 is a sinusoid,

$$r(t) = \sin \omega_0 t \qquad (6\text{-}67)$$

where $\omega_0 = 2$. Then

$$
\begin{aligned}
r_s(t) &= \sin \omega_0 t \\
\dot{r}_s(t) &= \omega_0 \cos \omega_0 t \\
\ddot{r}_s(t) &= -\omega_0^2 \sin \omega_0 t \\
\dddot{r}_s(t) &= -\omega_0^3 \cos \omega_0 t \\
&\;\;\vdots
\end{aligned}
\qquad (6\text{-}68)
$$

The error series can be written

$$e_s(t) = \left[C_0 - \frac{C_2}{2!}\omega_0^2 + \frac{C_4}{4!}\omega_0^4 - \ldots \right] \sin \omega_0 t + \left[C_1 \omega_0 - \frac{C_3}{3!}\omega_0^3 + \ldots \right] \cos \omega_0 t \qquad (6\text{-}69)$$

Because of the sinusoidal input, the error series is now an infinite series. The convergence of the series is important in arriving at a meaningful answer to the steady-state error. It is clear that the convergence of the error series depends on the value of ω_0 and K. Let us assign the value of K to be 100. Then

$$C_0 = \frac{1}{1 + K} = 0.0099$$

$$C_1 = \frac{K}{(1 + K)^2} = 0.0098$$

$$C_2 = -\frac{2K}{(1 + K)^3} = -0.000194$$

$$C_3 = \frac{6K}{(1 + K)^5} = 5.65 \times 10^{-8}$$

Thus, using only the first four error coefficients, Eq. (6-69) becomes

$$e_s(t) \simeq \left[0.0099 + \frac{0.000194}{2} \cdot 4 \right] \sin 2t + 0.0196 \cos 2t \qquad (6\text{-}70)$$

$$= 0.01029 \sin 2t + 0.0196 \cos 2t$$

or

$$e_s(t) \simeq 0.02215 \sin (2t + 62.3°) \qquad (6\text{-}71)$$

Therefore, the steady-state error in this case is also a sinusoid, as given by Eq. (6-71).

6.4 Time-Domain Performance of Control Systems—
Transient Response

The transient portion of the time response is that part which goes to zero as time becomes large. Of course, the transient response has significance only when a stable system is referred to, since for an unstable system the response does not diminish and is out of control.

The transient performance of a control system is usually characterized by the use of a unit step input. Typical performance criteria that are used to characterize the transient response to a unit step input include overshoot, delay time, rise time, and settling time. Figure 6-4 illustrates a typical unit step response of

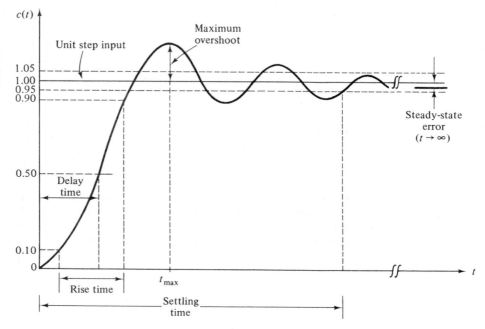

Fig. 6-4. Typical unit step response of a control system.

a linear control system. The above-mentioned criteria are defined with respect to the step response:

1. Maximum overshoot. The maximum overshoot is defined as the largest deviation of the output over the step input during the transient state. The amount of maximum overshoot is also used as a measure of the relative stability of the system. The maximum overshoot is often represented as a percentage of the final value of the step response; that is,

$$\text{per cent maximum overshoot} = \frac{\text{maximum overshoot}}{\text{final value}} \times 100\%$$

$$(6\text{-}72)$$

2. Delay time. The delay time T_d is defined as the time required for the step response to reach 50 per cent of its final value.

3. Rise time. The rise time T_r is defined as the time required for the step response to rise from 10 per cent to 90 per cent of its final value. Sometimes an alternative measure is to represent the rise time as a

reciprocal of the slope of the step response at the instant that the response is equal to 50 per cent of its final value.

4. Settling time. The settling time T_s is defined as the time required for the step response to decrease and stay within a specified percentage of its final value. A frequently used figure is 5 per cent.

The four quantities defined above give a direct measure of the transient characteristics of the step response. These quantities are relatively easy to measure when a step response is already plotted. However, analytically these quantities are difficult to determine except for the simple cases.

Performance Index

Since the general design objective of a control system is to have a small overshoot, fast rise time, short delay time, short settling time, and low steady-state error, it is advantageous to use a performance index that gives a measure of the overall quality of the response. Let us define the input signal of a system as $r(t)$ and the output as $c(t)$. The difference between the input and the output is defined as the error signal, as in Eq. (6-9). Sometimes $r(t)$ is referred to as the desired output.

In trying to minimize the error signal, time integrals of functions of the error signal may be used as performance indices. For example, the simplest integral function of the error is

$$I = \int_0^\infty e(t) \, dt \qquad (6\text{-}73)$$

where I is used to designate performance index. It is easy to see that Eq. (6-73) is not a practical performance index, since minimizing it is equivalent to minimizing the area under $e(t)$, and an oscillatory signal would yield a zero area and thus a zero I. Some of the practical integral performance indices are

$$\int_0^\infty |e(t)| \, dt \qquad \int_0^\infty te(t) \, dt \qquad \int_0^\infty e^2(t) \, dt$$

and there are many others. The subject of the design of control systems using performance indices is covered in Chapter 11.

6.5 Transient Response of a Second-Order System

Although true second-order control systems are rare in practice, their analysis generally helps to form a basis for the understanding of design and analysis techniques.

Consider that a second-order feedback control system is represented by the state diagram of Fig. 6-5. The state equations are written

$$\begin{bmatrix} \dot{x}_1(t) \\ \dot{x}_2(t) \end{bmatrix} = \begin{bmatrix} 0 & 1 \\ -\omega_n^2 & -2\zeta\omega_n \end{bmatrix} \begin{bmatrix} x_1(t) \\ x_2(t) \end{bmatrix} + \begin{bmatrix} 0 \\ 1 \end{bmatrix} r(t) \qquad (6\text{-}74)$$

where ζ and ω_n are constants.

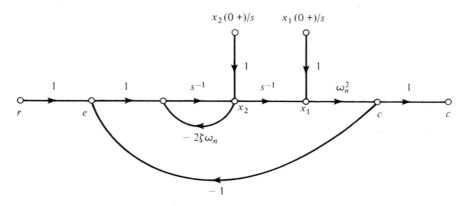

Fig. 6-5. State diagram of a second-order feedback control system.

The output equation is

$$c(t) = \omega_n^2 x_1(t) \tag{6-75}$$

Applying the gain formula to the state diagram of Fig. 6-5, the state transition equations are written

$$\begin{bmatrix} X_1(s) \\ X_2(s) \end{bmatrix} = \frac{1}{\Delta}\begin{bmatrix} s + 2\zeta\omega_n & 1 \\ -\omega_n^2 & s \end{bmatrix}\begin{bmatrix} x_1(0+) \\ x_2(0+) \end{bmatrix} + \frac{1}{\Delta}\begin{bmatrix} 1 \\ s \end{bmatrix}R(s) \tag{6-76}$$

where

$$\Delta = s^2 + 2\zeta\omega_n s + \omega_n^2 \tag{6-77}$$

The inverse Laplace transform of Eq. (6-76) is carried out with the help of the Laplace transform table. For a unit step function input, we have

$$\begin{bmatrix} x_1(t) \\ x_2(t) \end{bmatrix} = \frac{e^{-\zeta\omega_n t}}{\sqrt{1-\zeta^2}}\begin{bmatrix} \sin\left(\omega_n\sqrt{1-\zeta^2}\,t + \psi\right) & \frac{1}{\omega_n}\sin\omega_n\sqrt{1-\zeta^2}\,t \\ -\omega_n\sin\omega_n\sqrt{1-\zeta^2}\,t & \sin\left(\omega_n\sqrt{1-\zeta^2}\,t + \phi\right) \end{bmatrix}\begin{bmatrix} x_1(0+) \\ x_2(0+) \end{bmatrix}$$

$$+ \begin{bmatrix} \dfrac{1}{\omega_n^2}\left\{1 + \dfrac{1}{\sqrt{1-\zeta^2}}e^{-\zeta\omega_n t}\sin\left[\omega_n\sqrt{1-\zeta^2}\,t - \phi\right]\right\} \\ \dfrac{1}{\omega_n\sqrt{1-\zeta^2}}e^{-\zeta\omega_n t}\sin\omega_n\sqrt{1-\zeta^2}\,t \end{bmatrix} \quad t \geq 0 \tag{6-78}$$

where

$$\psi = \tan^{-1}\frac{\sqrt{1-\zeta^2}}{\zeta} \tag{6-79}$$

$$\phi = \tan^{-1}\frac{\sqrt{1-\zeta^2}}{-\zeta} \tag{6-80}$$

Although Eq. (6-78) gives the complete solution of the state variables in terms of the initial states and the unit step input, it is a rather formidable-looking expression, especially in view of the fact that the system is only of the second order. However, the analysis of control systems does not rely completely on the evaluation of the complete state and output responses. The development of linear control theory allows the study of control system performance by use of the transfer function and the characteristic equation. We shall show that a great deal can be learned about the system's behavior by studying the location of the roots of the characteristic equation.

The closed-loop transfer function of the system is determined from Fig. 6-5.

$$\frac{C(s)}{R(s)} = \frac{\omega_n^2}{s^2 + 2\zeta\omega_n s + \omega_n^2} \tag{6-81}$$

The characteristic equation of the system is obtained by setting Eq. (6-77) to zero; that is,

$$\Delta = s^2 + 2\zeta\omega_n s + \omega_n^2 = 0 \tag{6-82}$$

For a unit step function input, $R(s) = 1/s$, the output response of the system is determined by taking the inverse Laplace transform of

$$C(s) = \frac{\omega_n^2}{s(s^2 + 2\zeta\omega_n s + \omega_n^2)} \tag{6-83}$$

Or, $c(t)$ is determined by use of Eqs. (6-75) and (6-78) with zero initial states:

$$c(t) = 1 + \frac{e^{-\zeta\omega_n t}}{\sqrt{1-\zeta^2}} \sin\left[\omega_n\sqrt{1-\zeta^2}\,t - \tan^{-1}\frac{\sqrt{1-\zeta^2}}{-\zeta}\right] \qquad t \geq 0 \tag{6-84}$$

It is interesting to study the relationship between the roots of the characteristic equation and the behavior of the step response $c(t)$. The two roots of Eq. (6-82) are

$$\begin{aligned} s_1, s_2 &= -\zeta\omega_n \pm j\omega_n\sqrt{1-\zeta^2} \\ &= -\alpha \pm j\omega \end{aligned} \tag{6-85}$$

The physical significance of the constants ζ, ω_n, α, and ω is now described as follows:

As seen from Eq. (6-85), $\alpha = \zeta\omega_n$, and α appears as the constant that is multiplied to t in the exponential term of Eq. (6-84). Therefore, α controls the rate of rise and decay of the time response. In other words, α controls the "damping" of the system and is called the *damping constant* or the *damping factor*. The inverse of α, $1/\alpha$, is proportional to the time constant of the system.

When the two roots of the characteristic equation are real and identical we call the system *critically damped*. From Eq. (6-85) we see that *critical damping occurs when $\zeta = 1$.* Under this condition the damping factor is simply $\alpha = \omega_n$. Therefore, we can regard ζ as the *damping ratio*, which is the ratio between the actual damping factor and the damping factor when the damping is critical.

ω_n is defined as the *natural undamped frequency.* As seen from Eq. (6-85), when the damping is zero, $\zeta = 0$, the roots of the characteristic equation are imaginary, and Eq. (6-84) shows that the step response is purely sinusoidal. Therefore, ω_n corresponds to the frequency of the undamped sinusoid.

Equation (6-85) shows that

$$\omega = \omega_n\sqrt{1-\zeta^2} \tag{6-86}$$

However, since unless $\zeta = 0$, the response of Eq. (6-84) is not a periodic function. Therefore, strictly, ω is not a frequency. For the purpose of reference ω is sometimes defined as *conditional frequency.*

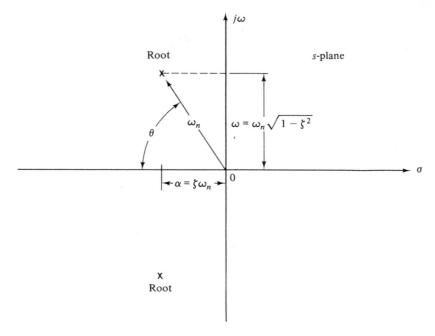

Fig. 6-6. Relationship between the characteristic equation roots of a second-order system and α, ζ, ω_n, and ω.

Figure 6-6 illustrates the relationship between the location of the characteristic equation roots and α, ζ, ω_n, and ω. For the complex-conjugate roots shown, ω_n is the radial distance from the roots to the origin of the s-plane. The damping factor α is the real part of the roots; the conditional frequency is the imaginary part of the roots, and the damping ratio ζ is equal to the cosine of the angle between the radial line to the roots and the negative real axis; that is,

$$\zeta = \cos \theta \qquad (6\text{-}87)$$

Figure 6-7 shows the constant-ω_n loci, the constant-ζ loci, the constant-α loci, and the constant-ω loci. Note that the left-half of the s-plane corresponds to positive damping (i.e., the damping factor or ratio is positive), and the right-half of the s-plane corresponds to negative damping. The imaginary axis corresponds to zero damping ($\alpha = 0$, $\zeta = 0$). As shown by Eq. (6-84), when the damping is positive, the step response will settle to its constant final value because of the negative exponent of $e^{-\zeta \omega_n t}$. Negative damping will correspond to a response that grows without bound, and zero damping gives rise to a sustained sinusoidal oscillation. These last two cases are defined as *unstable* for linear systems. Therefore, we have demonstrated that the location of the characteristic equation roots plays a great part in the dynamic behavior of the transient response of the system.

The effect of the characteristic-equation roots on the damping of the second-

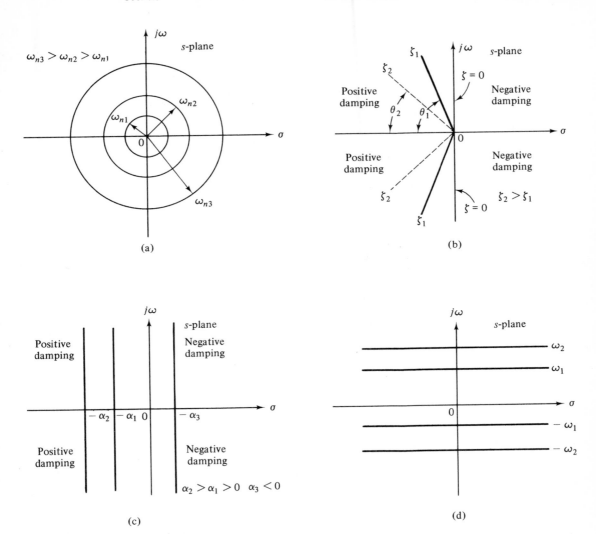

Fig. 6-7. (a) Constant natural undamped frequency loci. (b) Constant damping ratio loci. (c) Constant damping factor loci. (d) Constant conditional frequency loci.

order system is further illustrated by Figs. 6-8 and 6-9. In Fig. 6-8, ω_n is held constant while the damping ratio ζ is varied from $-\infty$ to $+\infty$. The following classification of the system dynamics with respect to the value of ζ is given:

$$0 < \zeta < 1: \quad s_1, s_2 = -\zeta\omega_n \pm j\omega_n\sqrt{1 - \zeta^2} \quad \text{underdamped case}$$
$$\zeta = 1: \quad s_1, s_2 = -\omega_n \quad \text{critically damped case}$$
$$\zeta > 1: \quad s_1, s_2 = -\zeta\omega_n \pm \omega_n\sqrt{\zeta^2 - 1} \quad \text{overdamped case}$$
$$\zeta = 0: \quad s_1, s_2 = \pm j\omega_n \quad \text{undamped case}$$
$$\zeta < 0: \quad s_1, s_2 = -\zeta\omega_n \pm j\omega_n\sqrt{1 - \zeta^2} \quad \text{negatively damped case}$$

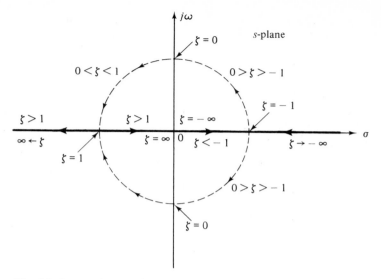

Fig. 6-8. Locus of roots of Eq. (6-82) when ω_n is held constant while the damping ratio is varied from $-\infty$ to $+\infty$.

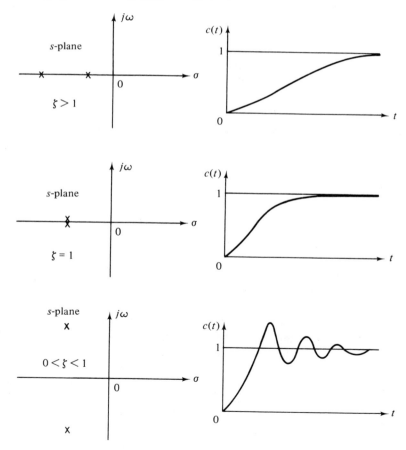

Fig. 6-9. Response comparison for various root locations in the s-plane.

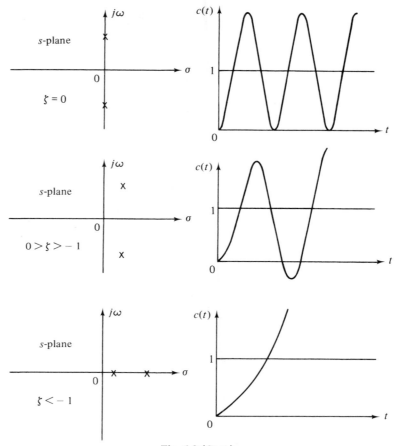

Fig. 6-9 (Cont.).

Figure 6-9 illustrates typical step responses that correspond to the various root locations.

In practical applications only stable systems are of interest. Therefore, the cases when ζ is positive are of particular interest. In Fig. 6-10 is plotted the variation of the unit step response described by Eq. (6-84) as a function of the normalized time $\omega_n t$, for various values of the damping ratio ζ. It is seen that the response becomes more oscillatory as ζ decreases in value. When $\zeta \geq 1$ there is no overshoot in the step response; that is, the output never exceeds the value of the reference input.

The exact relation between the damping ratio and the amount of overshoot can be obtained by taking the derivative of Eq. (6-84) and setting the result to zero. Thus

$$\frac{dc(t)}{dt} = -\frac{\zeta \omega_n e^{-\zeta \omega_n t}}{\sqrt{1 - \zeta^2}} \sin(\omega t - \phi)$$

$$+ \frac{e^{-\zeta \omega_n t}}{\sqrt{1 - \zeta^2}} \omega_n \sqrt{1 - \zeta^2} \cos(\omega t - \phi) \qquad t \geq 0 \tag{6-88}$$

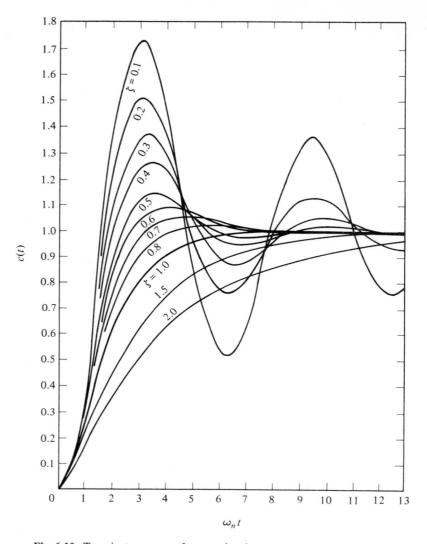

Fig. 6-10. Transient response of a second-order system to a unit step function input.

where

$$\phi = \tan^{-1}\frac{\sqrt{1-\zeta^2}}{-\zeta} \tag{6-89}$$

Equation (6-88) is simplified to

$$\frac{dc(t)}{dt} = \frac{\omega_n}{\sqrt{1-\zeta^2}}e^{-\zeta\omega_n t}\sin\omega_n\sqrt{1-\zeta^2}t \qquad t \geq 0 \tag{6-90}$$

Therefore, setting Eq. (6-90) to zero, we have $t = \infty$ and

$$\omega_n\sqrt{1-\zeta^2}t = n\pi \qquad n = 0, 1, 2, \ldots \tag{6-91}$$

or

$$t = \frac{n\pi}{\omega_n\sqrt{1 - \zeta^2}} \tag{6-92}$$

The first maximum value of the step response $c(t)$ occurs at $n = 1$. Therefore, the time at which the maximum overshoot occurs is given by

$$t_{max} = \frac{\pi}{\omega_n\sqrt{1 - \zeta^2}} \tag{6-93}$$

In general, for all odd values of n, that is, $n = 1, 3, 5, \ldots$, Eq. (6-92) gives the times at which the overshoots occur. For all even values of n, Eq. (6-92) gives the times at which the undershoots occur, as shown in Fig. 6-11. It is interesting to note that, although the maxima and the minima of the response occur at periodic intervals, the response is a damped sinusoid and is not a periodic function.

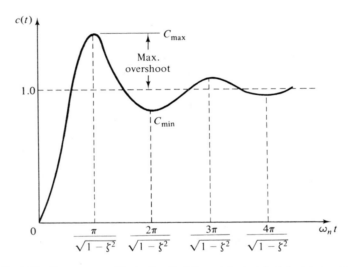

Fig. 6-11. Step response illustrating that the maxima and minima occur at periodic intervals.

The magnitudes of the overshoots and the undershoots can be obtained by substituting Eq. (6-92) into Eq. (6-84). Thus

$$c(t)\big|_{max\ or\ min} = 1 + \frac{e^{-n\pi\zeta/\sqrt{1-\zeta^2}}}{\sqrt{1 - \zeta^2}} \sin\left(n\pi - \tan^{-1}\frac{\sqrt{1 - \zeta^2}}{-\zeta}\right) \quad n = 1, 2, 3, \ldots \tag{6-94}$$

or

$$c(t)\big|_{max\ or\ min} = 1 + (-1)^{n-1}e^{-n\pi\zeta/\sqrt{1-\zeta^2}} \tag{6-95}$$

The maximum overshoot is obtained by letting $n = 1$ in Eq. (6-95). Therefore,

$$\text{maximum overshoot} = c_{max} - 1 = e^{-\pi\zeta/\sqrt{1-\zeta^2}} \tag{6-96}$$

and

$$\text{per cent maximum overshoot} = 100e^{-\pi\zeta/\sqrt{1-\zeta^2}} \tag{6-97}$$

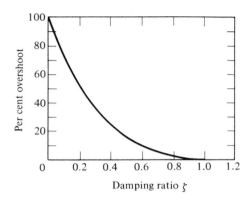

Fig. 6-12. Per cent overshoot as a function of damping ratio for the step response of a second-order system.

Note that for the second-order system, the maximum overshoot of the step response is only a function of the damping ratio. The relationship between the per cent maximum overshoot and damping ratio for the second-order system is shown in Fig. 6-12.

From Eqs. (6-93) and (6-94) it is seen that for the second-order system under consideration, the maximum overshoot and the time at which it occurs are all exactly expressed in terms of ζ and ω_n. For the delay time, rise time, and settling time, however, the relationships are not so simple. It would be difficult to determine the exact expressions for these quantities. For instance, for the delay time, we would have to set $c(t) = 0.5$ in Eq. (6-84) and solve for t. An easier way would be to plot $\omega_n t_d$ versus ζ as shown in Fig. 6-13. Then, over the range of $0 < \zeta < 1.0$ it is possible to approximate the curve by a straight line,

$$\omega_n t_d \simeq 1 + 0.7\zeta \qquad (6\text{-}98)$$

Thus the delay time is

$$t_d \simeq \frac{1 + 0.7\zeta}{\omega_n} \qquad (6\text{-}99)$$

For a wider range of ζ, a second-order equation should be used. Then

$$t_d \cong \frac{1 + 0.6\zeta + 0.15\zeta^2}{\omega_n} \qquad (6\text{-}100)$$

For the rise time t_r, which is the time for the step response to reach from 10 per cent to 90 per cent of its final value, the exact values can again be obtained directly from the responses of Fig. 6-10. The plot of $\omega_n t_r$ versus ζ is shown in Fig. 6-14. In this case the rise time versus ζ relation can again be approximated by a straight line over a limited range of ζ. Therefore,

$$t_r \cong \frac{0.8 + 2.5\zeta}{\omega_n} \qquad 0 < \zeta < 1 \qquad (6\text{-}101)$$

A better approximation may be obtained by using a second-order equation;

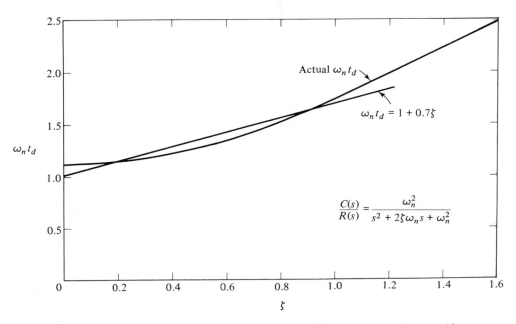

Fig. 6-13. Normalized delay time versus ζ for a second-order control system.

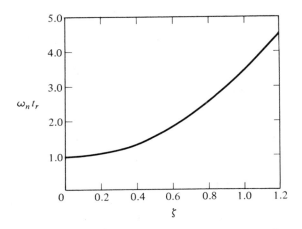

Fig. 6-14. Normalized rise time versus ζ for a second-order system.

then

$$t_r \cong \frac{1 + 1.1\zeta + 1.4\zeta^2}{\omega_n} \tag{6-102}$$

From the definition of settling time, it is clear that the expression for the settling time is the most difficult to determine. However, we can obtain an approximation for the case of $0 < \zeta < 1$ by using the envelope of the damped sinusoid, as shown in Fig. 6-15.

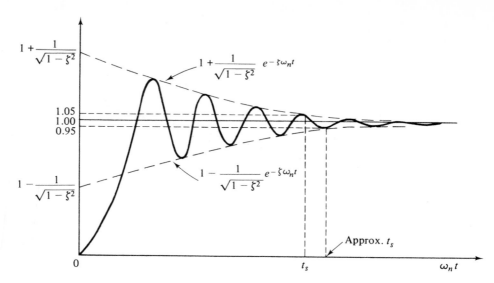

Fig. 6-15. Approximation of settling time using the envelope of the decaying step response of a second-order system $(0 < \zeta < 1)$.

From the figure it is clear that the same result is obtained with the approximation whether the upper envelope or the lower envelope is used. Therefore,

$$c(t) = 1 + \frac{e^{-\zeta \omega_n t_s}}{\sqrt{1 - \zeta^2}} = 1.05 \tag{6-103}$$

Solving for $\omega_n t_s$ from the last equation, we have

$$\omega_n t_s = -\frac{1}{\zeta} \ln [0.05\sqrt{1 - \zeta^2}] \tag{6-104}$$

For small values of ζ, Eq. (6-104) is simplified to

$$\omega_n t_s \cong \frac{3}{\zeta} \tag{6-105}$$

or

$$t_s \cong \frac{3}{\zeta \omega_n} \qquad 0 < \zeta < 1 \tag{6-106}$$

Now reviewing the relationships for the delay time, rise time, and settling time, it is seen that small values of ζ would yield short rise time and short delay time. However, a fast settling time requires a large value for ζ. Therefore, a compromise in the value of ζ should be made when all these criteria are to be satisfactorily met in a design problem. Together with the consideration on maximum overshoot, a generally accepted range of damping ratio for satisfactory all-around performance is between 0.5 and 0.8.

6.6 Time Response of a Positional Control System

In this section we shall study the time-domain performance of a control system whose objective is to control the position of a load that has viscous friction and inertia. The schematic diagram of the system is shown in Fig. 6-16.

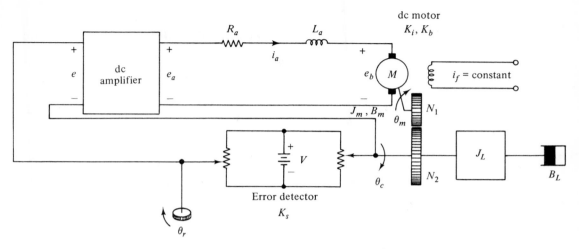

Fig. 6-16. Direct current positional control system.

A set of potentiometers form the error detector with sensitivity K_s. The error detector sends a signal to the dc amplifier which is proportional to the difference between the angular positions of the reference input shaft and the output shaft. The output of the dc amplifier is used to control the armature of a dc motor. The current in the field of the dc motor is held constant.

The parameters of the system are given as follows:

Sensitivity of error detector	$K_s = 1/57.3$ volt/deg = 1 volt/rad
Gain of dc amplifier	A (variable)
Resistance of armature of motor	$R_a = 5\ \Omega$
Inductance of armature of motor	L_a = negligible
Inertia of rotor of motor	$J_m = 10^{-3}$ lb-ft-sec^2
Friction of motor shaft	B_m = negligible
Friction of load shaft	$B_L = 0.1$ lb-ft-sec
Inertia of load	$J_L = 0.1$ lb-ft-sec^2
Gear ratio	$n = N_1/N_2 = \frac{1}{10}$
Torque constant of motor	$K_i = 0.5$ lb-ft/amp

The first step in the analysis is to write the equations for the system. This is done by proceeding from the input to the output following through the cause-and-effect relationship.

1. Error detector:

$$\theta_e(t) = \theta_r(t) - \theta_c(t) \tag{6-107}$$

$$e(t) = K_s\theta_e(t) \tag{6-108}$$

2. DC amplifier:

$$e_a(t) = Ae(t) \tag{6-109}$$

3. Armature-controlled dc motor:

$$L_a \frac{di_a(t)}{dt} = -R_a i_a(t) + e_a(t) - e_b(t) \tag{6-110}$$

$$e_b(t) = K_b \omega_m(t) \tag{6-111}$$

where K_b is the back emf constant of the motor:

$$T_m(t) = K_i i_a(t) \tag{6-112}$$

$$J_{me} \frac{d\omega_m(t)}{dt} = -B_{me} \omega_m(t) + T_m(t) \tag{6-113}$$

where J_{me} and B_{me} are the equivalent inertia and viscous frictional coefficients seen by the motor, respectively, and

$$J_{me} = J_m + n^2 J_L = 10^{-3} + 0.01(0.1) = 2 \times 10^{-3} \text{ lb-ft-sec}^2 \tag{6-114}$$

$$B_{me} = B_m + n^2 B_L = 10^{-3} \text{ lb-ft-sec} \tag{6-115}$$

4. Output:

$$\frac{d\theta_m(t)}{dt} = \omega_m(t) \tag{6-116}$$

$$\theta_c(t) = n\theta_m(t) \tag{6-117}$$

The value of the back emf constant, K_b, is not given originally, but a definite relationship exists between K_b and K_i. In the British unit system, K_i is given in lb-ft/amp and the units of the back emf constant are volts/rad/sec. With these units, K_b is related to K_i through a constant ratio. The mechanical power developed in the motor armature is (*see* Sec. 5.7)

$$p(t) = e_b(t)i_a(t) \qquad \text{watts}$$
$$= \frac{1}{746} e_b(t)i_a(t) \qquad \text{hp} \tag{6-118}$$

Substituting Eqs. (6-111) and (6-112) into Eq. (6-118), we have

$$p(t) = \frac{K_b}{746 K_i} T_m(t)\omega_m(t) \qquad \text{hp} \tag{6-119}$$

Also, it is known that

$$p(t) = \frac{1}{550} T_m(t)\omega_m(t) \qquad \text{hp} \tag{6-120}$$

Therefore, equating Eq. (6-119) to Eq. (6-120) gives

$$K_i = \frac{550}{746} K_b = 0.737 K_b \tag{6-121}$$

or

$$K_b = 1.36 K_i \tag{6-122}$$

Thus, given K_i to be 0.5 lb-ft/amp, K_b is found to be 0.68 volt/rad/sec.

Using Eqs. (6-107) through (6-116), the state equations of the system are written in the matrix form as follows:

$$
\begin{bmatrix} \dfrac{di_a(t)}{dt} \\[2ex] \dfrac{d\omega_m(t)}{dt} \\[2ex] \dfrac{d\theta_m(t)}{dt} \end{bmatrix} = \begin{bmatrix} -\dfrac{R_a}{L_a} & -\dfrac{K_b}{L_a} & -\dfrac{nAK_s}{L_a} \\[2ex] \dfrac{K_i}{J_{me}} & -\dfrac{B_{me}}{J_{me}} & 0 \\[2ex] 0 & 1 & 0 \end{bmatrix} \begin{bmatrix} i_a(t) \\[2ex] \omega_m(t) \\[2ex] \theta_m(t) \end{bmatrix} + \begin{bmatrix} \dfrac{AK_s}{L_a} \\[2ex] 0 \\[2ex] 0 \end{bmatrix} \theta_r(t) \quad (6\text{-}123)
$$

The output equation is given by Eq. (6-117). The state diagram of the system is drawn as shown in Fig. 6-17.

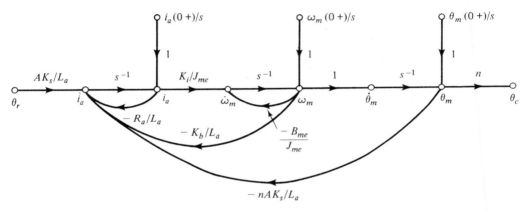

Fig. 6-17. State diagram of the positional control system in Fig. 6-16.

When the initial states are assumed to be zero, the closed-loop transfer function of the system is obtained as

$$
\frac{\theta_c(s)}{\theta_r(s)} = \frac{K_s AK_i n}{R_a B_{me} s(1 + \tau_a s)(1 + \tau_{me} s) + K_b K_i s + K_s AK_i n} \quad (6\text{-}124)
$$

where

$$
\tau_a = \frac{L_a}{R_a} = \text{negligible}
$$

$$
\tau_{me} = \frac{J_{me}}{B_{me}} = 2 \text{ sec}
$$

The error between the reference input and the output is defined as

$$
\theta_e(t) = \theta_r(t) - \theta_c(t) \quad (6\text{-}125)
$$

The open-loop transfer function of the system is obtained by use of Eq. (6-124). Therefore,

$$
G(s) = \frac{\theta_c(s)}{\theta_e(s)} = \frac{K_s AK_i n}{R_a B_{me} s(1 + \tau_a s)(1 + \tau_{me} s) + K_b K_i s} \quad (6\text{-}126)
$$

The state transition equations of the system can be obtained from the state diagram by use of the gain formula, in the usual fashion. However, the main objective of this problem is to demonstrate the behavior of the time response of the positional control system with respect to the system parameters, and it is sufficient to assume that all the initial states of the system are zero.

Since L_a is negligible, $\tau_a = 0$, and the closed-loop transfer function in Eq. (6-124) is simplified to

$$\frac{\theta_c(s)}{\theta_r(s)} = \frac{K_s A K_i n}{R_a J_{me} s^2 + (K_b K_i + R_a B_{me}) s + K_s A K_i n} \tag{6-127}$$

Equation (6-127) is of the second order; thus it can be written in the standard form of Eq. (6-81). The natural undamped frequency of the system is

$$\omega_n = \pm \sqrt{\frac{K_s A K_i n}{R_a J_{me}}} \tag{6-128}$$

The damping ratio is

$$\zeta = \frac{K_b K_i + R_a B_{me}}{2 R_a J_{me} \omega_n} = \frac{K_b K_i + R_a B_{me}}{2 \sqrt{K_s A K_i R_a J_{me} n}} \tag{6-129}$$

When the values of the system parameters are substituted into Eq. (6-127), the closed-loop transfer function of the system becomes

$$\frac{\theta_c(s)}{\theta_r(s)} = \frac{5A}{s^2 + 34.5s + 5A} \tag{6-130}$$

Suppose that the gain of the dc amplifier is arbitrarily set at 200. The natural undamped frequency and the damping ratio are, respectively,

$$\omega_n = \pm 31.6 \text{ rad/sec} \tag{6-131}$$

$$\zeta = 0.546 \tag{6-132}$$

The characteristic equation of the system is

$$s^2 + 34.5s + 1000 = 0 \tag{6-133}$$

whose roots are

$$s_1, s_2 = -17.25 \pm j26.5 \tag{6-134}$$

Time Response to a Step Input

Consider that the reference input is a unit step function, $\theta_r(t) = u_s(t)$ rad; then the output of the system, under zero initial state condition, is

$$\theta_c(t) = \mathcal{L}^{-1} \left[\frac{1000}{s(s^2 + 34.5s + 1000)} \right] \tag{6-135}$$

or

$$\theta_c(t) = 1 + 1.2 e^{-0.546 \omega_n t} \sin (0.837 \omega_n t + \pi + \tan^{-1} 1.53)$$
$$= 1 + 1.2 e^{-17.25t} \sin (26.4t + 236.8°) \tag{6-136}$$

The output response of the system is plotted in Fig. 6-18 as a function of the normalized time $\omega_n t$.

It is interesting to study the effects on the time response when the value of the gain A is varied. Note that in Eq. (6-128) and in Eq. (6-129), an increase in the gain A increases the natural undamped frequency ω_n but decreases the damping ratio ζ. For $A = 1500$, $\zeta = 0.2$, and $\omega_n = 86.2$ rad/sec. The output response to a unit step displacement input for $A = 1500$ is plotted as shown in

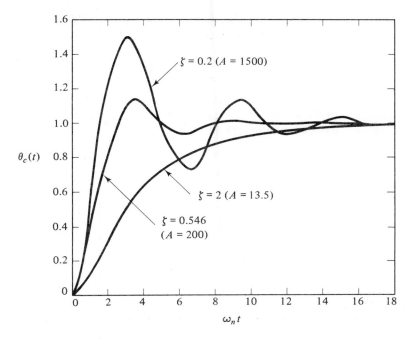

Fig. 6-18. Time response of the positional control system in Fig. 6-16 when the input is a unit step displacement.

Fig. 6-18. In this case the overshoot is very high, although the rise time and delay time are apparently reduced. The settling time is also shortened by the increase in A, although the responses in Fig. 6-18 do not show this because the time axis is normalized by ω_n. In fact, in Fig. 6-18 the step response for $A = 1500$ seems to take a longer time to reach the steady state than that for $A = 200$. This is distorted by the normalized time axis; for $A = 1500$, $\omega_n = 86.2$ rad/sec, as compared to $\omega_n = 31.6$ rad/sec for $A = 200$.

When A is set at a relatively low value, 13.5, the damping ratio and the natural undamped frequency are $\zeta = 2.1$ and $\omega_n = 8.22$ rad/sec, respectively. Since ζ is greater than one, the step response corresponding to $A = 13.5$ is overdamped. Table 6-2 gives the comparison of the time responses of the system for the three different values of A used.

Table 6-2   Comparison of Transient Response of a Second-Order Positional Control System When the Gain Varies

Gain A	Damping Ratio ζ	ω_n	Maximum Overshoot	T_d	T_r	T_s	t_{max}
13.5	2.1	8.22	0	0.325	1.02	1.51	—
200	0.546	31.6	0.13	0.043	0.064	0.174	0.119
1500	0.2	86.6	0.52	0.013	0.015	0.17	0.037

When the value of A is set at 13.5 and 200, the roots of the characteristic equation are determined and listed as follows:

$$A = 13.5 \qquad s_1, s_2 = -2.1, -32.4$$
$$A = 200 \qquad s_1, s_2 = -17.25 \pm j26.5$$

These roots, together with the ones for $A = 1500$, are located in the s-plane as shown in Fig. 6-19. In general, for any value of A, the roots of the characteristic equation are given by

$$s_1, s_2 = -17.25 \pm \frac{1}{2}\sqrt{1190 - 20A} \qquad (6\text{-}137)$$

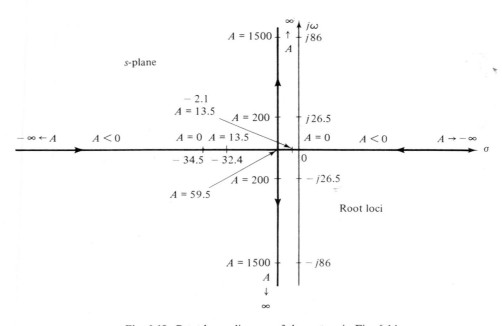

Fig. 6-19. Root locus diagram of the system in Fig. 6-14.

Therefore, for values of A between zero and 59.5, the two roots are real and lie on the negative real axis in the s-plane, and the system is overdamped. For values of A greater than 59.5, the roots are complex conjugate; the real parts of the roots are equal to -17.25 and are not affected by the value of A. Therefore, as A approaches infinity, the damping factor of the second-order system is always equal to 17.25 sec^{-1}. When A varies continuously between $-\infty$ and ∞, the two roots of the characteristic equation trace out two continuous loci in the s-plane, as shown in Fig. 6-19. In general, these are called the *root loci* of the characteristic equation or of the control system. In control system studies a root-locus diagram can be very useful for analysis and design once the relationship between the root location and the transient behavior has been established. In this case Fig. 6-19 shows that the second-order system is always stable for all finite positive values of A. When A is negative, one of the roots is positive, which corresponds to a time response that increases monotonically with time, and the system is said to be unstable. The dynamic characteristics of the transient response as determined from the root-locus diagram of Fig. 6-19 are summarized as follows:

Amplifier Gain	Characteristic Equation Roots	System Dynamics
$0 < A < 59.5$	Two negative distinct real roots	Overdamped ($\zeta > 1$)
$A = 59.5$	Two negative equal real roots	Critically damped ($\zeta = 1$)
$59.5 < A < \infty$	Two complex conjugate roots with negative real parts	Underdamped ($\zeta < 1$)
$-\infty < A < 0$	Two distinct real roots, one positive and one negative	Unstable system ($\zeta < 0$)

Since the system under consideration is of type 1, the steady-state error of the system is zero for all positive values of A, when the input is a step function. In other words, for a step input, the positional error constant K_p is to be used. Substituting the open-loop transfer function $G(s)$ for the system and $H(s) = 1$ into Eq. (6-21), we have

$$K_p = \lim_{s \to 0} \frac{5A}{s(s + 34.5)} = \infty \tag{6-138}$$

Therefore, the steady-state error is given by Eq. (6-22) as

$$e_{ss} = \frac{R}{1 + K_p} = 0 \tag{6-139}$$

The unit step responses of Fig. 6-18 verify this fact.

Time Response to a Ramp Input

When a unit ramp function input $\theta_r(t) = tu_s(t)$ is applied to the control system of Fig. 6-16, the output response is described by

$$\theta_c(t) = \mathcal{L}^{-1}\left[\frac{\omega_n^2}{s^2(s^2 + 2\zeta\omega_n s + \omega_n^2)}\right] \tag{6-140}$$

From the Laplace transform table, Eq. (6-140) is written

$$\theta_c(t) = t - \frac{2\zeta}{\omega_n} + \frac{1}{\omega_n\sqrt{1 - \zeta^2}}e^{-\zeta\omega_n t} \sin [\omega_n\sqrt{1 - \zeta^2}\, t - \phi] \tag{6-141}$$

where

$$\phi = 2 \tan^{-1}\frac{\sqrt{1 - \zeta^2}}{-\zeta} \tag{6-142}$$

The ramp responses for $A = 13.5$, 200, and 1500 are sketched as shown in Fig. 6-20. Notice that in this case the steady-state error of the system is nonzero. As seen from Eq. (6-141), the last term will decay to zero as time approaches infinity. Therefore, the steady-state response of the system due to a unit ramp

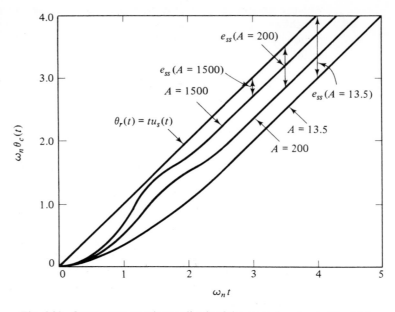

Fig. 6-20. Output response (normalized) of the control system in Fig. 6-16 when the input is a unit ramp function.

input is

$$\lim_{t \to \infty} \theta_c(t) = \lim_{t \to \infty} \left[t - \frac{2\zeta}{\omega_n} \right] \tag{6-143}$$

It is simple to see that the steady-state error due to the unit ramp input is

$$e_{ss} = \frac{2\zeta}{\omega_n} = \frac{34.5}{5A} \tag{6-144}$$

which is a constant.

A more systematic method of determining the steady-state error to a ramp input is to use the velocity error constant K_v. From Eq. (6-26),

$$K_v = \lim_{s \to 0} sG(s)H(s) = \lim_{s \to 0} \frac{5A}{s + 34.5} = \frac{5A}{34.5} \tag{6-145}$$

Therefore, from Eq. (6-27),

$$e_{ss} = \frac{1}{K_v} = \frac{34.5}{5A} \tag{6-146}$$

which agrees with the result of Eq. (6-144).

Equation (6-146) shows that the steady-state error is inversely proportional to the magnitude of A. However, if we choose to improve the steady-state accuracy of the system by increasing the forward gain, the transient response becomes more oscillatory. This phenomenon is rather typical in all control systems. For higher-order systems, if the loop gain of the system is too high the system may become unstable.

Transient Response of a Third-Order System

It was shown in the last section that if the armature inductance L_a of the dc motor is neglected, the control system is of the second order and is stable for all positive values of A. Suppose now that we let $L_a = 0.1$ henry in the system in Fig. 6-16 and keep the other parameters unchanged. The armature time constant τ_a is now 0.02 sec. The closed-loop transfer function given by Eq. (6-124) is now

$$\frac{\theta_c(s)}{\theta_r(s)} = \frac{0.05A}{0.005s(1 + 0.02s)(1 + 2s) + 0.34s + 0.05A} \tag{6-147}$$

or

$$\frac{\theta_c(s)}{\theta_r(s)} = \frac{250A}{s^3 + 50.5s^2 + 1725s + 250A} \tag{6-148}$$

The open-loop transfer function is

$$G(s) = \frac{\theta_c(s)}{\theta_e(s)} = \frac{250A}{s(s^2 + 50.5s + 1725)} \tag{6-149}$$

The characteristic equation of the system is

$$s^3 + 50.5s^2 + 1725s + 250A = 0 \tag{6-150}$$

It is apparent that when the armature inductance is restored, the system is now of the third order.

If we let $A = 13.5$, the closed-loop transfer function of Eq. (6-147) becomes

$$\frac{\theta_c(s)}{\theta_r(s)} = \frac{1}{(1 + 0.48s)(1 + 0.0298s + 0.000616s^2)} \tag{6-151}$$

The characteristic equation has a real root at $s = -2.08$ and two complex roots at $-24.2 \pm j\,32.2$. For a unit step function input the output response is

$$\theta_c(t) = 1 - 1.06e^{-2.08t} + 0.0667e^{-24.2t} \sin(32.2t + 1.88) \tag{6-152}$$

In Eq. (6-152), since the time constant of the pure exponential term is more than 10 times greater than that of the damped sinusoid term, the transient response of $\theta_c(t)$ is predominantly governed by the characteristic root at $s = -2.08$. The response due to the last term of Eq. (6-152) decreases to zero very rapidly with the increase of t. Comparing Eq. (6-152) with Eq. (6-136) we see that the damping factor for the second-order system ($L_a = 0$) is 17.25, whereas for the third-order system ($L_a = 0.1$) the output response is governed by the exponential term, which has an equivalent damping factor of 2.08. Thus the third-order system will have a slower rise time. This result is expected, since the presence of inductance will slow down the buildup of the armature current, thus slowing down the torque development of the motor. However, higher inductance will cause a higher overshoot in the step response, as shown in Fig. 6-21. With $L_a = 0.1\ H$, and $A = 348$, the characteristic equation of Eq. (6-150) is factored as

$$(s + 50.5)(s^2 + 1725) = 0 \tag{6-153}$$

Thus the characteristic equation has two imaginary roots at $s = \pm j41.5$. The response corresponding to these imaginary roots is an undamped sinusoid. The

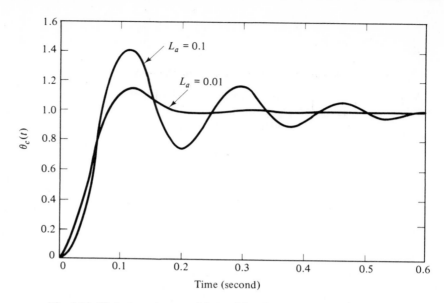

Fig. 6-21. Unit step response of the positional control system of Fig. 6-16 with varying inductance, $A = 200$.

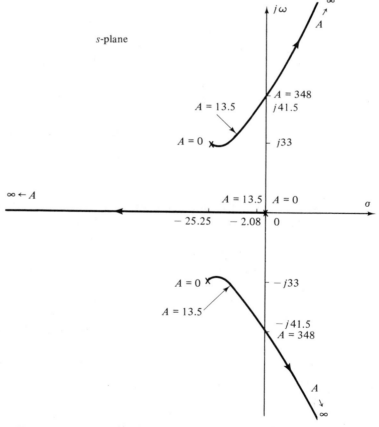

Fig. 6-22. Root loci of the characteristic equation of Eq. (6-150) when A is varied from zero to infinity.

frequency of the sinusoidal response is 41.5 rad/sec. When the roots of a characteristic equation lie on the imaginary axis of the *s*-plane, such as in the present situation, the linear system is said to be on the verge of instability.

Figure 6-22 shows the root loci of the characteristic equation of Eq. (6-150) when *A* is varied from zero to infinity. For all values of *A* greater than 348, the two complex roots are found in the right-half of the *s*-plane, and, with time, the step response of the third-order system will increase without bound.

From this illustrative example we have learned that a second-order system is always stable as long as the loop gain is finite and positive; third- and higher-order systems may become unstable if the loop gain becomes high.

6.7 Effects of Derivative Control on the Time Response of Feedback Control Systems

The control systems considered thus far in this chapter are all of the *proportional type*, in that the system develops a correcting effort that is proportional to the magnitude of the actuating signal only. The illustrative example given in Section 6.6 shows that a proportional type of control system has the limitation or disadvantage that it is often difficult to find a proper forward path gain so that the steady-state and the transient responses satisfy their respective requirements. Often, in practice, a single gain parameter is seldom sufficient to meet the design requirements on two performance criteria.

It is logical to perform other operations, in addition to the proportional control, on the actuating signal. In terms of signal processing, we may perform a time derivative of the actuating signal. Figure 6-23 shows the block diagram

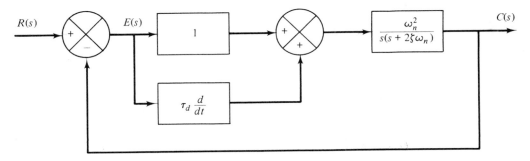

Fig. 6-23. Feedback control system with derivative control.

of a typical second-order feedback control system with derivative control added to the proportional control. In this case the constant τ_d represents the amount of derivative control used in proportion to the ordinary proportional control. The open-loop transfer function of the system is now

$$\frac{C(s)}{E(s)} = \frac{\omega_n^2(1 + \tau_d s)}{s(s + 2\zeta\omega_n)} \tag{6-154}$$

Analytically, Eq. (6-154) shows that the derivative control is equivalent to the addition of a simple zero at $s = -1/\tau_d$ to the open-loop transfer function.

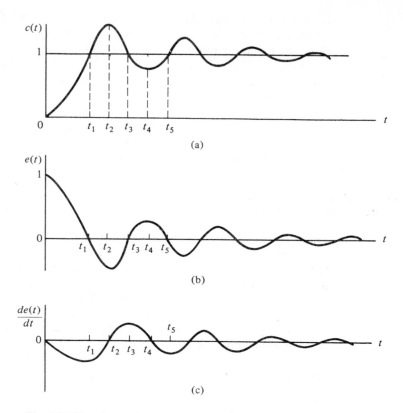

Fig. 6-24. Waveforms of $c(t)$, $e(t)$, and $de(t)/dt$ showing the effect of derivative control. (a) Step response. (b) Error signal. (c) Time rate of change of error signal.

The effect of the derivative control on the transient response of a feedback control system can be investigated by referring to Fig. 6-24. Let us assume that the unit step response of a proportional type of system is shown in Fig. 6-24(a). The corresponding error signal $e(t)$ and the time derivative of $e(t)$ are as shown in Fig. 6-24(b) and (c), respectively. Notice that under the assumed case the step response exhibits a high peak overshoot. For a system that is driven by a motor of some kind, this large overshoot is due to the excessive amount of torque developed by the motor in the time interval $0 < t < t_1$, during which the error signal is positive. For the time interval $t_1 < t < t_3$, the error signal is negative, and the corresponding motor torque is negative. This negative torque tends to reverse the direction of motion of the output, thus causing $c(t)$ to undershoot during $t_3 < t < t_5$. During the time interval $t_3 < t < t_5$ the motor torque is again positive, thus tending to reduce the undershoot in the response caused by the negative torque in the previous interval. Since the system is assumed to be stable, the error amplitude is reduced with each oscillation, and the output eventually is settled to its final desired value.

Considering the explanation given above, we can say that the contributing factors to a high overshoot are as follows: (1) The positive correcting torque in the interval $0 < t < t_1$ is too large, and (2) the retarding torque in the time interval $t_1 < t < t_2$ is inadequate. Therefore, in order to reduce the overshoot in the step response, a logical approach is to decrease the amount of positive correcting torque and to increase the retarding torque. Similarly, in the time interval $t_2 < t < t_4$, the negative corrective torque should be reduced, and the retarding torque, which is now in the positive direction, should be increased in order to improve the undershoot.

The derivative control as represented by the system of Fig. 6-23 gives precisely the compensation effect described in the last paragraph. Let us consider that the proportional type of control system whose signals are described in Fig. 6-24 is now modified so that the torque developed by the motor is proportional to the signal $e(t) + \tau_d de(t)/dt$. In other words, in addition to the error signal, a signal that is proportional to the time rate of change of error is applied to the motor. As shown in Fig. 6-24(c), for $0 < t < t_1$, the time derivative of $e(t)$ is negative; this will reduce the original torque developed due to $e(t)$ alone. For $t_1 < t < t_2$, both $e(t)$ and $de(t)/dt$ are negative, which means that the negative retarding torque developed will be greater than that of the proportional case. Therefore, all these effects will result in a smaller overshoot. It is easy to see that $e(t)$ and $de(t)/dt$ have opposite signs in the time interval $t_2 < t < t_3$; therefore, the negative torque that originally contributes to the undershoot is reduced also.

Since $de(t)/dt$ represents the slope of $e(t)$, the derivative control is essentially an anticipatory type of control. Normally, in a linear system, if the slope of $e(t)$ or $c(t)$ due to a step input is large, a high overshoot will subsequently occur. The derivative control measures the instantaneous slope of $e(t)$, predicts the large overshoot ahead of time, and makes a proper correcting effort before the overshoot actually occurs.

It is apparent that the derivative control will affect the steady-state error of a system only if the steady-state error varies with time. If the steady-state error of a system is constant with respect to time, the time derivative of this error is zero, and the derivative control has no effect on the steady-state error. But if the steady-state error increases with time, a torque is again developed in proportion to $de(t)/dt$, which will reduce the magnitude of the error.

Consider that the positional control system of Fig. 6-16 is modified by replacing the dc amplifier with derivative control so that the open-loop transfer function is now

$$\frac{\theta_c(s)}{\theta_e(s)} = \frac{5A(1 + \tau_d s)}{s(s + 34.5)} \qquad (6\text{-}155)$$

Figure 6-25 illustrates the unit step responses of the closed-loop system when $A = 13.5$ and $\tau_d = 0.01$, 0.1, and 1.0. Notice that in the case of the low value for A, an increasing value of τ_d has the effect of slowing down the response, and the damping is increased. Figure 6-26 illustrates the unit step responses when $A = 1500$. In this case the damping is noticeably improved by the derivative control. When $\tau_d = 0$, Fig. 6-18 shows that the overshoot is 52 per cent for $A = 1500$, whereas with $\tau_d = 0.01$, the overshoot is reduced to approximately 14 per cent. When $\tau_d = 0.1$, the overshoot is completely eliminated, but the step

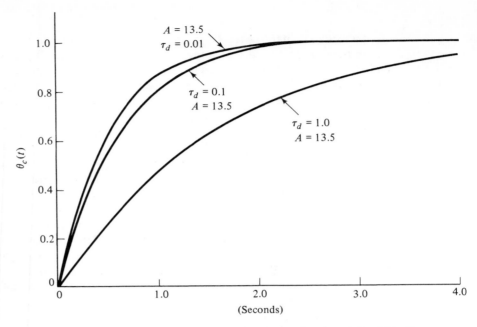

Fig. 6-25. Step responses of the positional control system of Fig. 6-16 with derivative control; $A = 13.5$.

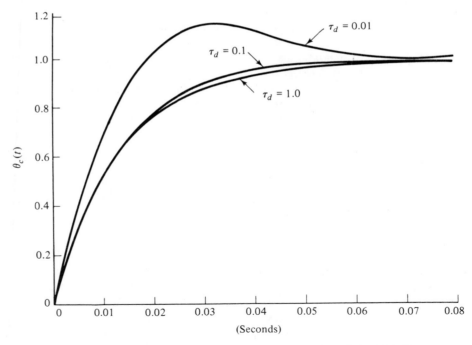

Fig. 6-26. Step responses of the positional control system of Fig. 6-16 with derivative control; $A = 1500$.

response is slow in reaching its final value. Figure 6-26 also shows the step response for $\tau_d = 1.0$.

The effect of derivative control on the transient response of a feedback control system can also be studied by referring to the open-loop transfer function of Eq. (6-154). The corresponding closed-loop transfer function of the system is

$$\frac{C(s)}{R(s)} = \frac{(1 + \tau_d s)\omega_n^2}{s^2 + (2\zeta\omega_n + \tau_d\omega_n^2)s + \omega_n^2} \tag{6-156}$$

The characteristic equation of the system is

$$s^2 + (2\zeta\omega_n + \tau_d\omega_n^2)s + \omega_n^2 = 0 \tag{6-157}$$

Notice that the derivative control has the effect of increasing the coefficient of the s term by the quantity $\tau_d\omega_n^2$. This means that the damping of the system is increased. Using the values as represented by Eq. (6-155), the characteristic equation becomes

$$s^2 + (34.5 + 5A\tau_d)s + 5A = 0 \tag{6-158}$$

Figure 6-27 shows the loci of the roots of Eq. (6-158) when $A = 13.5$ and τ_d is varied from 0 to ∞. The improvement on the system damping due to the derivative control is illustrated by the root loci of Fig. 6-28 with A set at 1500. Notice that for small values of τ_d, the roots of the characteristic equation are still complex, although they are farther away from the imaginary axis than those when $\tau_d = 0$. For large values of τ_d, the roots become real.

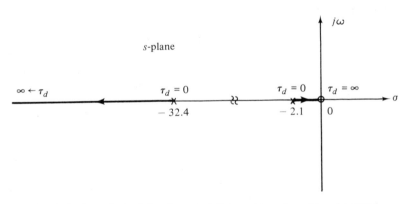

Fig. 6-27. Root loci of the characteristic equation of positional control system with derivative control, $s^2 + (34.5 + 5A\tau_d)s + 5A = 0$, $A = 13.5$.

It should be pointed out that although the derivative control as fashioned by the scheme shown in Fig. 6-23 generally improves the damping of a feedback control system, no considerations have been given here on the practicability of the configuration. In practice, however, the transfer function of $(1 + \tau_d s)$ cannot be physically realized by passive elements. Furthermore, derivative control has the characteristics of a high-pass filter which tends to propagate noise and disturbances through the system. Practical controllers which are used to improve

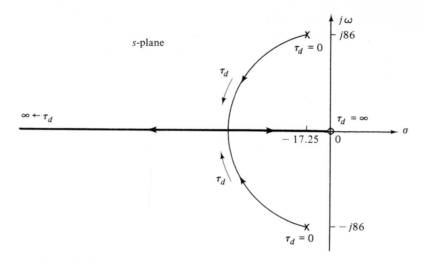

Fig. 6-28. Root loci of the characteristic equation of positional control system with derivative control; $s^2 + (34.5 + 5A\tau_d)s + 5A = 0$, $A = 1500$.

the performance of control systems usually have more complex transfer functions than that of the derivative controller illustrated here.

6.8 Effects of Integral Control on the Time Response of Feedback Control Systems

The counterpart of derivative control is the integral control that introduces a signal in proportion to the time integral of the error. Figure 6-29 illustrates the basic scheme for the integral control applied to a second-order system. The signal applied to the process consists of two components: one proportional to the instantaneous error signal, the other, to the time integral of the error. The parameter K_1 is a proportional constant. The open-loop transfer function of the system is

$$\frac{C(s)}{E(s)} = \frac{\omega_n^2(s + K_1)}{s^2(s + 2\zeta\omega_n)} \qquad (6\text{-}159)$$

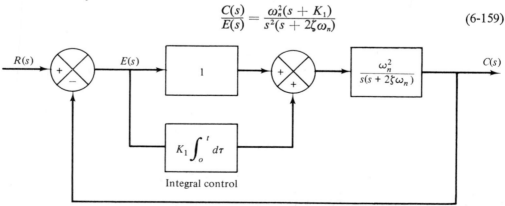

Fig. 6-29. Feedback control system with integral control.

One obvious effect of the integral control is that it increases the order of the system by 1. More important is that it increases the type of the system by 1. Therefore, the steady-state error of the original system without integral control is improved by an order of 1. In other words, if the steady-state error to a given input is constant, the integral control reduces it to zero. In the case of Eq. (6-159), the system will now have a zero steady-state error when the input is a ramp function. However, because the system is now of the third order it tends to be less stable than the original second-order system. In fact, if the loop gain of the system is high, the system may be unstable.

Let us consider the positional control system of Fig. 6-16 again, with $L_a = 0$. With the integral control, the open-loop transfer function of the system becomes

$$\frac{\theta_c(s)}{\theta_e(s)} = \frac{5A(s + K_1)}{s^2(s + 34.5)} \tag{6-160}$$

The characteristic equation is

$$s^3 + 34.5s^2 + 5As + 5AK_1 = 0 \tag{6-161}$$

The effect of the value of K_1 on the transient behavior of the system may be investigated by plotting the roots of Eq. (6-161) as a function of A. Figure 6-30

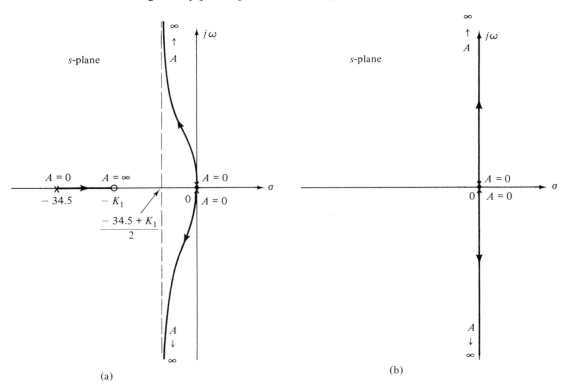

(a)　　　　　　　　　　　　　　　　　　　　　　　(b)

Fig. 6-30. (a) Root loci of the characteristic equation of a feedback control system with integral control, $s^3 + 34.5s^2 + 5As + 5AK_1 = 0$, $K_1 < 34.5$. (b) Root loci of the characteristic equation of a feedback control system with integral control, $s^3 + 34.5s^2 + 5As + 5AK_1 = 0$, $K_1 = 34.5$.

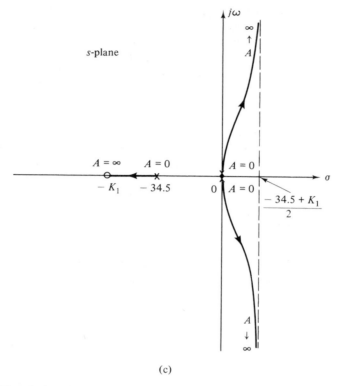

(c)

Fig. 6-30 (Cont.). (c) Root loci of the characteristic equation of a feedback control system with integral control, $s^3 + 34.5s^2 + 5As + 5AK_1 = 0$, $K_1 > 34.5$.

illustrates the root loci of Eq. (6-161) for three different values of K_1 and when A varies between zero and infinity. Notice that when K_1 lies between 0 and 34.5, the roots of the characteristic equation all lie in the left half s-plane. When $K_1 = 34.5$, the system becomes second order and the two roots lie on the $j\omega$-axis for all values of A between 0 and ∞, and the system is on the verge of instability. When the value of K_1 exceeds 34.5, the system with the integral control is unstable for all values of A.

As an alternative, we can fix the value of A and show the effect of varying K_1 on the roots of the characteristic equation. Figure 6-31 shows the root loci of Eq. (6-161) with $A = 13.5$ and $0 \le K_1 < \infty$.

To verify the analytical results obtained in the preceding sections, Fig. 6-32 shows the step responses of the system with $A = 13.5$ and $K_1 = 1, 10$, and 34.5. As predicted by the root loci of Figs. 6-30 and 6-31, the response for $K_1 = 34.5$ is a pure sinusoid.

6.9 Rate Feedback or Tachometer Feedback Control

The philosophy of using the derivative of the actuating signal to improve the damping of a system can be applied to the output signal to achieve a similar

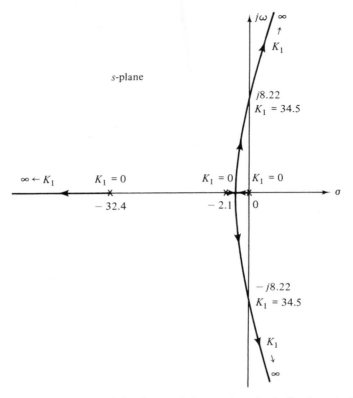

Fig. 6-31. Root loci of the characteristic equation of a feedback control system with integral control, $s^3 + 34.5s^2 + 5As + 5AK_1 = 0$, $A = 13.5$.

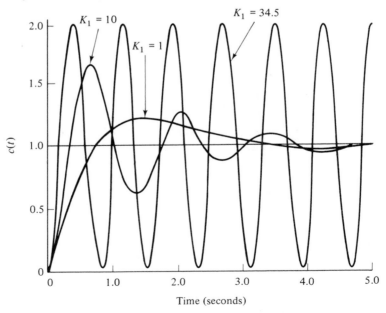

Fig. 6-32. Step responses of the positional control system of Fig. 6-16 with $L_a = 0$ and integral control, $A = 13.5$.

effect. In other words, the derivative of the output signal is fed back and compared with the reference input. In practice, if the output variable is mechanical displacement, a tachometer may be used which converts the mechanical displacement into an electrical signal that is proportional to the derivative of the displacement. Figure 6-33 shows the block diagram of a second-order system with a secondary path that feeds back the derivative of the output. The transfer function of the tachometer is denoted by $K_t s$, where K_t is the tachometer constant, usually expressed in volts/unit velocity.

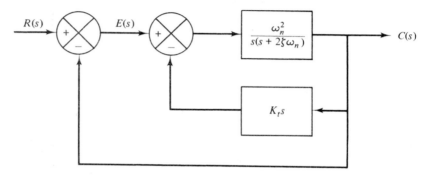

Fig. 6-33. Second-order system with tachometer feedback.

The closed-loop transfer function of the system of Fig. 6-33 is

$$\frac{C(s)}{R(s)} = \frac{\omega_n^2}{s^2 + (2\zeta\omega_n + K_t\omega_n^2)s + \omega_n^2} \qquad (6\text{-}162)$$

and the characteristic equation is

$$s^2 + (2\zeta\omega_n + K_t\omega_n^2)s + \omega_n^2 = 0 \qquad (6\text{-}163)$$

Comparing Eq. (6-163) with Eq. (6-157), which is the characteristic equation of the second-order system with derivative control, we see that the two equations are of the same form. In fact, they are identical if τ_d is interchanged with K_t. Therefore, we conclude that the rate or tachometer feedback also improves the damping of a feedback control system. However, it should be noted that the closed-loop transfer function of Eq. (6-162) does not have a zero, and thus the responses of the two systems will not be identical even if K_t equals τ_d.

The open-loop transfer function of the system with tachometer feedback is obtained from Fig. 6-33:

$$\frac{C(s)}{E(s)} = \frac{\omega_n^2}{s(s + 2\zeta\omega_n + K_t\omega_n^2)} \qquad (6\text{-}164)$$

The system is still of type 1, so the basic characteristic of the steady-state error is not altered. However, for a unit ramp function input, the steady-state error to the system of Fig. 6-23 which has the derivative control is $2\zeta/\omega_n$, whereas that of the system in Fig. 6-33 is $(2\zeta + K_t\omega_n)/\omega_n$.

6.10 Control by State-Variable Feedback[4, 5]

One of the design techniques in modern control theory is that instead of using controllers that have dynamics and feedback from the output variable, flexibility can be gained by feeding back some or all of the state variables to control the process. In the system with tachometer feedback, shown in Fig. 6-33, if we decompose the process in the forward path by direct decomposition, we can show that the system is actually equivalent to having state feedback. Figure 6-34(a) shows the state diagram of the process

$$\frac{C(s)}{E(s)} = \frac{\omega_n^2}{s(s + 2\zeta\omega_n)} \tag{6-165}$$

which is decomposed by direct decomposition. If the states x_1 and x_2 are physically accessible, we may feed back these variables through individual gains, as shown in Fig. 6-34(b), to give closed-loop control of the process.

The closed-loop transfer function of the system in Fig. 6-34(b) is

$$\frac{C(s)}{R(s)} = \frac{\omega_n^2}{s^2 + (2\zeta\omega_n + g_2)s + g_1} \tag{6-166}$$

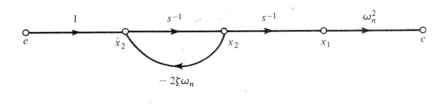

(a)

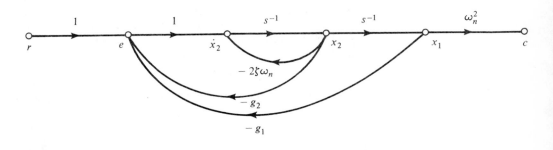

(b)

Fig. 6-34. Control of a second-order system by state feedback.

Comparing this transfer function with that of the system with tachometer feedback, Eq. (6-162), we notice that the two transfer functions would be identical if $g_1 = \omega_n^2$ and $g_2 = K_t\omega_n^2$. In fact, in selecting the feedback gains g_1 and g_2, in order to have zero steady-state error for a step input, g_1 should equal ω_n^2. The value of g_2 is selected to satisfy the damping requirements of the system.

The significance of this illustration is that, if all the state variables are available for feedback, we can achieve at least the same or better control with state feedback as with dynamic controllers and feeding back only the output. Note that the system with tachometer feedback shown in Fig. 6-33 has only the output variable available for feedback. If we regard the output as one of the states, it is fed back through a unity gain. The second state needed for feedback is actually "reconstructed" by the dynamics of the tachometer. In other words, the tachometer acts as an "observer" (see Chapter 11), which recovers the state variable from the output variable.

In modern control theory, certain types of design algorithm, such as the linear regulator theory, naturally lead to state-variable feedback. Since the eigenvalues (or the poles of the closed-loop transfer function) of a linear system directly control the transient response of the system, it would be desirable if the designer is able to place the eigenvalues according to the performance specifications. It is shown in Chapter 11 that if a system is completely controllable, the eigenvalues of the system can be placed arbitrarily.

The following example gives an illustration on how the state feedback and eigenvalue assignment affect the time response of a linear control system.

EXAMPLE 6-3   Consider that the transfer function of a linear process is

$$G(s) = \frac{C(s)}{E(s)} = \frac{20}{s^2(s + 1)} \tag{6-167}$$

Figure 6-35(a) shows the state diagram of $G(s)$, and Fig. 6-35(b) shows the state diagram with feedback from all three states. The closed-loop transfer function of the system is

$$\frac{C(s)}{R(s)} = \frac{20}{s^3 + (g_3 + 1)s^2 + g_2 s + g_1} \tag{6-168}$$

Let us assume that we desire to have zero steady-state error with the input being a unit step function, and in addition, two of the closed-loop poles must be at $s = -1 + j$ and $s = -1 - j$. The steady-state requirement fixes the value of g_1 at 20, and only g_2 and g_3 need to be determined from the eigenvalue location.

The characteristic equation of the system is

$$s^3 + (g_3 + 1)s^2 + g_2 s + 20 = (s + 1 - j)(s + 1 + j)(s + a) \tag{6-169}$$

Equating the coefficients of the corresponding terms in the last equation, we get

$$g_2 = 22 \quad \text{and} \quad g_3 = 11$$

and the third pole is at $s = -10$.

Since the complex closed-loop poles have a damping ratio of 0.707, and the third pole is quite far to the left of these poles, the system acts like a second-order system. Figure 6-36 shows that the unit step response has an overshoot of 4 per cent.

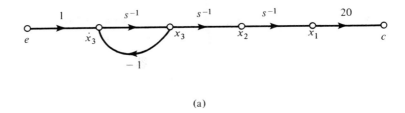

(a)

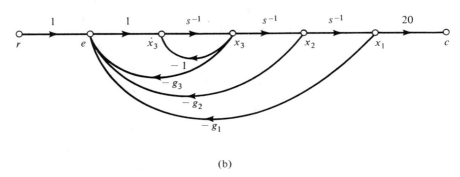

(b)

Fig. 6-35. Control of a third-order system by state feedback.

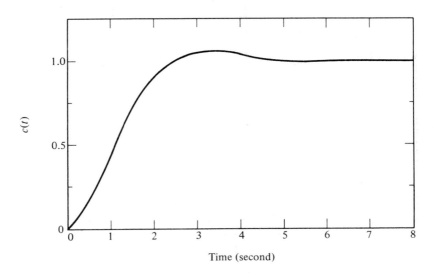

Fig. 6-36. Unit step response of the control system in Example 6-3.

REFERENCES

Time-Domain Analysis

1. O. L. R. JACOBS, "The Damping Ratio of an Optimal Control System," *IEEE Trans. Automatic Control*, Vol. AC-10, pp. 473–476, Oct. 1965.

2. G. A. JONES, "On the Step Response of a Class of Third-Order Linear Systems," *IEEE Trans. Automatic Control*, Vol. AC-12, p. 341, June 1967.

3. R. A. MONZINGO, "On Approximating the Step Response of a Third-Order Linear System by a Second-Order Linear System," *IEEE Trans. Automatic Control*, Vol. AC-13, p. 739, Dec. 1968.

State-Variable Feedback

4. W. M. WONHAM, "On Pole Assignment in Multi-input Controllable Linear Systems," *IEEE Trans. Automatic Control*, Vol. AC-12, pp. 660–665, Dec. 1967.

5. J. C. WILLEMS and S. K. MITTER, "Controllability, Observability, Pole Allocation, and State Reconstruction," *IEEE Trans. Automatic Control*, Vol. AC-16, pp. 582–595, Dec. 1971.

PROBLEMS

6.1. A pair of complex-conjugate poles in the s-plane is required to meet the various specifications below. For each specification, sketch the region in the s-plane in which the poles may be located.
(a) $\zeta \geq 0.707$, $\omega_n \geq 2$ rad/sec   (positive damping)
(b) $0 \leq \zeta \leq 0.707$ $\omega \leq 2$ rad/sec   (positive damping)
(c) $\zeta \leq 0.5$, $2 \leq \omega_n \leq 4$ rad/sec   (positive damping)
(d) $0.5 \leq \zeta \leq 0.707$, $\omega_n \leq 2$ rad/sec   (positive and negative damping)

6.2. Determine the position, velocity, and acceleration error constants for the following control systems with unity feedback. The open-loop transfer functions are given by

(a) $G(s) = \dfrac{50}{(1 + 0.1s)(1 + 2s)}$

(b) $G(s) = \dfrac{K}{s(1 + 0.1s)(1 + 0.5s)}$

(c) $G(s) = \dfrac{K}{s(s^2 + 4s + 200)}$

(d) $G(s) = \dfrac{K(1 + 2s)(1 + 4s)}{s^2(s^2 + 2s + 10)}$

6.3. For the systems in Problem 6.2, determine the steady-state error for a unit step input, a unit ramp input, and an acceleration input $t^2/2$.

6.4. The open-loop transfer function of a control system with unity feedback is

$$G(s) = \frac{500}{s(1 + 0.1s)}$$

Evaluate the error series for the system. Determine the steady-state error of the system when the following inputs are applied:

(a) $r(t) = t^2 u_s(t)/2$

(b) $r(t) = (1 + 2t + t^2)u_s(t)$

Show that the steady-state error obtained from the error series is equal to the inverse Laplace transform of $E(s)$ with the terms generated by the poles of $E(s)/R(s)$ discarded.

6.5. In Problem 6.4, if a sinusoidal input $r(t) = \sin \omega t$ is applied to the system at $t = 0$, determine the steady-state error of the system by use of the error series for $\omega = 5$ rad/sec. What are the limitations in the error series when $r(t)$ is sinusoidal?

6.6. A machine-tool contouring control system is to cut the piecewise-linear contour shown in Fig. P6-6 with a two-axis control system. Sketch the reference input of each of the two-axis systems as a function of time.

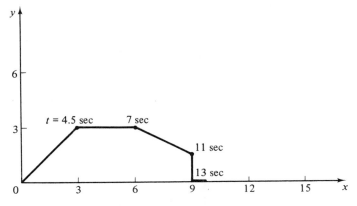

Figure P6-6.

6.7. A machine-tool contouring control system is to cut a perfect circular arc with a two-axis control system. Determine the reference inputs of the two systems that will accomplish this.

6.8. A step motor gives a single step response shown in Fig. P6-8 after a pulse exci-

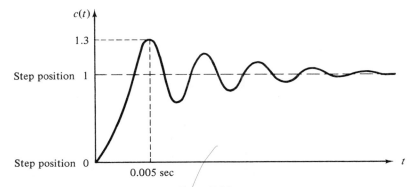

Figure P6-8.

tation is applied. Find a linear second-order transfer function to model the motor for this operation.

6.9. The attitude control of the missile shown in Fig. P5-11 is accomplished by thrust vectoring. The transfer function between the thrust angle δ and the angle of attack θ can be represented by

$$G_p(s) = \frac{\theta(s)}{\delta(s)} = \frac{K}{s^2 - a}$$

(refer to Problem 5.11) where K and a are positive constants. The attitude control system is represented by the block diagram in Fig. P6-9.

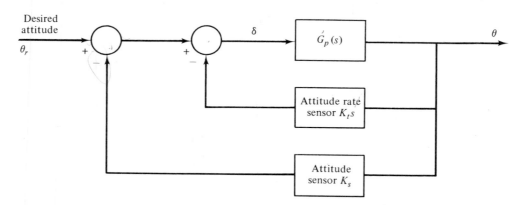

Figure P6-9.

(a) In Fig. P6-9, consider that only the attitude sensor loop is in operation ($K_t = 0$). Determine the performance of the overall system with respect to the relative values of K, K_s, and a.

(b) Consider that both loops are in operation. Determine the minimum values of K_t and K_s in terms of K and a so that the missile will not tumble.

(c) It is desired that the closed-loop system should have a damping ratio of ζ and a natural undamped frequency of ω_n. Find the values of K_s and K_t in terms of ζ, ω_n, a, and K.

6.10. A controlled process is represented by the following state equations:

$$\dot{x}_1 = x_1 - 3x_2$$
$$\dot{x}_2 = 5x_1 + u$$

The control is obtained from state feedback such that

$$u = -g_1 x_1 - g_2 x_2$$

where g_1 and g_2 are real constants.

(a) Find the locus in the g_1 versus g_2 plane on which the overall system has a natural undamped frequency of $\sqrt{2}$ rad/sec.

(b) Find the locus in the g_1 versus g_2 plane on which the overall system has a damping ratio of 70.7 per cent.

(c) Find the values of g_1 and g_2 such that $\zeta = 0.707$ and $\omega_n = \sqrt{2}$ rad/sec.

6.11. Given a linear time-invariant system which is described by the state equations

$$\dot{\mathbf{x}} = \mathbf{A}\mathbf{x} + \mathbf{B}u$$

where

$$
\mathbf{A} = \begin{bmatrix} 0 & 1 & 0 \\ 0 & 0 & 1 \\ -1 & -2 & -3 \end{bmatrix} \quad \mathbf{B} = \begin{bmatrix} 0 \\ 0 \\ 1 \end{bmatrix}
$$

The input u is a unit step function. Determine the equilibrium state which satisfies $\mathbf{Ax} + \mathbf{B}u = 0$.

6.12. Repeat Problem 6.11 when

$$
\mathbf{A} = \begin{bmatrix} 0 & 1 & 0 \\ 0 & 0 & 1 \\ 0 & -2 & -3 \end{bmatrix} \quad \mathbf{B} = \begin{bmatrix} 0 \\ 0 \\ 1 \end{bmatrix}
$$

6.13. The block diagram of a missile attitude control system is shown in Fig. P6-13; control is represented by $u(t)$, and the dynamics of the missile are represented by

$$
\frac{\theta_0(s)}{U(s)} = \frac{L}{Js^2}
$$

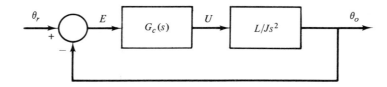

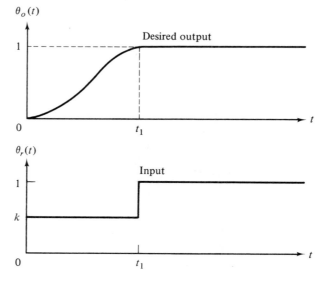

Figure P6-13.

The attitude controller is represented by $G_c(s)$, and $\theta_0(s)$ is the actual heading or output.

(a) With $G_c(s) = 1$, determine the response of the system, $\theta_0(t)$, when the input $\theta_r(t)$ is a unit step function. Assume zero initial conditions. Discuss the effects of L and J on this response.

(b) Let $G_c(s) = (1 + T_d s)$, $L = 10$, and $J = 1000$. Determine the value of T_d so that the system is critically damped.

(c) It is desired to obtain an output response with no overshoot (see Fig. P6-13); the response time may not be minimum. Let $G_c(s) = 1$, and the system is controlled through the input $\theta_r(t)$, which is chosen to be of the form shown. Determine the values of k and t_1 so that the desired output is obtained. Use the same values of L and J as given before.

6.14. The block diagram of a simple servo system is shown in Fig. P6-14.

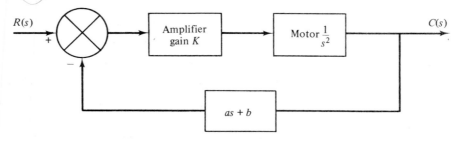

Figure P6-14.

(a) For $K = 10$, determine the values of a and b to give an overshoot of 16 per cent and a time constant of 0.1 sec of the system response to a unit step input. Time constant is defined here as the inverse of the damping factor.

(b) If the value of K is decreased slightly, how does it affect the damping ratio of the system?

skip ⟶ (c) Plot several points on the loci of roots of the system characteristic equation as K is varied from zero to infinity.

6.15. The parameters of the positioning servo system shown in Fig. P6-15 are given below:

J_L = load inertia		1 ft-lb/rad/sec²
B_L = load viscous friction		0.00143 ft-lb/rad/sec
J_m = motor inertia		8×10^{-4} ft-lb/rad/sec²
B_m = motor viscous friction		negligible
R_f = generator field resistance		50 Ω
L_f = generator field inductance		5 henry
R_a = total armature resistance of generator and		
motor		48.8 Ω
K_i = motor torque constant		0.812 ft-lb/A
K_g = generator constant		200 volts/amp
L_a = total armature inductance of generator and		
motor		negligible

(a) For an amplifier gain of $K = 100$, find the roots of the characteristic equation of the system. Locate these roots in the s-plane.

(b) For $K = 100$, evaluate the output response $\theta_L(t)$ when $\theta_r(t)$ is a unit step displacement input. Sketch the waveform of $\theta_L(t)$.

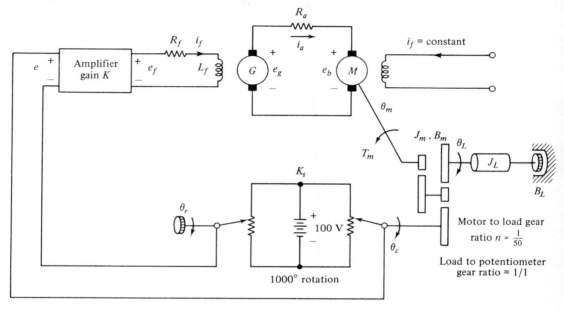

Figure P6-15.

(c) Repeat parts (a) and (b) for $K = 60.7$.

(d) Repeat parts (a) and (b) for $K = 50$. How does the steady-state response of $\theta_L(t)$ compare with the reference input $\theta_r(t)$?

6.16. The following parameters are given for the servo system shown in Fig. P6-16.

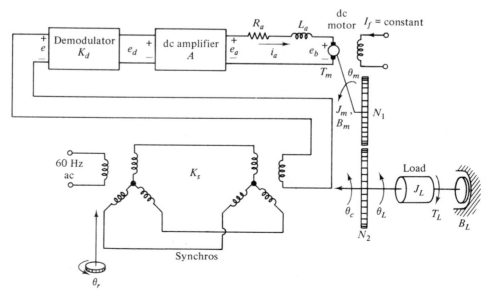

Figure P6-16.

K_s = sensitivity of error detector   1 volt/rad
J_L = load inertia   0.05 ft-lb/rad/sec²
B_L = load viscous friction   0.005 ft-lb/rad/sec
J_m = motor inertia   0.05 ft-lb/rad/sec²
B_i = motor viscous friction   negligible
K_i = motor torque constant   1 ft-lb/amp
L_a = motor armature inductance   negligible
R_a = motor armature resistance   10 Ω
K_d = gain of demodulator   1 volt/volt
Gear ratio $n = N_1/N_2$ 1 : 1

(a) Write the characteristic equation of the system and determine the value of
the dc amplifier gain A for a critically damped system.
(b) For a unit ramp function input $\theta_r(t) = t u_s(t)$, what should be the minimum
value of A so that the steady-state value of the response $\theta_L(t)$ will follow
the reference input with a positional error not exceeding 0.0186 rad? With
this gain setting, evaluate the output response $\theta_L(t)$.

6.17. In the feedback control system shown in Fig. P6-17, the sensitivity of the synchro error detector is 1 V/deg. After the entire system is set up, the transfer
function of the two-phase motor is determined experimentally as

$$\frac{\theta_m(s)}{E_2(s)} = \frac{K_m}{s(1 + T_m s)}$$

where $K_m = 10$ volts/sec, and $T_m = 0.1$ sec.

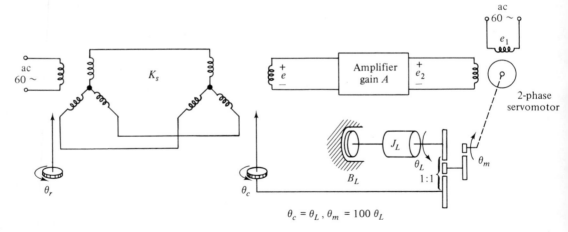

$$\theta_c = \theta_L, \; \theta_m = 100 \, \theta_L$$

Figure P6-17.

(a) If the load on the output shaft is to be driven in its steady state at a constant
speed of 30 rpm, what is the minimum value of gain A of the amplifier in
order that the deviation between output and input positions will not exceed
3° when the steady state is reached?
(b) The gain of the amplifier is given by $A = 35$. Determine the damping ratio
ζ and the undamped natural frequency of the system.

(c) The amplifier is modified to differentiate the error signal so that the output of the amplifier is written

$$e_2(t) = Ae(t) + AT_d\frac{de(t)}{dt}$$

where $A = 35$. Determine the value of T_d so that the damping ratio is 40 per cent. Repeat part (a) with the modified amplifier.

7

Stability of Control Systems

7.1 Introduction

It was shown in Chapter 6 that the transient response of a linear feedback control system is governed by the location of the roots of the characteristic equation. Basically, the design of linear feedback control systems may be regarded as a problem of arranging the location of the characteristic equation roots in such a way that the corresponding system will perform according to the prescribed specifications. We shall learn in Chapter 11 that there are conditions the system must satisfy before its characteristic equation roots can be *arbitrarily* assigned.

Among the many forms of performance specifications used in the design of control systems, the most important requirement is that the system must be stable at all times. However, there are various ways of defining stability, especially when we include all types of systems, such as linear and nonlinear.

In any case, the notion of stability is used to distinguish two classes of systems: *useful* and *useless*. From a practical standpoint, a stable system may be useful, whereas an unstable system is useless.

In essence, the analysis and design of linear control systems is centered on stability studies. The notion of stability may be expanded to include *absolute stability* and *relative stability*. Absolute stability refers to the condition of stable or unstable. Once the system is found to be stable, it is of interest to determine how stable it is, and the degree of stability is measured by relative stability. Parameters such as overshoot and damping ratio used in relation to the transient response in Chapter 6 provide indications of the relative stability of a linear time-invariant system in the time domain.

316

In this chapter we are concerned with the subject of absolute stability of control systems, which is simply a *yes* or *no* proposition.

From the illustrative examples of Chapter 6 we may summarize the relation between the transient response and the characteristic equation roots as follows:

1. When all the roots of the characteristic equation are found in the left half of the *s*-plane, the system responses due to the initial conditions will decrease to zero as time approaches infinity.
2. If one or more pairs of simple roots are located on the imaginary axis of the *s*-plane, but there are no roots in the right half of the *s*-plane, the responses due to initial conditions will be undamped sinusoidal oscillations.
3. If one or more roots are found in the right half of the *s*-plane, the responses will increase in magnitude as time increases.

In linear system theory, the last two categories are usually defined as *unstable* conditions. Note that the responses referred to in the above conditions are due to initial conditions only, and they are often called the *zero-input responses*.

7.2 Stability, Characteristic Equation, and the State Transition Matrix

We can show from a more rigorous approach that the zero-input stability of a linear time-invariant system is determined by the location of the roots of the characteristic equation or the behavior of the state transition matrix $\boldsymbol{\phi}(t)$.

Let a linear time-invariant system be described by the state equation

$$\dot{\mathbf{x}}(t) = \mathbf{A}\mathbf{x}(t) + \mathbf{B}\mathbf{u}(t) \tag{7-1}$$

where $\mathbf{x}(t)$ is the state vector and $\mathbf{u}(t)$ the input vector. For zero input, $\mathbf{x}(t) = \mathbf{0}$ satisfies the homogeneous state equation $\dot{\mathbf{x}}(t) = \mathbf{A}\mathbf{x}(t)$ and is defined as the *equilibrium state* of the system. The zero-input stability is defined as follows: *If the zero-input response* $\mathbf{x}(t)$, *subject to finite initial state* $\mathbf{x}(t_0)$, *returns to the equilibrium state* $\mathbf{x}(t) = \mathbf{0}$ *as t approaches infinity, the system is said to be stable; otherwise, the system is unstable. This type of stability is also known as the asymptotic stability.*

In a more mathematical manner, the foregoing definition may be stated: *A linear time-invariant system is said to be stable (zero input) if for any finite initial state* $\mathbf{x}(t_0)$ *there is a positive number M [which depends on* $\mathbf{x}(t_0)$*] such that*

(1) $$\|\mathbf{x}(t)\| < M \qquad \text{for all } t \geq t_0 \tag{7-2}$$

and

(2) $$\lim_{t \to \infty} \|\mathbf{x}(t)\| = 0 \tag{7-3}$$

where $||\mathbf{x}(t)||$ represents the norm* of the state vector $\mathbf{x}(t)$, or

$$||\mathbf{x}(t)|| = \left[\sum_{i=1}^{n} x_i^2(t) \right]^{1/2} \tag{7-4}$$

The condition stated in Eq. (7-2) implies that the transition of state for any $t > t_0$ as represented by the norm of the vector $\mathbf{x}(t)$ must be bounded. Equation (7-3) states that the system must reach its equilibrium state as time approaches infinity.

The interpretation of the stability criterion in the state space is illustrated by the second-order case shown in Fig. 7-1. The state trajectory represents the transition of $\mathbf{x}(t)$ for $t > t_0$ from a finite initial state $\mathbf{x}(t_0)$. As shown in the figure, $\mathbf{x}(t_0)$ is represented by a point that is the tip of the vector obtained from the vector sum $x_1(t_0)$ and $x_2(t_0)$. A cylinder with radius M forms the upper bound for the trajectory points for all $t > t_0$, and as t approaches infinity, the system reaches its equilibrium state $\mathbf{x}(t) = \mathbf{0}$.

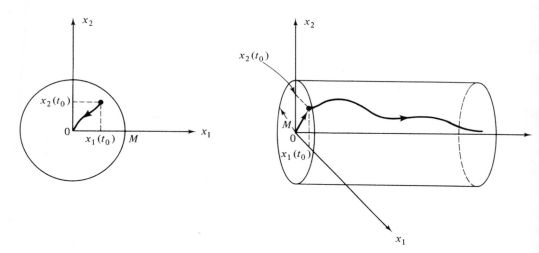

Fig. 7-1. Stability concept illustrated in the state space.

Next we shall show that the definition of stability of linear time-invariant systems given above leads to the same conclusion on the condition of the roots of the characteristic equation. For the zero-input condition, the state transition equation of the system is

$$\mathbf{x}(t) = \boldsymbol{\phi}(t - t_0)\mathbf{x}(t_0) \tag{7-5}$$

where $\boldsymbol{\phi}(t - t_0)$ is the state transition matrix.

Taking the norm on both sides of Eq. (7-5) gives

$$||\mathbf{x}(t)|| = ||\boldsymbol{\phi}(t - t_0)\mathbf{x}(t_0)|| \tag{7-6}$$

*The norm of a vector is the generalization of the idea of length. $||\mathbf{x}||$ is always a real number.

An important property of the norm of a vector is

$$\| \mathbf{x}(t) \| \le \| \boldsymbol{\phi}(t - t_0) \| \, \| \mathbf{x}(t_0) \| \tag{7-7}$$

which is analogous to the relation between lengths of vectors.

Then the condition in Eq. (7-2) requires that $\| \boldsymbol{\phi}(t - t_0) \| \, \| \mathbf{x}(t_0) \|$ be finite. Thus, if $\| \mathbf{x}(t_0) \|$ is finite as postulated, $\| \boldsymbol{\phi}(t - t_0) \|$ must also be finite for $t > t_0$. Similarly, Eq. (7-3) leads to the condition that

$$\lim_{t \to \infty} \| \boldsymbol{\phi}(t - t_0) \| = 0 \tag{7-8}$$

or

$$\lim_{t \to \infty} \phi_{ij}(t - t_0) = 0 \tag{7-9}$$

$i, j = 1, 2, \ldots, n$, where $\phi_{ij}(t - t_0)$ is the ijth element of $\boldsymbol{\phi}(t - t_0)$.

In Eq. (4-42) the state transition matrix is written

$$\boldsymbol{\phi}(t) = \mathcal{L}^{-1}[(s\mathbf{I} - \mathbf{A})^{-1}] \tag{7-10}$$

or

$$\boldsymbol{\phi}(t) = \mathcal{L}^{-1}\left[\frac{\text{adj}\,(s\mathbf{I} - \mathbf{A})}{|s\mathbf{I} - \mathbf{A}|}\right] \tag{7-11}$$

Since $|s\mathbf{I} - \mathbf{A}| = 0$ is the characteristic equation of the system, Eq. (7-11) implies that the time response of $\boldsymbol{\phi}(t)$ is governed by the roots of the characteristic equation. Thus the condition in Eq. (7-8) requires that the roots of the characteristic equation must all have negative real parts.

7.3 Stability of Linear Time-Invariant Systems with Inputs

Although the stability criterion for linear time-invariant systems given in the preceding section is for the zero-input condition, we can show that, in general, the stability condition for this class of systems is *independent* of the inputs. An alternative way of defining stability of linear time-invariant systems is as follows: *A system is stable if its output is bounded for any bounded input.*

In other words, let $c(t)$ be the output and $r(t)$ the input of a linear system with a single input–output. If

$$|r(t)| \le N < \infty \qquad \text{for } t \ge t_0 \tag{7-12}$$

then

$$|c(t)| \le M < \infty \qquad \text{for } t \ge t_0 \tag{7-13}$$

However, there are a few exceptions to the foregoing definition. A differentiator gives rise to an impulse response at $t = t_0$ when it is subjected to a unit step function input. In this case the amplitude of the input is bounded, but the amplitude of the output is not, since an impulse is known to have an infinite amplitude. Also, when a unit step function is applied to a perfect integrator, the output is an unbounded ramp function. However, since a differentiator and an integrator are all useful systems, they are defined as stable systems, and are exceptions to the stability condition defined above.

We shall show that the bounded input–bounded output definition of stability again leads to the requirement that the roots of the characteristic equation be located in the left half of the s-plane.

Let us express the input–output relation of a linear system by the convolution integral

$$c(t) = \int_0^\infty r(t - \tau)g(\tau)\, d\tau \tag{7-14}$$

where $g(\tau)$ is the impulse response of the system.

Taking the absolute value on both sides of Eq. (7-14), we get

$$|c(t)| = \left| \int_0^\infty r(t - \tau)g(\tau)\, d\tau \right| \tag{7-15}$$

Since the absolute value of an integral is not greater than the integral of the absolute value of the integrand, Eq. (7-15) is written

$$|c(t)| \leq \int_0^\infty |r(t - \tau)|\,|g(\tau)|\, d\tau \tag{7-16}$$

Now if $r(t)$ is a bounded signal, then, from Eq. (7-12),

$$|c(t)| \leq \int_0^\infty N\,|g(\tau)|\, d\tau = N \int_0^\infty |g(\tau)|\, d\tau \tag{7-17}$$

Therefore, if $c(t)$ is to be a bounded output,

$$N \int_0^\infty |g(\tau)|\, d\tau \leq M < \infty \tag{7-18}$$

or

$$\int_0^\infty |g(\tau)|\, d\tau \leq P < \infty \tag{7-19}$$

A physical interpretation of Eq. (7-19) is that the area under the absolute-value curve of the impulse response $g(t)$, evaluated from $t = 0$ to $t = \infty$, must be finite.

We shall now show that the requirement on the impulse response for stability can be linked to the restrictions on the characteristic equation roots. By definition, the transfer function $G(s)$ of the system and the impulse response $g(t)$ are related through the Laplace transform integral

$$G(s) = \int_0^\infty g(t)e^{-st}\, dt \tag{7-20}$$

Taking the absolute value on the left side of Eq. (7-20) gives

$$|G(s)| \leq \int_0^\infty |g(t)|\,|e^{-st}|\, dt \tag{7-21}$$

The roots of the characteristic equation are the poles of $G(s)$, and when s takes on these values, $|G(s)| = \infty$. Also, $s = \sigma + j\omega$; the absolute value of e^{-st} is $|e^{-\sigma t}|$. Equation (7-21) becomes

$$\infty \leq \int_0^\infty |g(t)|\,|e^{-\sigma t}|\, dt \tag{7-22}$$

If one or more roots of the characteristic equation are in the right half or on the imaginary axis of the s-plane, $\sigma \geq 0$, and thus $|e^{-\sigma t}| \leq N = 1$. Thus

Eq. (7-22) is written

$$\infty \leq \int_{0}^{\infty} N\,|g(t)|\,dt = \int_{0}^{\infty} |g(t)|\,dt \qquad (7\text{-}23)$$

for $\text{Re}(s) = \sigma \geq 0$.

Since Eq. (7-23) contradicts the stability criterion given in Eq. (7-19), we conclude that *for the system to be stable, the roots of the characteristic equation must all lie inside the left half of the s-plane.*

The discussions conducted in the preceding sections lead to the conclusions that the stability of linear time-invariant systems can be determined by checking whether any roots of the characteristic equation are in the right half or on the imaginary axis of the s-plane. The regions of stability and instability in the s-plane are illustrated in Fig. 7-2. The imaginary axis, excluding the origin, is included in the unstable region.

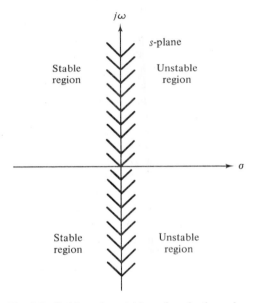

Fig. 7-2. Stable and unstable regions in the s-plane.

7.4 Methods of Determining Stability of Linear Control Systems

Although the stability of linear time-invariant systems may be checked by investigating the impulse response, the state transition matrix, or by finding the roots of the characteristic equation, these criteria are difficult to implement in practice. For instance, the impulse response is obtained by taking the inverse Laplace transform of the transfer function, which is not always a simple task; a similar process is required to evaluate the state transition matrix $\phi(t)$. The solving of the roots of a high-order polynomial can only be carried out by a digital computer. In practice, therefore, the stability analysis of a linear system is seldom carried out by working with the impulse response or the state transition matrix, or even by finding the exact location of the roots of the characteristic equation. In general, we are interested in algorithms that are straightforward to apply and which can provide answers to stability or instability, without excessive computations. The methods outlined below are frequently used for the stability studies of linear time-invariant systems.

1. *Routh–Hurwitz criterion*[1]: an algebraic method that provides information on the absolute stability of a linear time-invariant system. The criterion tests whether any roots of the characteristic equation lie in the right half of the *s*-plane. The number of roots that lie on the imaginary axis and in the right half of the *s*-plane is also indicated.

2. *Nyquist criterion*[6]: a semigraphical method that gives information on the difference between the number of poles and zeros of the closed loop transfer function by observing the behavior of the Nyquist plot of the loop transfer function. The poles of the closed-loop transfer function are the roots of the characteristic equation. This method requires that we know the relative location of the zeros of the closed-loop transfer function.

3. *Root locus plot* (Chapter 8): represents a diagram of loci of the characteristic equation roots when a certain system parameter varies. When the root loci lie in the right half of the *s*-plane, the closed-loop system is unstable.

4. *Bode diagram* (Appendix A): the Bode plot of the loop transfer function $G(s)H(s)$ may be used to determine the stability of the closed-loop system. However, the method can be used only if $G(s)H(s)$ has no poles and zeros in the right-half *s*-plane.

5. *Lyapunov's stability criterion*: a method of determining the stability of nonlinear systems, although it can also be applied to linear systems. The stability of the system is determined by checking on the properties of the *Lyapunov function* of the system.

7.5 Routh–Hurwitz Criterion[1-5]

The Routh–Hurwitz criterion represents a method of determining the location of zeros of a polynomial with constant real coefficients with respect to the left half and the right half of the *s*-plane, without actually solving for the zeros. The method can be applied to systems with single input and output and with multiple inputs and outputs, as well as single- or multiple-loop systems.

Consider that the characteristic equation of a linear time-invariant system is of the form

$$F(s) = a_0 s^n + a_1 s^{n-1} + a_2 s^{n-2} + \ldots + a_{n-1} s + a_n = 0 \qquad (7\text{-}24)$$

where all the coefficients are real numbers.

In order that there be no roots of the last equation with positive real parts, it is necessary but not sufficient that*

1. All the coefficients of the polynomial have the same sign.
2. None of the coefficients vanishes.

These two necessary conditions can easily be checked by inspection. However, these conditions are not sufficient; it is quite possible that a polynomial with all its coefficients nonzero and of the same sign still have zeros in the right half of the s-plane.

The necessary and sufficient condition that all the roots of Eq. (7-24) lie in the left half of the s-plane is that the polynomial's Hurwitz determinants, D_k, $k = 1, 2, \ldots, n$, must be all positive.

The Hurwitz determinants of Eq. (7-24) are given by

$$D_1 = a_1 \qquad D_2 = \begin{vmatrix} a_1 & a_3 \\ a_0 & a_2 \end{vmatrix} \qquad D_3 = \begin{vmatrix} a_1 & a_3 & a_5 \\ a_0 & a_2 & a_4 \\ 0 & a_1 & a_3 \end{vmatrix}$$

$$\cdots \qquad D_n = \begin{vmatrix} a_1 & a_3 & a_5 & \cdots & a_{2n-1} \\ a_0 & a_2 & a_4 & \cdots & a_{2n-2} \\ 0 & a_1 & a_3 & \cdots & a_{2n-3} \\ 0 & a_0 & a_2 & \cdots & a_{2n-4} \\ 0 & 0 & a_1 & \cdots & a_{2n-5} \\ \cdots & \cdots & \cdots & \cdots & \cdots \\ 0 & 0 & 0 & \cdots & a_n \end{vmatrix} \qquad (7\text{-}25)$$

where the coefficients with indices larger than n or with negative indices are replaced by zeros.

At first glance the application of the Hurwitz determinants may seem to be formidable for high-order polynomials, because of the labor involved in evaluating the determinants in Eq. (7-25). Fortunately, the rule was simplified by Routh into a tabulation, so one does not have to work with the determinants of Eq. (7-25).

---

*   From the basic laws of algebra, the following relations are observed for the polynomial in Eq. (7-24):

$$\frac{a_1}{a_0} = -\Sigma \text{ all roots}$$

$$\frac{a_2}{a_0} = \Sigma \text{ products of the roots taken 2 at a time}$$

$$\frac{a_3}{a_0} = -\Sigma \text{ products of the roots taken 3 at a time}$$

$$\vdots$$

$$\frac{a_n}{a_0} = (-1)^n \text{ products of all roots}$$

All the ratios must be positive and nonzero unless at least one of the roots has a positive real part.

The first step in the simplification of the Routh–Hurwitz criterion is to arrange the polynomial coefficients into two rows. The first row consists of the first, third, fifth, ... coefficients, and the second row consists of the second, the fourth, sixth, ... coefficients, as shown in the following tabulation:

$$a_0 \quad a_2 \quad a_4 \quad a_6 \quad a_8 \quad \cdots$$
$$a_1 \quad a_3 \quad a_5 \quad a_7 \quad a_9 \quad \cdots$$

The next step is to form the following array of numbers by the indicated operations (the example shown is for a sixth-order system):

s^6	a_0	a_2	a_4	a_6
s^5	a_1	a_3	a_5	0
s^4	$\dfrac{a_1 a_2 - a_0 a_3}{a_1} = A$	$\dfrac{a_1 a_4 - a_0 a_5}{a_1} = B$	$\dfrac{a_1 a_6 - a_0 \times 0}{a_1} = a_6$	0
s^3	$\dfrac{A a_3 - a_1 B}{A} = C$	$\dfrac{A a_5 - a_1 a_6}{A} = D$	$\dfrac{A \times 0 - a_1 \times 0}{A} = 0$	0
s^2	$\dfrac{BC - AD}{C} = E$	$\dfrac{C a_6 - A \times 0}{C} = a_6$	$\dfrac{C \times 0 - A \times 0}{C} = 0$	0
s^1	$\dfrac{ED - C a_6}{E} = F$	0	0	0
s^0	$\dfrac{F a_6 - E \times 0}{F} = a_6$	0	0	0

The array of numbers and operations given above is known as the *Routh tabulation* or the *Routh array*. The column of *s*s on the left side is used for identification purpose.

Once the Routh tabulation has been completed, the last step in the Routh–Hurwitz criterion is to investigate the signs of the numbers in the first column of the tabulation. The following conclusions are drawn: *The roots of the polynomial are all in the left half of the s-plane if all the elements of the first column of the Routh tabulation are of the same sign. If there are changes of signs in the elements of the first column, the number of sign changes indicates the number of roots with positive real parts.*

The reason for the foregoing conclusion is simple, based on the requirements on the Hurwitz determinants. The relations between the elements in the first column of the Routh tabulation and the Hurwitz determinants are:

$$
\begin{aligned}
s^6 \qquad & a_0 = a_0 \\
s^5 \qquad & a_1 = D_1 \\
s^4 \qquad & A = \frac{D_2}{D_1} \\
s^3 \qquad & C = \frac{D_3}{D_2} \\
s^2 \qquad & E = \frac{D_4}{D_3} \\
s^1 \qquad & F = \frac{D_5}{D_4} \\
s^0 \qquad & a_6 = \frac{D_6}{D_5}
\end{aligned}
$$

Therefore, if all the Hurwitz determinants are positive, the elements in the first column would also be of the same sign.

The following two examples illustrate the application of the Routh–Hurwitz criterion to simple problems.

EXAMPLE 7-1   Consider the equation

$$(s - 2)(s + 1)(s - 3) = s^3 - 4s^2 + s + 6 = 0 \tag{7-26}$$

which has one negative coefficient. Thus, from the necessary condition, we know without applying the Routh–Hurwitz test that the equation has at least one root with positive real parts. But for the purpose of illustrating the Routh–Hurwitz criterion, the Routh tabulation is formed as follows:

s^3	1	1
s^2	-4	6
s^1	$\dfrac{(-4)(1) - (6)(1)}{-4} = 2.5$	0
s^0	$\dfrac{(2.5)(6) - (-4)(0)}{2.5} = 6$	0

Sign change (between s^3 and s^2)
Sign change (between s^2 and s^1)

Since there are two sign changes in the first column of the tabulation, the polynomial has two roots located in the right half of the s-plane. This agrees with the known result, as Eq. (7-26) clearly shows that the two right-half plane roots are at $s = 2$ and $s = 3$.

EXAMPLE 7-2   Consider the equation

$$2s^4 + s^3 + 3s^2 + 5s + 10 = 0 \tag{7-27}$$

Since the equation has no missing terms and the coefficients are all of the same sign, it satisfies the necessary condition for not having roots in the right half of the s-plane or on the imaginary axis. However, the sufficient condition must still be checked. The Routh tabulation is made as follows:

s^4	2	3	10
s^3	1	5	0
s^2	$\dfrac{(1)(3) - (2)(5)}{1} = -7$	10	0
s^1	$\dfrac{(-7)(5) - (1)(10)}{-7} = 6.43$	0	0
s^0	10		

Sign change (between s^3 and s^2)
Sign change (between s^2 and s^1)

Since there are two changes in sign in the first column, the equation has two roots in the right half of the s-plane.

Special Cases

The two illustrative examples given above are designed so that the Routh–Hurwitz criterion can be carried out without any complications. However, depending upon the equation to be tested, the following difficulties may occur occasionally when carrying out the Routh test:

1. The first element in any one row of the Routh tabulation is zero, but the other elements are not.
2. The elements in one row of the Routh tabulation are all zero.

In the first case, if a zero appears in the first position of a row, the elements in the next row will all become infinite, and the Routh test breaks down. This situation may be corrected by multiplying the equation by the factor $(s + a)$, where a is any number,* and then carry on the usual tabulation.

EXAMPLE 7-3 Consider the equation

$$(s - 1)^2(s + 2) = s^3 - 3s + 2 = 0 \tag{7-28}$$

Since the coefficient of the s^2 term is zero, we know from the necessary condition that at least one root of the equation is located in the right half of the s-plane. To determine how many of the roots are in the right-half plane, we carry out the Routh tabulation as follows:

$$
\begin{array}{ccc}
s^3 & 1 & -3 \\
s^2 & 0 & 2 \\
s^1 & \infty &
\end{array}
$$

Because of the zero in the first element of the second row, the first element of the third row is infinite. To correct this situation, we multiply both sides of Eq. (7-28) by the factor $(s + a)$, where a is an arbitrary number. The simplest number that enters one's mind is 1. However, for reasons that will become apparent later, we do not choose a to be 1 or 2. Let $a = 3$; then Eq. (7-28) becomes

$$(s - 1)^2(s + 2)(s + 3) = s^4 + 3s^3 - 3s^2 - 7s + 6 = 0 \tag{7-29}$$

The Routh tabulation of Eq. (7-29) is

$$
\begin{array}{cccc}
s^4 & 1 & -3 & 6 \\
s^3 & 3 & -7 & 0 \\
\text{Sign change} & & & \\
s^2 & \dfrac{-9 + 7}{3} = -\dfrac{2}{3} & 6 & 0 \\
\text{Sign change} & & & \\
s^1 & \dfrac{(-\frac{2}{3})(-7) - 18}{-\frac{2}{3}} = 20 & 0 & \\
s^0 & 6 & &
\end{array}
$$

Since there are two changes in sign in the first column of the Routh tabulation, the equation has two roots in the right half of the s-plane.

As an alternative to the remedy of the situation described above, we may replace the zero element in the Routh tabulation by an arbitrary small positive

---

*If one chooses to use a negative number for a, the $(s + a)$ term will contribute a root in the right half of the s-plane, and this root must be taken into account when interpreting the Routh tabulation.

number ϵ and then proceed with the Routh test. For instance, for the equation given in Eq. (7-28), we may replace the zero element in the second row of the Routh tabulation by ϵ; then we have

	s^3	1	-3
	s^2	ϵ	2
Sign change			
	s^1	$\dfrac{-3\epsilon - 2}{\epsilon}$	0
Sign change			
	s^0	2	

Since ϵ is postulated to be a small positive number, $(-3\epsilon-2)/\epsilon$ approaches $-2/\epsilon$, which is a negative number; thus the first column of the last tabulation has two sign changes. This agrees with the result obtained earlier. On the other hand, we may assume ϵ to be negative, and we can easily verify that the number of sign changes is still two, but they are between the first three rows.

In the second special case, when all the elements in one row of the Routh tabulation are zeros, it indicates that one or more of the following conditions may exist:

1. Pairs of real roots with opposite signs.
2. Pairs of imaginary roots.
3. Pairs of complex-conjugate roots forming symmetry about the origin of the s-plane.

The equation that is formed by using the coefficients of the row just above the row of zeros is called the auxiliary equation. The order of the auxiliary equation is always even, and it indicates the number of root pairs that are equal in magnitude but opposite in sign. For instance, if the auxiliary equation is of the second order, there are two equal and opposite roots. For a fourth-order auxiliary equation, there must be two pairs of equal and opposite roots. *All these roots of equal magnitude can be obtained by solving the auxiliary equation.*

When a row of zeros appear in the Routh tabulation, again the test breaks down. The test may be carried on by performing the following remedies:

1. Take the derivative of the auxiliary equation with respect to s.
2. Replace the row of zeros with the coefficients of the resultant equation obtained by taking the derivative of the auxiliary equation.
3. Carry on the Routh test in the usual manner with the newly formed tabulation.

EXAMPLE 7-4   Consider the same equation, Eq. (7-28), which is used in Example 7-3. In multiplying this equation by a factor $(s + a)$, logically the first number that comes into one's mind would be $a = 1$. Multiplying both sides of Eq. (7-28) by $(s + 1)$, we have

$$(s - 1)^2(s + 2)(s + 1) = s^4 + s^3 - 3s^2 - s + 2 = 0 \qquad (7\text{-}30)$$

The Routh tabulation is made as follows:

$$
\begin{array}{cccc}
s^4 & 1 & -3 & 2 \\
s^3 & 1 & -1 & 0 \\
s^2 & \dfrac{(1)(-3)-(1)(-1)}{1}=-2 & 2 & \\
s^1 & \dfrac{2-2}{-2}=0 & 0 &
\end{array}
$$

Since the s^1 row contains all zeros, the Routh test terminates prematurely. The difficulty in this case is due to the multiplication of the original equation, which already has a root at $s = 1$, by the factor $(s + 1)$. This makes the new equation fit special case (2). To remedy this situation, we form the auxiliary equation using the coefficients contained in the s^2 row, that is, the row preceding the row of zeros. Thus the auxiliary equation is written

$$A(s) = -2s^2 + 2 = 0 \tag{7-31}$$

Taking the derivative of $A(s)$ with respect to s gives

$$\frac{dA(s)}{ds} = -4s \tag{7-32}$$

Now, the row of zeros in the Routh tabulation is replaced by the coefficients of Eq. (7-32), and the new tabulation reads as follows:

$$
\begin{array}{clcl}
& s^4 & 1 \quad -3 \quad 2 & \\
& s^3 & 1 \quad -1 & \\
\text{Sign change} & & & \\
& s^2 & -2 \quad 2 & \text{(coefficients of auxiliary equations)} \\
& s^1 & -4 \quad 0 & [\text{coefficients of } dA(s)/ds] \\
\text{Sign change} & & & \\
& s^0 & 2 \quad 0 &
\end{array}
$$

Since the preceding tabulation has two sign changes, the equation of Eq. (7-28) has two roots in the right-half plane. By solving the roots of the auxiliary equation in Eq. (7-31), we have

$$s^2 = 1 \qquad \text{or} \qquad s = \pm 1$$

These are also the roots of the equation in Eq. (7-30). It should be remembered that the roots of the auxiliary equation are also roots of the original equation, which is under the Routh test.

EXAMPLE 7-5   In this example we shall consider equations with imaginary roots. Consider

$$(s + 2)(s - 2)(s + j)(s - j)(s^2 + s + 1)$$
$$= s^6 + s^5 - 2s^4 - 3s^3 - 7s^2 - 4s - 4 = 0 \tag{7-33}$$

which is known to have two pairs of equal roots with opposite signs at $s = \pm 2$ and $s = \pm j$. The Routh tabulation is

$$
\begin{array}{ccccc}
s^6 & 1 & -2 & -7 & -4 \\
s^5 & 1 & -3 & -4 & \\
s^4 & 1 & -3 & -4 & \\
s^3 & 0 & 0 & 0 &
\end{array}
$$

Since a row of zeros appears prematurely, we form the auxiliary equation using the coefficients of the s^4 row. The auxiliary equation is

$$A(s) = s^4 - 3s^2 - 4 = 0 \tag{7-34}$$

The derivative of $A(s)$ with respect to s is

$$\frac{dA(s)}{ds} = 4s^3 - 6s = 0 \tag{7-35}$$

from which the coefficients 4 and -6 are used to replace the row of zeros in the Routh tabulation. The new Routh tabulation is

s^6	1	-2	-7	-4
s^5	1	-3	-4	
s^4	1	-3	-4	
s^3	4	-6	$0 \left[\text{coefficients of } \frac{dA(s)}{ds}\right]$	
s^2	-1.5	-4	0	
s^1	-16.7	0		
s^0	-4	0		

Sign change

Since there is one change in sign in the first column of the new Routh tabulation, Eq. (7-33) has one root in the right half of the s-plane. The equal, but opposite roots that caused the all-zero row to occur are solved from the auxiliary equation. From Eq. (7-34) these roots are found to be

$$s = +2, -2, +j, \text{ and } -j.$$

A frequent use of the Routh–Hurwitz criterion is for a quick check of the stability and the simple design of a linear feedback control system. For example, the control system with integral control shown in Fig. 6-29 has the characteristic equation

$$s^3 + 34.5s^2 + 7500s + 7500K_1 = 0 \tag{7-36}$$

The Routh–Hurwitz criterion may be used to determine the range of K_1 for which the closed-loop system is stable. The Routh tabulation of Eq. (7-36) is

s^3	1	7500
s^2	34.5	$7500K_1$
s^1	$\dfrac{258,750 - 7500K_1}{34.5}$	0
s^0	$7500K_1$	

For the system to be stable, all the coefficients in the first column of the last tabulation should be of the same sign. This leads to the following conditions:

$$\frac{258,750 - 7500K_1}{34.5} > 0 \tag{7-37}$$

$$7500K_1 > 0 \tag{7-38}$$

From the condition in Eq. (7-37) we have

$$K_1 < 34.5 \tag{7-39}$$

and the condition in Eq. (7-38) gives

$$K_1 > 0 \qquad (7\text{-}40)$$

Therefore, the condition for stability is that K_1 must satisfy

$$0 < K_1 < 34.5 \qquad (7\text{-}41)$$

EXAMPLE 7-6   Consider the characteristic equation of a certain closed-loop control system,

$$s^3 + 3Ks^2 + (K + 2)s + 4 = 0 \qquad (7\text{-}42)$$

It is desired to determine the range of K so that the system is stable. The Routh tabulation of Eq. (7-42) is

s^3	1	$(K + 2)$
s^2	$3K$	4
s^1	$\dfrac{3K(K + 2) - 4}{3K}$	0
s^0	4	

From the s^2 row, the condition of stability is

$$K > 0$$

and from the s^1 row, the condition of stability is

$$3K^2 + 6K - 4 > 0 \quad \text{or} \quad K < -2.528 \quad \text{or} \quad K > 0.528$$

When the conditions of $K > 0$ and $K > 0.528$ are compared, it is apparent that the latter limitation is the more stringent one. Thus for the closed-loop system to be stable, K must satisfy

$$K > 0.528$$

The requirement of $K < -2.528$ is disregarded since K cannot be negative.

It should be reiterated that the Routh–Hurwitz criterion is valid only if the characteristic equation is algebraic and all the coefficients are real. If any one of the coefficients of the characteristic equation is a complex number, or if the equation contains exponential functions of s, such as in the case of a system with time delays, the Routh–Hurwitz criterion cannot be applied.

Another limitation of the Routh–Hurwitz criterion is that it offers information only on the absolute stability of the system. If a system is found to be stable by the Routh's test, one still does not know how good the system is— in other words, how closely the roots of the characteristic equation roots are located to the imaginary axis of the s-plane. On the other hand, if the system is shown to be unstable, the Routh–Hurwitz criterion gives no indication of how the system can be stabilized.

7.6 Nyquist Criterion[6-15]

Thus far, two methods of determining stability by investigating the location of the roots of the characteristic equation have been indicated:

1. The roots of the characteristic equation are actually solved. This

step can be carried out on a digital computer by means of a root-finding subroutine.

2. The relative location of the roots with respect to the imaginary axis of the s-plane is determined by means of the Routh–Hurwitz criterion.

These two methods are not very useful for design purposes. It is necessary to devise stability criteria that will enable the analyst to determine the relative stability of the system. It would be desirable if the criterion indicates how the stability of the system might be improved.

The Nyquist criterion possesses the following features that make it desirable for the analysis as well as the design of control systems:

1. It provides the same amount of information on the absolute stability of a control system as the Routh–Hurwitz criterion.
2. In addition to absolute system stability, the Nyquist criterion indicates the degree of stability of a stable system and gives an indication of how the system stability may be improved, if needed.
3. It gives information on the frequency-domain response of the system.
4. It can be used for a stability study of systems with time delay.
5. It can be modified for nonlinear systems.

We may formulate the Nyquist criterion by first using modern control notation, that is, state equations. Let a linear control system with a single input be represented by the following set of equations:

$$\dot{\mathbf{x}}(t) = \mathbf{A}\mathbf{x}(t) + \mathbf{B}e(t) \tag{7-43}$$

$$c(t) = \mathbf{D}\mathbf{x}(t) \tag{7-44}$$

$$e(t) = r(t) - c(t) \tag{7-45}$$

A block diagram that gives the transfer relations of the system is shown in Fig. 7-3. This is known as a *closed-loop system with unity feedback*. The open-loop transfer function of the system is defined as

$$G(s) = \mathbf{D}(s\mathbf{I} - \mathbf{A})^{-1}\mathbf{B} \tag{7-46}$$

When the feedback is nonunity, with the feedback dynamics represented by the

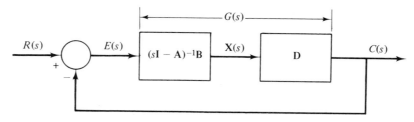

Fig. 7-3. Block diagram of a closed-loop control system with unity feedback.

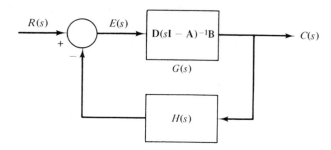

Fig. 7-4. Block diagram of a closed-loop control system.

transfer function $H(s)$, the system block diagram is modified as shown in Fig. 7-4.

The closed-loop transfer function of the system of Fig. 7-4 is

$$\frac{C(s)}{R(s)} = \frac{G(s)}{1 + G(s)H(s)} \tag{7-47}$$

Let the denominator of the closed-loop transfer function be represented by $F(s)$. Then

$$F(s) = 1 + G(s)H(s) = 1 + \mathbf{D}(s\mathbf{I} - \mathbf{A})^{-1}\mathbf{B}H(s) \tag{7-48}$$

It is easy to see that the zeros of $F(s)$ are the roots of the characteristic equation of the system, if there is no pole-zero cancellation in $G(s)H(s)$.

Let us assume that $G(s)$ and $H(s)$ are rational functions; that is, $G(s)H(s)$ is a quotient of two polynomials with constant coefficients. In general, $F(s)$ can be written

$$F(s) = \frac{K(s + z_1)(s + z_2)\ldots(s + z_m)}{s^j(s + p_1)(s + p_2)\ldots(s + p_n)} \tag{7-49}$$

where z_i $(i = 1, 2, \ldots, m)$ and p_k $(k = 1, 2, \ldots, n)$ are either real or in complex-conjugate pairs. Then the roots of the characteristic equations are $s = -z_1$, $-z_2, \ldots, -z_m$. For the closed-loop system to be stable, it is required that none of these roots has a positive real part. There is no particular restriction on the location of the poles of $F(s)$, which are at $s = 0, -p_1, -p_2, \ldots, -p_n$. It is important to note that the poles of $F(s)$ are the same as those of $G(s)H(s)$. If any one of the poles of $G(s)H(s)$ lies in the right half of the s-plane, the open-loop system is said to be unstable; however, the closed-loop system can still be stable, if all the zeros of $F(s)$ are found in the left half of the s-plane. This is a very important feature of a feedback control system. In preceding chapters it has been pointed out that a high forward-path gain generally reduces the steady-state error of a feedback control system. Consequently, it is desirable to use high gain in multiple-loop control systems. Although this practice may result in an unstable inner-loop system, the entire closed-loop system can be made stable by proper design.

Before embarking on the fundamentals of the Nyquist criterion, it is essential to summarize on the pole–zero relationships with respect to the system functions.

1. Identification of poles and zeros:
   loop transfer function zeros = zeros of $G(s)H(s)$
   loop transfer function poles = poles of $G(s)H(s)$
   closed-loop poles = poles of $C(s)/R(s)$
   $\qquad\qquad\quad$ = zeros of $F(s) = 1 + G(s)H(s)$
   $\qquad\qquad\quad$ = roots of the characteristic equation
2. The poles of $F(s)$ are the same as the poles of the loop transfer function $G(s)H(s)$.
3. For a closed-loop system to be stable, there is no restriction on the location of the poles and zeros of the loop transfer function $G(s)H(s)$, but the poles of the closed-loop transfer function must all be located in the left half of the s-plane.

"Encircled" versus "Enclosed"

It is important to distinguish between the concepts *encircled* and *enclosed*, which are used frequently with the Nyquist criterion.

Encircled. *A point is said to be encircled by a closed path if it is found inside the path.* For example, point A in Fig. 7-5 is encircled by the closed path Γ, since A is found inside the closed path, whereas point B is not encircled. In the application of the Nyquist criterion, the closed path also has a direction associated with it. As shown in Fig. 7-5, point A is said to be encircled by Γ in the counterclockwise direction.

When considering all the points inside the closed path, we can say that the region inside the closed path Γ is encircled in the indicated direction.

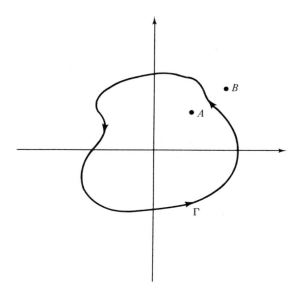

Fig. 7-5. Illustration of the definition of encirclement.

Enclosed. A point or region is said to be enclosed by a closed path if it is found to lie to the left of the path when the path is traversed in the prescribed direction. Putting it another way, *enclosure* is equivalent to *counterclockwise encirclement*. For instance, the shaded regions shown in Fig. 7-6(a) and (b) are considered to be enclosed by the closed path Γ. In other words, point A in Fig. 7-6(a) is enclosed by Γ, but point A in Fig. 7-6(b) is not. However, in Fig. 7-6(b), point B and all the points in the region outside Γ are considered to be enclosed.

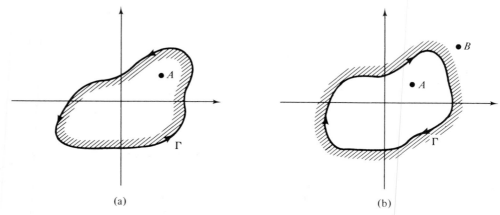

(a)   (b)

Fig. 7-6. Definition of enclosed points and regions. (a) Point A is enclosed by Γ. (b) Point A is not enclosed but B is enclosed by the locus Γ.

Number of encirclement and enclosure. When point A is encircled or enclosed by a closed path, a number N may be assigned to the number of encirclement or enclosure, as the case may be. The value of N may be determined by drawing a vector from A to any arbitrary point s_1 on the closed path Γ and then let s_1 follow the path in the prescribed direction until it returns to the starting point. The total *net* number of revolutions traversed by this vector is N. For example, point A in Fig. 7-7(a) is encircled *once* by Γ, and point B is

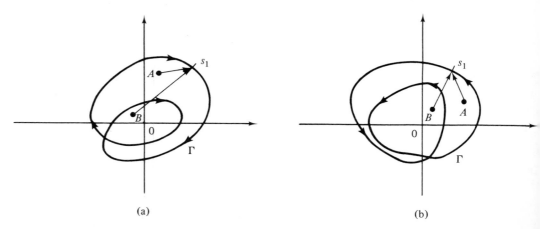

(a)   (b)

Fig. 7-7. Definition of the number of encirclement and enclosure.

encircled twice, all in the clockwise direction. Point A in Fig. 7-7(b) is enclosed once; point B is enclosed twice.

Principle of the Argument

The Nyquist criterion was originated as an engineering application of the well-known principle of the argument in complex variable theory. The principle is stated as follows, in a heuristic manner. Let $F(s)$ be a single-valued rational function that is analytic everywhere in a specific region except at a finite number of points in the s-plane. For each point at which $F(s)$ is analytic in the specified region in the s-plane, there is a corresponding point in the $F(s)$-plane.

Suppose that a continuous closed path Γ_s is arbitrarily chosen in the s-plane, as shown in Fig. 7-8(a). If all the points on Γ_s are in the specified region

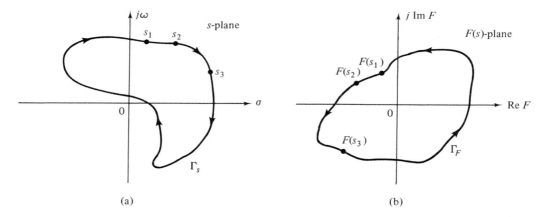

(a)   (b)

Fig. 7-8. (a) Arbitrarily chosen closed path in the s-plane. (b) Corresponding locus Γ_F in the $F(s)$-plane.

in which $F(s)$ is analytic, then curve Γ_F mapped by the function $F(s)$ into the $F(s)$-plane is also a closed one, as shown in Fig. 7-8(b). If, corresponding to point s_1 in the s-plane, point $F(s_1)$ is located in the $F(s)$-plane, then as the Γ_s locus is traversed starting from point s_1 in the arbitrarily chosen direction (clockwise) and then returning to s_1 after going through all the points on the Γ_s locus [as shown in Fig. 7-8(a)], the corresponding Γ_F locus will start from point $F(s_1)$ and go through points $F(s_2)$ and $F(s_3)$, which correspond to s_2 and s_3, respectively, and return to the starting point, $F(s_1)$. The direction of traverse of Γ_F may be either clockwise or counterclockwise; that is, in the same direction or the opposite direction as that of Γ_s, depending on the particular function $F(s)$. In Fig. 7-8(b) the direction of Γ_F is shown, for illustration purposes, to be counterclockwise.

It should be pointed out that, although the mapping from the s-plane to the $F(s)$-plane is one to one for a rational function $F(s)$, the reverse process is usually not a one-to-one mapping. For example, consider the function

$$F(s) = \frac{K}{s(s+1)(s+2)} \qquad (7\text{-}50)$$

which is analytic in the finite s-plane except at the points $s = 0$, -1, and -2. For each value of s in the finite s-plane other than the three points 0, -1, and -2, there is only one corresponding point in the $F(s)$-plane. However, for each point in the $F(s)$-plane, the function maps into three corresponding points in the s-plane. The simplest way to illustrate this is to write Eq. (7-50) as

$$s(s + 1)(s + 2) = \frac{K}{F(s)} \qquad (7\text{-}51)$$

The left side of Eq. (7-51) is a third-order equation, which has three roots when $F(s)$ is chosen to be a constant.

The principle of the argument can be stated: *Let $F(s)$ be a single-valued rational function that is analytic in a given region in the s-plane except at a finite number of points. Suppose that an arbitrary closed path Γ_s is chosen in the s-plane so that $F(s)$ is analytic at every point on Γ_s; the corresponding $F(s)$ locus mapped in the $F(s)$-plane will encircle the origin as many times as the difference between the number of the zeros and the number of poles of $F(s)$ that are encircled by the s-plane locus Γ_s.*

In equation form, this statement can be expressed as

$$N = Z - P \qquad (7\text{-}52)$$

where

$N =$ number of encirclement of the origin made by the $F(s)$-plane locus Γ_F

$Z =$ number of zeros of $F(s)$ encircled by the s-plane locus Γ_s in the s-plane

$P =$ number of poles of $F(s)$ encircled by the s-plane locus Γ_s in the s-plane

In general, N can be positive $(Z > P)$, zero $(Z = P)$, or negative $(Z < P)$. These three situations will now be described.

1. $N > 0 \, (Z > P)$. If the s-plane locus encircles more zeros than poles of $F(s)$ in a certain prescribed direction (clockwise or counterclockwise), N is a positive integer. In this case the $F(s)$-plane locus will encircle the origin of the $F(s)$-plane N times in the same direction as that of Γ_s.
2. $N = 0 \, (Z = P)$. If the s-plane locus encircles as many poles as zeros, or no poles and zeros, of $F(s)$, the $F(s)$-plane locus Γ_F will not encircle the origin of the $F(s)$-plane.
3. $N < 0 \, (Z < P)$. If the s-plane locus encircles more poles than zeros of $F(s)$ in a certain direction, N is a negative integer. In this case the $F(s)$-plane locus, Γ_F, will encircle the origin N times in the opposite direction from that of Γ_s.

A convenient way of determining N with respect to the origin (or any other point) of the $F(s)$ plane is to draw a line from the point in any direction to

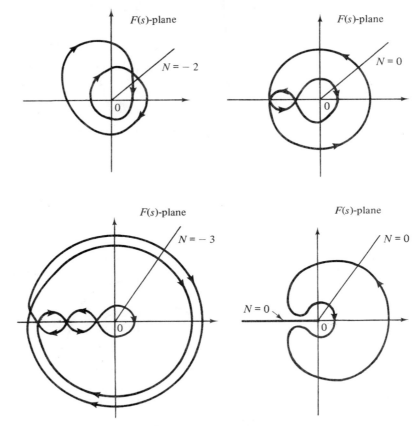

Fig. 7-9. Examples of the determination of N in the $F(s)$-plane.

infinity; the number of net intersections of this line with the $F(s)$ locus gives the magnitude of N. Figure 7-9 gives several examples of this method of determining N. It is assumed that the Γ_s locus has a counterclockwise sense.

A rigorous proof of the principle of the argument is not given here. The following illustration may be considered as a heuristic explanation of the principle. Let us consider the function $F(s)$ given by

$$F(s) = \frac{K(s + z_1)}{(s + p_1)(s + p_2)} \tag{7-53}$$

where K is a positive number. The poles and zero of $F(s)$ are assumed to be as shown in Fig. 7-10(a).

The function $F(s)$ can be written

$$F(s) = |F(s)| \underline{/F(s)}$$
$$= \frac{K|s + z_1|}{|s + p_1||s + p_2|}(\underline{/s + z_1} - \underline{/s + p_1} - \underline{/s + p_2}) \tag{7-54}$$

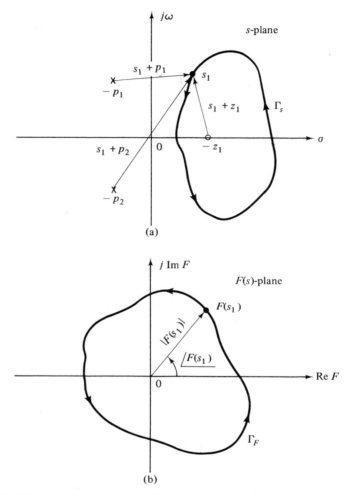

Fig. 7-10. (a) Pole–zero configuration of $F(s)$ in Eq. (7-53) and the s-plane trajectory Γ_s. (b) $F(s)$-plane locus, Γ_F, which corresponds to the Γ_s locus of (a) through the mapping of Eq. (7-53).

Figure 7-10(a) shows an arbitrarily chosen trajectory Γ_s in the s-plane, with the arbitrary point s_1 on the path. The function $F(s)$ evaluated at $s = s_1$ is given by

$$F(s_1) = \frac{K(s_1 + z_1)}{(s_1 + p_1)(s_1 + p_2)} \qquad (7-55)$$

The factor $s_1 + z_1$ can be represented graphically by the vector drawn from $-z_1$ to s_1. Similar vectors can be defined for $(s_1 + p_1)$ and $(s_1 + p_2)$. Thus $F(s_1)$ is represented by the vectors drawn from the given poles and zero to the point s_1, as shown in Fig. 7-10(a). Now, if the point s_1 moves along the locus Γ_s in the prescribed counterclockwise direction until it returns to the starting point, the angles generated by the vectors drawn from the poles (and zeros if there were any) that are not encircled by Γ_s when s_1 completes one round trip are zero; whereas the vector $(s_1 + z_1)$ drawn from the zero at $-z_1$, which is encircled by Γ_s, generates a positive angle (counterclockwise sense) of 2π rad. Then, in Eq. (7-54), the net angle or argument of $F(s)$ as the point s_1 travels around Γ_s once is equal to 2π, which means that the corresponding $F(s)$ plot must go around the origin 2π radians or one revolution in a counterclockwise direction, as shown in Fig. 7-10(b). This is why only the poles and zeros of $F(s)$, which are inside the Γ_s path in the s-plane, would contribute to the value of N of Eq. (7-52). Since poles of $F(s)$ correspond to negative phase angles and zeros correspond to positive phase angles, the value of N depends only on the difference between Z and P.

In the present case,

$$Z = 1 \qquad P = 0$$

Thus

$$N = Z - P = 1$$

which means that the $F(s)$-plane locus should encircle the origin once in the *same* direction as the s-plane locus. It should be kept in mind that Z and P refer only to the zeros and poles, respectively, of $F(s)$ that are encircled by Γ_s, and not the total number of zeros and poles of $F(s)$.

In general, if there are N more zeros than poles of $F(s)$, which are encircled by the s-plane locus Γ_s in a prescribed direction, the net angle traversed by the $F(s)$-plane locus as the s-plane locus is traversed once is equal to

$$2\pi(Z - P) = 2\pi N \qquad (7\text{-}56)$$

This equation implies that the $F(s)$-plane locus will encircle the origin N times in the same direction as that of Γ_s. Conversely, if N more poles than zeros are encircled by Γ_s, in a given prescribed direction, N in Eq. (7-56) will be negative, and the $F(s)$-plane locus must encircle the origin N times in the opposite direction to that of Γ_s.

A summary of all the possible outcomes of the principle of the argument is given in Table 7-1.

Table 7-1   Summary of All Possible Outcomes of the Principle of the Argument

| | | F(s)-Plane Locus | |
$N = Z - P$	Sense of the s-Plane Locus	Number of Encirclements of the Origin	Direction of Encirclement
$N > 0$	Clockwise Counterclockwise	N	Clockwise Counterclockwise
$N < 0$	Clockwise Counterclockwise	N	Counterclockwise Clockwise
$N = 0$	Clockwise Counterclockwise	0	No encirclement No encirclement

Nyquist Path

At this point the reader should place himself in the position of Nyquist many years ago, confronted with the problem of determining whether or not the rational function $1 + G(s)H(s)$ has zeros in the right half of the s-plane. Apparently, Nyquist discovered that the principle of the argument could be used to solve the stability problems, if the s-plane locus, Γ_s, is taken to be one that encircles the entire right half of the s-plane. Of course, as an alternative, Γ_s can be chosen to encircle the entire left-half s-plane, as the solution is a relative one. Figure 7-11 illustrates a Γ_s locus, with a counterclockwise sense, which encircles the entire right half of the s-plane. This path is often called the *Nyquist path.*

Since the Nyquist path must not pass through any singularity of $F(s)$, the small semicircles shown along the $j\omega$ axis in Fig. 7-11 are used to indicate that the path should go around these singular points of $F(s)$. It is apparent that

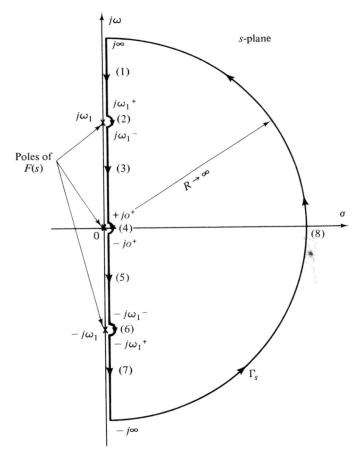

Fig. 7-11. Nyquist path.

if any pole or zero of $F(s)$ lies inside the right half of the s-plane, it will be encircled by this Nyquist path.

For the convenience of analysis, the Nyquist path is divided into a minimum of three sections. The exact number of sections depends upon how many of those small semicircles are necessary on the imaginary axis. For the situation illustrated in Fig. 7-11, a total of eight sections needs to be defined. The order of numbering these sections is entirely arbitrary. The notations $j\omega_1^+, j\omega_1^-, +j0^+$, $-j0^+, -j\omega_1^+$, and $-j\omega_1^-$ are used to identify the starting and ending points of the small semicircles only.

> Section 1: from $s = +j\infty$ to $+j\omega_1^+$ along the $j\omega$ axis.
> Section 2: from $+j\omega_1^+$ to $+j\omega_1^-$ along the small semicircle around $s = j\omega_1$.
> Section 3: from $s = j\omega_1^-$ to $+j0^+$ along the $j\omega$ axis.
> Section 4: from $+j0^+$ to $-j0^+$ along the small semicircle around $s = 0$
> Section 5: from $s = -j0^+$ to $-j\omega_1^-$ along the $j\omega$ axis (mirror image of section 3).
> Section 6: from $s = -j\omega_1^-$ to $-j\omega_1^+$ along the semicircle around $s = -j\omega_1$ (mirror image of section 2).
> Section 7: from $s = -j\omega_1^+$ to $s = -j\infty$ along the $j\omega$ axis (mirror image of section 1).
> Section 8: from $s = -j\infty$ to $s = +j\infty$ along the semicircle of infinite radius.

Nyquist Criterion and the G(s)H(s) Plot

The Nyquist criterion is a direct application of the principle of the argument when the s-plane locus is the Nyquist path. In principle, once the Nyquist path is specified, the stability of a closed-loop system can be determined by plotting the $F(s) = 1 + G(s)H(s)$ locus when s takes on values along the Nyquist path, and investigating the behavior of the $F(s)$ plot with respect to the origin of the $F(s)$-plane. This is called the *Nyquist plot of F(s)*. However, since usually the function $G(s)H(s)$ is given, a simpler approach is to construct the Nyquist plot of $G(s)H(s)$, and the same result of $F(s)$ can be determined from the behavior of the $G(s)H(s)$ plot with respect to the $(-1, j0)$ point in the $G(s)H(s)$-plane. This is because the origin of the $F(s)$-plane corresponds to the $(-1, j0)$ point (or the -1 point on the real axis) of the $G(s)H(s)$-plane. In addition, if the location of the poles of $G(s)H(s)$ is not known, the stability of the open-loop system can be determined by investigating the behavior of the Nyquist plot of $G(s)H(s)$ with respect to the origin of the $G(s)H(s)$-plane.

Let us define the origin of the $G(s)H(s)$-plane or the origin of the $1 + G(s)H(s)$-plane, as the problem requires, as the *critical point*. We are interested basically in two types of stability: stability of the open-loop system, and stability of the closed-loop system.

It is important to clarify that closed-loop stability implies that $1 + G(s)H(s)$ has *zeros* only in the left half of the s-plane. Open-loop stability implies that $G(s)H(s)$ has *poles* only in the left half of the s-plane.

With this added dimension to the stability problem, it is necessary to define two sets of N, Z, and P, as follows:

$N_0 =$ number of encirclements of the origin made by $G(s)H(s)$

$Z_0 =$ number of zeros of $G(s)H(s)$ that are encircled by the Nyquist path, or in the right half of the s-plane

$P_0 =$ number of poles of $G(s)H(s)$ that are encircled by the Nyquist path, or in the right half of the s-plane

$N_{-1} =$ number of encirclements of the $(-1, j0)$ point made by $G(s)H(s)$

$Z_{-1} =$ number of zeros of $1 + G(s)H(s)$ that are encircled by the Nyquist path, or in the right half of the s-plane

$P_{-1} =$ number of poles of $G(s)H(s)$ that are encircled by the Nyquist path, or in the right half of the s-plane

Several facts become clear and should be remembered at this point:

$$P_0 = P_{-1} \qquad (7\text{-}57)$$

since $G(s)H(s)$ and $1 + G(s)H(s)$ always have the same poles. Closed-loop stability implies or requires that

$$Z_{-1} = 0 \qquad (7\text{-}58)$$

but open-loop stability requires that

$$P_0 = 0 \qquad (7\text{-}59)$$

The crux of the matter is that closed-loop stability is determined by the properties of the Nyquist plot of the open-loop transfer function $G(s)H(s)$.

The procedure of applying the principle of the argument for stability studies is summarized as follows:

1. Given the control system, which is represented by the closed-loop transfer function of Eq. (7-47), the Nyquist path is defined according to the pole–zero properties of $G(s)H(s)$.
2. The Nyquist plot of $G(s)H(s)$ is constructed.
3. The values of N_0 and N_{-1} are determined by observing the behavior of the Nyquist plot of $G(s)H(s)$ with respect to the origin and the $(-1, j0)$ point, respectively.
4. Once N_0 and N_{-1} are determined, the value of P_0 (if it is not already known) is determined from

$$N_0 = Z_0 - P_0 \qquad (7\text{-}60)$$

if Z_0 is given. Once P_0 is determined, $P_{-1} = P_0$ [Eq. (7-57)], and Z_{-1} is determined from

$$N_{-1} = Z_{-1} - P_{-1} \qquad (7\text{-}61)$$

Since it has been established that for a stable closed-loop system Z_{-1} must be zero, Eq. (7-61) gives

$$N_{-1} = -P_{-1} \qquad (7\text{-}62)$$

Therefore, the Nyquist criterion may be formally stated: For a closed-loop system to be stable, *the Nyquist plot of $G(s)H(s)$ must encircle the $(-1, j0)$ point as many times as the number of poles of $G(s)H(s)$ that are in the right half of the s-plane, and the encirclement, if any, must be made in the clockwise direction.*

In general, it is not always necessary to construct the Nyquist plot which corresponds to the entire Nyquist path. In fact, a great majority of the practical problems can be solved simply by sketching the part of $G(s)H(s)$ that corresponds to the $+j\omega$-axis of the s-plane. The condition when this simplified procedure is possible is when $P_0 = P_{-1} = 0$; that is, $G(s)H(s)$ has no poles in the right half of the s-plane. In this case Eq. (7-61) becomes

$$N_{-1} = Z_{-1} \tag{7-63}$$

which means that the Nyquist plot of $G(s)H(s)$ can encircle the $(-1, j0)$ point only in the counterclockwise direction, since N_{-1} in Eq. (7-63) can only be zero or positive. Since counterclockwise encirclement is equivalent to enclosure, we can determine closed-loop stability by checking whether the $(-1, j0)$ critical point is enclosed by the $G(s)H(s)$ plot. Furthermore, if we are interested only in whether N_{-1} is zero, we need not sketch the entire Nyquist plot for $G(s)H(s)$, only the portion from $s = +j\infty$ to $s = 0$ along the imaginary axis of the s-plane.

The Nyquist criterion for the $P_{-1} = 0$ case may be stated: *If the function $G(s)H(s)$ has no poles in the right half of the s-plane, for the closed-loop system to be stable, the Nyquist plot of $G(s)H(s)$ must not enclose the critical point $(-1, j0)$.*

Furthermore, if we are not interested in the number of roots of the characteristic equation that are in the right-half plane, but only the closed-loop stability, only the $G(s)H(s)$ plot that corresponds to the positive imaginary axis of the s-plane is necessary. Figure 7-12 illustrates how the $s = j\infty$ to $s = 0$

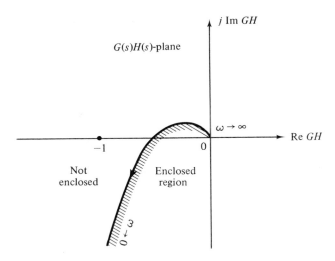

Fig. 7-12. Nyquist plot of $G(s)H(s)$, which corresponds to $s = j\omega$ to $s = 0$, to indicate whether the critical point is enclosed.

portion of the Nyquist plot may be used to determine whether the critical point at $(-1, j0)$ is enclosed.

7.7 Application of the Nyquist Criterion

The following examples serve to illustrate the practical application of the Nyquist criterion to the stability of control systems.

EXAMPLE 7-7   Consider a single-loop feedback control system with the loop transfer function given by

$$G(s)H(s) = \frac{K}{s(s + a)} \tag{7-64}$$

where K and a are positive constants. It is apparent that $G(s)H(s)$ does not have any pole in the right-half s-plane; thus, $P_0 = P_{-1} = 0$. To determine the stability of the closed-loop system, it is necessary only to sketch the Nyquist plot of $G(s)H(s)$ that corresponds to $s = j\infty$ to $s = 0$ on the Nyquist path and see if it encloses the $(-1, j0)$ point in the $G(s)H(s)$-plane. However, for the sake of illustration, we shall construct the entire $G(s)H(s)$ plot for this problem.

The Nyquist path necessary for the function of Eq. (7-64) is shown in Fig. 7-13.

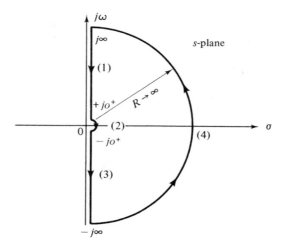

Fig. 7-13. Nyquist path for the system in Example 7-7.

Since $G(s)H(s)$ has a pole at the origin, it is necessary that the Nyquist path includes a small semicircle around $s = 0$. The entire Nyquist path is divided into four sections, as shown in Fig. 7-13.

Section 2 of the Nyquist path is magnified as shown in Fig. 7-14(a). The points on this section may be represented by the phasor

$$s = \epsilon e^{j\theta} \tag{7-65}$$

where $\epsilon(\epsilon \longrightarrow 0)$ and θ denote the magnitude and phase of the phasor, respectively. As the Nyquist path is traversed from $+j0^+$ to $-j0^+$ along section 2, the phasor of

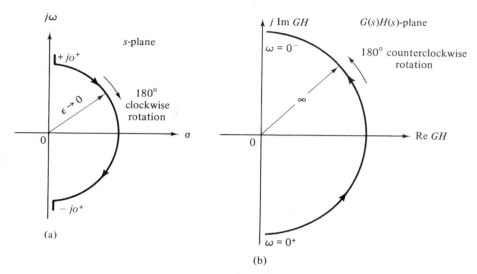

Fig. 7-14. (a) Section 2 of the Nyquist path of Fig. 7-13. (b) Nyquist plot of $G(s)H(s)$ that corresponds to section 2.

Eq. (7-65) rotates in the clockwise direction through 180°. Also, in going from $+j0^+$ to $-j0^+$, θ varies from $+90°$ to $-90°$ through 0°. The corresponding Nyquist plot of $G(s)H(s)$ can be determined simply by substituting Eq. (7-65) into Eq. (7-64). Thus

$$G(s)H(s)\Big|_{s=\epsilon e^{j0}} = \frac{K}{\epsilon e^{j\theta}(\epsilon e^{j\theta} + a)} \tag{7-66}$$

Since $\epsilon \longrightarrow 0$, the last expression is simplified to

$$G(s)H(s)\Big|_{s=\epsilon e^{j0}} \simeq \frac{K}{a\epsilon e^{j\theta}} = \infty e^{-j\theta} \tag{7-67}$$

which indicates that all points on the Nyquist plot of $G(s)H(s)$ that correspond to section 2 of the Nyquist path have an infinite magnitude, and the corresponding phase is opposite that of the s-plane locus. Since the phase of the Nyquist path varies from $+90°$ to $-90°$ in the clockwise direction, the minus sign in the phase relation of Eq. (7-67) indicates that the corresponding $G(s)H(s)$ plot should have a phase that varies from $-90°$ to $+90°$ in the counterclockwise direction, as shown in Fig. 7-14(b).

In general, when one has acquired proficiency in the sketching of the Nyquist plots, the determination of the behavior of $G(s)H(s)$ at $s = 0$ and $s = \infty$ may be carried out by inspection. For instance, in the present problem the behavior of $G(s)H(s)$ corresponding to section 2 of the Nyquist path is determined from

$$\lim_{s \to 0} G(s)H(s) = \lim_{s \to 0} \frac{K}{s(s + a)} = \lim_{s \to 0} \frac{K}{sa} \tag{7-68}$$

From this equation it is clear that the behavior of $G(s)H(s)$ at $s = 0$ is inversely proportional to s. As the Nyquist path is traversed by a phasor with infinitesimally small magnitude, from $+j0^+$ to $-j0^+$ through a clockwise rotation of 180°, the corresponding $G(s)H(s)$ plot is traced out by a phasor with an infinite magnitude, 180° in the opposite or counterclockwise direction. It can be concluded that, in general, if the limit of

$G(s)H(s)$ as s approaches zero assumes the form

$$\lim_{s \to 0} G(s)H(s) = \lim_{s \to 0} Ks^{\pm n} \qquad (7\text{-}69)$$

the Nyquist plot of $G(s)H(s)$ that corresponds to section 2 of Fig. 7-14(a) is traced out by a phasor of infinitesimally small magnitude $n \times 180$ degrees in the clockwise direction if the plus sign is used, and by a phasor of infinite magnitude $n \times 180°$ in the counterclockwise direction if the negative sign is used.

The technique described above may also be used to determine the behavior of the $G(s)H(s)$ plot, which corresponds to the semicircle with infinite radius on the Nyquist path. The large semicircle referred to as section 4 in Fig. 7-13 is isolated, as shown in Fig. 7-15(a). The points on the semicircle may be represented by the phasor

$$s = Re^{j\phi} \qquad (7\text{-}70)$$

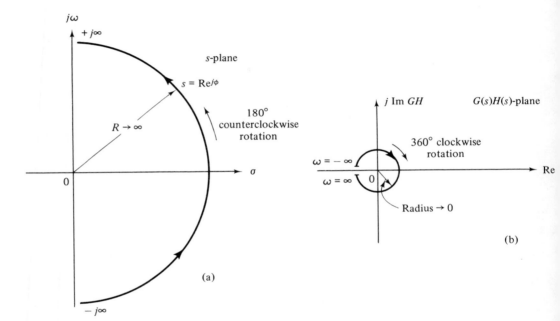

Fig. 7-15. (a) Section 4 of the Nyquist path of Fig. 7-13. (b) Nyquist plot of $G(s)H(s)$ that corresponds to section 4.

where $R \longrightarrow \infty$. Substituting Eq. (7-70) into Eq. (7-64) yields

$$G(s)H(s)\Big|_{s=Re^{j\phi}} = \frac{K}{R^2 e^{j2\phi}} = 0e^{-j2\phi} \qquad (7\text{-}71)$$

which implies that the behavior of the $G(s)H(s)$ plot at infinite frequency is described by a phasor with infinitesimally small magnitude which rotates around the origin $2 \times 180°$ $= 360°$ in the clockwise direction. Thus the $G(s)H(s)$ plot that corresponds to section 4 of the Nyquist path is sketched as shown in Fig. 7-15(b).

Now to complete the Nyquist plot of the transfer function of Eq. (7-64) we must consider sections 1 and 3. Section 1 is usually constructed by substituting $s = j\omega$ into Eq. (7-64) and solving for the possible crossing points on the real axis of the $G(s)H(s)$-

plane. Equation (7-64) becomes

$$G(j\omega)H(j\omega) = \frac{K}{j\omega(j\omega + a)} \tag{7-72}$$

which is rationalized by multiplying the numerator and denominator by the complex conjugate of the denominator. Thus

$$G(j\omega)H(j\omega) = \frac{K(-\omega^2 - ja\omega)}{\omega^4 + a^2\omega^2} \tag{7-73}$$

The intersect of $G(j\omega)H(j\omega)$ on the real axis is determined by equating the imaginary part of $G(j\omega)H(j\omega)$ to zero. Thus the frequency at which $G(j\omega)H(j\omega)$ intersects the real axis is found from

$$\text{Im } G(j\omega)H(j\omega) = \frac{-Ka\omega}{\omega^4 + a^2\omega^2} = \frac{-Ka}{\omega(\omega^2 + a^2)} = 0 \tag{7-74}$$

which gives $\omega = \infty$. This means that the only intersect on the real axis in the $G(s)H(s)$-plane is at the origin with $\omega = \infty$. Since the Nyquist criterion is not concerned with the exact shape of the $G(s)H(s)$ locus but only the number of encirclements, it is not necessary to obtain an exact plot of the locus. The complete Nyquist plot of the function of Eq. (7-64) is now sketched in Fig. 7-16 by connecting the terminal points of the loci

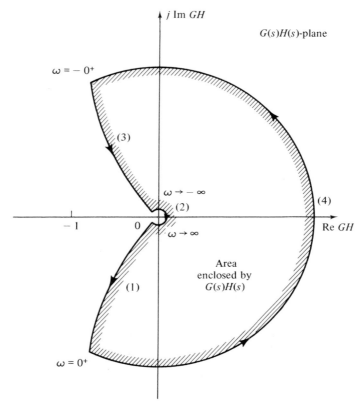

Fig. 7-16. Complete Nyquist plot of $G(s)H(s) = K/[s(s + a)]$.

that correspond to sections 2 and 4, without intersecting any finite part of the real axis.

It is of interest to check all the pertinent data that can be obtained from the Nyquist plot of Fig. 7-16. First, $N_0 = N_{-1} = 0$. By inspection of Eq. (7-64), $Z_0 = 0$ and $P_0 = 0$, which satisfy Eq. (7-60). Since $P_0 = P_{-1}$, Eq. (7-61) leads to

$$Z_{-1} = N_{-1} + P_{-1} = 0 \qquad (7\text{-}75)$$

Therefore, the closed-loop system is stable. This solution should have been anticipated, since for the second-order system, the characteristic equation is simply

$$s^2 + as + K = 0 \qquad (7\text{-}76)$$

whose roots will always lie in the left half of the s-plane for positive a and K.

Figure 7-16 also shows that for this problem it is necessary only to sketch the portion of $G(s)H(s)$ that corresponds to section 1 of the Nyquist path. It is apparent that the $(-1, j0)$ point will never be enclosed by the $G(s)H(s)$ plot for all positive values of K.

EXAMPLE 7-8   Consider that a control system with single feedback loop has the loop transfer function

$$G(s)H(s) = \frac{K(s - 1)}{s(s + 1)} \qquad (7\text{-}77)$$

The characteristic equation of the system is

$$s^2 + (1 + K)s - K = 0 \qquad (7\text{-}78)$$

which has one root in the right half of the s-plane for all positive K. The Nyquist path of Fig. 7-13 is applicable to this case.

Section 2. $s = \epsilon e^{j\theta}$:

$$\lim_{s \to 0} G(s)H(s) = \lim_{s \to 0} \frac{-K}{s} = \infty e^{-j(\theta + \pi)} \qquad (7\text{-}79)$$

This means that the Nyquist plot of $G(s)H(s)$ which corresponds to section 2 of the Nyquist path is traced by a phasor with infinite magnitude. This phasor starts at an angle of $+90°$ and ends at $-90°$ and goes around the origin of the $G(s)H(s)$-plane counterclockwise a total of $180°$.

Section 4. $s = Re^{j\phi}$:

$$\lim_{s \to \infty} G(s)H(s) = \lim_{s \to \infty} \frac{K}{s} = 0 e^{-j\phi} \qquad (7\text{-}80)$$

Thus the Nyquist plot of $G(s)H(s)$ corresponding to section 4 of the Nyquist path goes around the origin $180°$ in the clockwise direction with zero magnitude.

Section 1. $s = j\omega$:

$$G(j\omega)H(j\omega) = \frac{K(j\omega - 1)}{j\omega(j\omega + 1)} = \frac{K(j\omega - 1)(-\omega^2 - j\omega)}{\omega^4 + \omega^2} = K\frac{2\omega + j(1 - \omega^2)}{\omega(\omega^2 + 1)} \qquad (7\text{-}81)$$

Setting the imaginary part of $G(j\omega)H(j\omega)$ to zero, we have

$$\omega = \pm 1 \text{ rad/sec} \qquad (7\text{-}82)$$

which are frequencies at which the $G(j\omega)H(j\omega)$ locus crosses the real axis. Then

$$G(j1)H(j1) = K \qquad (7\text{-}83)$$

Based on the information gathered in the last three steps, the complete Nyquist

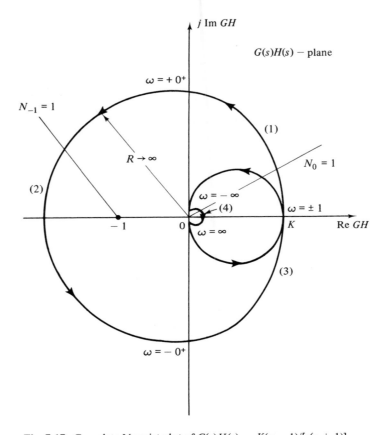

Fig. 7-17. Complete Nyquist plot of $G(s)H(s) = K(s - 1)/[s(s + 1)]$.

plot of $G(s)H(s)$ is sketched as shown in Fig. 7-17. We can conclude that, by inspection,

$$Z_0 = 1$$
$$P_0 = P_{-1} = 0$$

Figure 7-17 indicates that $N_0 = 1$, which is in agreement with the Nyquist criterion. Figure 7-17 also gives $N_{-1} = 1$. Then

$$Z_{-1} = N_{-1} + P_{-1} = 1 \tag{7-84}$$

which means that the closed-loop system is unstable, since the characteristic equation has one root in the right half of the s-plane. The Nyquist plot of Fig. 7-17 further indicates that the system cannot be stabilized by changing only the value of K.

EXAMPLE 7-9   Consider the control system shown in Fig. 7-18. It is desired to determine the range of K for which the system is stable. The open-loop transfer function of the system is

$$\frac{C(s)}{E(s)} = G(s) = \frac{10K(s + 2)}{s^3 + 3s^2 + 10} \tag{7-85}$$

Since this function does not have any pole or zero on the $j\omega$ axis, the Nyquist path

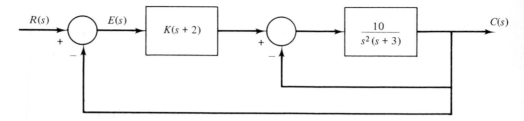

Fig. 7-18. Block diagram of the control system for Example 7-9.

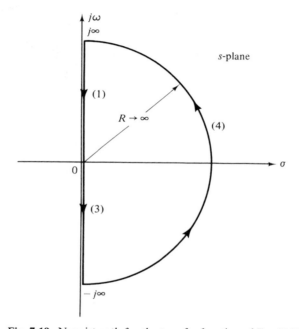

Fig. 7-19. Nyquist path for the transfer function of Eq. (7-85).

can consist of only three sections, as shown in Fig. 7-19. The construction of the Nyquist plot of $G(s)$ is outlined as follows:

Section 4. $s = Re^{j\phi}$:

$$\lim_{s \to \infty} G(s) = \frac{10K}{s^2} = 0e^{-j2\phi} \tag{7-86}$$

As the phasor for section 4 of the Nyquist path is traversed from $-90°$ to $+90°$ through $180°$ counterclockwise, Eq. (7-86) indicates that the corresponding Nyquist plot of $G(s)$ is traced by a phasor of practically zero length from $+180°$ to $-180°$ through a total of $360°$ in the clockwise sense.

Section 1. $s = j\omega$:

$$G(j\omega) = \frac{10K(j\omega + 2)}{(10 - 3\omega^2) - j\omega^3} \tag{7-87}$$

Rationalizing, the last equation becomes

$$G(j\omega) = \frac{10K[2(10 - 3\omega^2) - \omega^4 + j\omega(10 - 3\omega^2) + j2\omega^3]}{(10 - 3\omega^2)^2 + \omega^6} \tag{7-88}$$

Setting the imaginary part of $G(j\omega)$ to zero gives

$$\omega = 0 \text{ rad/sec}$$

and

$$\omega = \pm\sqrt{10} \text{ rad/sec}$$

which correspond to the frequencies at the intersects on the real axis of the $G(s)$-plane. In this case it is necessary to determine the intersect of the $G(s)$ plot on the imaginary axis. Setting the real part of $G(j\omega)$ to zero in Eq. (7-88), we have

$$\omega^4 + 6\omega^2 - 20 = 0 \tag{7-89}$$

which gives

$$\omega = \pm 1.54 \text{ rad/sec}$$

Therefore, the intersects on the real axis of the $G(s)$-plane are at

$$G(j0) = 2K$$

and

$$G(j\sqrt{10}) = -K$$

and the intersect on the imaginary axis is

$$G(j\sqrt{2}) = j10\sqrt{2}K/3$$

With the information gathered in the preceding steps, the Nyquist plot for $G(s)$ of Eq. (7-85) is sketched as shown in Fig. 7-20. The information on the imaginary axis is needed so that the direction of section 1 may be determined without actually plotting the locus point by point.

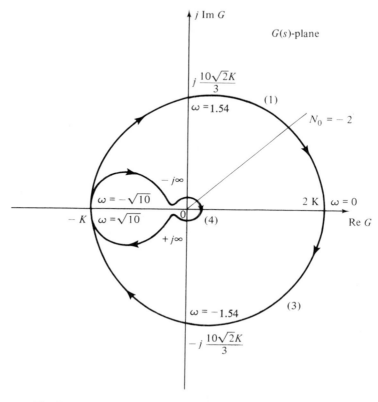

Fig. 7-20. Nyquist plot of $G(s) = 10K(s + 2)/(s^3 + 3s^2 + 10)$.

Inspection of the Nyquist plot of Fig. 7-20 reveals that $N_0 = -2$. Since Eq. (7-85) shows that $G(s)$ has no zeros inside the right half of the s-plane, $Z_0 = 0$; this means that $P_0 = 2$. Thus $P_{-1} = 2$. Now, applying the Nyquist criterion, we have

$$N_{-1} = Z_{-1} - P_{-1} = Z_{-1} - 2 \qquad (7\text{-}90)$$

Thus for the closed-loop system to be stable, $Z_{-1} = 0$, which requires that $N_{-1} = -2$. With reference to Fig. 7-20, the stability criterion requires that the $(-1, j0)$ point must be encircled twice in the clockwise direction. In other words, the critical point should be to the right of the crossover point at $-K$. Thus, for stability,

$$K > 1 \qquad (7\text{-}91)$$

The reader can easily verify this solution by applying the Routh–Hurwitz criterion to the characteristic equation of the system.

It should be reiterated that although the Routh–Hurwitz criterion is much simpler to use in stability problems such as the one stated in this illustrative example, in general the Nyquist criterion leads to a more versatile solution, which also includes information on the relative stability of the system.

7.8 Effects of Additional Poles and Zeros of $G(s)H(s)$ on the Shape of the Nyquist Locus

Since the performance and the stability of a feedback control system are often influenced by adding and moving poles and zeros of the transfer functions, it is informative to illustrate how the Nyquist locus is affected when poles and zeros are added to a typical loop transfer function $G(s)H(s)$. This investigation will also be helpful to gain further insight on the quick sketch of the Nyquist locus of a given function.

Let us begin with a first-order transfer function

$$G(s)H(s) = \frac{K}{1 + sT_1} \qquad (7\text{-}92)$$

The Nyquist locus of $G(j\omega)H(j\omega)$ for $0 \leq \omega < \infty$ is a semicircle, as shown in Fig. 7-21.

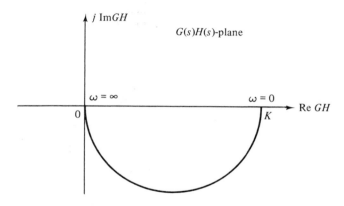

Fig. 7-21. Nyquist plot of $G(s)H(s) = K/(1 + T_1 s)$.

Addition of poles at s = 0. Consider that a pole at $s = 0$ is added to the transfer function of Eq. (7-92); then we have

$$G(s)H(s) = \frac{K}{s(1 + sT_1)} \tag{7-93}$$

The effect of adding this pole is that the phase of $G(j\omega)H(j\omega)$ is reduced by $-90°$ at both zero and infinite frequencies.

In other words, the Nyquist locus of $G(j\omega)H(j\omega)$ is rotated by $-90°$ from that of Fig. 7-21 at $\omega = 0$ and $\omega = \infty$, as shown in Fig. 7-22. In addition, the

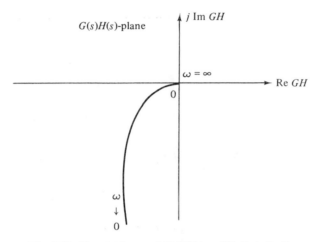

Fig. 7-22. Nyquist locus of $G(s)H(s) = K/[s(1 + T_1s)]$.

magnitude of $G(j\omega)H(j\omega)$ at $\omega = 0$ becomes infinite. In general, adding a pole of multiplicity j at $s = 0$ to the transfer function of Eq. (7-92) will give the following properties to the Nyquist locus of $G(s)H(s)$:

$$\lim_{\omega \to \infty} \underline{/G(j\omega)H(j\omega)} = -(j + 1)\frac{\pi}{2} \tag{7-94}$$

$$\lim_{\omega \to 0} \underline{/G(j\omega)H(j\omega)} = -j\frac{\pi}{2} \tag{7-95}$$

$$\lim_{\omega \to \infty} |G(j\omega)H(j\omega)| = 0 \tag{7-96}$$

$$\lim_{\omega \to 0} |G(j\omega)H(j\omega)| = \infty \tag{7-97}$$

Figure 7-23 illustrates the Nyquist plots of

$$G(s)H(s) = \frac{K}{s^2(1 + T_1s)} \tag{7-98}$$

and

$$G(s)H(s) = \frac{K}{s^3(1 + T_1s)} \tag{7-99}$$

In view of these illustrations it is apparent that addition of poles at $s = 0$ will affect the stability adversely, and systems with a loop transfer function of more than one pole at $s = 0$ are likely to be unstable.

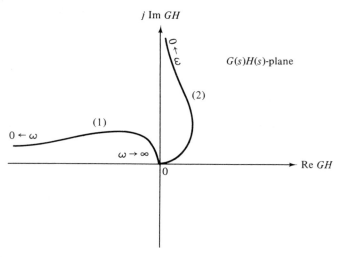

Fig. 7-23. Nyquist loci for (1) $G(s)H(s) = K/[s^2(1 + T_1s)]$. (2) $G(s)H(s) = K/[s^3(1 + T_1s)]$.

Addition of finite poles. When a pole at $s = -1/T_2$ is added to the $G(s)$ $H(s)$ function of Eq. (7-92), we have

$$G(s)H(s) = \frac{K}{(1 + T_1s)(1 + T_2s)} \qquad (7\text{-}100)$$

The Nyquist locus of $G(s)H(s)$ at $\omega = 0$ is not affected by the addition of the pole, since

$$\lim_{\omega \to 0} G(j\omega)H(j\omega) = K \qquad (7\text{-}101)$$

The Nyquist locus at $\omega = \infty$ is

$$\lim_{\omega \to \infty} G(j\omega)H(j\omega) = \lim_{\omega \to \infty} \frac{-K}{T_1T_2\omega^2} = 0\underline{/-180^\circ} \qquad (7\text{-}102)$$

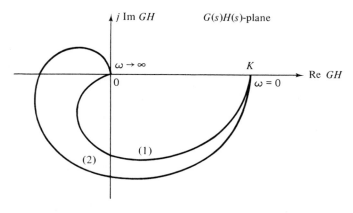

Fig. 7-24. Nyquist loci for (1) $G(s)H(s) = K/[(1 + T_1s)(1 + T_2s)]$. (2) $G(s)H(s) = K/[(1 + T_1s)(1 + T_2s)(1 + T_3s)]$.

Thus the effect of adding a pole at $s = -1/T_2$ to the transfer function of Eq. (7-92) is to shift the phase of the Nyquist locus by $-90°$ at infinite frequency, as shown in Fig. 7-24. This figure also shows the Nyquist locus of

$$G(s)H(s) = \frac{K}{(1 + T_1 s)(1 + T_2 s)(1 + T_3 s)} \qquad (7\text{-}103)$$

These examples show the adverse effects on stability that result from the addition of poles to the loop transfer function.

Addition of zeros. It was pointed out in Chapter 6 that the effect of the derivative control on a closed-loop control system is to make the system more stable. In terms of the Nyquist plot, this stabilization effect is easily shown, since the multiplication of the factor $(1 + T_d s)$ to the loop transfer function increases the phase of $G(s)H(s)$ by $90°$ at $\omega = \infty$.

Consider that the loop transfer function of a closed-loop system is given by

$$G(s)H(s) = \frac{K}{s(1 + T_1 s)(1 + T_2 s)} \qquad (7\text{-}104)$$

It can be shown that the closed-loop system is stable for $0 \leq K < (T_1 + T_2)/T_1 T_2$. Suppose that a zero at $s = -1/T_d$ is added to the transfer function of Eq. (7-104), such as with a derivative control. Then

$$G(s)H(s) = \frac{K(1 + T_d s)}{s(1 + T_1 s)(1 + T_2 s)} \qquad (7\text{-}105)$$

The Nyquist loci of the two transfer functions of Eqs. (7-104) and (7-105) are sketched as shown in Fig. 7-25. The effect of the zero in Eq. (7-105) is to add

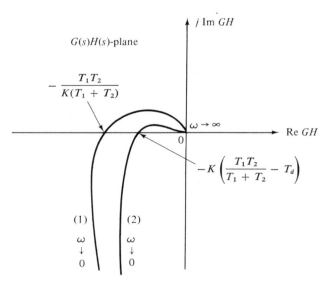

Fig. 7-25. Nyquist loci for (1) $G(s)H(s) = K/[s(1 + T_1 s)(1 + T_2 s)]$. (2) $G(s)H(s) = K(1 + T_d s)/[s(1 + T_1 s)(1 + T_2 s)]$.

$90°$ to the phase of $G(j\omega)H(j\omega)$ at $\omega = \infty$ while not affecting the locus at $\omega = 0$. The crossover point on the real axis is moved from $-K(T_1 + T_2)/T_1T_2$ to $-K(T_1 + T_2)/T_1T_2(1 + T_d)$, which is closer to the origin of the $G(j\omega)H(j\omega)$-plane.

7.9 Stability of Multiloop Systems

The stability analyses conducted in the preceding sections are all centered toward systems with a single feedback loop, with the exception of the Routh–Hurwitz criterion, which apparently can be applied to systems of any configuration, as long as the characteristic equation is known. We shall now illustrate how the Nyquist criterion is applied to a linear system with multiple feedback loops.

In principle, all feedback systems with single input and output can be reduced to the basic system configuration of Fig. 7-4. Then it would seem that the Nyquist criterion can be applied in a straightforward manner to the equivalent loop transfer function $G(s)H(s)$. However, owing to the multiloop nature of the original system, the poles and/or the zeros of $G(s)H(s)$ may be unknown, and a systematic approach to the problem may be adopted.

Let us illustrate the procedure of applying Nyquist criterion to a multiloop control system by means of a specific example. Figure 7-26 gives the block diagram of a control system with two loops. The transfer function of each block is indicated in the diagram. In this case it is simple to derive the open-loop transfer function of the system as

$$\frac{C(s)}{E(s)} = G(s) = \frac{G_1(s)G_2(s)}{1 + G_2(s)H(s)}$$

$$= \frac{K(s + 2)}{(s + 10)[s(s + 1)(s + 2) + 5]} \tag{7-106}$$

The stability of the overall system can be investigated by sketching the Nyquist

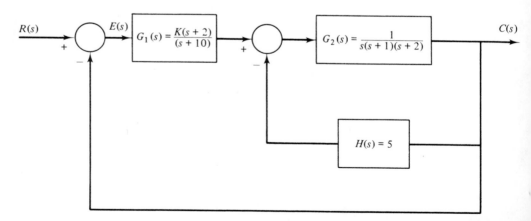

Fig. 7-26. Multiloop feedback control system.

locus of $G(s)$, except that the poles of $G(s)$ are not entirely known. To avoid the construction of the entire Nyquist locus of $G(s)$, we can attack the problem in two stages, as there are two feedback loops. First, consider only the inner loop, whose loop transfer function is $G_2(s)H(s)$. We shall first sketch the Nyquist locus of $G_2(s)H(s)$ for $0 \leq \omega < \infty$. The property of the $G_2(s)H(s)$ plot with respect to the $(-1, j0)$ point gives an indication of the number of zeros of $1 + G_2(s)H(s)$ that are in the right half of the s-plane. Having found this information, we then proceed to sketch the Nyquist locus of $G(s)$ of Eq. (7-106) only for $0 \leq \omega < \infty$ to determine the stability of the overall system.

Figure 7-27 shows the Nyquist locus of

$$G_2(s)H(s) = \frac{5}{s(s + 1)(s + 2)} \tag{7-107}$$

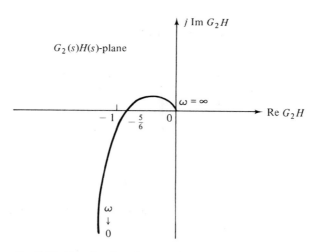

Fig. 7-27. Nyquist plot of $G_2(s)H(s) = 5/[s(s + 1)(s + 2)]$.

Since the $(-1, j0)$ point is not enclosed by the locus, the inner loop is stable by itself, and the zeros of $1 + G_2(s)H(s)$ are all in the left half of the s-plane. Next, the Nyquist locus of $G(s)$ of Eq. (7-106) is sketched as shown in Fig. 7-28. Since all the poles and zeros of $G(s)$ are found to be in the left half of the s-plane, we only have to investigate the crossover point of the $G(s)$ locus with respect to the $(-1, j0)$ point to determine the requirement on K for the overall system to be stable. In this case the range of K for stability is $0 \leq K < 50$.

Now, let us consider a system that is a slightly modified version of Fig. 7-26. Use the same block diagram, but with

$$G_1(s) = \frac{s + 2}{s + 1} \tag{7-108}$$

$$G_2(s) = \frac{K}{s(s + 1)(s + 2)} \tag{7-109}$$

$$H(s) = 5 \tag{7-110}$$

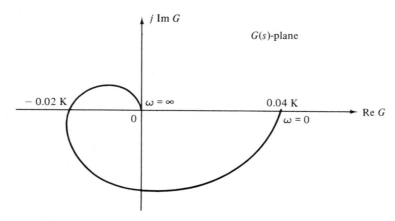

Fig. 7-28. Nyquist plot of $G(s) = [K(s + 2)]/\{(s + 10)[s(s + 1)(s + 2) + 5]\}$.

In this case we cannot use the method outline above, since the unknown gain parameter is in the inner loop. However, we may still use the Nyquist criterion to solve this problem. The open-loop transfer function of the system is

$$G(s) = \frac{G_1(s)G_2(s)}{1 + G_2(s)H(s)}$$

$$= \frac{s + 2}{(s + 10)[s(s + 1)(s + 2) + 5K]} \tag{7-111}$$

Since the unknown parameter K does not appear as a gain factor of $G(s)$, it would be of no avail to sketch the Nyquist locus of $G(s)/K$. However, we can write the characteristic equation of the overall system as

$$s(s + 10)(s + 1)(s + 2) + s + 2 + 5K(s + 10) = 0 \tag{7-112}$$

In order to create an equivalent open-loop transfer function with K as a multiplying factor, we divide both sides of Eq. (7-112) by terms that do not contain K. We have

$$1 + \frac{5K(s + 10)}{s(s + 10)(s + 1)(s + 2) + s + 2} = 0 \tag{7-113}$$

Since this equation is of the form $1 + G_3(s) = 0$, the roots of the characteristic equation may be investigated by sketching the Nyquist locus of $G_3(s)$. However, the poles of $G_3(s)$ are not known, since the denominator of $G_3(s)$ is not in factored form. The zeros of the polynominal $s(s + 10)(s + 1)(s + 2) + s + 2$ may be studied by investigating the Nyquist plot of still another function $G_4(s)$, which we create as follows:

$$G_4(s) = \frac{s + 2}{s(s + 10)(s + 1)(s + 2)} \tag{7-114}$$

Figure 7-29 shows that the Nyquist locus of $G_4(s)$ intersects the real axis to the right of the $(-1, j0)$ point. Thus all the poles of $G_3(s)$ are in the left half of the s-plane. The Nyquist plot of $G_3(s)$ is sketched as shown in Fig. 7-30. Since

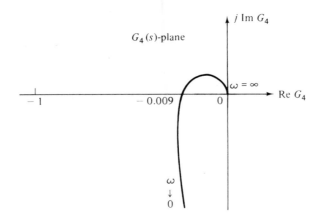

Fig. 7-29. Nyquist locus of $G_4(s) = (s + 2)/[s(s + 10)(s + 1)(s + 2)]$.

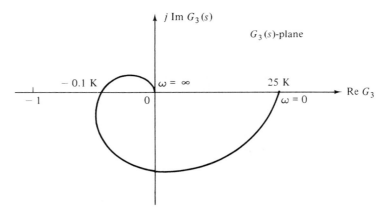

Fig. 7-30. Nyquist locus of $G_3(s) = [5K(s + 10)]/[s(s + 10)(s + 1)(s + 2) + s + 2]$.

the intersect of the locus on the real axis is at $-0.1K$, the range of K for stability is $0 \leq K < 10$.

In this section we have investigated the application of the Nyquist criterion to multiloop control systems. For analysis problems, the stability of the system can be investigated by applying the Nyquist criterion in a systematic fashion from the inner loop toward the outer loops. For design problems when a system parameter K is to be determined for stability, it is sometimes necessary to start with the characteristic equation, which may be written in the form

$$P(s) + KQ(s) = 0 \tag{7-115}$$

where $P(s)$ and $Q(s)$ are polynominals. The stability of the system is studied by sketching the Nyquist plot of the equivalent open-loop transfer function

$$G(s) = \frac{KQ(s)}{P(s)} \tag{7-116}$$

Thus we have also indicated a method of applying the Nyquist criterion to design a stable system by using the characteristic equation. The Nyquist locus represents a more convenient approach than the Routh–Hurwitz criterion, since the latter often involves the solution of an inequality equation in the high order of K, whereas in the Nyquist approach K can be determined from the intersection of the $G(s)$ locus with the real axis.

The stability analysis discussed thus far in this chapter can be applied to linear multivariable systems. Since the characteristic equation of a multivariable is still written

$$|s\mathbf{I} - \mathbf{A}| = 0 \qquad (7\text{-}117)$$

the Routh–Hurwitz criterion is applied in the usual manner. We can also apply the Nyquist criterion to Eq. (7-117) in the manner outlined in Eqs. (7-115) and (7-116). On the other hand, since the stability of a linear system is independent of the inputs and outputs of the system, we can apply the Nyquist criterion to a multivariable system between any input–output pair while setting all other inputs to zero, as long as the system is completely controllable and observable.

7.10 Stability of Linear Control Systems with Time Delays

Systems with time delays and their modeling have been discussed in Section 5.12. In general, closed-loop systems with time delays in the loops will be subject to more stability problems than systems without delays. Since a pure time delay T_d is modeled by the transfer function relationship $e^{-T_d s}$, the characteristic equation of the system will no longer have constant coefficients. Therefore, the Routh–Hurwitz criterion is *not* applicable. However, we shall show in the following that the Nyquist criterion is readily applicable to a system with a pure time delay.

Let us consider that the loop transfer function of a feedback control system with pure time delay is represented by

$$G(s)H(s) = e^{-T_d s}G_1(s)H_1(s) \qquad (7\text{-}118)$$

where $G_1(s)H_1(s)$ is a rational function with constant coefficients; T_d is the pure time delay in seconds. Whether the time delay occurs in the forward path or the feedback path of the system is immaterial from the stability standpoint.

In principle, the stability of the closed-loop system can be investigated by sketching the Nyquist locus of $G(s)H(s)$ and then observing its behavior with reference to the $(-1, j0)$ point of the complex function plane.

The effect of the exponential term in Eq. (7-118) is that it rotates the phasor $G_1(j\omega)H_1(j\omega)$ at each ω by an angle of ωT_d radians in the clockwise direction. The amplitude of $G_1(j\omega)H_1(j\omega)$ is not affected by the time delay, since the magnitude of $e^{-j\omega T_d}$ is unity for all frequencies.

In control systems the magnitude of $G_1(j\omega)H_1(j\omega)$ usually approaches zero as ω approaches infinity. Thus the Nyquist locus of the transfer function of Eq. (7-118) will usually spiral toward the origin as ω approaches infinity, and there are infinite number of intersects on the negative real axis of the $G(s)H(s)$-plane. For the closed-loop system to be stable, all the intersects of the $G(j\omega)H(j\omega)$ locus with the real axis must occur to the right of the $(-1, j0)$ point.

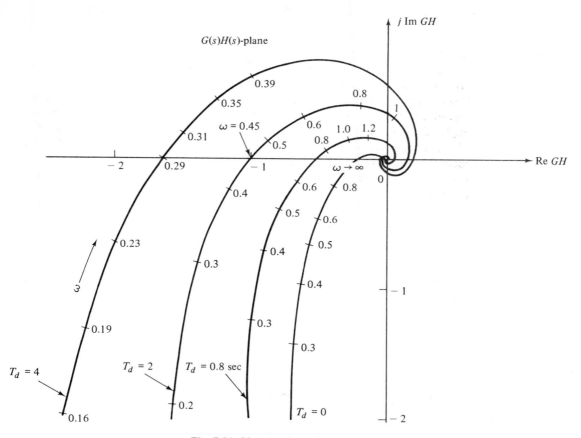

Fig. 7-31. Nyquist plots of $G(s)H(s) = e^{-T_d s}/[s(s+1)(s+2)]$.

Figure 7-31 shows an example of the Nyquist plot of

$$G(s)H(s) = e^{-T_d s}G_1(s)H_1(s) = \frac{e^{-T_d s}}{s(s+1)(s+2)} \qquad (7\text{-}119)$$

for several values of T_d. It is observed from this diagram that the closed-loop system is stable when the time delay T_d is zero, but the stability condition deteriorates as T_d increases. The system is on the verge of becoming unstable when $T_d = 2$ sec. This is shown with the Nyquist plot passing through the $(-1, j0)$ point.

Unlike the rational function case, the analytical solution of the crossover points on the real axis of the $G(s)H(s)$-plane is not trivial, since the equations that govern the crossings are no longer algebraic. For instance, the loop transfer function of Eq. (7-119) may be rationalized in the usual manner by multiplying its numerator and denominator by the complex conjugate of the denominator. The result is

$$G(j\omega)H(j\omega) = \frac{(\cos \omega T_d - j \sin \omega T_d)[-3\omega^2 - j\omega(2 - \omega^2)]}{9\omega^4 + \omega^2(2 - \omega^2)^2} \qquad (7\text{-}120)$$

The condition for crossings on the real axis of the $G(s)H(s)$-plane is

$$3\omega^2 \sin \omega T_d - \omega(2 - \omega^2) \cos \omega T_d = 0$$

which is not easily solved, given T_d.

Since the term $e^{-j\omega T_d}$ always has a magnitude of 1 for all frequencies, the crossover problem is readily solved in the Bode plot domain (Appendix A). Since the time-delay term affects only the phase but not the magnitude of $G(j\omega)H(j\omega)$, the phase of the latter is obtained in the Bode plot by adding a negative angle of $-\omega T_d$ to the phase curve of $G_1(j\omega)H_1(j\omega)$. The frequency at which the phase curve of $G(j\omega)H(j\omega)$ crosses the $180°$ axis is the place where the Nyquist locus intersects the negative real axis. In general, analysis and design problems involving pure time delays are more easily carried out graphically in the Bode diagram.

If the time delay is small, it is possible to approximate the time delay transfer function relation by a truncated power series; that is,

$$e^{-T_d s} = 1 - T_d s + \frac{T_d^2 s^2}{2!} - \frac{T_d^3 s^3}{3!} + \dots \tag{7-121}$$

Figure 7-32 shows the Nyquist plots of the transfer function of Eq. (7-119)

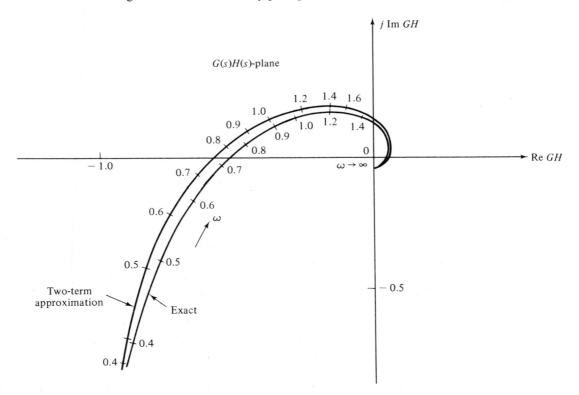

Fig. 7-32. Approximation of Nyquist plot of system with time delay by truncated power-series expansion.

with $T_d = 0.8$ sec, and

$$G(s)H(s) = \frac{1 - T_d s}{s(s + 1)(s + 2)} \tag{7-122}$$

which is the result of truncating the series of Eq. (7-121) after two terms.

7.11 Stability of Nonlinear Systems—Popov's Criterion[16-26]

It was mentioned in Section 7.1 that the stability analysis of nonlinear systems is a complex subject, and unlike the linear time-invariant case, no single method can be applied to all nonlinear systems. Although the major emphasis of this book is on linear systems, we shall show in the following that a certain class of nonlinear systems can be studied with stability criteria that are quite similar to the Nyquist criterion.

When a system is nonlinear, it does not have a characteristic equation, and therefore there are no eigenvalues to speak of. The Routh–Hurwitz criterion becomes useless in this case. The kind of stability that we are concerned with here is asymptotic stability, which means that the equilibrium state $\mathbf{x}_e$ is asymptotically stable if every motion starting at an initial state $\mathbf{x}_0$ will converge to $\mathbf{x}_e$ as time approaches infinity.

Popov's stability criterion applies to a closed-loop control system that contains one nonlinear element, as shown in Fig. 7-33. The nonlinearity is described

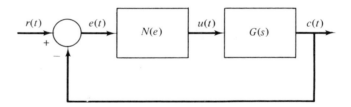

Fig. 7-33. Nonlinear closed-loop control system.

by a functional relation that must lie in the first and third quadrants, as shown in Fig. 7-34.

Many nonlinear control systems in practice can be modeled by the block diagram and nonlinear characteristic of Figs. 7-33 and 7-34. For instance, Fig 7-35 illustrates several common-type nonlinearities, which are encountered often in control systems that fit the characteristic specified by Fig. 7-34.

Popov's stability theorem is based on the following assumptions:

1. The linear part of the system is described by the transfer function $G(s)$, which has more poles than zeros, and there are no cancellations of poles and zeros.
2. The nonlinear characteristic is bound by k_1 and k_2 as shown in Fig. 7-34; or

$$k_1 \leq N(e) \leq k_2 \tag{7-123}$$

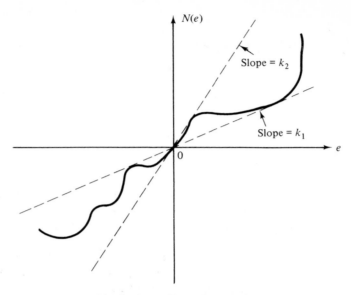

Fig. 7-34. Nonlinear characteristics.

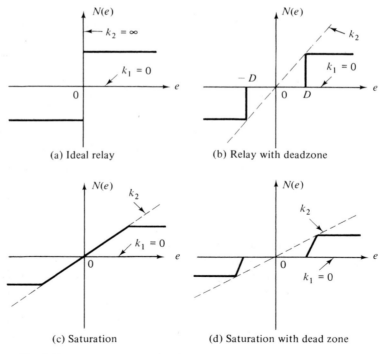

(a) Ideal relay

(b) Relay with deadzone

(c) Saturation

(d) Saturation with dead zone

Fig. 7-35. Several common nonlinearities encountered in control systems. (a) Ideal relay. (b) Relay with dead zone. (c) Saturation. (d) Saturation with dead zone.

The theorem is stated in the following without giving the proof: *The closed-loop system is asymptotically stable if the Nyquist plot of G(jω) does not intersect or enclose the circle that is described by*

$$\left[x + \frac{k_1 + k_2}{2k_1 k_2}\right]^2 + y^2 = \left[\frac{k_2 - k_1}{2k_1 k_2}\right]^2 \tag{7-124}$$

where x and y denote the real and imaginary coordinates of the $G(j\omega)$-plane, respectively.

It should be noted that the Popov stability criterion is sufficient but is not necessary. In other words, if the $G(j\omega)$ locus does not intersect or enclose the circle, the system is stable, although *the criterion is usually overly conservative.* On the other hand, if the above condition is violated, it does not necessarily mean that the system is unstable. The possible implications are illustrated in Fig. 7-36.

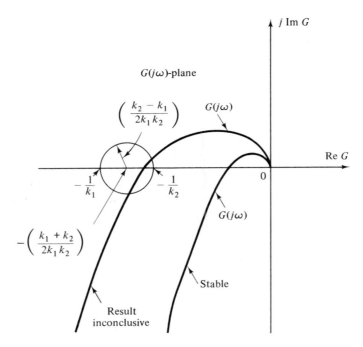

Fig. 7-36. Interpretation of Popov's stability criterion.

It is interesting to observe that if the system is linear, $k_1 = k_2 = k$, the circle degenerates to the $(-1, j0)$ point, and Popov's criterion becomes Nyquist's criterion. Figure 7-35 shows that a great majority of the nonlinearities in practice are with $k_1 = 0$. In these cases, the circle of Eq. (7-124) becomes a straight line

$$x = -\frac{1}{k_2} \tag{7-125}$$

For stability, the Nyquist locus of $G(j\omega)$ must not intersect this line.

The following example serves to illustrate the practical application of the Popov stability criterion.

EXAMPLE 7-10   Consider that the block diagram shown in Fig. 7-37 represents a feedback control system that has a saturation element in the forward path. The system could be a position control system using a motor as an actuator. The saturation is in the power amplifier of the motor controller, with K representing the gain of the linear portion of the amplifier characteristic.

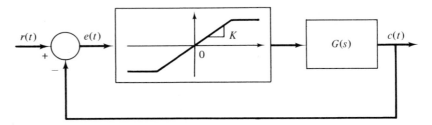

Fig. 7-37. Nonlinear system with a saturation characteristic.

The linear transfer function of the system is

$$G(s) = \frac{1}{s(s + 1)(s + 2)} \tag{7-126}$$

It can be shown that if the saturation characteristic were absent, the closed-loop system would be stable for $0 \le K < 6$. It is desired to determine the value of K so that the nonlinear system is assured of stability. Notice that the Popov criterion is of such a nature that for the saturation nonlinearity, the result of the criterion is independent of the level of saturation, which is in a way a qualitativeness of the test.

The Nyquist locus of $G(j\omega)$ is sketched as shown in Fig. 7-38. For stability of the closed-loop system, the $G(j\omega)$ locus must not intersect the vertical line, which passes through the point $(-1/K, j0)$.

The maximum value of K is determined by finding the maximum magnitude of the real part of $G(j\omega)$. Rationalizing $G(j\omega)$, the real part is written

$$\text{Re } G(j\omega) = \frac{-3}{9\omega^2 + (2 - \omega^2)^2} \tag{7-127}$$

The frequency at which Re $G(j\omega)$ is a maximum is determined from

$$\frac{d \text{ Re } G(j\omega)}{d\omega} = \frac{\omega[-18 + 4(2 - \omega^2)]}{[9\omega^2 + (2 - \omega^2)^2]^2} = 0 \tag{7-128}$$

which gives $\omega = 0$. Thus

$$\text{Max. Re } G(j\omega) = -\tfrac{3}{4} \tag{7-129}$$

and the closed-loop system is stable for

$$K < \tfrac{4}{3} \tag{7-130}$$

which is considerably less than the critical value of $K = 6$ for the linear system.

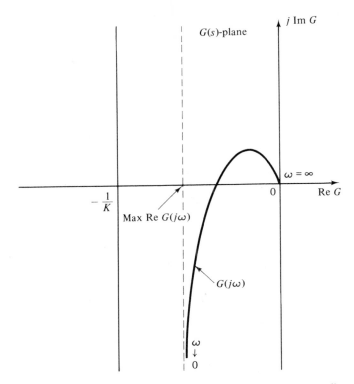

Fig. 7-38. Application of the Popov stability criterion to the nonlinear system in Fig. 7-37.

REFERENCES

Routh–Hurwitz Criterion

1. E. J. ROUTH, *Dynamics of a System of Rigid Bodies*, Chap. 6, Part II, Macmillan & Co. Ltd., London, 1905.

2. N. N. PURI and C. N. WEYGANDT, "Second Method of Liapunov and Routh's Canonical Form," *J. Franklin Inst.*, Vol. 76, pp. 365–384, Nov. 1963.

3. G. V. S. S. RAJU, "The Routh Canonical Form," *IEEE Trans. Automatic Control*, Vol. AC-12, pp. 463–464, Aug. 1967.

4. V. KRISHNAMURTHI, "Correlation Between Routh's Stability Criterion and Relative Stability of Linear Systems," *IEEE Trans. Automatic Control*, Vol. AC-17, pp. 144–145, Feb. 1972.

5. V. KRISHNAMURTHI, "Gain Margin of Conditionally Stable Systems from Routh's Stability Criterion," *IEEE Trans. Automatic Control*, Vol. AC-17, pp. 551–552, Aug. 1972.

General Linear Stability—Nyquist

6. H. NYQUIST, "Regeneration Theory," *Bell System Tech. J.*, Vol. 11, pp. 126–147, Jan. 1932.

7. C. H. HOFFMAN, "How To Check Linear Systems Stability: I. Solving the Characteristic Equation," *Control Engineering*, pp. 75–80, Aug. 1964.

8. C. H. HOFFMAN, "How To Check Linear Systems Stability: II. Locating the Roots by Algebra," *Control Engineering*, pp. 84–88, Feb. 1965.

9. C. H. HOFFMAN, "How To Check Linear Systems Stability: III. Locating the Roots Graphically," *Control Engineering*, pp. 71–78, June 1965.

10. M. R. STOJIĆ and D. D. ŠILJAK, "Generalization of Hurwitz, Nyquist, and Mikhailov Stability Criteria," *IEEE Trans. Automatic Control*, Vol. AC-10, pp. 250–254, July 1965.

11. R. W. BROCKETT and J. L. WILLEMS, "Frequency Domain Stability Criteria—Part I," *IEEE Trans. Automatic Control*, Vol. AC-10, pp. 255–261, July 1965.

12. R. W. BROCKETT and J. L. WILLEMS, "Frequency Domain Stability Criteria—Part II," *IEEE Trans. Automatic Control*, Vol. AC-10, pp. 407–413, Oct. 1965.

13. D. D. ŠILJAK, "A Note on the Generalized Nyquist Criterion," *IEEE Trans. Automatic Control*, Vol. AC-11, p. 317, April 1966.

14. L. EISENBERG, "Stability of Linear Systems with Transport Lag," *IEEE Trans. Automatic Control*, Vol. AC-11, pp. 247–254, April 1966.

15. T. R. NATESAN, "A Supplement to the Note on the Generalized Nyquist Criterion," *IEEE Trans. Automatic Control*, Vol. AC-12, pp. 215–216, Apr. 1967.

Popov's Criterion

16. V. M. POPOV, "Absolute Stability of Nonlinear Systems of Automatic Control," *Automation and Remote Control*, Vol. 22, pp. 961–978, Aug. 1961.

17. Z. V. REKASUIS, "A Stability Criterion for Feedback Systems with One Nonlinear Element," *IEEE Trans. Automatic Control*, Vol. AC-9, pp. 46–50, Jan. 1964.

18. C. A. DESOER, "A Generalization of the Popov Criterion," *IEEE Trans. Automatic Control*, Vol. AC-10, pp. 182–185, Apr. 1965.

19. S. C. PINCURA, "On the Inapplicability of the Popov Stability Criterion in Certain Classes of Control Systems," *IEEE Trans. Automatic Control*, Vol. AC-12, pp. 465–466, Aug. 1967.

20. Y. S. CHO and K. S. NARENDIA, "An Off-Axis Circle Criterion for the Stability of Feedback Systems with a Monotonic Nonlinearity," *IEEE Trans. Automatic Control*, Vol. AC-13, No. 4, pp. 413–416, Aug. 1968.

21. J. B. MOORE, "A Circle Criterion Generalization for Relative Stability," *IEEE Trans. Automatic Control*, Vol. AC-13, No. 1, pp. 127–128, Feb. 1968.

22. C. A. DESOER, "An Extension to the Circle Criterion," *IEEE Trans. Automatic Control*, Vol. AC-13, pp. 587–588, Oct. 1968.

23. E. Y. SHAPIRO, "Circle Criterion for Nonlinear System Stability," *Control Engineering*, Vol. 17, No. 3, pp. 81–83, Mar. 1970.

24. A. C. Tsoi and H. M. Power, "Equivalent Predictions of the Circle Criterion and an Optimum Quadratic Form for a Second-Order System," *IEEE Trans. Automatic Control*, Vol. AC-17, pp. 565–566, Aug. 1972.

25. C. E. Zimmerman and G. J. Thaler, "Application of the Popov Criterion to Design of Nonlinear Systems," *IEEE Trans. Automatic Control*, Vol. AC-16, pp. 76–79, Feb. 1971.

26. H. M. Power and A. C. Tsoi, "Improving the Predictions of the Circle Criterion by Combining Quadratic Forms," *IEEE Trans. Automatic Control*, Vol. AC-18, pp. 65–67, Feb. 1973.

PROBLEMS

7.1. By means of the Routh–Hurwitz criterion, determine the stability of systems that have the following characteristic equations. In each case, determine the number of roots of the equation which are in the right-half s-plane.
(a) $s^3 + 20s^2 + 9s + 100 = 0$
(b) $s^3 + 20s^2 + 9s + 200 = 0$
(c) $3s^4 + 10s^3 + 5s^2 + s + 2 = 0$
(d) $s^4 + 2s^3 + 6s^2 + 8s + 8 = 0$
(e) $s^6 + 2s^5 + 8s^4 + 12s^3 + 20s^2 + 16s + 16 = 0$

7.2. The characteristic equations for certain feedback control systems are given below. In each case, determine the values of K that correspond to a stable system.
(a) $s^4 + 22s^3 + 10s^2 + 2s + K = 0$
(b) $s^4 + 20Ks^3 + 5s^2 + (10 + K)s + 15 = 0$
(c) $s^3 + (K + 0.5)s^2 + 4Ks + 50 = 0$

7.3. The conventional Routh–Hurwitz criterion gives only the location of the roots of a polynomial with respect to the right half and the left half of the s-plane. The open-loop transfer function of a unity feedback control system is given as

$$G(s) = \frac{K}{s(1 + Ts)}$$

It is desired that all the roots of the system's characteristic equation lie in the region to the left of the line $s = -a$. This will assure not only that a stable system is obtained, but also that the system has a minimum amount of damping. Extend the Routh–Hurwitz criterion to this case, and determine the values of K and T required so that there are no roots to the right of the line $s = -a$.

7.4. The loop transfer function of a feedback control system is given by

$$G(s)H(s) = \frac{K(s + 1)}{s(1 + Ts)(1 + 2s)}$$

The parameters K and T may be represented in a plane with K as the horizontal axis and T as the vertical axis. Determine the region in which the closed-loop system is stable.

7.5. The open-loop transfer function of a unity feedback control system is given by

$$G(s) = \frac{K(s + 5)(s + 40)}{s^3(s + 200)(s + 1000)}$$

Discuss the stability of the closed-loop system as a function of K. Determine the values of K that will cause sustained oscillations in the closed-loop system. What are the frequencies of oscillations?

7.6. A controlled process is represented by the following state equations:

$$\dot{x}_1 = x_1 - 3x_2$$
$$\dot{x}_2 = 5x_1 + u$$

The control is obtained from state feedback such that

$$u = g_1 x_1 + g_2 x_2$$

where g_1 and g_2 are real constants. Determine the region in the g_2 versus g_1 plane in which the overall system is stable.

7.7. Given a linear time-invariant system that is described by the following state equations:

$$\dot{\mathbf{x}}(t) = \mathbf{A}\mathbf{x}(t) + \mathbf{B}u(t)$$

where

$$\mathbf{A} = \begin{bmatrix} 0 & 1 & 0 \\ 0 & 0 & 1 \\ 0 & -3 & -2 \end{bmatrix} \quad \mathbf{B} = \begin{bmatrix} 0 \\ 0 \\ 1 \end{bmatrix}$$

The closed-loop system is implemented by state feedback so that

$$u(t) = -\mathbf{G}\mathbf{x}(t)$$

where $\mathbf{G}$ is the feedback matrix, $\mathbf{G} = [g_1 \quad g_2 \quad g_3]$ with g_1, g_2, and g_3 equal to real constants. Determine the constraints on the elements of $\mathbf{G}$ so that the overall system is asymptotically stable.

7.8. Given the system $\dot{\mathbf{x}} = \mathbf{A}\mathbf{x} + \mathbf{B}u$, where

$$\mathbf{A} = \begin{bmatrix} 1 & 0 & 0 \\ 0 & -2 & 0 \\ 0 & 0 & 3 \end{bmatrix} \quad \mathbf{B} = \begin{bmatrix} 2 \\ 0 \\ 1 \end{bmatrix}$$

Consider that state feedback may be implemented. Is the system stabilizable by state feedback?

7.9. For the following loop gain functions, sketch the Nyquist diagrams that correspond to the entire Nyquist path. In each case check the values of N, P, and Z with respect to the origin in the GH-plane. Determine the values of N, P, and Z with respect to the -1 point, and determine if the closed-loop system is stable. Specify in which case it is necessary to sketch only the Nyquist plot for $\omega = 0$ to ∞ (section 1) on the Nyquist path to investigate the stability of the closed-loop system.

(a) $G(s)H(s) = \dfrac{50}{s(1 + 0.1s)(1 + 0.2s)}$

(b) $G(s)H(s) = \dfrac{10}{s^2(1 + 0.25s)(1 + 0.5s)}$

(c) $G(s)H(s) = \dfrac{100(1 + s)}{s(1 + 0.1s)(1 + 0.5s)(1 + 0.8s)}$

(d) $G(s)H(s) = \dfrac{5(1 - 0.5s)}{s(1 + 0.1s)(1 - 0.25s)}$

(e) $G(s)H(s) = \dfrac{10}{s(1 + 0.2s)(s - 1)}$

(f) $G(s)H(s) = \dfrac{2.5(1 + 0.2s)}{1 + 2s + s^3}$

7.10. Sketch Nyquist diagrams for the following loop transfer functions. Sketch only the portion that is necessary to determine the stability of the closed-loop system. Determine the stability of the systems.

(a) $G(s)H(s) = \dfrac{100}{s(s^2 + 2s + 2)(s + 1)}$

(b) $G(s)H(s) = \dfrac{50}{s(s + 2)(s^2 + 4)}$

(c) $G(s)H(s) = \dfrac{s}{1 - 0.2s}$

7.11. Figure P7-11 shows the entire Nyquist plots of the loop gains $G(s)H(s)$ of some feedback control systems. It is known that in each case, the zeros of $G(s)H(s)$

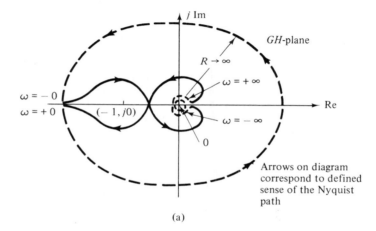

(a)

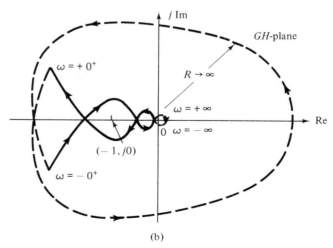

(b)

Figure P7-11.

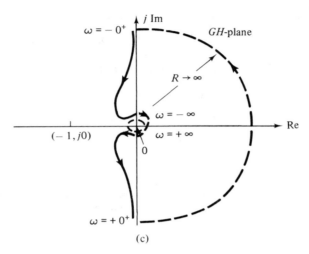

Figure P7-11 (Cont.).

are all located in the left half of the s-plane; that is, $Z = 0$ with respect to the origin in the GH-plane. Determine the number of poles of $G(s)H(s)$ that are in the right half of the s-plane. State the stability of the open-loop systems. State whether the closed-loop system is stable; if not, give the number of roots of the characteristic equation that are in the right half of the s-plane.

7.12. The characteristic equation of a feedback control system is given by

$$s^3 + 5Ks^2 + (2K + 3)s + 10 = 0$$

Apply the Nyquist criterion to determine the values of K for a stable closed-loop system. Check the answer by means of the Routh–Hurwitz criterion.

7.13. The Nyquist criterion was originally devised to investigate the absolute stability of a closed-loop system. By sketching the Nyquist plot of $G(s)H(s)$ that corresponds to the Nyquist path, it is possible to tell whether the system's characteristic equation has roots in the right half of the s-plane.

(a) Define a new Nyquist path in the s-plane that may be used to ensure that all the complex roots of the characteristic equation have damping ratios greater than a value ζ_1.

(b) Define a new Nyquist path in the s-plane that may be used to ensure that all the characteristic equation roots are in the left half of the s-plane with real parts greater than α_1.

7.14. The block diagram of a feedback control system is shown in Fig. P7-14.

(a) Determine the values of a so that the system is stable.

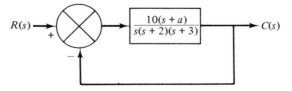

Figure P7-14.

(b) Determine the values of a so that the eigenvalues of the system all lie to the left of the $\text{Re}(s) = -1$ line in the s-plane.

7.15. The block diagram of a multivariable control system is shown in Fig. P7-15.

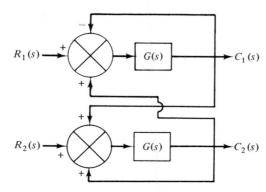

Figure P7-15.

The transfer function $G(s)$ is given by

$$G(s) = \frac{K}{(s+1)(s+2)}$$

and K is a positive constant. Determine the range of K so that the system is asymptotically stable.

(a) Use the Routh–Hurwitz criterion.

(b) Use the Nyquist criterion.

7.16. Figure P7-16 shows the block diagram of a control system in which the amplifier

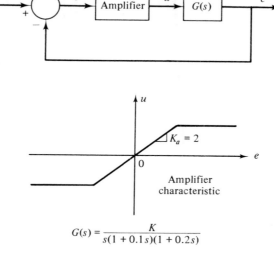

$$G(s) = \frac{K}{s(1+0.1s)(1+0.2s)}$$

Figure P7-16.

has a nonlinear saturation characteristic. Determine the maximum value of K such that the overall system will be absolutely stable.

7.17. Figure P7-17 shows a control system that has a synchro control transformer as an error transducer. The output of the synchro is proportional to the sine of the input shaft position. Assuming that the synchro operation is limited to $-\pi/2 \leq \theta_e \leq \pi/2$, the input–output characteristic of the device may be represented by the sine wave shown in Fig. P7-17. Determine the limiting value of K so that the system is absolutely stable.

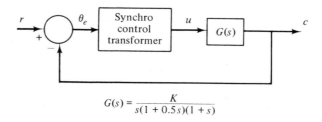

$$G(s) = \frac{K}{s(1 + 0.5s)(1 + s)}$$

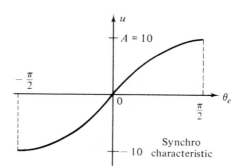

Figure P7-17.

8

Root Locus Techniques

8.1 Introduction

In the design of control systems it is often necessary to investigate the performance of a system when one or more parameters of the system varies over a given range. Since the characteristic equation plays an important role in the dynamic behavior of linear systems, an important problem in linear control systems theory is the investigation of the trajectories of the roots of the characteristic equation, or simply, the root loci, when a certain system parameter varies. In fact, the examples of Chapter 6 already have illustrated the importance of the root loci in the study of linear control systems.

The root locus technique is not confined to the inclusive study of control systems. The equation under investigation does not necessarily have to be the characteristic equation of a linear system. The technique can also be used to assist in the determination of roots of high-order algebraic equations.

In general, the root locus problem for one variable parameter can be defined by referring to equations of the following form:

$$F(s) = s^n + a_1 s^{n-1} + \ldots + a_{n-1}s + a_n$$
$$+ K(s^m + b_1 s^{m-1} + \ldots + b_{m-1}s + b_m) = 0 \tag{8-1}$$

where K is the parameter considered to vary between $-\infty$ and ∞. The coefficients $a_1, \ldots, a_n, b_1, \ldots, b_{m-1}, b_m$ are assumed to be fixed. These coefficients can be real or complex, although our main interest here is in real coefficients. The root locus problem of multiple-variable parameters will also be considered in this chapter.

Although the loci of the roots of Eq. (8-1) when K varies between $-\infty$ and

∞ are generally referred to as the *root loci* in control system literature, the following categories are defined:

1. Root loci: the portion of the root loci when K assumes positive values; that is, $0 \leq K < \infty$.
2. Complementary root loci: the portion of the root loci when K assumes negative values; that is, $-\infty < K \leq 0$.
3. Root contours: loci of roots when more than one parameter varies.

The *complete root loci* refers to the combination of the root loci and the complementary root loci.

8.2 Basic Conditions of the Root Loci

Since our main interest is in control systems, let us consider that Eq. (8-1) is the characteristic equation of a linear control system that has the closed-loop transfer function

$$\frac{C(s)}{R(s)} = \frac{G(s)}{1 + G(s)H(s)} \tag{8-2}$$

The characteristic equation is obtained by setting the numerator of $1 + G(s)H(s)$ to zero. In other words, the left side of Eq. (8-1) should correspond to the numerator of $1 + G(s)H(s)$. Or, the roots of the characteristic equation must also satisfy

$$1 + G(s)H(s) = 0 \tag{8-3}$$

The reader should have a special understanding of the relationship between the characteristic equation, the equation $1 + G(s)H(s) = 0$, and the function $G(s)H(s)$. Although these relationships are very simple to understand, one can easily overlook these important concepts when deeply entangled in the intricacies of the construction of the root loci.

If we divide both sides of Eq. (8-1) by the terms that do not contain K, we get

$$1 + \frac{K(s^m + b_1 s^{m-1} + \ldots + b_{m-1}s + b_m)}{s^n + a_1 s^{n-1} + \ldots + a_{n-1}s + a_n} = 0 \tag{8-4}$$

Comparing Eqs. (8-3) and (8-4) we see that the following relation can be established:

$$G(s)H(s) = \frac{K(s^m + b_1 s^{m-1} + \ldots + b_{m-1}s + b_m)}{s^n + a_1 s^{n-1} + \ldots + a_{n-1}s + a_n} \tag{8-5}$$

where $G(s)H(s)$ is known as the loop transfer function of the control system. Since we have mentioned that the root locus technique is not limited to control systems, in general, given Eq. (8-1), we can regard $G(s)H(s)$ of Eq. (8-5) as the loop transfer function of an equivalent control system. The control system is equivalent in the sense that it has Eq. (8-1) as its characteristic equation. Later we shall discover that the step in obtaining Eq. (8-4) from Eq. (8-1) is a very useful one in many other analysis and design situations. For lack of a better name, we shall refer to the procedure of dividing both sides of the characteristic equation by the terms that do not contain K as the *Golden Rule*.

We can now define the complete root loci as the loci of the points in the s-plane that satisfy Eq. (8-3) as K is varied between $-\infty$ and ∞.

Now we are ready to establish the conditions under which Eq. (8-3) is satisfied. Let us express $G(s)H(s)$ as

$$G(s)H(s) = KG_1(s)H_1(s) \tag{8-6}$$

where $G_1(s)H_1(s)$ no longer contains the variable parameter K. Then Eq. (8-3) is written

$$G_1(s)H_1(s) = -\frac{1}{K} \tag{8-7}$$

To satisfy this equation, the following conditions must be met simultaneously:

$$|G_1(s)H_1(s)| = \frac{1}{|K|} \qquad\qquad -\infty < K < \infty \tag{8-8}$$

$$\underline{/G_1(s)H_1(s)} = (2k+1)\pi \qquad\qquad K \geq 0 \tag{8-9}$$

$$\underline{/G_1(s)H_1(s)} = 2k\pi \qquad\qquad K \leq 0 \tag{8-10}$$

where $k = 0, \pm 1, \pm 2, \ldots$ (all integers).

In practice, the complete root loci are constructed by finding all points in the s-plane that satisfy Eqs. (8-9) and (8-10), and then the values of K along the loci are determined using Eq. (8-8).

The construction of the root loci is basically a graphical problem, although some of the rules of construction are arrived at analytically. The starting point of the graphical construction of the root loci is based on knowledge of the poles and zeros of the function $G(s)H(s)$. In other words, Eq. (8-5) must first be written*

$$G(s)H(s) = \frac{K(s + z_1)(s + z_2)\ldots(s + z_m)}{(s + p_1)(s + p_2)\ldots(s + p_n)} \tag{8-11}$$

$$= KG_1(s)H_1(s)$$

where the poles and zeros of $G(s)H(s)$ are real or complex-conjugate numbers.

Using Eq. (8-11), the conditions stated in Eqs. (8-8), (8-9), and (8-10) become

$$|G_1(s)H_1(s)| = \frac{\prod_{i=1}^{m} |s + z_i|}{\prod_{j=1}^{n} |s + p_j|} = \frac{1}{|K|} \qquad -\infty < K < \infty \tag{8-12}$$

and

$$\underline{/G_1(s)H_1(s)} = \sum_{i=1}^{m} \underline{/s + z_i} - \sum_{j=1}^{n} \underline{/s + p_j}$$
$$= (2k+1)\pi \qquad 0 \leq K < \infty \tag{8-13}$$

---

*We shall first consider that $G(s)H(s)$ is a rational function of s. For systems with time delays, $G(s)H(s)$ will contain exponential terms such as e^{-Ts}.

$$\underline{/G_1(s)H_1(s)} = \sum_{i=1}^{m} \underline{/s + z_i} - \sum_{j=1}^{n} \underline{/s + p_j}$$

$$= 2k\pi \qquad -\infty < K \leq 0 \tag{8-14}$$

for $k = 0, \pm 1, \pm 2, \ldots$.

It was mentioned earlier that Eqs. (8-13) and (8-14) may be used for the construction of the complete root loci in the s-plane. In other words, Eq. (8-13) implies that for any positive value of K, a point (e.g., s_1) in the s-plane is a point on the root loci if the difference between the sums of the angles of the vectors drawn from the zeros and the poles of $G(s)H(s)$ to s_1 is an odd multiple of $180°$. Similarly, for negative values of K, Eq. (8-14) shows that any point on the complementary root loci must satisfy the condition that the difference between the sums of the angles of the vectors drawn from the zeros and the poles to the point is an even multiple of $180°$, or $0°$.

To illustrate the use of Eqs. (8-13) and (8-14) for the construction of the root loci, let us consider

$$G(s)H(s) = \frac{K(s + z_1)}{s(s + p_2)(s + p_3)} \tag{8-15}$$

The locations of the poles and the zero of $G(s)H(s)$ are arbitrarily assumed, as shown in Fig. 8-1. Next, we select an arbitrary point, s_1, in the s-plane and draw vectors directing from the poles and the zero of $G(s)H(s)$ to the point s_1. If s_1

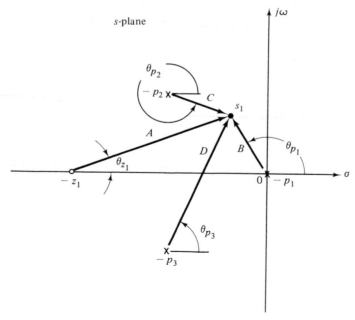

Fig. 8-1. Pole–zero configuration of $G(s)H(s) = [K(s + z_1)]/[s(s + p_2)(s + p_3)]$.

is indeed a point on the root loci ($0 \le K < \infty$) [remember that the root loci represent the loci of zeros of $1 + G(s)H(s)$], it must satisfy the following two conditions simultaneously: from Eq. (8-12),

$$\frac{|s_1 + z_1|}{|s_1||s_1 + p_2||s_1 + p_3|} = \frac{1}{|K|} \tag{8-16}$$

and from Eq. (8-13),

$$\underline{/s_1 + z_1} - (\underline{/s_1} + \underline{/s_1 + p_2} + \underline{/s_1 + p_3}) = (2k + 1)\pi \qquad k = 0, \pm 1, \pm 2, \ldots \tag{8-17}$$

Similarly, if s_1 is to be a point on the complementary root loci ($-\infty < K \le 0$), it must satisfy Eq. (8-14); that is,

$$\underline{/s + z_1} - (\underline{/s_1} + \underline{/s_1 + p_2} + \underline{/s_1 + p_3}) = 2k\pi \tag{8-18}$$

for $k = 0, \pm 1, \pm 2, \ldots$.

As shown in Fig. 8-1, the angles θ_{z_1}, θ_{p_1}, θ_{p_2}, and θ_{p_3} are the angles of the vectors measured with the positive real axis as zero reference. Equations (8-17) and (8-18) become

$$\theta_{z_1} - (\theta_{p_1} + \theta_{p_2} + \theta_{p_3}) = (2k + 1)\pi \qquad 0 \le K < \infty \tag{8-19}$$

and

$$\theta_{z_1} - (\theta_{p_1} + \theta_{p_2} + \theta_{p_3}) = 2k\pi \qquad -\infty < K \le 0 \tag{8-20}$$

respectively.

If s_1 is found to satisfy either Eq. (8-19) or Eq. (8-20), Eq. (8-16) is used to determine the value of K at the point. Rewriting Eq. (8-16), we have

$$|K| = \frac{|s_1||s_1 + p_2||s_1 + p_3|}{|s_1 + z_1|} \tag{8-21}$$

where the factor $|s_1 + z_1|$ is the length of the vector drawn from the zero z_1 to the point s_1. If, as in Fig. 8-1, the vector lengths are represented by A, B, C, and D, Eq. (8-21) becomes

$$|K| = \frac{BCD}{A} \tag{8-22}$$

The sign of K, of course, depends on whether s_1 is on the root loci or the complementary root loci. Consequently, given the pole–zero configuration of $G(s)H(s)$, the construction of the complete root locus diagram involves the following two steps:

1. A search for all the s_1 points in the s-plane that satisfy Eqs. (8-9) and (8-10).
2. The determination of the values of K at points on the root loci and the complementary root loci by use of Eq. (8-8).

From the basic principles of the root locus plot discussed thus far, it may seem that the search for all the s_1 points in the s-plane that satisfy Eqs. (8-9) and (8-10) is a very tedious task. However, aided with the properties of the root loci that we are going to assemble in the next section, the actual sketching of the root locus diagram in most cases is not so formidably complex. Normally, with some experience on the part of the analyst, the root loci can be sketched by following through the "rules" of construction. A special graphical aid, called the *Spirule*, can also be used to help plot the root locus diagram. However, since the Spirule is simply a device that assists in adding and subtracting angles of vectors quickly, according to Eq. (8-13) or Eq. (8-14), it can be used effectively only if we already know the general proximity of the roots. Indeed, when we set out to find a point s_1 on the root loci, we must first have some general knowledge of the location of this point; then we select a trial point and test it in Eqs. (8-13) and (8-14). If the trial point is not close to any point on the root loci, the search procedure can be frustrating, even with the aid of a Spirule.

Digital and analog computer programs[32-34] can be prepared for the plotting of the root locus diagram. Analytical procedures[26-29] have also been developed for obtaining the root loci. In fact, a computer or the analytical method must be used if the poles and zeros of $G(s)H(s)$ are not known a priori.

The material presented in this chapter will emphasize the principle of construction of the root loci, since one must obtain a thorough understanding of the basic principles before any engineering implementation or numerical methods can be applied successfully.

8.3 Construction of the Complete Root Loci

The following rules of construction are developed from the relation between the poles and zeros of $G(s)H(s)$ and the zeros of $1 + G(s)H(s)$. These rules should be regarded only as an aid to the construction of the root loci and the complementary root loci, as they do not give the exact plots.

K = 0 Points

Theorem 8-1. The $K = 0$ points on the complete root loci are at the poles of $G(s)H(s)$.

Proof: From Eq. (8-12),

$$|G_1(s)H_1(s)| = \frac{\prod_{i=1}^{m} |s + z_i|}{\prod_{j=1}^{n} |s + p_j|} = \frac{1}{|K|} \tag{8-23}$$

As K approaches zero, the value of Eq. (8-23) approaches infinity, and, correspondingly, s approaches the poles of $G_1(s)H_1(s)$ or of $G(s)H(s)$; that is, s approaches $-p_j$ $(j = 1, 2, \ldots, n)$. It is apparent that this property applies to both the root loci and the complementary root loci, since the sign of K has no bearing in Eq. (8-23).

EXAMPLE 8-1   Consider the following equation:

$$s(s + 2)(s + 3) + K(s + 1) = 0 \qquad (8\text{-}24)$$

When $K = 0$, the three roots of the equation are at $s = 0$, $s = -2$, and $s = -3$. These three points are also the poles of the function $G(s)H(s)$ if we divide both sides of Eq. (8-24) by the terms that do not contain K (the Golden Rule) and establish the relationship

$$1 + G(s)H(s) = 1 + \frac{K(s + 1)}{s(s + 2)(s + 3)} = 0 \qquad (8\text{-}25)$$

Thus

$$G(s)H(s) = \frac{K(s + 1)}{s(s + 2)(s + 3)} \qquad (8\text{-}26)$$

The three $K = 0$ points on the complete root loci are as shown in Fig. 8-2.

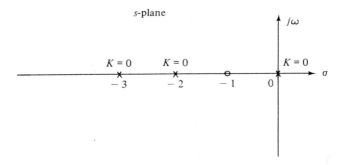

Fig. 8-2. Points at which $K = 0$ on the complete root loci of $s(s + 2)$ $(s + 3) + K(s + 1) = 0$.

K = ±∞ Points

Theorem 8-2. The $K = \pm\infty$ points on the complete root loci are at the zeros of $G(s)H(s)$.

Proof: Referring again to Eq. (8-23), as K approaches $\pm\infty$, the equation approaches zero. This corresponds to s approaching the zeros of $G(s)H(s)$; or s approaching $-z_i$ $(i = 1, 2, \ldots, m)$.

EXAMPLE 8-2   Consider again the equation

$$s(s + 2)(s + 3) + K(s + 1) = 0 \qquad (8\text{-}27)$$

It is apparent that when K is very large, the equation can be approximated by

$$K(s + 1) = 0 \qquad (8\text{-}28)$$

which has the root $s = -1$. Notice that this is also the zero of $G(s)H(s)$ in Eq. (8-26). Therefore, Fig. 8-3 shows the point $s = -1$ at which $K = \pm\infty$. However, $G(s)H(s)$ in this case also has two other zeros located at infinity, because for a rational function,

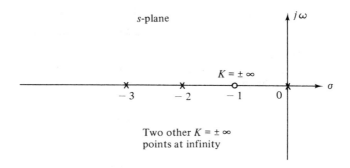

Fig. 8-3. Points at which $K = \pm\infty$ on the complete root loci of $s(s + 2)$ $(s + 3) + K(s + 1) = 0$.

the total number of poles and zeros must be equal if the poles and zeros at infinity are included. Therefore, for the equation in Eq. (8-27), the $K = \pm\infty$ points are at $s = -1$, ∞, and ∞.

Number of Branches on the Complete Root Loci

A branch of the complete root loci is the locus of one root when K takes on values between $-\infty$ and ∞. Since the number of branches of the complete root loci must equal the number of roots of the equation, the following theorem results:

> *Theorem 8-3. The number of branches of the root loci of Eq. (8-1) is equal to the greater of n and m.*

EXAMPLE 8-3   The number of branches on the complete root loci of

$$s(s + 2)(s + 3) + K(s + 1) = 0 \qquad (8\text{-}29)$$

is three, since $n = 3$ and $m = 1$. Or, since the equation is of the third order in s, it must have three roots, and therefore three root loci.

Symmetry of the Complete Root Loci

> *Theorem 8-4. The complete root loci are symmetrical with respect to the real axis of the s-plane. In general, the complete root loci are symmetrical with respect to the axes of symmetry of the poles and zeros of G(s)H(s).*

Proof: The proof of the first statement is self-evident, since, for real coefficients in Eq. (8-1), the roots must be real or in complex-conjugate pairs.

The reasoning behind the second statement is also simple, since if the poles and zeros of $G(s)H(s)$ are symmetrical to an axis other than the real axis in the s-plane, we can regard this axis of symmetry as if it were the real axis of a new complex plane obtained through a linear transformation.

EXAMPLE 8-4   Let us consider the equation

$$s(s + 1)(s + 2) + K = 0 \qquad (8\text{-}30)$$

Dividing both sides of the equation by the terms that do not contain K leads to

$$G(s)H(s) = \frac{K}{s(s + 1)(s + 2)} \qquad (8\text{-}31)$$

The complete root loci of Eq. (8-30) are sketched as shown in Fig. 8-4. Notice that since the poles of $G(s)H(s)$ are symmetrical with respect to the $s = -1$ axis (in addition to being always symmetrical with respect to the real axis), the complete root loci are symmetrical to the $s = -1$ axis and the real axis.

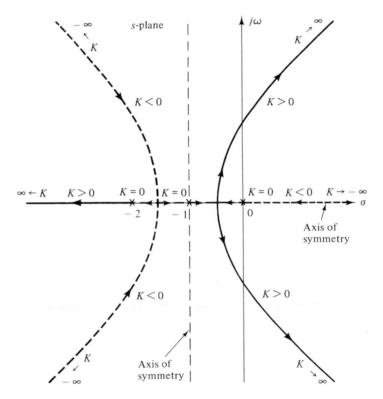

Fig. 8-4. Root loci of $s(s + 1)(s + 2) + K = 0$, showing the properties of symmetry.

EXAMPLE 8-5   When the pole–zero configuration of $G(s)H(s)$ is symmetrical with respect to a point in the s-plane, the complete root loci will also be symmetrical to that point. This is illustrated by the root locus plot of

$$s(s + 2)(s + 1 + j)(s + 1 - j) + K = 0 \qquad (8\text{-}32)$$

as shown in Fig. 8-5.

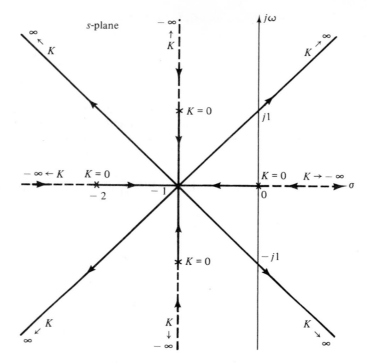

Fig. 8-5. Root loci of $s(s + 2)(s + 1 + j)(s + 1 - j) + K = 0$, showing the properties of symmetry.

Asymptotes of the Complete Root Loci (Behavior of Root Loci at $s = \infty$)

The properties of the complete root loci near infinity in the s-plane are important, since when $n \neq m$, $2|n - m|$ of the loci will approach infinity in the s-plane.

> *Theorem 8-5. For large values of s, the root loci for $K \geq 0$ are asymptotic to straight lines or asymptotes with angles given by*
>
> $$\theta_k = \frac{(2k + 1)\pi}{n - m} \qquad (8\text{-}33)$$
>
> *where $k = 0, 1, 2, \ldots, |n - m| - 1^*$ and n and m are defined in Eq. (8-1).*
> *For the complementary root loci, $K \leq 0$, the angles of the asymptotes are*
>
> $$\theta_k = \frac{2k\pi}{n - m} \qquad (8\text{-}34)$$
>
> *where $k = 0, 1, 2, \ldots, |n - m| - 1$.*

*According to the defining equations of the root loci, $k = 0, \pm 1, \pm 2, \ldots$. However, since there are only $|n - m|$ asymptotes for each type of root loci, we need to assign only $|n - m|$ values to k.

Equations (8-33) and (8-34) imply that the asymptotes of the complementary root loci are linear extensions of the asymptotes of the root loci, and vice versa. When the asymptotes of one type of root loci are determined, those of the other type are found without further calculation.

Proof: Let us divide both sides of Eq. (8-1) by

$$s^m + b_1 s^{m-1} + \ldots + b_{m-1}s + b_m$$

We then have

$$\frac{s^n + a_1 s^{n-1} + \ldots + a_{n-1}s + a_n}{s^m + b_1 s^{m-1} + \ldots + b_{m-1}s + b_m} + K = 0 \tag{8-35}$$

Carrying out the fraction of the left side of Eq. (8-35) by the process of long division, and for large s neglecting all but the first two terms, we have

$$s^{n-m} + (a_1 - b_1)s^{n-m-1} \cong -K \tag{8-36}$$

or

$$s\left(1 + \frac{a_1 - b_1}{s}\right)^{1/(n-m)} \cong (-K)^{1/(n-m)} \tag{8-37}$$

The factor $[1 + (a_1 - b_1)/s]^{1/(n-m)}$ in Eq. (8-37) is expanded by binomial expansion, and Eq. (8-37) becomes

$$s\left[1 + \frac{a_1 - b_1}{(n-m)s} + \cdots\right] \cong (-K)^{1/(n-m)} \tag{8-38}$$

Again, if only the first two terms in the last series are retained, we get

$$s + \frac{a_1 - b_1}{n - m} \cong (-K)^{1/(n-m)} \tag{8-39}$$

Now let $s = \sigma + j\omega$, and, using DeMoivre's algebraic theorem, Eq. (8-39) is written

$$\sigma + j\omega + \frac{a_1 - b_1}{n - m} \cong K^{1/(n-m)}\left[\cos\frac{(2k+1)\pi}{n - m} + j\sin\frac{(2k+1)\pi}{n - m}\right] \tag{8-40}$$

for $0 \leq K < \infty$, and

$$\sigma + j\omega + \frac{a_1 - b_1}{n - m} \cong |K^{1/(n-m)}|\left[\cos\frac{2k\pi}{n - m} + j\sin\frac{2k\pi}{n - m}\right] \tag{8-41}$$

for $-\infty < K \leq 0$, and $k = 0, \pm1, \pm2, \ldots$.

Equating the real and imaginary parts of both sides of Eq. (8-40), we have, for $0 \leq K < \infty$,

$$\sigma + \frac{a_1 - b_1}{n - m} \cong K^{1/(n-m)} \cos\frac{(2k+1)\pi}{n - m} \tag{8-42}$$

and

$$\omega \cong K^{1/(n-m)} \sin\frac{(2k+1)\pi}{n - m} \tag{8-43}$$

Solving for $K^{1/(n-m)}$ from Eq. (8-43), we have

$$K^{1/(n-m)} \cong \frac{\omega}{\sin\dfrac{2k+1}{n-m}\pi} \cong \frac{\sigma + \dfrac{a_1 - b_1}{n - m}}{\cos\dfrac{2k+1}{n-m}\pi} \tag{8-44}$$

or

$$\omega \cong \tan \frac{(2k+1)\pi}{n-m}\left(\sigma + \frac{a_1 - b_1}{n-m}\right) \qquad (8\text{-}45)$$

Equation (8-45) represents a straight line in the s-plane, and the equation is of the form

$$\omega \cong M(\sigma - \sigma_1) \qquad (8\text{-}46)$$

where M represents the slope of the straight line or the asymptote, and σ_1 is the intersect with the σ axis.

From Eqs. (8-45) and (8-46) we have

$$M = \tan \frac{2k+1}{n-m}\pi \qquad (8\text{-}47)$$

$k = 0, 1, 2, \ldots, |n-m| - 1$, and

$$\sigma_1 = -\frac{a_1 - b_1}{n-m} \qquad (8\text{-}48)$$

Note that these properties of the asymptotes are for the root loci ($0 \leq K < \infty$) only.

Similarly, from Eq. (8-41) we can show that for the complementary root loci ($-\infty < K \leq 0$),

$$M = \tan \frac{2k\pi}{n-m} \qquad (8\text{-}49)$$

$k = 0, 1, 2, \ldots, |n-m| - 1$, and the same expression as in Eq. (8-48) is obtained for σ_1. Therefore, the angular relations for the asymptotes, given by Eqs. (8-33) and (8-34), have been proved. This proof also provided a by-product, which is the intersect of the asymptotes with the real axis of the s-plane, and therefore resulted in the following theorem.

Intersection of the Asymptotes (Centroid)

Theorem 8-6. (a) The intersection of the $2|n - m|$ asymptotes of the complete root loci lies on the real axis of the s-plane.
(b) The intersection of the asymptotes is given by

$$\sigma_1 = \frac{b_1 - a_1}{n-m} \qquad (8\text{-}50)$$

where a_1, b_1, n, and m are defined in Eq. (8-1).

Proof: The proof of (a) is straightforward, since it has been established that the complete root loci are symmetrical to the real axis.

The proof of (b) is a consequence of Eq. (8-48). Furthermore, if we define a function $G(s)H(s)$ as in Eq. (8-5), Eq. (8-50) may be written as

$$\sigma_1 = \frac{b_1 - a_1}{n-m}$$

$$= \frac{\sum \text{finite poles of } G(s)H(s) - \sum \text{finite zeros of } G(s)H(s)}{\text{number of finite poles of } G(s)H(s)} \qquad (8\text{-}51)$$
$$- \text{ number of finite zeros of } G(s)H(s)$$

since from the law of algebra,

$$-a_1 = \text{sum of the roots of } s^n + a_1 s^{n-1} + \ldots + a_{n-1} s + a_n = 0$$

$$= \text{sum of the finite poles of } G(s)H(s) \tag{8-52}$$

$$-b_1 = \text{sum of the roots of } s^m + b_1 s^{m-1} + \ldots + b_{m-1} s + b_m = 0$$

$$= \text{sum of the finite zeros of } G(s)H(s) \tag{8-53}$$

Since the poles and zeros are either real or complex-conjugate pairs, the imaginary parts always cancel each other. Thus in Eq. (8-51) the terms in the summations may be replaced by the real parts of the poles and zeros of $G(s)H(s)$, respectively.

It should be noted that Eq. (8-51) is valid for the root loci as well as the complementary root loci.

EXAMPLE 8-6   Consider the equation

$$s(s + 4)(s^2 + 2s + 2) + K(s + 1) = 0 \tag{8-54}$$

This equation corresponds to the characteristic equation of a feedback control system with the loop transfer function

$$G(s)H(s) = \frac{K(s + 1)}{s(s + 4)(s^2 + 2s + 2)} \tag{8-55}$$

The pole–zero configuration of $G(s)H(s)$ is shown in Fig. 8-6. From the six theorems on the construction of the complete root loci described so far, the following information concerning the root loci and the complementary root loci of Eq. (8-54) is obtained:

1. $K = 0$: The $K = 0$ points on the complete root loci are at the poles of $G(s)H(s)$: $s = 0$, $s = -4$, $s = -1 + j1$, and $s = -1 - j1$.
2. $K = \pm\infty$: The $K = \pm\infty$ points on the complete root loci are at the zeros of $G(s)H(s)$: $s = -1$, $s = \infty$, $s = \infty$, and $s = \infty$.
3. Since Eq. (8-54) is of the fourth order, there are four complete root loci.
4. The complete root loci are symmetrical to the real axis.
5. For large values of s, the root loci are asymptotic to straight lines with angles measured from the real axis: Root loci ($K \geq 0$), Eq. (8-33):

$$k = 0 \qquad \theta_0 = \frac{180°}{3} = 60°$$

$$k = 1 \qquad \theta_1 = \frac{540°}{3} = 180°$$

$$k = 2 \qquad \theta_2 = \frac{900°}{3} = 300°$$

The angles of the asymptotes of the complementary root loci are given by Eq. (8-34):

$$k = 0 \qquad \theta_0 = \frac{0°}{3} = 0°$$

$$k = 1 \qquad \theta_1 = \frac{360°}{3} = 120°$$

$$k = 2 \qquad \theta_2 = \frac{720°}{3} = 240°$$

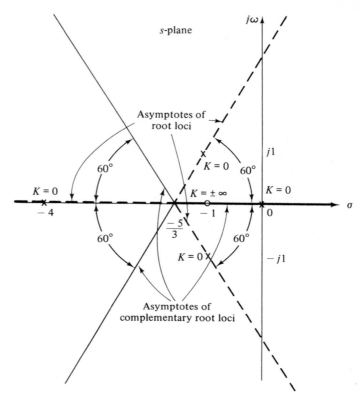

Fig. 8-6. Asymptotes of the complete root loci of $s(s + 4)(s^2 + 2s + 2)$ $+ K(s + 1) = 0$.

The asymptotes of the complementary root loci may be obtained by extending the asymptotes of the root loci.

6. The six asymptotes of the complete root loci intersect at

$$\sigma_1 = \frac{\sum \text{finite poles of } G(s)H(s) - \sum \text{finite zeros of } G(s)H(s)}{n - m}$$

$$= \frac{(0 - 4 - 1 + j1 - 1 - j1) - (-1)}{4 - 1} = -\frac{5}{3} \tag{8-56}$$

The asymptotes are sketched as shown in Fig. 8-6.

EXAMPLE 8-7   The asymptotes of the complete root loci for several different equations are shown in Fig. 8-7.

Root Loci on the Real Axis

Theorem 8-7. (a) Root loci: On a given section of the real axis, root loci $(K \geq 0)$ may be found in the section only if the total number of real poles and zeros of $G(s)H(s)$ to the right of the section is odd.

(b) Complementary root loci: On a given section of the real axis, complementary root loci $(K \leq 0)$ may be found in the section only if the total number of real poles and zeros of $G(s)H(s)$ to the right of the section is even. Alternatively,

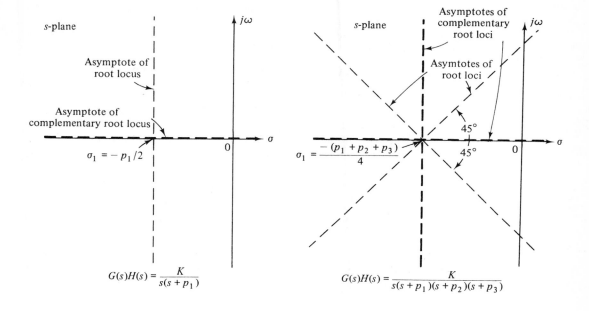

$$G(s)H(s) = \frac{K}{s(s+p_1)}$$

$$G(s)H(s) = \frac{K}{s(s+p_1)(s+p_2)(s+p_3)}$$

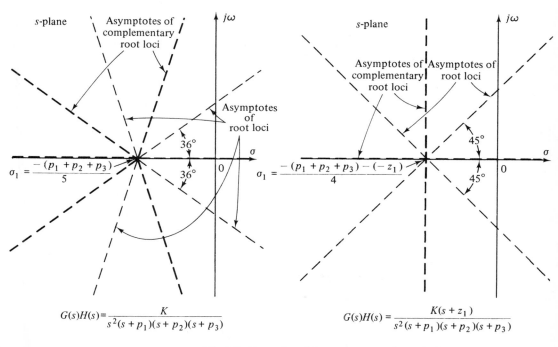

$$G(s)H(s) = \frac{K}{s^2(s+p_1)(s+p_2)(s+p_3)}$$

$$G(s)H(s) = \frac{K(s+z_1)}{s^2(s+p_1)(s+p_2)(s+p_3)}$$

Fig. 8-7. Examples of the asymptotes of root loci.

we can state that *complementary root loci will be found in sections on the real axis not occupied by the root loci.*

In all cases the complex poles and zeros of $G(s)H(s)$ do not affect the existence properties of the root loci on the real axis.

Proof: The proof of the theorem is based on the following observations:

1. At any point (e.g., s_1) on the real axis, the angles of the vectors drawn from the complex-conjugate poles and zeros of $G(s)H(s)$ add up to be zero. Therefore, the only contribution to the angular relations in Eqs. (8-13) and (8-14) is from the real poles and zeros of $G(s)H(s)$.
2. Only the real poles and zeros of $G(s)H(s)$ that lie to the right of the point s_1 may contribute to Eqs. (8-13) and (8-14), since real poles and zeros that lie to the left of the point contribute zero degrees.
3. Each real pole of $G(s)H(s)$ to the right of the point s_1 contributes $-180°$ and each zero to the right of the point contributes $180°$ to Eqs. (8-13) and (8-14).

The last observation shows that for s_1 to be a point on the root loci, there must be an odd number of poles and zeros of $G(s)H(s)$ to the right of s_1, and for s_1 to be a point on the complementary root loci the total number of poles and

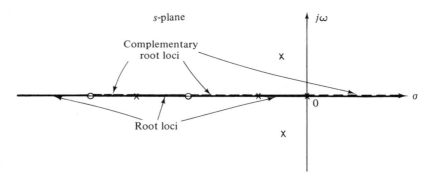

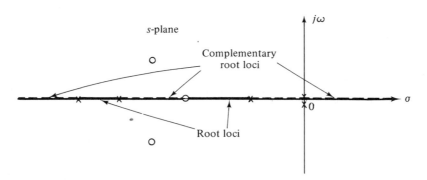

Fig. 8-8. Properties of root loci on the real axis.

zeros of $G(s)H(s)$ to the right of s_1 must be even. The following example illustrates the properties of the complete root loci on the real axis of the s-plane.

EXAMPLE 8-8   The complete root loci on the real axis in the s-plane are shown in Fig. 8-8 for two different pole–zero configuraions of $G(s)H(s)$. Notice that the entire real axis is occupied by either the root loci or the complementary root loci.

Angles of Departure (from Poles) and the Angles of Arrival (at Zeros) of the Complete Root Loci

The angle of departure (arrival) of the complete root locus at a pole (zero) of $G(s)H(s)$ denotes the behavior of the root loci near that pole (zero). For the root loci ($K \geq 0$) these angles can be determined by use of Eq. (8-13). For instance, in the pole–zero configuration of $G(s)H(s)$ given in Fig. 8-9, it is desired to deter-

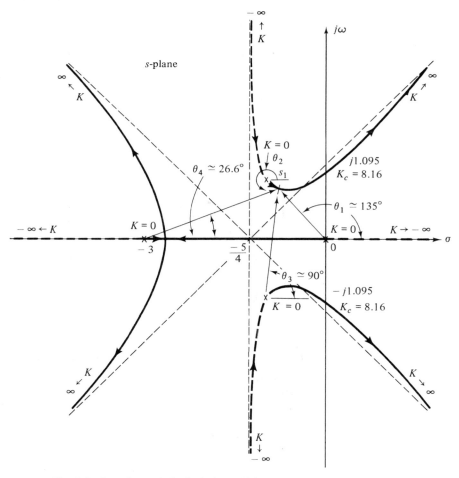

Fig. 8-9. Complete root loci of $s(s + 3)(s^2 + 2s + 2) + K = 0$ to illustrate the angles of departure or arrival.

mine the angle at which the root locus leaves the pole at $-1 + j1$. Notice that the unknown angle θ_2 is measured with respect to the real axis. Let us assume that s_1 is a point on the root locus leaving the pole at $-1 + j1$ and is very near the pole. Then s_1 must satisfy Eq. (8-13). Thus

$$\underline{/G(s_1)H(s_1)} = -(\theta_1 + \theta_2 + \theta_3 + \theta_4) = (2k + 1)180° \qquad (8\text{-}57)$$

Since s_1 is very close to the pole at $-1 + j1$, the angles of the vectors drawn from the other three poles are determined from Fig. 8-9, and Eq. (8-57) becomes

$$-(135° + \theta_2 + 90° + 26.6°) = (2k + 1)180° \qquad (8\text{-}58)$$

We can simply set k equal to zero, since the same result is obtained for all other values. Therefore,

$$\theta_2 = -431.6°$$

which is the same as $-71.6°$.

When the angle of the root locus at a pole or zero of $G(s)H(s)$ is determined, the angle of the complementary root loci at the same point differs from this angle by 180°, since Eq. (8-14) must now be used.

Intersection of the Root Loci with the Imaginary Axis

The points where the complete root loci intersect the imaginary axis of the s-plane, and the corresponding values of K, may be determined by means of the Routh–Hurwitz criterion.

For complex situations with multiple intersections, the critical values of K and ω may be more easily determined approximately from the Bode diagram.

EXAMPLE 8-9   The complete root loci of the equation

$$s(s + 3)(s^2 + 2s + 2) + K = 0 \qquad (8\text{-}59)$$

are drawn in Fig. 8-9. The root loci intersect the $j\omega$-axis at two conjugate points. Applying the Routh–Hurwitz criterion to Eq. (8-59), we have, by solving the auxiliary equation, $K_c = 8.16$ and $\omega_c = \pm 1.095$ rad/sec.

Breakaway Points (Saddle Points) on the Complete Root Loci

Breakaway points or saddle points on the root loci of an equation correspond to multiple-order roots of the equation. Figure 8-10(a) illustrates a case in which two branches of the root loci meet at the breakaway point on the real axis and then depart from the axis in opposite directions. In this case the breakaway point represents a double root of the equation to which the root loci belong. Figure 8-10(b) shows another common situation where a breakaway point may occur.

In general a breakaway point may involve more than two root loci. Figure 8-11 illustrates a situation where the breakaway point represents a fourth-order root.

A root locus diagram can, of course, have more than one breakaway point. Moreover, the breakaway points need not always be on the real axis. However, because of the conjugate symmetry of the root loci, the breakaway points must either be real or in complex-conjugate pairs.

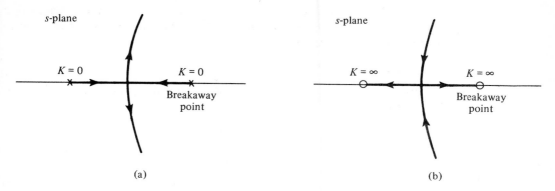

(a)   (b)

Fig. 8-10. Examples of breakaway points on the real axis in the s-plane.

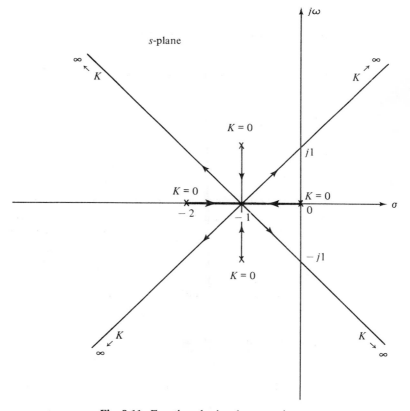

Fig. 8-11. Fourth-order breakaway point.

Because of the symmetry of the root loci, it is easy to see that the root loci in Fig. 8-10(a) and (b) break away at 180° apart, whereas in Fig. 8-11 the four root loci depart with angles 90° apart. In general, if n root loci $(-\infty < K < \infty)$ approach or leave a breakaway point, they must be $180/n$ degrees apart.

Several graphical and analytical methods are available for the determination of the location of the breakaway points. Two analytical methods that seem to be the most general are presented below.

Method 1.

Theorem 8-8. The breakaway points on the complete root loci of $1 + KG_1(s)H_1(s) = 0$ must satisfy

$$\frac{dG_1(s)H_1(s)}{ds} = 0 \tag{8-60}$$

Proof: Let Eq. (8-1) be written

$$Q(s) + KP(s) = 0 \tag{8-61}$$

Then Eq. (8-5) may be written

$$G(s)H(s) = \frac{KP(s)}{Q(s)} \tag{8-62}$$

If we consider that K is varied by an increment ΔK, Eq. (8-61) becomes

$$Q(s) + (K + \Delta K)P(s) = 0 \tag{8-63}$$

Dividing both sides of Eq. (8-63) by $Q(s) + KP(s)$, we have

$$1 + \frac{\Delta K P(s)}{Q(s) + KP(s)} = 0 \tag{8-64}$$

which can be written

$$1 + \Delta K F(s) = 0 \tag{8-65}$$

where

$$F(s) = \frac{P(s)}{Q(s) + KP(s)} \tag{8-66}$$

Since the denominator of $F(s)$ is the same as the left-hand side of Eq. (8-61), at points very close to an nth-order root $s = s_i$ of Eq. (8-61), which corresponds to a breakaway point of n loci, $F(s)$ can be approximated by

$$F(s) \simeq \frac{A_i}{(s - s_i)^n} = \frac{A_i}{(\Delta s)^n} \tag{8-67}$$

where A_i is a constant.

Substituting Eq. (8-67) into Eq. (8-65) gives

$$1 + \frac{\Delta K A_i}{(\Delta s)^n} = 0 \tag{8-68}$$

from which we obtain

$$\frac{\Delta K}{\Delta s} = -\frac{(\Delta s)^{n-1}}{A_i} \tag{8-69}$$

Taking the limit on both sides of the last equation as ΔK approaches zero, we have

$$\lim_{\Delta K \to 0} \frac{\Delta K}{\Delta s} = \frac{dK}{ds} = 0 \tag{8-70}$$

We have shown that at a breakaway point on the root loci, dK/ds is zero.*

---

*The quantity $(ds/s)/(dK/K)$ is defined as the root sensitivity[44-46] of an equation with respect to incremental variation of the parameter K. In this case it is proved that at the breakaway points of the root loci, the roots have infinite sensitivity.

Now, since the roots of Eq. (8-61) must also satisfy

$$1 + KG_1(s)H_1(s) = 0 \tag{8-71}$$

or

$$K = -\frac{1}{G_1(s)H_1(s)} \tag{8-72}$$

it is simple to see that $dK/ds = 0$ is equivalent to

$$\frac{dG_1(s)H_1(s)}{ds} = 0 \tag{8-73}$$

It is important to point out that the condition for the breakaway point given by Eq. (8-73) is necessary but not sufficient. In other words, all breakaway points must satisfy Eq. (8-73), but not all solutions of Eq. (8-73) are breakaway points. To be a breakaway point, the solution of Eq. (8-73) must also satisfy Eq. (8-71); or, Eq. (8-73) must be a factor of Eq. (8-71) for some real K.

In general, the following conclusions can be made with regard to the solutions of Eq. (8-73):

1. All real solutions of Eq. (8-73) are breakaway points on the root loci $(-\infty < K < \infty)$, since the entire real axis of the s-plane is occupied by the complete root loci.
2. The complex-conjugate solutions of Eq. (8-73) are breakaway points only if they also satisfy Eq. (8-1). This uncertainty does not cause difficulty in the effective use of Eq. (8-73), since the other properties of the root loci are usually sufficient to provide information on the general proximity of the breakaway points. This information may also be helpful in solving for the roots of a high-order equation, as a result of Eq. (8-60), by trial and error.

The following examples are devised to illustrate the application of Eq. (8-60) for the determination of breakaway points on the root loci.

EXAMPLE 8-10   Consider that it is desired to sketch the root loci of the second-order equation

$$s(s + 2) + K(s + 4) = 0 \tag{8-74}$$

Based on some of the theorems on root loci, the root loci of Eq. (8-74) are easily sketched, as shown in Fig. 8-12. It can be proven that the complex part of the loci is described by a circle. The two breakaway points are all on the real axis, one between 0 and -2 and the other between -4 and $-\infty$.

When we divide both sides of Eq. (8-74) by $s(s + 2)$ (the Golden Rule), we obtain the identity

$$G_1(s)H_1(s) = \frac{(s + 4)}{s(s + 2)} \tag{8-75}$$

Therefore, from Eq. (8-60), the breakaway points must satisfy

$$\frac{dG_1(s)H_1(s)}{ds} = \frac{s(s + 2) - 2(s + 1)(s + 4)}{s^2(s + 2)^2} = 0 \tag{8-76}$$

or

$$s^2 + 8s + 8 = 0 \tag{8-77}$$

Solving Eq. (8-77), we find that the two breakaway points of the root loci are at $s =$

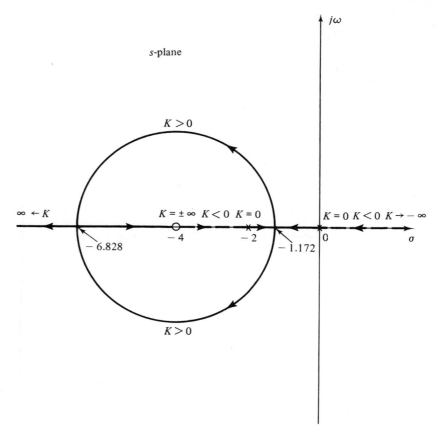

Fig. 8-12. Root loci of $s(s+2) + K(s+4) = 0$.

-1.172 and $s = -6.828$. Note also that the breakaway points happen to occur all on the root loci ($K > 0$).

EXAMPLE 8-11   Consider the equation

$$s^2 + 2s + 2 + K(s+2) = 0 \qquad (8\text{-}78)$$

The equivalent $G(s)H(s)$ is obtained by dividing both sides of Eq. (8-78) by $s^2 + 2s + 2$,

$$G(s)H(s) = \frac{K(s+2)}{s^2 + 2s + 2} \qquad (8\text{-}79)$$

Based on the poles and zeros of $G(s)H(s)$, the complete root loci of Eq. (8-78) are sketched as shown in Fig. 8-13. The diagram shows that both the root loci and the complementary root loci possess a breakaway point. These breakaway points are determined from

$$\frac{dG_1(s)H_1(s)}{ds} = \frac{d}{ds}\frac{(s+2)}{s^2 + 2s + 2}$$

$$= \frac{s^2 + 2s + 2 - 2(s+1)(s+2)}{(s^2 + 2s + 2)^2} = 0 \qquad (8\text{-}80)$$

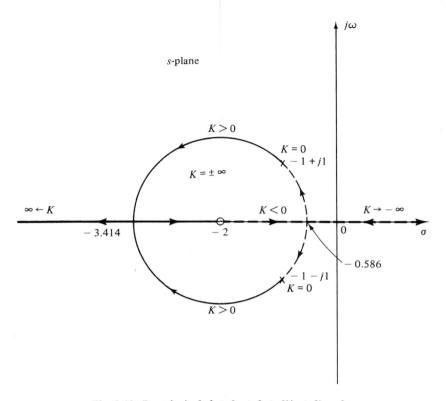

Fig. 8-13. Root loci of $s^2 + 2s + 2 + K(s + 2) = 0$.

or

$$s^2 + 4s + 2 = 0 \tag{8-81}$$

Upon solving Eq. (8-81), the breakaway points are found to be at $s = -0.586$ and $s = -3.414$. Notice that in this case $s = -3.414$ is a breakaway point on the root loci, whereas $s = -0.586$ is a breakaway point on the complementary root loci.

EXAMPLE 8-12   Figure 8-14 shows the complete root loci of the equation

$$s(s + 4)(s^2 + 4s + 20) + K = 0 \tag{8-82}$$

Dividing both sides of Eq. (8-82) by the terms that do not contain K, we have

$$1 + G(s)H(s) = 1 + \frac{K}{s(s + 4)(s^2 + 4s + 20)} = 0 \tag{8-83}$$

Since the poles of $G(s)H(s)$ are symmetrical about the axes $\sigma = -2$ and $\omega = 0$ in the s-plane, the complete root loci of the equation are also symmetrical with respect to these two axes.

Taking the derivative of $G_1(s)H_1(s)$ with respect to s, we get

$$\frac{dG_1(s)H_1(s)}{ds} = -\frac{4s^3 + 24s^2 + 72s + 80}{[s(s + 4)(s^2 + 4s + 20)]^2} = 0 \tag{8-84}$$

or

$$s^3 + 6s^2 + 18s + 20 = 0 \tag{8-85}$$

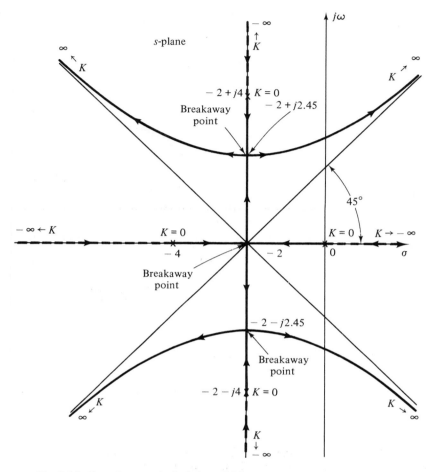

Fig. 8-14. Complete root loci of $s(s + 4)(s^2 + 4s + 20) + K = 0$.

Because of the symmetry of the poles of $G(s)H(s)$, one of the breakaway points is easily determined to be at $s = -2$. The other two breakaway points are found by solving Eq. (8-85) using this information; they are $s = -2 + j2.45$ and $s = -2 - j2.45$.

EXAMPLE 8-13   In this example we shall show that the solutions of Eq. (8-60) do not necessarily represent breakaway points on the root loci.
The complete root loci of the equation

$$s(s^2 + 2s + 2) + K = 0 \qquad (8\text{-}86)$$

are shown in Fig. 8-15; neither the root loci nor the complementary root loci have any breakaway point in this case. However, writing Eq. (8-86) as

$$1 + KG_1(s)H_1(s) = 1 + \frac{K}{s(s^2 + 2s + 2)} = 0 \qquad (8\text{-}87)$$

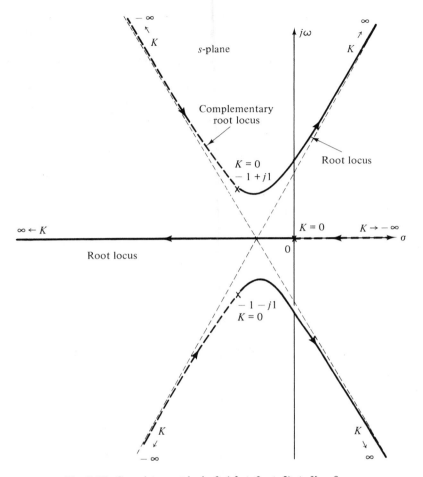

Fig. 8-15. Complete root loci of $s(s^2 + 2s + 2) + K = 0$.

and applying Eq. (8-60), we have

$$\frac{dG_1(s)H_1(s)}{ds} = \frac{d}{ds}\frac{1}{s(s^2 + 2s + 2)} = 0 \qquad (8\text{-}88)$$

which gives

$$3s^2 + 4s + 2 = 0 \qquad (8\text{-}89)$$

The roots of Eq. (8-89) are $s = -0.677 + j0.471$ and $s = -0.677 - j0.471$. These two roots do not represent breakaway points on the root loci, since they do not satisfy Eq. (8-86) for any real values of K. Another way of stating this is that Eq. (8-89) is not a factor of Eq. (8-86) for any real K.

Method 2. An algorithm for finding the breakaway points on the root loci $(-\infty < K < \infty)$ was introduced by Remec.[17] The method is derived from the theory of equations,[11] and the proofs of its necessary and sufficient conditions are given in the literature.

The breakaway-point algorithm using a tabulation that resembles the Routh tabulation for stability study is described below:

1. Let the equation for which the root loci are desired be written as

$$F(s) = A_0 s^n + A_1 s^{n-1} + \ldots + A_{n-1} s + A_n = 0 \qquad (8\text{-}90)$$

   where $A_0, A_1, \ldots, A_{n-1}, A_n$ are constant coefficients and the variable parameter K is considered to be contained in the coefficients.

2. Obtain the function $F'(s)$ which is the derivative of $F(s)$ with respect to s. Let $F'(s)$ be of the form

$$F'(s) = B_0 s^{n-1} + B_1 s^{n-2} + \ldots + B_{n-2} s + B_{n-1} = 0 \qquad (8\text{-}91)$$

   where $B_0 = nA_0$, $B_1 = (n-1)A_1, \ldots$, etc.

3. Arrange the coefficients of $F(s)$ and $F'(s)$ in two rows as follows:

$$
\begin{array}{cccccc}
A_0 & A_1 & A_2 & \ldots & A_{n-1} & A_n \\
B_0 & B_1 & B_2 & \ldots & B_{n-1} & 0
\end{array}
$$

4. Form the following array of numbers obtained by the indicated operations. Notice that this is *not* a Routh tabulation, although the cross-multiplication operation is the same as in the Routh tabulation. The illustrated tabulation is shown for a fourth-order equation, that is, for $n = 4$. Higher-order cases can be easily extended.

s^4	A_0	A_1	A_2	A_3	A_4
s^3	B_0	B_1	B_2	B_3	0
s^3	$C_0 = \dfrac{B_0 A_1 - B_1 A_0}{B_0}$	$C_1 = \dfrac{B_0 A_2 - B_2 A_0}{B_0}$	$C_2 = \dfrac{B_0 A_3 - B_3 A_0}{B_0}$	$C_3 = \dfrac{B_0 A_4 - A_0 0}{B_0} = A_4$	0
s^3	B_0	B_1	B_2	B_3	0
s^2	$D_0 = \dfrac{B_0 C_1 - B_1 C_0}{B_0}$	$D_1 = \dfrac{B_0 C_2 - B_2 C_0}{B_0}$	$D_2 = \dfrac{B_0 C_3 - B_3 C_0}{B_0}$	$D_3 = \dfrac{B_0 0 - 0 C_0}{B_0} = 0$	0
s^2	$E_0 = \dfrac{D_0 B_1 - D_1 B_0}{D_0}$	$E_1 = \dfrac{D_0 B_2 - D_2 B_0}{D_0}$	$E_2 = \dfrac{D_0 B_3 - D_3 B_0}{D_0}$	$E_3 = 0$	0
s^2	D_0	D_1	D_2	0	0
s^1	$F_0 = \dfrac{D_0 E_1 - D_1 E_0}{D_0}$	$F_1 = \dfrac{D_0 E_2 - D_2 E_0}{D_0}$	$F_2 = 0$	0	
s^1	$G_0 = \dfrac{F_0 D_1 - D_0 F_1}{F_0}$	$G_1 = \dfrac{F_0 D_2 - F_2 D_0}{F_0}$	0	0	
s^1	F_0	F_1	0		
s^0	$H_0 = \dfrac{F_0 G_1 - F_1 G_0}{F_0}$	0			

Several important features of this tabulation must be mentioned at this point.

1. The s^j $(j = 0, 1, 2, \ldots, n)$ terms assigned to each row of the tabulation are used for reference purposes.

2. The s^j terms repeat for three consecutive rows for $j = (n-1), \ldots,$ 1. There is only one s^0 row and one s^n row.
3. If we regard the rows of coefficients in the tabulation that have the same s^j as a *group*, then it is noticed that the coefficients in the last row of a group are always a repeat of those in the first row of the same group.

If $F(s)$ has multiple-order roots, which means that the root loci will have breakaway points, a row of the tabulation shown above will contain all zero elements. In other words, the tabulation process will terminate prematurely. The multiple-order roots, which are the breakaway points, are obtained by solving the equation formed by using the row of coefficients just preceding the row of zeros.

Let us use the following numerical examples to illustrate the application of the tabulation method of finding the breakaway points.

EXAMPLE 8-14   Consider the equation

$$F(s) = (s+1)^2(s+2)^2 = s^4 + 6s^3 + 13s^2 + 12s + 4 = 0 \qquad (8\text{-}92)$$

We have stated the problem in such a way that the equation is known to have two double roots at $s = -1$ and $s = -2$. We are merely going to demonstrate the properties of the tabulation when the multiple-order-root situation occurs.

Taking the derivative on both sides of Eq. (8-92) with respect to s, we have

$$F'(s) = 4s^3 + 18s^2 + 26s + 12 = 0 \qquad (8\text{-}93)$$

The tabulation using the coefficients of $F(s)$ and $F'(s)$ is made in the following manner:

s^4	1	6	13	12	4
s^3	4	18	26	12	0
s^3	$\dfrac{(4)(6)-(1)(18)}{4} = \dfrac{3}{2}$	$\dfrac{(4)(13)-(1)(26)}{4} = \dfrac{13}{2}$	$\dfrac{(4)(12)-(1)(12)}{4} = 9$	4	0
s^3	4	18	26	12	0
s^2	$-\frac{1}{4}$	$-\frac{3}{4}$	$-\frac{1}{2}$		
s^2	6	18	12		
s^2	$-\frac{1}{4}$	$-\frac{3}{4}$	$-\frac{1}{2}$		
s^1	0	0	0		

Since there is a row of zero elements in the tabulation before the tabulation process is completed, this indicates that the equation $F(s)$ has multiple-order roots. The equation that needs to be solved is formed with the coefficients taken from the row just above the row of zeros. The order of the equation is given by the power of s in the reference column. Therefore, we have

$$-\tfrac{1}{4}s^2 - \tfrac{3}{4}s - \tfrac{1}{2} = 0 \qquad (8\text{-}94)$$

or

$$s^2 + 3s + 2 = 0 \qquad (8\text{-}95)$$

The roots of Eq. (8-95) are $s = -1$ and $s = -2$, which are the roots of $F(s)$ with multiple order.

EXAMPLE 8-15   Let us consider a root locus problem, that is, the determination of the breakaway points on the root loci. The equation, Eq. (8-74), considered in Example 8-10 will be used here. The equation is rewritten

$$F(s) = s^2 + (2 + K)s + 4K = 0 \qquad (8\text{-}96)$$

Then

$$F'(s) = 2s + (2 + K) = 0 \qquad (8\text{-}97)$$

The following tabulation is made:

s^2	1	$2 + K$	$4K$
s^1	2	$2 + K$	0
s^1	$\dfrac{2 + K}{2}$	$4K$	0
s^1	2	$2 + K$	0
s^0	$4K - \dfrac{(2 + K)^2}{4}$	0	

None of the rows can be made to contain all zeros except the last one. Then, we set

$$4K - \frac{(2 + K)^2}{4} = 0 \qquad (8\text{-}98)$$

or

$$-K^2 + 12K - 4 = 0 \qquad (8\text{-}99)$$

Solving for K from the last equation, we have

$$K = 0.344$$

and

$$K = 11.656$$

When K equals either one of these two values, the s^0 row of the tabulation will contain all zeros, thus signifying a multiple-order root for $F(s)$. The equation from which the roots or breakaway points can be obtained is formed by using the coefficients in the row preceding the s^0 row. The equation thus formed is

$$2s + (2 + K) = 0 \qquad (8\text{-}100)$$

Now substituting $K = 0.344$ and $K = 11.656$ into Eq. (8-100) in turn, we find the two breakaway points on the root loci at

$$s = -1.172 \qquad K = 0.344$$

and

$$s = -6.828 \qquad K = 11.656$$

It is apparent that these answers agree with those obtained in Example 8-10. Furthermore, a by-product of the tabulation method is that the values of K at the breakaway points are also determined. In fact, the values of K are always determined first before the breakaway points are found.

EXAMPLE 8-16   Now consider the root locus problem of Example 8-12. Equation (8-82) is written as

$$F(s) = s^4 + 8s^3 + 36s^2 + 80s + K = 0 \qquad (8\text{-}101)$$

The root loci of Eq. (8-101) have three breakaway points at $s = -2, -2 + j2.45$, and $-2 - j2.45$. Let us now determine the breakaway points by the tabulation method.

We have

$$F'(s) = 4s^3 + 24s^2 + 72s + 80 = 0 \qquad (8\text{-}102)$$

or

$$s^3 + 6s^2 + 18s + 20 = 0 \qquad (8\text{-}103)$$

The following tabulation is made:

s^4	1	8	36	80	K
s^3	1	6	18	20	
s^3	2	18	60	K	
s^3	1	6	18	20	
s^2	6	24	$K - 40$	0	
s^2	2	$18 - \dfrac{K - 40}{6}$	20	0	
s^2	6	24	$K - 40$	0	
s^1	$10 - \dfrac{K - 40}{6}$	$20 - \dfrac{K - 40}{3}$	0	0	

We would like to interrupt the tabulation at this point to remark that the elements in the first row of the s^1 group can be made zero by setting $K = 100$. Therefore, $F(s)$ has multiple-order roots when $K = 100$. The breakaway points are found from the equation

$$6s^2 + 24s + (K - 40) = 0 \qquad K = 100 \qquad (8\text{-}104)$$

or

$$s^2 + 4s + 10 = 0 \qquad (8\text{-}105)$$

Therefore, the two breakaway points occur at $s = -2 + j2.45$ and $s = -2 - j2.45$, which are the solutions of Eq. (8-105).

In order to complete the tabulation, we must now consider that $K \neq 100$, so that the coefficients in the subsequent row (the second row of the s^1 group) are finite. In fact, since the multiple-order roots at $K = 100$ are already determined, should there be any additional multiple-order roots, they would have to occur at different values of K than 100. Therefore, in the remaining three rows of tabulation it is implied that $K \neq 100$. Resuming the tabulation, we have

s^1	12	$K - 40$	
s^1	$10 - \dfrac{K - 40}{6}$	$20 - \dfrac{K - 40}{3}$	$K \neq 100$
s^0	$K - 64$	0	

Now the only row that can be all zero is the s^0 row, and only when $K = 64$. Therefore, the breakaway point is found by substituting $K = 64$ into

$$\left(10 - \frac{K - 40}{6}\right)s + \left(20 - \frac{K - 40}{3}\right) = 0 \qquad (8\text{-}106)$$

which gives

$$s = -2$$

An alternative way of completing the tabulation is to factor out the common factor $10 - (K - 40)/6$ in the last row of the s^1 group, and the tabulation is as follows:

$$\begin{array}{c|cc} s^1 & 12 & K-40 \\ s^1 & 1 & 2 \\ \\ s^0 & K-64 & 0 \end{array}$$

Therefore, the same results, $K = 64$ and $s = -2$, are obtained.

EXAMPLE 8-17   In this example we shall show that the tabulation method actually indicates explicitly that the root locus diagram in Fig. 8-15, Example 8-13, does not have any breakaway points.

Equation (8-86) is written

$$F(s) = s^3 + 2s^2 + 2s + K = 0 \tag{8-107}$$

Then

$$F'(s) = 3s^2 + 4s + 2 = 0 \tag{8-108}$$

The following tabulation is made:

$$\begin{array}{c|cccc} s^3 & 1 & 2 & 2 & K \\ \\ s^2 & 3 & 4 & 2 & \\ s^2 & \frac{2}{3} & \frac{4}{3} & K & \\ s^2 & 3 & 4 & 2 & \\ \\ s^1 & \frac{4}{9} & K - \frac{4}{9} & 0 & \\ s^1 & 1 - \dfrac{27K}{4} & 2 & 0 & \\ s^1 & \frac{4}{9} & K - \frac{4}{9} & 0 & \\ \\ s^0 & 2 - \frac{9}{4}\left(K - \frac{4}{9}\right)\left(1 - \frac{27K}{4}\right) & 0 & & \end{array}$$

It is apparent that only the elements of the last row of the tabulation can be made zero for any K. Therefore, we set

$$2 - \frac{9}{4}\left(K - \frac{4}{9}\right)\left(1 - \frac{27K}{4}\right) = 0 \tag{8-109}$$

which is simplified to

$$81K^2 - 48K + 16 = 0 \tag{8-110}$$

However, there are no real values of K that will satisfy this equation. Therefore, the conclusion is that $F(s)$ does not have any multiple-order roots, or the root loci do not have any breakaway points.

Comparison of the two methods of calculation of breakaway points.

1. The condition stated in Eq. (8-60) for the breakaway point is necessary but not sufficient. The method involves solving for the *possible* breakaway points as roots of Eq. (8-60). For higher-order systems with a large number of breakaway points, the amount of work involved in solving a higher-order equation in Eq. (8-60) may be excessive.

2. The tabulation method gives a necessary and sufficient condition for the breakaway points. In general, the procedure still involves the solving of the roots of an equation, but the order of the equation may be lower than that involved in the first method. The tabulation method also gives the values of K at the breakaway points.

These methods also represent ways of solving for the multiple-order root of an equation.

Calculation of K on the Root Loci

Once the root loci have been constructed, the values of K at any point s_1 on the loci can be determined by use of the defining equation of Eq. (8-8); that is,

$$|K| = \frac{1}{|G_1(s_1)H_1(s_1)|} \tag{8-111}$$

If $G_1(s)H_1(s)$ is of the form shown in Eq. (8-11), Eq. (8-12) is written

$$|K| = \frac{\prod_{j=1}^{n} |s_1 + p_j|}{\prod_{i=1}^{m} |s_1 + z_i|} \tag{8-112}$$

or

$$|K| = \frac{\text{product of lengths of vectors drawn from the poles of } G_1(s)H_1(s) \text{ to } s_1}{\text{product of lengths of vectors drawn from the zeros of } G_1(s)H_1(s) \text{ to } s_1} \tag{8-113}$$

Equations (8-112) and (8-113) can be evaluated either graphically or analytically. Usually, if the root loci are already drawn accurately, the graphical method is more convenient. For example, the root loci of the equation

$$s^2 + 2s + 2 + K(s + 2) = 0 \tag{8-114}$$

are shown in Fig. 8-16. The value of K at the point s_1 is given by

$$K = \frac{A \cdot B}{C} \tag{8-115}$$

where A and B are the lengths of the vectors drawn from the poles of $G(s)H(s)$ $= K(s + 2)/(s^2 + 2s + 2)$ to the point s_1 and C is the length of the vector drawn from the zero of $G(s)H(s)$ to s_1. In the illustrated case s_1 is on the root loci, so K is positive. If s_1 is a point on the complementary root loci, K should have a negative sign.

The value of K at the point where the root loci intersect the imaginary axis can also be found by the method just described. However, the Routh–Hurwitz criterion usually represents a more direct method of computing this critical value of K.

The eleven rules on the construction of root locus diagrams described above should be regarded only as important properties of the root loci. Remember that earlier it was pointed out [Eq. (8-11)] that the usefulness of most of these rules of construction depends on first writing Eq. (8-1) in the form

$$(s + p_1)(s + p_2) \ldots (s + p_n) + K(s + z_1)(s + z_2) \ldots (s + z_m) = 0 \tag{8-116}$$

Then, except for extremely complex cases, these rules are usually adequate for the analyst to make a reasonably accurate sketch of the root loci just short of plotting them point by point. In complicated situations, one has to rely on a computer as a more practical means of constructing the root loci.

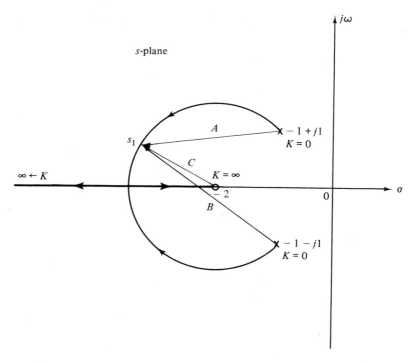

Fig. 8-16. Graphical method of finding the values of K on the root loci.

The following example serves as an illustration on the application of the rules of construction of the root loci.

EXAMPLE 8-18   Consider the equation

$$s(s + 5)(s + 6)(s^2 + 2s + 2) + K(s + 3) = 0 \qquad (8\text{-}117)$$

The complete root loci ($-\infty < K < \infty$) of the system are to be constructed. Using the rules of construction, the following properties of the root loci are determined:

1. The $K = 0$ points on the complete root loci are at $s = 0, -5, -6, -1 + j1$, and $-1 - j1$. Notice that these points are the poles of $G(s)H(s)$, where

$$G(s)H(s) = \frac{K(s + 3)}{s(s + 5)(s + 6)(s^2 + 2s + 2)} \qquad (8\text{-}118)$$

2. The $K = \pm\infty$ points on the complete root loci are at $s = -3, \infty, \infty, \infty, \infty$, which are the zeros of $G(s)H(s)$.
3. There are five separate branches on the complete root loci.
4. The complete root loci are symmetrical with respect to the real axis of the s-plane.
5. The angles of the asymptotes of the root loci at infinity are given by [Eq. (8-33)]

$$\theta_k = \frac{(2k + 1)\pi}{n - m} = \frac{(2k + 1)\pi}{5 - 1} \qquad 0 \le K < \infty \qquad (8\text{-}119)$$

for $k = 0, 1, 2, 3$. Thus the four root loci that approach infinity in the s-plane as K approaches $+\infty$ should approach asymptotes with angles of 45°, −45°, 135°, and −135°, respectively. The angles of the asymptotes of the complementary root loci at infinity are given by [Eq. (8-34)]

$$\theta_k = \frac{2k\pi}{n-m} = \frac{2k\pi}{5-1} \qquad -\infty < K \leq 0 \tag{8-120}$$

Therefore, as K approaches $-\infty$, four complementary root loci should approach infinity along asymptotes with angles of 0°, 90°, 180°, and 270°.

6. The intersection of the asymptotes is given by [Eq. (8-51)]

$$\sigma_1 = \frac{b_1 - a_1}{n-m} = \frac{\sum \text{poles of } G(s)H(s) - \sum \text{zeros of } G(s)H(s)}{n-m}$$

$$= \frac{(0 - 5 - 6 - 1 + j1 - 1 - j1) - (-3)}{4} = -2.5 \tag{8-121}$$

The results from these six steps are illustrated in Fig. 8-17.

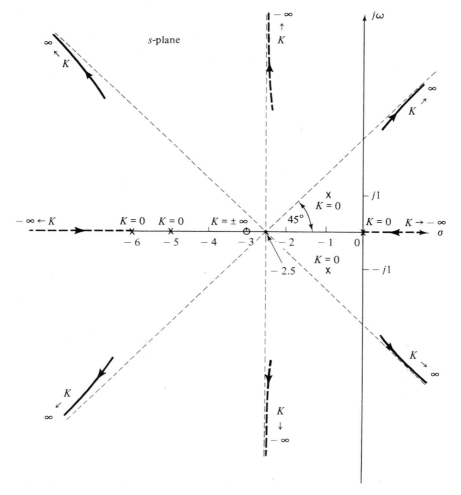

Fig. 8-17. Preliminary calculations of the root loci of $s(s+5)(s+6)(s^2 + 2s + 2) + K(s+3) = 0$.

In general, the rules on the asymptotes do not indicate on which side of the asymptote the root locus will lie. Therefore, the asymptotes indicate nothing more than the behavior of the root loci as $s \rightarrow \infty$. The complete root loci portions can be correctly sketched as shown in Fig. 8-17 only if the entire solution to the root loci problem is known.

7. Complete root loci on the real axis: There are root loci $(0 \leq K < \infty)$ on the real axis between $s = 0$ and $s = -3$, $s = -5$, and $s = -6$. There are complementary root loci $(-\infty < K \leq 0)$ on the remaining portion of the real axis, that is, between $s = -3$ and $s = -5$, and $s = -6$ and $s = -\infty$ (see Fig. 8-18).

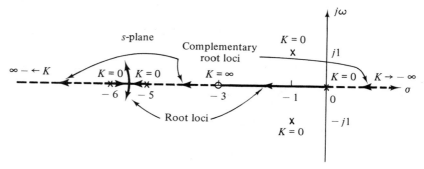

Fig. 8-18. Complete root loci on the real axis of $s(s + 5)(s + 6)(s^2 + 2s + 2) + K(s + 3) = 0$.

8. Angles of departure: The angle of departure, θ, of the root locus leaving the pole at $-1 + j1$ is determined using Eq. (8-13). If s_1 is a point on the root locus leaving $-1 + j1$, and s_1 is very close to $-1 + j1$ as shown in Fig. 8-19, Eq. (8-13) gives

$$\underline{/s_1 + 3} - (\underline{/s_1} + \underline{/s_1 + 1 + j1} + \underline{/s_1 + 5} + \underline{/s_1 + 6}$$
$$+ \underline{/s_1 + 1 - j1}) = (2k + 1)180° \quad (8\text{-}122)$$

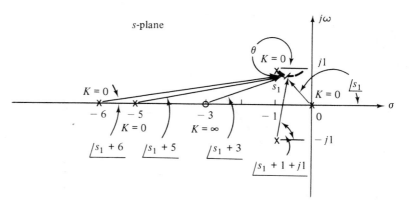

Fig. 8-19. Computation of angle of departure of the root loci of $s(s + 5)$ $(s + 6)(s^2 + 2s + 2) + K(s + 3) = 0$.

or

$$26.6° - (135° + 90° + 14° + 11.4° + \theta) \cong (2k + 1)180° \quad (8\text{-}123)$$

for $k = 0, \pm1, \pm2, \ldots$.
Therefore,

$$\theta \cong -43.8° \quad (8\text{-}124)$$

Similarly, Eq. (8-14) is used to determine the angle of arrival of the complementary root locus arriving at the point $-1 + j1$. If this angle is designated as θ', it is easy to see that θ' differs from θ by 180°; that is,

$$\theta' = 180° - 43.8° = 136.2° \quad (8\text{-}125)$$

9. The intersection of the root loci with the imaginary axis is determined by the Routh–Hurwitz criterion. Equation (8-117) is rewritten

$$s^5 + 13s^4 + 54s^3 + 82s^2 + (60 + K)s + 3K = 0 \quad (8\text{-}126)$$

The Routh tabulation is

s^5	1	54	$60 + K$
s^4	13	82	$3K$
s^3	47.7	$60 + 0.769K$	0
s^2	$65.6 - 0.212K$	$3K$	0
s^1	$\dfrac{3940 - 105K - 0.163K^2}{65.6 - 0.212K}$	0	0
s^0	$3K$	0	

For Eq. (8-126) to have no roots in the right half of the s-plane, the quantities in the first column of the Routh tabulation should be of the same sign. Therefore, the following inequalities must be satisfied:

$$65.6 - 0.212K > 0 \quad \text{or} \quad K < 309 \quad (8\text{-}127)$$

$$3940 - 105K - 0.163K^2 > 0 \quad \text{or} \quad K < 35 \quad (8\text{-}128)$$

$$K > 0 \quad (8\text{-}129)$$

Hence all the roots of Eq. (8-126) will stay in the left half of the s-plane if K lies between 0 and 35, which means that the root loci of Eq. (8-126) cross the imaginary axis when $K = 35$ and $K = 0$. The coordinate at the crossover point on the imaginary axis that corresponds to $K = 35$ is determined from the auxiliary equation

$$A(s) = (65.6 - 0.212K)s^2 + 3K = 0 \quad (8\text{-}130)$$

Substituting $K = 35$ into Eq. (8-130), we have

$$58.2s^2 + 105 = 0 \quad (8\text{-}131)$$

which yields

$$s = \pm j1.34$$

10. Breakaway points: Based on the information obtained from the preceding nine steps, a trial sketch of the root loci indicates that there can be only one breakaway point on the entire root loci, and the point lies between the two poles of $G(s)H(s)$ at $s = -5$ and -6. In this case, since there is only one breakaway point for this fifth-order system, the value of K at the point is obtained from the s^0 row of the breakaway-point tabulation, which would contain a total of 14 rows if carried out. In this case

it is actually easier to solve for the breakaway point by trial and error, since we know that the desired root is between -5 and -6.

Applying $dK/ds = 0$ to Eq. (8-126) gives

$$s^5 + 13.5s^4 + 66s^3 + 142s^2 + 123s + 45 = 0$$

After a few trial-and-error calculations, the root of the last equation that corresponds to the breakaway point is found to be $s = -5.53$.

From the information obtained in these 10 steps, the complete root locus diagram is sketched as shown in Fig. 8-20.

In this section we have described 11 important properties of the root loci. These properties have been regarded as rules when they are used in aiding the

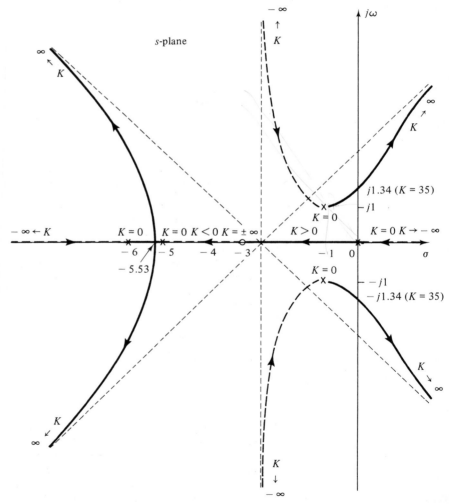

Fig. 8-20. Complete root loci of $s(s + 5)(s + 6)(s^2 + 2s + 2) + K(s + 3) = 0$.

construction of the root locus diagram. Of course, there are other minor properties of the root loci which are not mentioned here. However, in general, it is found that these 11 rules are adequate in helping to obtain a reasonably accurate sketch of the complete root loci just short of actually plotting them.

For easy reference, the 11 rules of construction are tabulated in Table 8-1.

Table 8-1   Rules of Construction of Root Loci

1. $K = 0$ points	The $K = 0$ points on the complete root loci are at the poles of $G(s)H(s)$. (The poles include those at infinity.)
2. $K = \pm\infty$ points	The $K = \pm\infty$ points on the complete root loci are at the zeros of $G(s)H(s)$. (The zeros include those at infinity.)
3. Number of separate root loci	The total number of root loci is equal to the order of the equation $F(s) = 0$.
4. Symmetry of root loci	The complete root loci of systems with rational transfer functions with constant coefficients are symmetrical with respect to the real axis of the s-plane.
5. Asymptotes of root loci as $s \longrightarrow \infty$	For large values of s, the root loci ($K > 0$) are asymptotic to straight lines with angles given by $$\theta_k = \frac{(2k+1)\pi}{n-m}$$ and for the complementary root loci ($K < 0$) $$\theta_k = \frac{2k\pi}{n-m}$$ where $k = 0, 1, 2, \ldots, \lvert n-m \rvert - 1$.
6. Intersection of the asymptotes (centroids)	(a) The intersection of the asymptotes lies only on the real axis in the s-plane. (b) The point of intersection of the asymptotes on the real axis is given by (for all values of K) $$\sigma_1 = \frac{\sum \text{real parts of poles of } G(s)H(s) - \sum \text{real parts of zeros of } G(s)H(s)}{n-m}$$
7. Root loci on the real axis	On a given section on the real axis in the s-plane, root loci are found for $K \geq 0$ in the section only if the total number of real poles and real zeros of $G(s)H(s)$ to the right of the section is *odd*. If the total number of real poles and zeros to the right of a given section is *even*, complementary root loci ($K \leq 0$) are found in the section.
8. Angles of departure and arrival	The angle of departure of the root locus ($K \geq 0$) from a pole or the angle of arrival at a zero of $G(s)H(s)$ can be determined by assuming a point s_1 that is on the root locus associated with the pole, or zero, and which is very close to the pole, or zero,

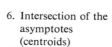

$n = \pm 1, \pm 2, \ldots$

$\alpha = $

$\dfrac{(n)180}{z-p}$

$P - z$

Table 8-1 (Cont.)

and applying the equation

$$\underline{/G(s_1)H(s_1)} = \sum_{i=1}^{m} \underline{/s_1 + z_i} - \sum_{j=1}^{n} \underline{/s_1 + p_j}$$

$$= (2k + 1)\pi \qquad k = 0, \pm1, \pm2, \ldots$$

The angle of departure or arrival of a complementary root locus is determined from

$$\underline{/G(s_1)H(s_1)} = \sum_{i=1}^{m} \underline{/s_1 + z_i} - \sum_{j=1}^{n} \underline{/s_1 + p_j}$$

$$= 2k\pi \qquad k = 0, \pm1, \pm2, \ldots$$

9. Intersection of the root loci with the imaginary axis	The values of ω and K at the crossing points of the root loci on the imaginary axis of the s-plane may be obtained by use of the Routh–Hurwitz criterion. The Bode plot of $G(s)H(s)$ may also be used.
10. Breakaway points (saddle points)	The breakaway points on the complete root loci are determined by finding the roots of $dK/ds = 0$, or $dG(s)H(s)/ds = 0$. These are necessary conditions only. Alternatively, the breakaway points are determined from a tabulation using the coefficients of the characteristic equations $F(s) = 0$ and $F'(s) = 0$. The conditions are necessary and sufficient.
11. Calculation of the values of K on the root loci	The absolute value of K at any point s_1 on the complete root loci is determined from the equation $$\|K\| = \frac{1}{\|G(s_1)H(s_1)\|}$$ $$= \frac{\text{product of lengths of vectors drawn from the poles of } G(s)H(s) \text{ to } s_1}{\text{product of lengths of vectors drawn from the zeros of } G(s)H(s) \text{ to } s_1}$$

8.4 Application of the Root Locus Technique to the Solution of Roots of a Polynomial

Although the root locus diagram is primarily intended for the portrayal of the trajectory of roots of a polynomial when a parameter, K, varies, the technique may also be used for the solution of the roots of a polynomial when all the coefficients are known. The principle is best illustrated by an example. Consider that it is desired to find the roots of the polynomial

$$F(s) = s^3 + 3s^2 + 4s + 20 = 0 \qquad (8\text{-}132)$$

To formulate this as a root locus problem, we should first convert Eq. (8-132) into the form of Eq. (8-4). Since Eq. (8-132) does not have a variable parameter K, the step of conversion is generally not unique. In other words, we can regard any one of the four coefficients of the polynomial as the parameter K. However, in general, the selection of a number of these coefficients as K will result in simpler construction of the root loci. In this case it is not difficult to see that it is more preferable to divide both sides of Eq. (8-132) by $s^3 + 3s^2$. Thus Eq. (8-132) leads to

$$1 + \frac{4(s + 5)}{s^2(s + 3)} = 0 \tag{8-133}$$

which is of the form of Eq. (8-4), with $K = 4$. Furthermore, it is apparent that the roots of Eq. (8-132) must also satisfy Eq. (8-133). Now the problem of solving Eq. (8-132) becomes that of a root locus problem, based on the pole–zero configuration of

$$G(s)H(s) = \frac{K(s + 5)}{s^2(s + 3)} \tag{8-134}$$

The desired roots are then found by setting $K = 4$. From a logistic standpoint, we have embedded a specific problem in a more general one by first solving for the solution to the general problem.

The root locus diagram based on the function $G(s)H(s)$ of Eq. (8-134) for positive K is shown in Fig. 8-21. When $K = 4$, the real root of Eq. (8-132) lies between -3 and -5, while the other two roots are complex with positive real

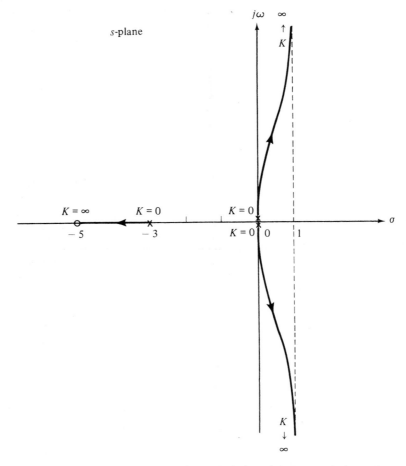

Fig. 8-21. Root locus diagram for the solution of the roots of $s^3 + 3s^2 + 4s + 20 = 0$.

parts. A few trial-and-error steps easily established the real root to be at $s = -3.495$. Thus the complex roots are found to be at $s = 0.2475 + j2.381$ and $s = 0.2475 - j2.381$.

EXAMPLE 8-19   Consider the problem of finding the roots of

$$s^4 + 5s^3 + 2s^2 + s + 10 = 0 \qquad (8\text{-}135)$$

Since this is a fourth-order polynomial, it is expected that the problem of solving for its roots will be more difficult than that of the last example. Let us first divide both sides of Eq. (8-135) by the first two terms of the equation; we have

$$1 + \frac{2(s^2 + 0.5s + 5)}{s^3(s + 5)} = 0 \qquad (8\text{-}136)$$

or

$$G(s)H(s) = \frac{K(s^2 + 0.5s + 5)}{s^3(s + 5)} \qquad K = 2 \qquad (8\text{-}137)$$

The root locus diagram based on the poles and zeros of $G(s)H(s)$ is constructed as shown in Fig. 8-22 for $\infty > K \geq 0$. However, from this root locus diagram it is not

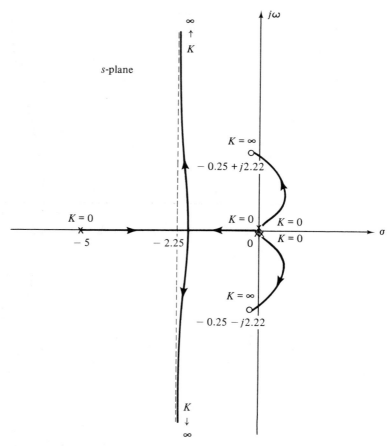

Fig. 8-22. Root locus diagram for the solution of the roots of $s^4 + 5s^3 + 2s^2 + s + 10 = 0$.

clear where the roots are when $K = 2$. Next let us divide both sides of Eq. (8-135) by the first three terms of the equation; we have

$$1 + \frac{s + 10}{s^2(s^2 + 5s + 2)} = 0 \qquad (8\text{-}138)$$

The root locus plot of

$$G(s)H(s) = \frac{K(s + 10)}{s^2(s^2 + 5s + 2)} \qquad K = 1 \qquad (8\text{-}139)$$

with $\infty > K \geq 0$ is shown in Fig. 8-23. The purpose of constructing the second root locus diagram is to establish a cross reference between the two root locus diagrams, since the roots of Eq. (8-135) must be found at the same points on both diagrams. Comparing the two root locus diagrams, we may conclude that if Eq. (8-135) has real roots, there must be two, and they must lie between -0.44 and -4.56 on the real axis. Once this range has been established, the real roots can be found by trial and error from Eq. (8-135). One real root is found to be at $s = -1.575$. Dividing both sides of Eq. (8-135) by $(s + 1.575)$, we have the remainder,

$$s^3 + 3.425s^2 - 3.394s + 6.346 = 0 \qquad (8\text{-}140)$$

Now we repeat the procedure to find the roots of Eq. (8-140). Since one of the three roots must be real, we can reason that it must be to the right of $s = -4.56$ on

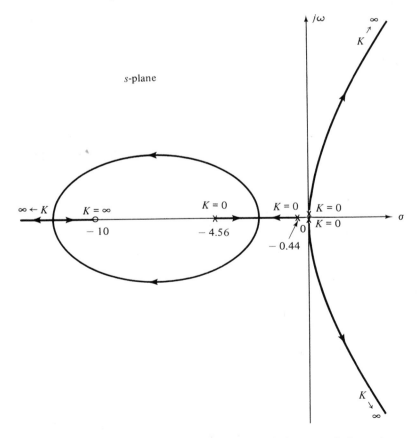

Fig. 8-23. Root locus diagram for the solution of the roots of $s^4 + 5s^3 + 2s^2 + s + 10 = 0$.

the real axis. However, we may acquire more information on this root by dividing both sides of the equation by the first two terms. We have

$$1 + \frac{-3.394(s - 1.87)}{s^2(s + 3.425)} = 0$$

or

$$G(s)H(s) = \frac{K(s - 1.87)}{s^2(s + 3.425)} \qquad K = -3.394 \tag{8-141}$$

The complementary root loci of Eq. (8-140) with the $G(s)H(s)$ of Eq. (8-141) are sketched in Fig. 8-24 for $-\infty < K \le 0$. Investigation of this complementary root locus

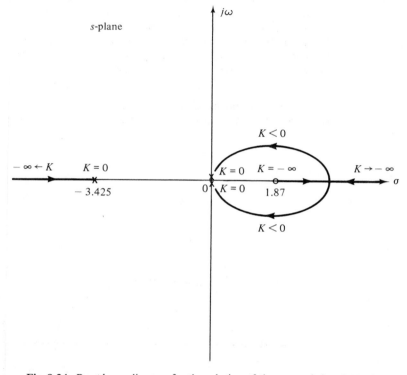

Fig. 8-24. Root locus diagram for the solution of the roots of $s^3 + 3.425s^2 - 3.394s + 6.346 = 0$.

diagram and the root loci of Fig. 8-24 reveals that the real root must lie between -3.425 and -4.56. To find the real root of Eq. (8-140), which is at $s = -4.493$, is a simple matter. The two complex roots are subsequently found to be at $s = 0.534 + j1.057$ and $s = 0.534 - j1.057$. Thus the roots of Eq. (8-135) are now all determined:

$$s = -1.575$$
$$s = -4.493$$
$$s = 0.534 + j1.057$$
$$s = 0.534 - j1.057$$

In summarizing, we have used two examples to illustrate the application of the root locus method to the solution of the roots of a high-order equation,

$$F(s) = 0 \tag{8-142}$$

where $F(s)$ is a polynomial in s. The general procedure is to convert the equation to be solved into the form $1 + G(s)H(s) = 0$ by dividing both sides of the equation by a certain appropriate portion of the equation.

In other words, we should express Eq. (8-142) as

$$P(s) + Q(s) = 0 \qquad (8\text{-}143)$$

Then, $G(s)H(s) = Q(s)/P(s)$, or $G(s)H(s) = P(s)/Q(s)$. The selection of the polynomials $P(s)$ and $Q(s)$ from $F(s)$ is more or less arbitrary, although in general the orders of $P(s)$ and $Q(s)$ should be close so that the root locus problem is made as simple as possible. In many circumstances it may be desirable to use more than one choice of division of $F(s)$ into $P(s)$ and $Q(s)$. This will usually provide additional information on the location of some of the roots of the equation, so that the trial-and-error procedure can be simplified. Example 8-19 illustrates how this is done.

In essence, the method of root finding presented here is that of utilizing the root locus technique to give an indication of the general location of some of the roots. The basic method of finding the exact roots is still cut-and-try. For high-order equations, and for equations only with complex roots, the root locus method of root finding may not be effective, and a computer solution is still the most preferable one.

8.5 Some Important Aspects of the Construction of the Root Loci

One of the important aspects of the root locus techniques is that for most control systems with moderate complexity, the analyst or designer may conduct a quick study of the system in the s-plane by making a sketch of the root loci using some or all of the rules of construction. In general, it is not necessary to make an exact plot of the root loci. Therefore, time may be saved by skipping some of the rules, and the sketching of the root locus diagram becomes an art that depends to some extent on the experience of the analyst.

In this section we shall present some of the important properties of the root loci which may be helpful in the construction of the root locus diagram.

Effect of Adding Poles and Zeros to G(s)H(s)

In Chapter 6 the effects of the derivative and integral control were illustrated by means of the root locus diagram. From the fundamental viewpoint we may investigate the effects to the root loci when poles and zeros are added to $G(s)H(s)$.

Addition of poles. In general we may state that adding a pole to the function $G(s)H(s)$ in the left half of the s-plane has the effect of pushing the original root loci toward the right-half plane. Although it is difficult to make a precise statement and provide the necessary proof, we can illustrate the point by several examples.

Let us consider the function

$$G(s)H(s) = \frac{K}{s(s + a)} \qquad a > 0 \qquad (8\text{-}144)$$

The zeros of $1 + G(s)H(s)$ are represented by the root locus diagram of Fig. 8-25(a). These root loci are constructed based on the poles of $G(s)H(s)$ at $s = 0$

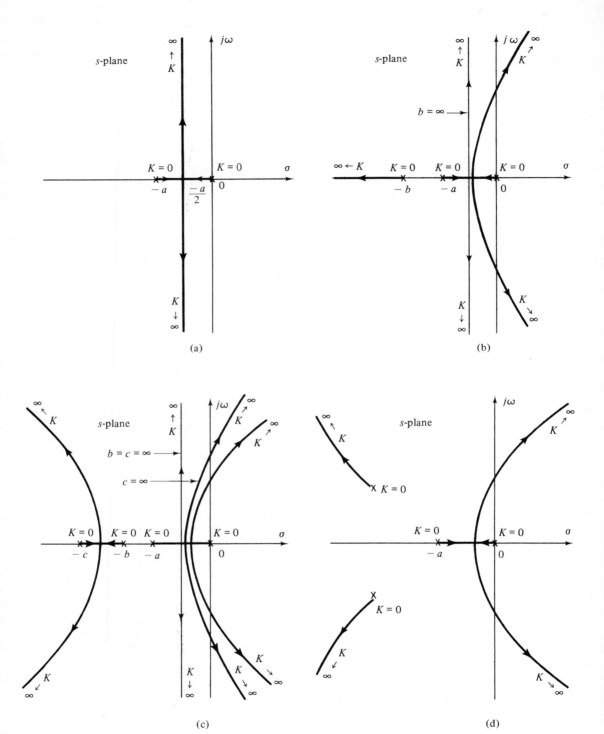

Fig. 8-25. Root locus diagrams that show the effects of adding poles to $G(s)H(s)$.

and $s = -a$. Now let us introduce a pole at $s = -b$ so that

$$G(s)H(s) = \frac{K}{s(s + a)(s + b)} \qquad b > a \qquad (8\text{-}145)$$

Figure 8-25(b) shows that the additional pole causes the complex part of the root loci to bend toward the right half of the s-plane. The angles of the asymptotes are changed from $\pm 90°$ to $\pm 60°$. The breakaway point is also moved to the right. For instance, if $a = 1$ and $b = 2$, the breakaway point is moved from -0.5 to -0.422 on the real axis. If $G(s)H(s)$ represents the loop transfer function of a feedback control system, the system with the root loci in Fig. 8-25(b) may become unstable if the value of K exceeds the critical value, whereas the system represented by the root loci of Fig. 8-25(a) is always stable. Figure 8-25(c) shows the root loci when another pole is added to $G(s)H(s)$ at $s = -c$. The system is now of the fourth order, and the two complex root loci are moved farther to the right. The angles of the asymptotes of these two loci are $\pm 45°$. For a feedback control system, the stability condition of the system becomes even more restricted. Figure 8-25(d) illustrates that the addition of a pair of complex-conjugate poles to the original two-pole configuration will result in a similar effect. Therefore, we may draw a general conclusion that the addition of poles to the function $G(s)H(s)$ has the effect of moving the root loci toward the right half of the s-plane.

Addition of zeros. Adding zeros to the function $G(s)H(s)$ has the effect of moving the root loci toward the left half of the s-plane. For instance, Fig. 8-26(a)

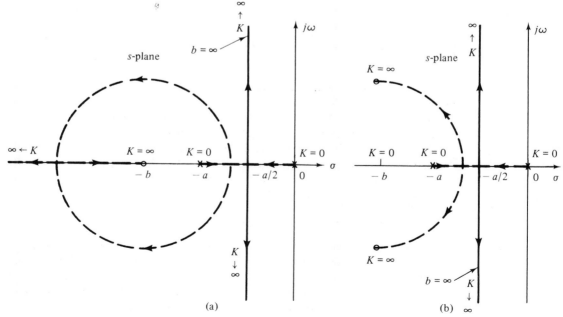

Fig. 8-26. Root locus diagrams that show the effects of adding a zero to $G(s)H(s)$.

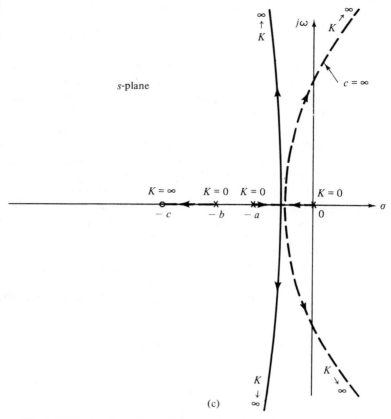

Fig. 8-26 (Cont.). Root locus diagrams that show the effects of adding a zero to $G(s)H(s)$.

shows the root locus diagram when a zero at $s = -b$ is added to the function $G(s)H(s)$ of Eq. (8-144), with $b > a$; the resultant root loci are bent toward the left and form a circle. Therefore, if $G(s)H(s)$ represents the loop transfer function of a feedback control system, the relative stability of the system is improved by the addition of the zero. Figure 8-26(b) illustrates that a similar effect will result if a pair of complex-conjugate zeros is added to the function of Eq. (8-144). Figure 8-26(c) shows the root locus diagram when a zero at $s = -c$ is added to the transfer function of Eq. (8-145).

Effects of Movements of Poles and Zeros

It was mentioned earlier that the construction of the root locus diagram depends greatly on the understanding of the principle of the technique rather than just the rigid rules of construction. In this section we show that in all cases the study of the effects of the movement of the poles and zeros of $G(s)H(s)$ on the root loci is an important and useful subject. Again, the best way to illustrate the subject is to use a few examples.

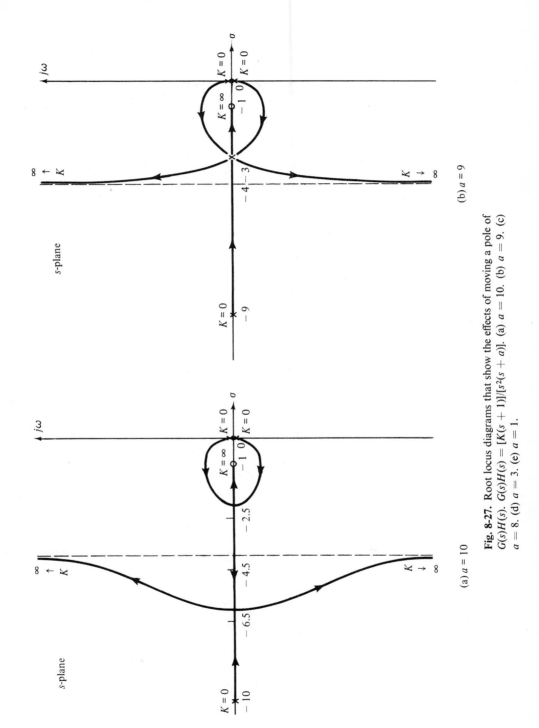

Fig. 8-27. Root locus diagrams that show the effects of moving a pole of $G(s)H(s)$. $G(s)H(s) = [K(s+1)]/[s^2(s+a)]$. (a) $a = 10$. (b) $a = 9$. (c) $a = 8$. (d) $a = 3$. (e) $a = 1$.

(a) $a = 10$

(b) $a = 9$

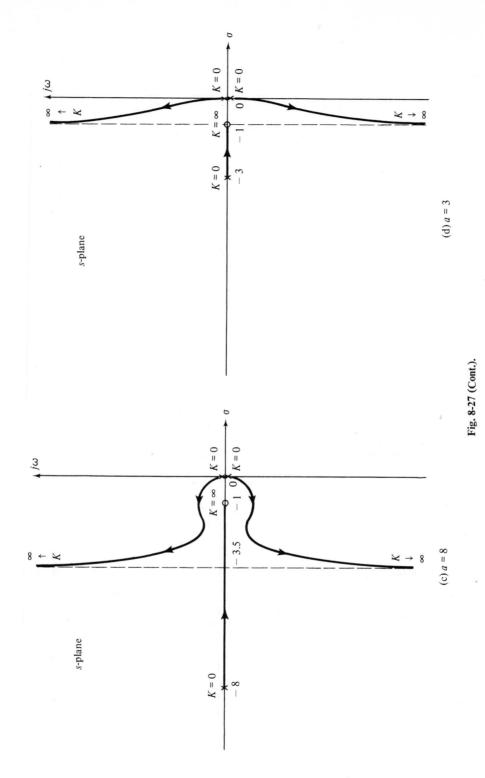

(c) $a = 8$

(d) $a = 3$

Fig. 8-27 (Cont.).

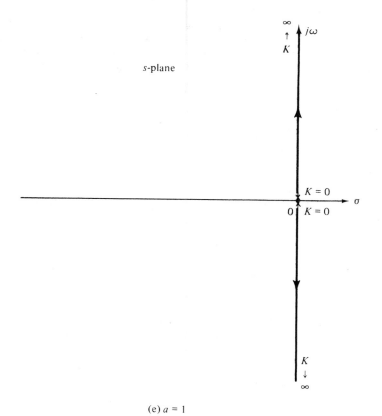

(e) $a = 1$

Fig. 8-27 (Cont.).

EXAMPLE 8-20   Consider the equation

$$s^2(s + a) + K(s + b) = 0 \qquad (8\text{-}146)$$

which is easily converted to the form of $1 + G(s)H(s) = 0$, with

$$G(s)H(s) = \frac{K(s + b)}{s^2(s + a)} \qquad (8\text{-}147)$$

Let us set $b = 1$ and investigate the root loci of Eq. (8-146) for several values of a.

Figure 8-27(a) illustrates the root loci of Eq. (8-146) with $a = 10$ and $b = 1$. The two breakaway points are found at $s = -2.5$ and -6.5. It can be shown that for arbitrary a the nonzero breakaway points are given by

$$s = -\frac{a + 3}{4} \pm \frac{1}{4}\sqrt{a^2 - 10a + 9} \qquad (8\text{-}148)$$

When $a = 9$, Eq. (8-148) indicates that the breakaway points converge to one point at $s = -3$, and the root locus diagram becomes that of Fig. 8-27(b). It is interesting to note that a change of the pole from -10 to -9 equals a considerable change to the root loci. For values of a less than 9, the values of s as given by Eq. (8-148) no longer satisfy the equation in Eq. (8-146), which means that there are no finite, nonzero, breakaway points. Figure 8-27(c) illustrates this case with $a = 8$. As the pole at $s = -a$ is moved farther to the right, the complex portion of the root loci is pushed farther toward the right-half plane. When $a = b$, the pole at $s = -a$ and the zero at $-b$ cancel each other, and the root loci degenerate into a second-order one and lie on the imaginary axis. These two cases are shown in Fig. 8-27(d) and (e), respectively.

EXAMPLE 8-21   Consider the equation

$$s(s^2 + 2s + a) + K(s + 2) = 0 \qquad (8\text{-}149)$$

which is converted to the form of $1 + G(s)H(s) = 0$, with

$$G(s)H(s) = \frac{K(s + 2)}{s(s^2 + 2s + a)} \qquad (8\text{-}150)$$

The objective is to study the complete root loci ($-\infty < K < \infty$) for various values of $a(> 0)$. As a start, let $a = 1$ so that the poles of $G(s)H(s)$ are at $s = 0, -1$, and -1. The complete root loci for this case are sketched in Fig. 8-28(a). By setting $dG(s)H(s)/ds$ to zero, the breakaway points are found at $s = -0.38, -1$, and -2.618.

As the value of a is increased from unity, the two double poles of $G(s)H(s)$ at $s = -1$ will move vertically up and down. The sketch of the root loci is governed mainly by the knowledge of the breakaway points. We can show that $dG(s)H(s)/ds$ leads to

$$s^3 + 4s^2 + 4s + a = 0 \qquad (8\text{-}151)$$

As the value of a increases, the breakaway points at $s = -0.38$ and $s = -2.618$ move to the left, whereas the breakaway point at $s = -1$ moves toward the right. Figure 8-28(b) shows the complete root loci with $a = 1.12$; that is,

$$G(s)H(s) = \frac{K(s + 2)}{s(s^2 + 2s + 1.12)} \qquad (8\text{-}152)$$

Since the real parts of the poles and zeros of $G(s)H(s)$ are not affected by the value of a, the intersect of the asymptotes is always at the origin of the s-plane. The breakaway points when $a = 1.12$ are at $s = -0.493, -0.857$, and -2.65. These are obtained by solving Eq. (8-151).

By solving for a double-root condition in Eq. (8-151) when $a = 1.185$, it can be shown that the two breakaway points that lie between $s = 0$ and $s = -1$ converge to a point. The root loci for this situation are sketched as shown in Fig. 8-28(c).

When a is greater than 1.185, Eq. (8-151) yields one real root and two complex-conjugate roots. Although complex breakaway points do occur quite often in root loci, we can easily show in the present case that these complex roots do not satisfy the original equation of Eq. (8-149) for any real K. Thus the root loci have only one breakaway point, as shown in Fig. 8-28(d) for $a = 3$. The transition between the cases in Fig. 8-28(c) and (d) should be apparent.

8.6 Root Contour—Multiple-Parameter Variation

The root locus technique discussed thus far is restricted to only one variable parameter in K. However, in many control systems problems, the effects of varying several parameters must be studied. For example, when designing a controller that is represented by a transfer function with poles and zeros, it is necessary to investigate the effects on the performance of the overall system when these poles and zeros take on various values.

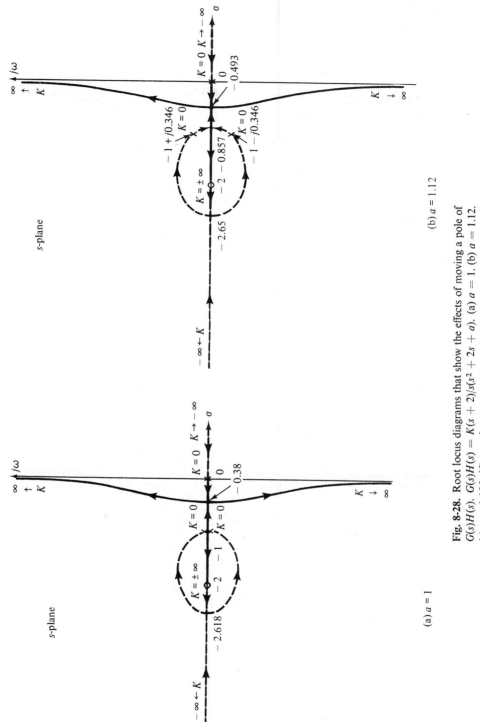

(a) $a = 1$

(b) $a = 1.12$

Fig. 8-28. Root locus diagrams that show the effects of moving a pole of $G(s)H(s)$. $G(s)H(s) = K(s + 2)/s(s^2 + 2s + a)$. (a) $a = 1$. (b) $a = 1.12$. (c) $a = 1.185$. (d) $a = 3$.

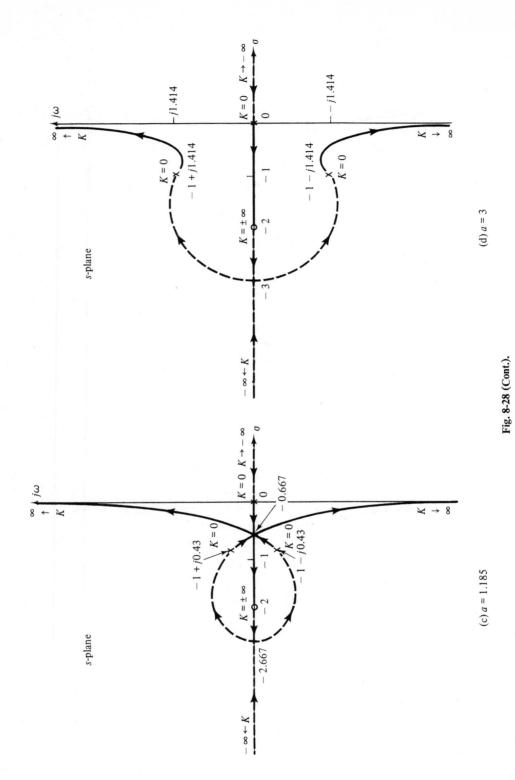

(c) $a = 1.185$

(d) $a = 3$

Fig. 8-28 (Cont.).

In Section 8.5 the root locus diagrams of equations with two variable parameters are studied by assigning different values to one of the parameters. In this section the multiparameter problem is investigated through a more systematic method of embedding. When more than one parameter varies continuously from $-\infty$ to ∞, the loci of the root are referred to as the *root contours*. It will be shown that the same conditions and rules of the root loci are still applicable to the construction of the root contours.

The principle of the root contours can be illustrated by considering the equation

$$Q(s) + K_1 P_1(s) + K_2 P_2(s) = 0 \qquad (8\text{-}153)$$

where K_1 and K_2 are the variable parameters and $Q(s)$, $P_1(s)$, and $P_2(s)$ are polynomials of s. The first step involves the setting of one of the parameters equal to zero. Let us set K_2 equal to zero. Then Eq. (8-153) becomes

$$Q(s) + K_1 P_1(s) = 0 \qquad (8\text{-}154)$$

The root loci of this equation may be obtained by dividing both sides of the equation (the Golden Rule) by $Q(s)$. Thus

$$1 + \frac{K_1 P_1(s)}{Q(s)} = 0 \qquad (8\text{-}155)$$

or

$$1 + G_1(s)H_1(s) = 0 \qquad (8\text{-}156)$$

The construction of the root loci depends on the pole–zero configuration of

$$G_1(s)H_1(s) = \frac{K_1 P_1(s)}{Q(s)} \qquad (8\text{-}157)$$

Next, we restore the value of K_2, and again apply the Golden Rule to Eq. (8-153), with K_2 as the variable parameter. We have

$$1 + \frac{K_2 P_2(s)}{Q(s) + K_1 P_1(s)} = 0 \qquad (8\text{-}158)$$

or

$$1 + G_2(s)H_2(s) = 0 \qquad (8\text{-}159)$$

Now the root contours of Eq. (8-153) are constructed based upon the poles and zeros of

$$G_2(s)H_2(s) = \frac{K_2 P_2(s)}{Q(s) + K_1 P_1(s)} \qquad (8\text{-}160)$$

However, one important feature is that the poles of $G_2(s)H_2(s)$ are identical to the roots of Eq. (8-156) or of Eq. (8-154). Thus the root contours of the original equation must all start ($K_2 = 0$) at the points that lie on the root loci of Eq. (8-156). This is the reason why one root contour problem is considered to be embedded in another. The same procedure may be extended to more than two variable parameters.

EXAMPLE 8-22   Consider the equation

$$s^3 + K_2 s^2 + K_1 s + K_1 = 0 \qquad (8\text{-}161)$$

where K_1 and K_2 are the variable parameters and with values that lie between 0 and ∞.

As a first step we let $K_2 = 0$; Eq. (8-161) becomes

$$s^3 + K_1 s + K_1 = 0 \qquad (8\text{-}162)$$

which is converted to

$$1 + \frac{K_1(s+1)}{s^3} = 0 \qquad (8\text{-}163)$$

The root loci of Eq. (8-162) are drawn from the poles and zeros of

$$G_1(s)H_1(s) = \frac{K_1(s+1)}{s^3} \qquad (8\text{-}164)$$

as shown in Fig. 8-29(a).

Next, we let K_2 vary between zero and infinity while holding K_1 at a constant nonzero value. Dividing both sides of Eq. (8-161) by the terms that do not contain K_2, we have

$$1 + \frac{K_2 s^2}{s^3 + K_1 s + K_1} = 0 \qquad (8\text{-}165)$$

Thus the root contours of Eq. (8-161) when K_2 varies may be drawn from the pole–zero configuration of

$$G_2(s)H_2(s) = \frac{K_2 s^2}{s^3 + K_1 s + K_1} \qquad (8\text{-}166)$$

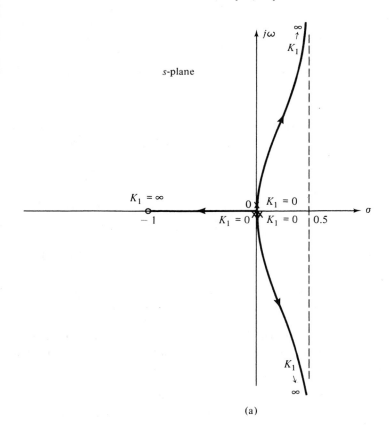

(a)

Fig. 8-29. (a) Root contours for $s^3 + K_2 s^2 + K_1 s + K_1 = 0$, $K_2 = 0$

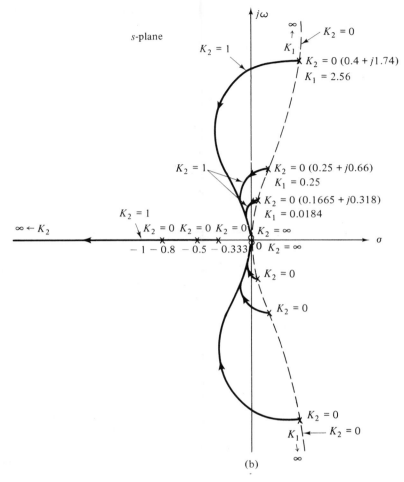

Fig. 8-29 (Cont.). (b) Root contours for $s^3 + K_2 s^2 + K_1 s + K_1 = 0$.
K_2 varies, $K_1 = $ constant.

The zeros of $G_2(s)H_2(s)$ are at $s = 0, 0$; but the poles are the zeros of $1 + G_1(s)H_1(s)$ which have been found on the contours of Fig. 8-29(a). Thus for fixed K_1 the root contours when K_2 varies must all emanate from the root contours of Fig. 8-29(a).

EXAMPLE 8-23   Consider the loop transfer function

$$G(s)H(s) = \frac{K}{s(1 + Ts)(s^2 + 2s + 2)} \qquad (8\text{-}167)$$

of a closed-loop control system. It is desired to construct the root contours of the characteristic equation with K and T as variable parameters.

The characteristic equation of the system is written

$$s(1 + Ts)(s^2 + 2s + 2) + K = 0 \qquad (8\text{-}168)$$

First, we shall set T equal to zero. The characteristic equation becomes

$$s(s^2 + 2s + 2) + K = 0 \qquad (8\text{-}169)$$

The root contours of this equation when K varies are drawn based on the poles and zeros of

$$G_1(s)H_1(s) = \frac{K}{s(s^2 + 2s + 2)} \qquad (8\text{-}170)$$

as shown in Fig. 8-30(a).

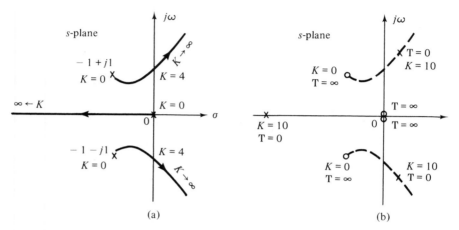

(a)   (b)

Fig. 8-30. (a) Root loci for $s(s^2 + 2s + 2) + K = 0$. (b) Pole–zero configuration of $G_2(s)H_2(s) = [Ts^2(s^2 + 2s + 2)]/s(s^2 + 2s + 2) + K]$.

For the root contours with T varying and K held constant, we write Eq. (8-168) as

$$1 + G_2(s)H_2(s) = 1 + \frac{Ts^2(s^2 + 2s + 2)}{s(s^2 + 2s + 2) + K} = 0 \qquad (8\text{-}171)$$

Therefore, the root contours when T varies are constructed from the pole–zero configuration of $G_2(s)H_2(s)$. When $T = 0$, the points on the root contours are at the poles of $G_2(s)H_2(s)$, which are the points on the root loci of Eq. (8-169), as shown in Fig. 8-30(b) for $K = 10$. The $T = \infty$ points on the root contours are at the zeros of $G_2(s)H_2(s)$, and these are at $s = 0, 0, -1 + j1$, and $-1 - j1$. The root contours for the system are sketched in Figs. 8-31, 8-32, and 8-33 for three different values of K; when $K = 0.5$ and $T = 0.5$, the characteristic equation has a quadruple root at $s = -1$.

EXAMPLE 8-24   As an example illustrating the effect of the variation of a zero of $G(s)H(s)$, consider

$$G(s)H(s) = \frac{K(1 + Ts)}{s(s + 1)(s + 2)} \qquad (8\text{-}172)$$

The problem may be regarded as a study of the effect of derivative control, as discussed in Section 6.7, on the root locations of the characteristic equation.

The characteristic equation of the system is

$$s(s + 1)(s + 2) + K(1 + Ts) = 0 \qquad (8\text{-}173)$$

Let us first consider the effect of varying the parameter K. Setting $T = 0$ in Eq. (8-173)

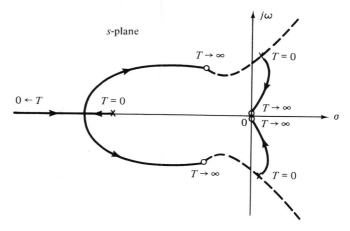

Fig. 8-31. Root contours for $s(1 + sT)(s^2 + 2s + 2) + K = 0$; $K > 4$.

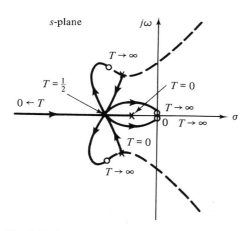

Fig. 8-32. Root contours for $s(1 + sT)(s^2 + 2s + 2) + K = 0$; $K = 0.5$.

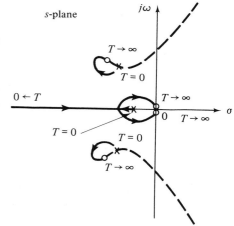

Fig. 8-33. Root contours for $s(1 + sT)(s^2 + 2s + 2) + K = 0$; $K < 0.5$.

yields

which leads to

$$s(s + 1)(s + 2) + K = 0 \tag{8-174}$$

$$1 + \frac{K}{s(s + 1)(s + 2)} = 0 \tag{8-175}$$

The root loci of Eq. (8-174) are sketched in Fig. 8-34, based on the pole–zero configuration of

$$G_1(s)H_1(s) = \frac{K}{s(s + 1)(s + 2)} \tag{8-176}$$

When T varies between zero and infinity, we write Eq. (8-173) as

$$1 + G_2(s)H_2(s) = 1 + \frac{TKs}{s(s + 1)(s + 2) + K} = 0 \tag{8-177}$$

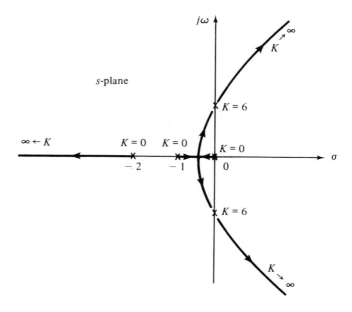

Fig. 8-34. Root loci for $s(s + 1)(s + 2) + K = 0$.

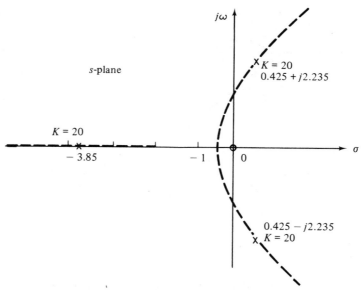

Fig. 8-35. Pole–zero configuration of $G_2(s)H_2(s) = TKs/[s(s + 1)(s + 2) + K]$, $K = 20$.

The points that correspond to $T = 0$ on the root contours are at the roots of $s(s + 1)(s + 2) + K = 0$, whose loci are sketched as shown in Fig. 8-34. If we choose $K = 20$, the pole–zero configuration of $G_2(s)H_2(s)$ is shown in Fig. 8-35. The root contours of Eq. (8-173) for $0 \leq T < \infty$ are sketched in Fig. 8-36 for three values of

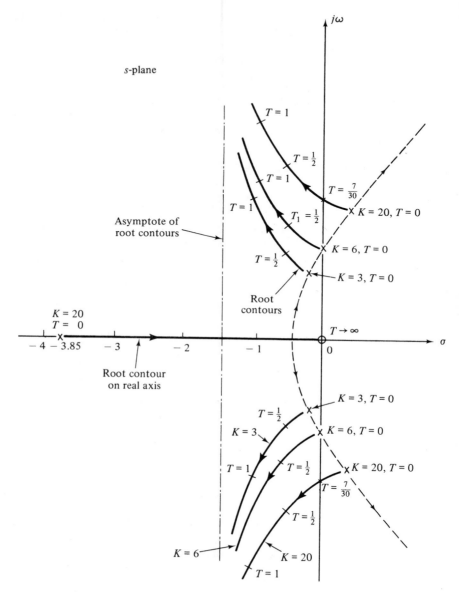

Fig. 8-36. Root contours of $s(s + 1)(s + 2) + K + KTs = 0$.

K. The intersection of the asymptotes of the root contours is obtained from Eq. (8-51); that is,

$$\sigma_1 = \frac{-3.85 + 0.425 + 0.425}{3 - 1} = -1.5 \qquad (8\text{-}178)$$

Therefore, the intersection of the asymptotes is always at $s = -1.5$ because the sum of the poles of $G_2(s)H_2(s)$ is always equal to -3, regardless of the value of K, and the sum of the zeros of $G_2(s)H_2(s)$ is zero.

The root contours shown in Fig. 8-36 verify the well-known fact that the derivative control generally improves the relative stability of the closed-loop system by moving the characteristic equation roots toward the left in the s-plane. The root contours also clearly indicate an important characteristic of the derivative control in that the bandwidth of the system is increased. In certain cases the contribution to the increase in bandwidth by increasing the value of T far exceeds the improvement made on the relative stability of the system. As shown in Fig. 8-36, for $K = 20$, the system is stabilized for all values of T greater than 0.2333. However, the largest damping ratio that the system can have by increasing T is approximately 30 per cent.

8.7 Root Loci of Systems with Pure Time Delay[51]

In Chapter 5 we investigated the modeling of systems with pure time delays and pointed out that the time delay between signals at various points of a system can be represented by the exponential term e^{-Ts} as its transfer function, where T is the delay in seconds. Therefore, we shall assume that the characteristic equation of a typical closed-loop system with pure time delay may be written

$$Q(s) + KP(s)e^{-Ts} = 0 \qquad (8\text{-}179)$$

where $Q(s)$ and $P(s)$ are polynomials of s. An alternative condition of Eq. (8-179) is

$$1 + KG_1(s)H_1(s)e^{-Ts} = 0 \qquad (8\text{-}180)$$

where

$$G_1(s)H_1(s) = \frac{P(s)}{Q(s)} \qquad (8\text{-}181)$$

Thus, similar to the development in Section 8.2, in order to satisfy Eq. (8-180), the following conditions must be met simultaneously:

$$e^{-T\sigma}|G_1(s)H_1(s)| = \frac{1}{|K|} \qquad\qquad -\infty < K < \infty \qquad (8\text{-}182)$$

$$\underline{/G_1(s)H_1(s)} = (2k + 1)\pi + \omega T \qquad\qquad K \geq 0 \qquad (8\text{-}183)$$

$$\underline{/G_1(s)H_1(s)} = 2k\pi + \omega T \qquad\qquad K \leq 0 \qquad (8\text{-}184)$$

where $s = \sigma + j\omega$ and $k = 0, \pm 1, \pm 2, \ldots$. Note that the condition for any point $s = s_1$ in the s-plane to be a point on the complete root loci is given in Eqs. (8-183) and (8-184), which differ from the conditions of Eqs. (8-9) and (8-10) by the term ωT. When $T = 0$, Eqs. (8-183) and (8-184) revert to Eqs. (8-9) and (8-10). Since ω is a variable in the s-plane, the angular conditions of Eqs. (8-183) and (8-184) are no longer constant in the s-plane but depend upon the point at which a root of Eq. (8-179) may lie. Viewing the problem from another standpoint, it is recognized that if $T = 0$, given a value of K, there are only n points in the s-plane that will satisfy either Eq. (8-183) or Eq. (8-184), for all possible values of k, where n is the highest order of $P(s)$ and $Q(s)$. However, for $T \neq 0$, the angular conditions in Eqs. (8-183) and (8-184) depend on ω, which varies along the vertical axis in the s-plane. Thus, for a given K, there may be more than n points which satisfy the angular conditions in the s-plane, as k takes on all possible integral values. In fact, there are an infinite number of these points, since Eq. (8-179), which is transcendental, is known to have an infinite number of roots.

The difficulty with the construction of the root loci of Eq. (8-179) is that many of the rules of construction developed originally for systems without time delay are no longer valid for the present case. It is of interest to investigate how some of the rules of construction given in Section 8.3 may be modified to apply to the time-delay case.

K = 0 Points

Theorem 8-9. The $K = 0$ points on the complete root loci of Eq. (8-180) are at the poles of $G_1(s)H_1(s)$ and $\sigma = -\infty$.

Proof: Equation (8-182) is repeated,

$$e^{-T\sigma}|G_1(s)H_1(s)| = \frac{1}{|K|} \tag{8-185}$$

Thus, if K equals zero, s approaches the poles of $G_1(s)H_1(s)$, or σ, which is the real part of s, approaches $-\infty$.

The K = ±∞ Points

Theorem 8-10. The $K = \pm\infty$ points on the complete root loci of Eq. (8-180) are at the zeros of $G_1(s)H_1(s)$ and $\sigma = \infty$.

Proof: Referring again to Eq. (8-185), the proof of this theorem becomes evident.

Number of Branches on the Complete Root Loci

The number of branches on the root loci of Eq. (8-179) is infinite, since the equation has an infinite number of roots.

Symmetry of the Complete Root Loci

The complete root loci are symmetrical with respect to the real axis of the s-plane. This is explained by expanding e^{-Ts} into an infinite series; then Eq. (8-179) again becomes a polynomial with a real coefficient but with infinite order.

Asymptotes of the Complete Root Loci

Theorem 8-11. The asymptotes of the root loci of Eq. (8-179) are infinite in number and all are parallel to the real axis of the s-plane. The intersects of the asymptotes with the imaginary axis are given by

$$\omega = \frac{N\pi}{T} \tag{8-186}$$

where N is tabulated in Table 8-2 for the various conditions indicated.

$$n = \text{number of finite poles of } G_1(s)H_1(s)$$

$$m = \text{number of finite zeros of } G_1(s)H_1(s)$$

Table 8-2

K	$n - m$	$K = 0$ *Asymptotes*	$K = \pm\infty$ *Asymptotes*
≥ 0	Odd	N = even integers $= 0, \pm 2, \pm 4, \ldots$	N = odd integers $= \pm 1, \pm 3, \pm 5, \ldots$
	Even	N = odd integers $= \pm 1, \pm 3, \pm 5, \ldots$	N = odd integers $= \pm 1, \pm 3, \pm 5, \ldots$
≤ 0	Odd	N = odd integers $= \pm 1, \pm 3, \pm 5, \ldots$	N = even integers $= 0, \pm 2, \pm 4, \ldots$
	Even	N = even integers $= 0, \pm 2, \pm 4, \ldots$	N = even integers $= 0, \pm 2, \pm 4, \ldots$

Proof: Since as $s \rightarrow \infty$ on the root loci, K either approaches zero or $\pm\infty$, Theorems 8-9 and 8-10 show that the asymptotes are at $\sigma = \infty$ ($K = \pm\infty$) and $\sigma = -\infty$ ($K = 0$). The intersections of the asymptotes with the $j\omega$-axis and the conditions given in Table 8-2 are arrived at by use of Eqs. (8-183) and (8-184).

Root Loci on the Real Axis

The property of the root loci of Eq. (8-179) on the real axis is the same as stated in Theorem 8-7, because on the real axis, $\omega = 0$, the angular conditions of Eqs. (8-183) and (8-184) revert to those of Eqs. (8-9) and (8-10), respectively.

Angles of Departure and Arrival

Angles of departure and arrival are determined by use of Eqs. (8-183) and (8-184).

Intersection of the Root Loci with the Imaginary Axis

Since Eq. (8-179) is not an algebraic equation of s, the intersection of its loci with the imaginary axis *cannot* be determined by use of the Routh–Hurwitz criterion. The determination of all the points of intersection of the root loci with the $j\omega$-axis is a difficult task, since the root loci have an infinite number of branches. However, we shall show in the following section that only the intersections nearest the real axis are of interest for stability studies.

Breakaway Points

Theorem 8-12. The breakaway points on the complete root loci of Eq. (8-179) must satisfy

$$\frac{dG_1(s)H_1(s)e^{-Ts}}{ds} = 0 \qquad (8\text{-}187)$$

Proof: The proof of this theorem is similar to that of Theorem 8-8.

Determination of the Values of K on the Root Loci

The value of K at any point $s = s_1$ on the root loci is determined from Eq. (8-182); that is,

$$|K| = \frac{e^{T\sigma_1}}{|G_1(s_1)H_1(s_1)|} \qquad (8\text{-}188)$$

where σ_1 is the real part of s_1.

EXAMPLE 8-25   Consider the equation

$$s + Ke^{-Ts} = 0 \qquad (8\text{-}189)$$

It is desired to construct the complete root loci of this equation for a fixed value of T. Dividing both sides of Eq. (8-189) by s, we get

$$1 + \frac{Ke^{-Ts}}{s} = 0 \qquad (8\text{-}190)$$

which is of the form of Eq. (8-180) with

$$G_1(s)H_1(s) = \frac{1}{s} \qquad (8\text{-}191)$$

The following properties of the root loci of Eq. (8-189) are determined:

1. The $K = 0$ points: From Theorem 8-9, $K = 0$ at $s = 0$ and at $\sigma = -\infty$. Using Theorem 8-11 and Table 8-2, we have

   $K \geq 0$: K approaches zero as σ approaches $-\infty$ at $\omega = 0$, $\pm 2\pi/T$, $\pm 4\pi/T, \ldots$

   $K \leq 0$: K approaches zero as σ approaches $-\infty$ at $\omega = \pm\pi/T$, $\pm 3\pi/T$, $\pm 5\pi/T, \ldots$

2. The $K = \pm\infty$ points: From Theorem 8-10, $K = \pm\infty$ at $\sigma = \infty$. Using Theorem 8-11 and Table 8-2, we have

   $K \geq 0$: K approaches $+\infty$ as σ approaches $+\infty$ at $\omega = \pm\pi/T$, $\pm 3\pi/T, \ldots$

   $K \leq 0$: K approaches $-\infty$ as σ approaches $+\infty$ at $\omega = 0$, $\pm 2\pi/T$, $\pm 4\pi/T, \ldots$

   The $K = 0$, $K = \pm\infty$ points, and the asymptotes of the root loci are shown in Fig. 8-37. The notation of 0^+ is used to indicate the asymptotes of the root loci, and 0^- is for the complementary root loci.

3. The root loci ($K \geq 0$) occupy the negative real axis, and the complementary root loci ($K \leq 0$) occupy the positive real axis.

4. The intersections of the root loci with the $j\omega$ axis are relatively easy to determine for this simple problem.

   Since $G_1(s)H_1(s)$ has only a simple pole at $s = 0$, for any point s_1 on

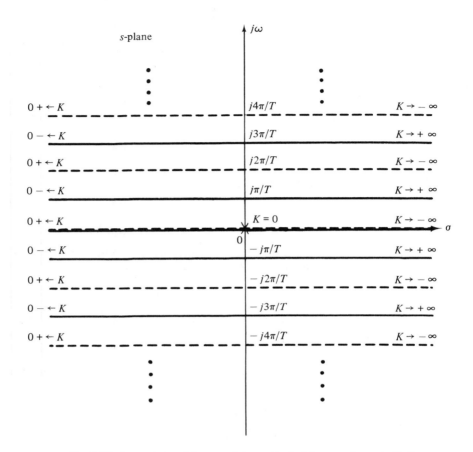

Fig. 8-37. Asymptotes of the complete root loci of the equation $s + Ke^{-Ts} = 0$.

the positive $j\omega$ axis,

$$\underline{/G_1(s_1)H_1(s_1)} = -\frac{\pi}{2} \tag{8-192}$$

and for any point s_1 on the negative $j\omega$ axis,

$$\underline{/G_1(s_1)H_1(s_1)} = \frac{\pi}{2} \tag{8-193}$$

Thus, for $K \geq 0$, Eq. (8-183) gives the condition of root loci on the $j\omega$ axis ($\omega > 0$),

$$-\frac{\pi}{2} = (2k + 1)\pi + \omega T \tag{8-194}$$

$k = 0, \pm 1, \pm 2, \ldots$ The values of ω that correspond to the points at which the root loci cross the positive $j\omega$ axis are

$$\omega = \frac{\pi}{2T}, \frac{5\pi}{2T}, \frac{9\pi}{2T}, \ldots \tag{8-195}$$

For $K \geq 0$, and $\omega < 0$,

$$\frac{\pi}{2} = (2k + 1)\pi + \omega T \tag{8-196}$$

and for $k = 0, \pm 1, \pm 2, \ldots$, the crossover points are found to be at

$$\omega = -\frac{\pi}{2T}, \; -\frac{5\pi}{2T}, \; -\frac{9\pi}{2T}, \cdots \tag{8-197}$$

Similarly, for $K \leq 0$, the conditions for the complementary root loci to cross the $j\omega$ axis are

$$-\frac{\pi}{2} = 2k\pi + \omega T \qquad \omega > 0 \tag{8-198}$$

$$\frac{\pi}{2} = 2k\pi + \omega T \qquad \omega < 0 \tag{8-199}$$

The crossover points are found by substituting $k = 0, \pm 1, \pm 2, \ldots$ into the last two equations. We have

$$\omega = \pm\frac{3\pi}{2T}, \; \pm\frac{7\pi}{2T}, \; \pm\frac{11\pi}{2T}, \cdots \tag{8-200}$$

5. Breakaway points: The breakaway points on the complete root loci are determined by the use of Eq. (8-187). Thus

$$\frac{dG_1(s)H_1(s)e^{-Ts}}{ds} = \frac{d}{ds}\left(\frac{e^{-Ts}}{s}\right) = 0 \tag{8-201}$$

or

$$\frac{-Te^{-Ts}s - e^{-Ts}}{s^2} = 0 \tag{8-202}$$

from which we have

$$e^{-Ts}(Ts + 1) = 0 \tag{8-203}$$

Therefore, the finite breakaway point is at $s = -1/T$.

6. The values of K at the crossover point on the $j\omega$ axis are found by using Eq. (8-188). Since $\sigma = 0$ on the $j\omega$ axis, we have

$$|K| = \frac{1}{|G_1(j\omega_c)H_1(j\omega_c)|} = |\omega_c| \tag{8-204}$$

where ω_c is a crossover point.

Based on the properties accumulated above, the complete root loci of Eq. (8-189) are sketched as shown in Fig. 8-38.

Although the equation of Eq. (8-179) has an infinite number of roots, and therefore the root loci have an infinite number of branches, from the system analysis standpoint, only the branches that lie between $-\pi/T \leq \omega \leq \pi/T$ are of interest. We shall refer to these as the *primary branches*. One reason is that the critical value of K at the crossover point on this portion of the root loci is equal to $\pi/2T$, whereas the critical value of K at the next branch at $\omega = \pm 5\pi/2T$ is $5\pi/2T$, which is much greater. Therefore, $K = \pi/T$ is the critical value for stability. Another reason for labeling the primary branches as the dominant loci is that for any value of K less than the critical value of $\pi/2T$, the corresponding roots on the other branches are all far to the left in the s-plane.

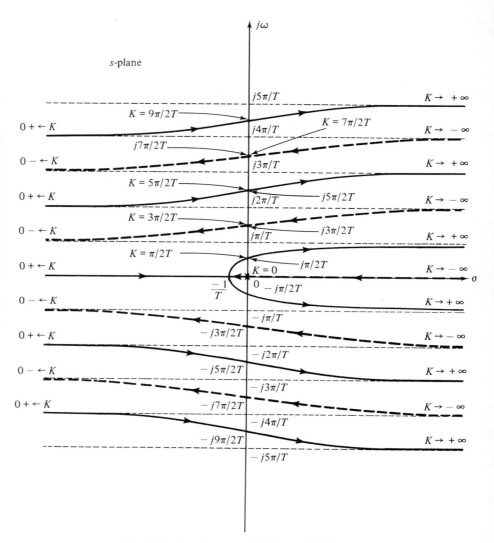

Fig. 8-38. Complete root loci for $s + Ke^{-Ts} = 0$.

Therefore, the transient response of the system, which has Eq. (8-189) as its characteristic equation, is predominantly controlled by the roots on the primary branches.

EXAMPLE 8-26   As a slightly more complex problem in the construction of the root loci of a system with pure time delay, let us consider the control system shown in Fig. 8-39. The loop transfer function of the system is

$$G(s)H(s) = \frac{Ke^{-Ts}}{s(s + 1)}$$

The characteristic equation is

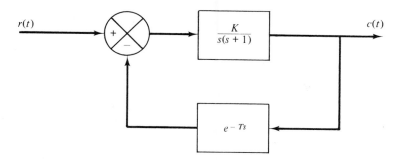

Fig. 8-39. Feedback control system with a pure time delay in the feedback path.

$$s^2 + s + Ke^{-Ts} = 0$$

In order to construct the complete root loci, the following properties are assembled by using the rules given in Theorems 8-9 through 8-12.

1. The $K = 0$ points: From Theorem 8-9, $K = 0$ at $s = 0$, $s = -1$, and $\sigma = -\infty$. Using Theorem 8-11 and Table 8-2, we have

   $K \geq 0$: K approaches zero as σ approaches $-\infty$ at $\omega = \pm\pi/T$, $\pm3/\pi T$, $\pm5\pi/T, \ldots$

   $K \leq 0$: K approaches zero as σ approaches $-\infty$ at $\omega = 0$, $\pm2\pi/T$, $\pm4\pi/T, \ldots$

2. The $K = \pm\infty$ points: From Theorem 8-10, $K = \pm\infty$ at $\sigma = \infty$. Using Theorem 8-11 and Table 8-2, we have

   $K \geq 0$: K approaches $+\infty$ as σ approaches $+\infty$ at $\omega = \pm\pi/T$, $\pm3\pi/T, \ldots$

   $K \leq 0$: K approaches $-\infty$ as σ approaches $+\infty$ at $\omega = 0$, $\pm2\pi/T$, $\pm4\pi/T, \ldots$

       Notice that the $K = 0$ asymptotes depend upon $n - m$, which is even in this case, but the $K = \pm\infty$ asymptotes depend only on the sign of K and not on $n - m$.

3. The root loci ($K \geq 0$) occupy the region between $s = 0$ and $s = -1$ on the real axis. The rest of the real axis is occupied by the complementary root loci ($K \leq 0$).

4. Breakaway points: The breakaway points of the complete root loci are found from Eq. (8-187); that is,

$$\frac{d}{ds}\left[\frac{e^{-Ts}}{s(s + 1)}\right] = 0$$

from which we have the two breakaway points at

$$s = \frac{1}{2T}[-(T + 2) \pm \sqrt{T^2 + 4}]$$

For $T = 1$ sec the two breakaway points are at

$$s = -0.382 \qquad s = -2.618$$

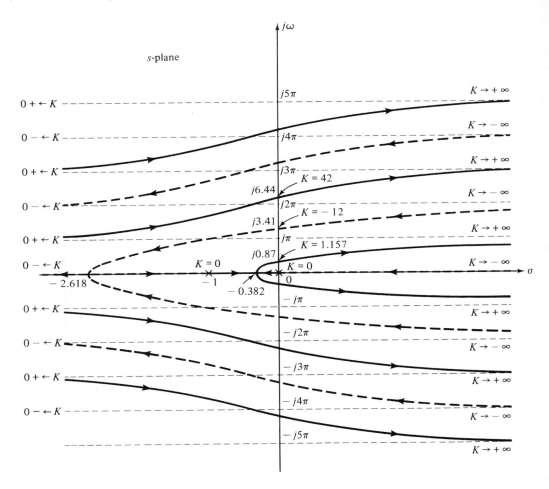

Fig. 8-40. Complete root loci for $s^2 + s + Ke^{-Ts} = 0$, $T = 1$.

where it is easily verified that one belongs to the root loci, and the other is on the complementary root loci.

The complete root loci of the system are sketched as shown in Fig. 8-40 for $T = 1$ sec. Notice that from the system analysis standpoint, only the portion of the root loci that lies between $\omega = \pi$ and $\omega = -\pi$ is of significance. The closed-loop system of Fig. 8-39 is stable for $0 \leq K < 1.157$. Therefore, the other root loci branches, including the complementary root loci, are perhaps only of academic interest.

Approximation of Systems with Pure Time Delay

Although the discussions given in the preceding section point to a systematic way of constructing the root loci of a closed-loop system with pure time delay, in general, for complex systems, the problem can still be quite difficult.

We shall investigate ways of approximating the time delay term, e^{-Ts}, by a polynomial or a rational function of s. One method is to approximate e^{-Ts}, as follows:

$$e^{-Ts} \cong \frac{1}{[1 + (Ts/n)]^n} \qquad (8\text{-}205)$$

Since e^{-Ts} has an infinite number of poles, the approximation is perfect when n becomes infinite. Figure 8-41 illustrates the effect of the approximation when the input to the pure time delay is a unit step function.

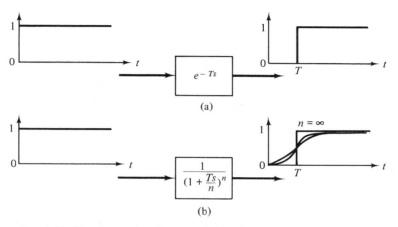

Fig. 8-41. Waveforms that illustrate the effect of approximating a pure time delay by finite number of poles.

If Eq. (8-205) is used as the approximation for the root locus problem, only the primary branches of the root loci will be realized. However, this will be adequate for the great majority of practical problems, since only the primary branches will contain the dominant eigenvalues of the system.

Let us approximate the exponential term of Eq. (8-189) by the right side of Eq. (8-205). Figure 8-42 illustrates the dominant root loci for $n = 2$, 3, and 4; $T = 1$, together with the primary branch of the exact root loci. The approximating root loci approach the exact ones as n becomes large.

Another way of approximating the pure time delay transfer relation is to use a power series; that is,

$$e^{-Ts} = 1 - Ts + \frac{T^2 s^2}{2!} - \frac{T^3 s^3}{3!} + \cdots \qquad (8\text{-}206)$$

The difference between this approximation and that of Eq. (8-205) is that in the former, the accuracy improves as the order n becomes larger, whereas in the present case, the validity of the approximation depends on the smallness of T. It is apparent that Eq. (8-206) can be conveniently applied if only a few terms of the series are used.

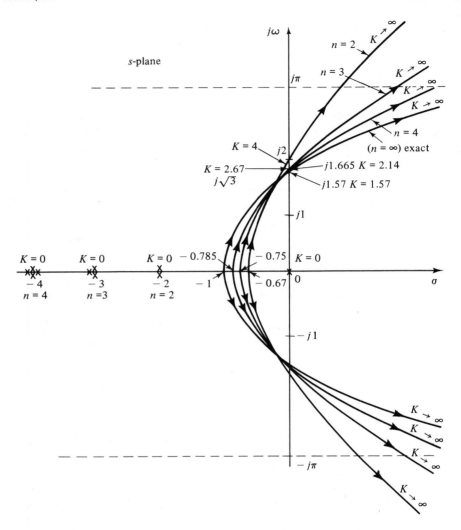

Fig. 8-42. Approximation of the root loci of $s + Ke^{-s} = 0$ by those of $(1 + s/n)^n + K = 0$.

8.8 Relationship Between Root Loci and the Polar Plot

In Chapter 7 the polar plot of a transfer function $G(s)H(s)$ is shown to be a mapping of the imaginary axis of the s-plane onto the $G(s)H(s)$ plane. In general, other trajectories in the s-plane can be mapped onto the $G(s)H(s)$-plane through the function $G(s)H(s)$, such as in the case of a system with pure time delay. Since the properties of the root loci depend upon the function $G(s)H(s)$, we can regard the root loci as a mapping from the $G(s)H(s)$ plane to the s-plane. Since the points on the complete root loci of the equation

$$1 + G(s)H(s) = 1 + KG_1(s)H_1(s) = 0 \tag{8-207}$$

satisfy the conditions

$$KG_1(s)H_1(s) = (2k + 1)\pi \qquad K \geq 0 \tag{8-208}$$

$$KG_1(s)H_1(s) = 2k\pi \qquad K \leq 0 \tag{8-209}$$

the root loci simply represent a mapping of the real axis of the $G(s)H(s)$ plane onto the s-plane. In fact, for the root loci, $K \geq 0$, the mapping points are on the negative real axis of the s-plane; and for the complementary root loci, $K \leq 0$, the mapping points are on the positive real axis of the s-plane.

It was shown in Chapter 7 that the mapping from the s-plane to the $G(s)H(s)$ plane via the polar plot or the Nyquist plot is single-valued. In other words, a point $\omega = \omega_1$ on the imaginary axis of the s-plane is mapped onto only one point, $G(j\omega_1)H(j\omega_1)$, in the $G(s)H(s)$-plane. However, in the root locus case, which is a reverse process, the mapping is multivalued. As an illustration, the polar plot of a type 1 transfer function of the third order is shown in Fig. 8-43.

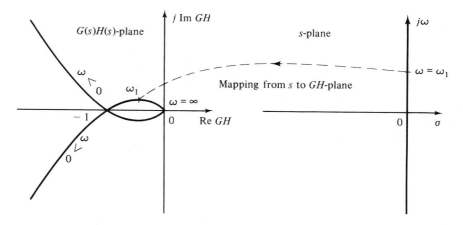

Fig. 8-43. Polar plot of $G(s)H(s) = K/[s(s + a)(s + b)]$ interpreted as a mapping of the $j\omega$ axis of the s-plane onto the $G(s)H(s)$-plane.

The entire polar plot in the $G(s)H(s)$ is interpreted as the mapping of the entire $j\omega$ axis of the s-plane. The complete root locus diagram for the same system is shown in Fig. 8-44 as a mapping of the real axis of the $G(s)H(s)$-plane onto the s-plane. Note that in this case each point of the $G(s)H(s)$-plane corresponds to three points in the s-plane. For instance, the $(-1, j0)$ point of the $G(s)H(s)$-plane corresponds to the two points where the root loci intersect the $j\omega$ axis and a point on the real axis.

The polar plot and the root loci each represents the mapping of only a very limited portion of one domain to the other. In general, it would be useful to consider the mapping of points other than those on the $j\omega$ axis of the s-plane and on the real axis of the $G(s)H(s)$-plane. For instance, we may use the mapping

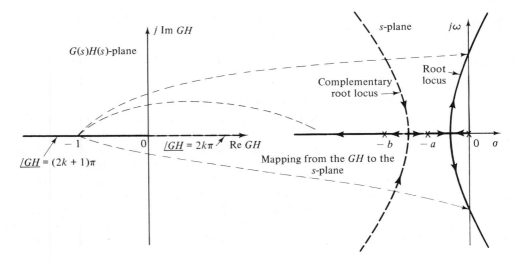

Fig. 8-44. Root locus diagram of $G(s)H(s) = K/[s(s + a)(s + b)]$ interpreted as a mapping of the real axis of the $G(s)H(s)$-plane onto the s-plane.

of the constant-damping-ratio lines onto the $G(s)H(s)$-plane for the purpose of determining relative stability of the closed-loop system. Figure 8-45 illustrates the $G(s)H(s)$ plots that correspond to different constant-damping-ratio lines in the s-plane. As shown by curve (3) in Fig. 8-45, when the $G(s)H(s)$ curve passes

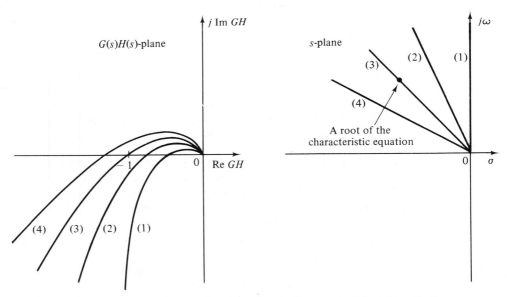

Fig. 8-45. $G(s)H(s)$ plots that correspond to constant-damping ratio lines in the s-plane.

through the $(-1, j0)$ point, it means that Eq. (8-207) is satisfied, and the corresponding trajectory in the s-plane passes through a root of the characteristic equation. Similarly, we may construct root loci that correspond to straight lines rotated at various angles from the real axis in the $G(s)H(s)$-plane, as shown by Fig. 8-46. Notice that these root loci now satisfy the condition of

$$\underline{/KG_1(s)H_1(s)} = (2k + 1)\pi - \theta \qquad K \geq 0. \tag{8-210}$$

Or, the root loci of Fig. 8-46 satisfy the equation

$$1 + G(s)H(s)e^{j\theta} = 0 \tag{8-211}$$

for the various values of θ indicated.

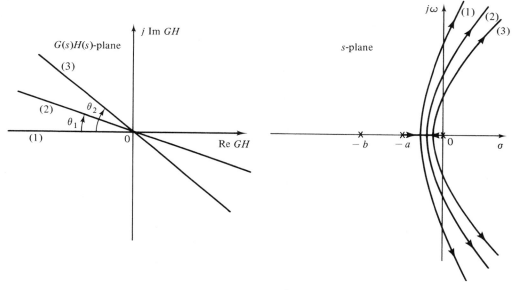

Fig. 8-46. Root loci that correspond to different phase-angle loci in the $G(s)H(s)$-plane. (The complementary root loci are not shown.)

It is of interest to note that Eq. (8-210) is very similar to the condition of Eq. (8-183), which is for the root loci of systems with pure time delays. In Eq. (8-183), the angle that is added to $(2k + 1)\pi$ is a function of the frequency ω.

8.9 Root Loci of Discrete-Data Control System[47-49]

The root locus technique for continuous-data systems can be readily applied to discrete-data systems without requiring any modifications. The characteristic equation roots of the discrete-data system having the block diagram of Fig. 8-47

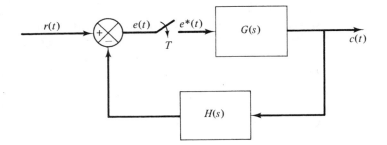

Fig. 8-47. Discrete-data control system.

must satisfy

$$1 + GH^*(s) = 0 \qquad (8\text{-}212)$$

if the roots are defined in the s-plane, or

$$1 + GH(z) = 0 \qquad (8\text{-}213)$$

if the z-plane is referred to.

Since

$$GH^*(s) = \frac{1}{T} \sum_{n=-\infty}^{\infty} G(s + jn\omega_s)H(s + jn\omega_s) \qquad (8\text{-}214)$$

which is an infinite series, the poles and zeros of $GH^*(s)$ in the s-plane will be infinite in number. This evidently makes the construction of the root loci of Eq. (8-212) more difficult. However, as an illustrative example of the difference between the characteristics of the root loci of continuous-data and discrete-data systems, let us consider that for the system of Fig. 8-47,

$$G(s)H(s) = \frac{K}{s(s + 1)} \qquad (8\text{-}215)$$

Using Eq. (8-214), we have

$$GH^*(s) = \frac{1}{T} \sum_{n=-\infty}^{\infty} \frac{K}{(s + jn\omega_s)(s + jn\omega_s + 1)} \qquad (8\text{-}216)$$

which has poles at $s = -jn\omega_s$ and $s = -1 - jn\omega_s$, where n takes on all integers between $-\infty$ and ∞. The pole configuration of $GH^*(s)$ is shown in Fig. 8-48(a). Using the rules of construction outlined earlier, the root loci of $1 + GH^*(s) = 0$ for positive K are drawn as shown in Fig. 8-48(b) for $T = 1$. The root loci contain an infinite number of branches, and these clearly indicate that the closed-loop system is unstable for all values of K greater than 4.32. In contrast, it is well known that the same system without sampling is stable for all positive values of K.

The root locus problem for discrete-data systems is simplified if the root

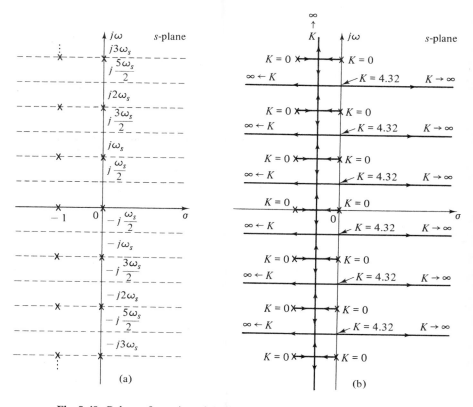

Fig. 8-48. Pole configuration of $GH^*(s)$ and the root locus diagram in the s-plane for the discrete-data system in Fig. 8-47 with $G(s)H(s) = K/[s(s+1)]$, $T = 1$ sec.

loci are constructed in the z-plane using Eq. (8-213). Since Eq. (8-213) is, in general, a polynomial in z with constant coefficients, the number of root loci is finite, and the same rules of construction for continuous-data systems are directly applicable.

As an illustrative example of the construction of root loci for discrete-data systems in the z-plane, let us consider the system of Fig. 8-47 with $T = 1$ sec, and $G(s)H(s)$ as given by Eq. (8-215). Taking the z-transform of Eq. (8-215), we have

$$GH(z) = \frac{0.632Kz}{(z - 1)(z - 0.368)} \qquad (8\text{-}217)$$

which has a zero at $z = 0$ and poles at $z = 1$ and $z = 0.368$. The root loci for the closed-loop system are constructed based on the pole–zero configuration of Eq. (8-217) and are shown in Fig. 8-49. Notice that when the value of K exceeds 4.32, one of the roots of the characteristic equation moves outside the unit circle in the z-plane, and the system becomes unstable.

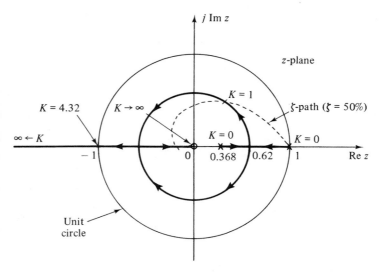

Fig. 8-49. Root locus diagram of discrete-data control system without zero-order hold. $G(s)H(s) = K/[s(s + 1)]$, $T = 1$ sec.

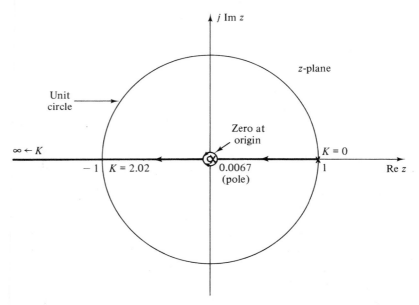

Fig. 8-50. Root locus diagram of discrete-data control system without zero-order hold. $G(s)H(s) = K/[s(s + 1)]$, $T = 5$ sec.

For the same system, if the sampling period is changed to $T = 5$ sec, the z-transform of $G(s)H(s)$ becomes

$$GH(z) = \frac{0.993Kz}{(z - 1)(z - 0.0067)} \tag{8-218}$$

The root loci for this case are drawn as shown in Fig. 8-50. It should be noted that although the complex part of the root loci for $T = 5$ sec takes the form of a smaller circle than that when $T = 1$ sec, the system is actually less stable. The marginal value of K for stability for $T = 5$ sec is 2.02 as compared to the marginal K of 4.32 for $T = 1$ sec.

The constant-damping-ratio path[48] may be superimposed on the root loci to determine the required value of K for a specified damping ratio. In Fig. 8-49 the constant-damping-ratio path for $\zeta = 0.5$ is drawn, and the intersection with the root loci gives the desired value of $K = 1$. Thus for all values of K less than 1, the damping ratio of the system will be greater than 50 per cent.

As another example, let us consider that a zero-order hold is inserted between the sampler and the controlled process $G(s)$ in the system of Fig. 8-47. For the loop transfer function of Eq. (8-215), the z-transform with the zero-order hold is

$$G_{h0}GH(z) = \frac{K[(T - 1 + e^{-T})z - Te^{-T} + 1 - e^{-T}]}{(z - 1)(z - e^{-T})} \qquad (8\text{-}219)$$

The root loci of the system with sample-and-hold for $T = 1$ sec and $T = 5$ sec are shown in Fig. 8-51(a) and (b), respectively. In this case the marginal

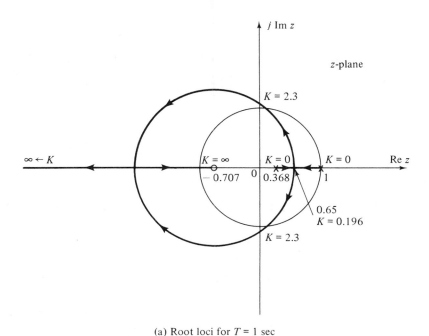

(a) Root loci for $T = 1$ sec

Fig. 8-51. Root locus diagrams of discrete-data control system with sample-and-hold. $G(s)H(s) = K/[s(s + 1)]$. (a) Root loci for $T = 1$ sec. (b) Root loci for $T = 5$ sec.

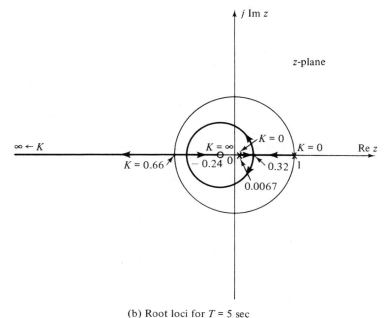

(b) Root loci for $T = 5$ sec

Fig. 8-51 (Cont.).

value of stability for K is 2.3 for $T = 1$ sec and 0.66 for $T = 5$ sec. This illustrates the well-established fact that the zero-order hold reduces the stability margin of a discrete-data system.

REFERENCES

General Subjects

1. W. R. EVANS, "Graphical Analysis of Control Sytems," *Trans. AIEE*, Vol. 67, pp. 547–551, 1948.

2. W. R. EVANS, "Control System Synthesis by Root Locus Method," *Trans. AIEE*, Vol. 69, pp. 66-69, 1950.

3. W. R. EVANS, *Control System Dynamics*, McGraw-Hill Book Company, New York, 1954.

4. Y. CHU and V. C. M. YEH, "Study of the Cubic Characteristic Equation by the Root Locus Method," *Trans. ASME*, Vol. 76, pp. 343–348, Apr. 1954.

5. J. G. TRUXAL, *Automatic Feedback Systems Synthesis*, McGraw-Hill Book Company, New York, 1955.

6. K. S. NARENDRA, "Inverse Root Locus, Reversed Root Locus, or Complementary Root Locus," *IRE Trans. Automatic Control*, pp. 359–360, Sept. 1961.

7. P. Mosner, J. Vullo, and K. Bley, "A Possible Source of Error in Root-Locus Analysis of Nonminimum Phase Systems," *IRE Trans. Automatic Control*, Vol. AC-7, pp. 87–88, Apr. 1962.

8. B. J. Matkowsky and A. H. Zemanian, "A Contribution to Root-Locus Techniques for the Analysis of Transient Responses," *IRE Trans. Automatic Control*, Vol. AC-7, pp. 69–73, Apr. 1962.

9. J. D. S. Muhlenberg, "Quartic Root Locus Types," *IEEE Trans. Automatic Control*, Vol. AC-12, pp. 228–229, Apr. 1967.

10. H. M. Power, "Root Loci Having a Total of Four Poles and Zeros," *IEEE Trans. Automatic Control*, Vol. AC-16, pp. 484–486, Oct. 1971.

Construction and Properties of The Root Locus

11. C. C. MacDuffe, *Theory of Equations*, John Wiley & Sons, Inc., New York, pp. 29–104, 1954.

12. G. A. Bendrikov and K. F. Teodorchik, "The Laws of Root Migration for Third and Fourth Power Linear Algebraic Equations as the Free Terms Change Continuously," *Automation and Remote Control*, Vol. 16, No. 3, 1955.

13. J. E. Gibson, "Build a Dynamic Root-Locus Plotter," *Control Eng.*, Feb. 1956.

14. C. S. Lorens and R. C. Titsworth, "Properties of Root Locus Asymptotes," *IRE Trans. Automatic Control*, AC-5, pp. 71–72, Jan. 1960.

15. K. F. Teodorchik and G. A. Bendrikov, "The Methods of Plotting Root Paths of Linear Systems and for Qualitative Determination of Path Type," *Proc. IFAC Cong.*, Vol. 1, pp. 8-12, 1960.

16. C. A. Stapleton, "On Root Locus Breakaway Points," *IRE Trans. Automatic Control*, Vol. AC-7, pp. 88–89, Apr. 1962.

17. M. J. Remec, "Saddle-Points of a Complete Root Locus and an Algorithm for Their Easy Location in the Complex Frequency Plane," *Proc. Natl. Electronics Conf.*, Vol. 21, pp. 605–608, 1965.

18. C. F. Chen, "A New Rule for Finding Breaking Points of Root Loci Involving Complex Roots," *IEEE Trans. Automatic Control*, Vol. AC-10, pp. 373–374, July 1965.

19. V. Krishnan, "Semi-analytic Approach to Root Locus," *IEEE Trans. Automatic Control*, Vol. AC-11, pp. 102–108, Jan. 1966.

20. R. H. LaBounty and C. H. Houpis, "Root Locus Analysis of a High-Gain Linear System with Variable Coefficients; Application of Horowitz's Method," *IEEE Trans. Automatic Control*, Vol. AC-11, pp. 255–263, Apr. 1966.

21. R. J. Fitzgerald, "Finding Root-Locus Breakaway Points with the Spirule," *IEEE Trans. Automatic Control*, Vol. AC-11, pp. 317–318, Apr. 1966.

22. J. D. S. Muhlenberg, "Synthetic Division for Gain Calibration on the Complex Root Locus," *IEEE Trans. Automatic Control*, Vol. AC-11, pp. 628–629, July 1966.

23. J. Feinstein and A. Fregosi, "Some Invariances of the Root Locus," *IEEE Trans. Automatic Control*, Vol. AC-14, pp. 102–103, Feb. 1969.

24. A. FREGOSI and J. FEINSTEIN, "Some Exclusive Properties of the Negative Root Locus," *IEEE Trans. Automatic Control*, Vol. AC-14, pp. 304–305, June 1969.

25. H. M. POWER, "Application of Bilinear Transformation to Root Locus Plotting," *IEEE Trans. Automatic Control*, Vol. AC-15, pp. 693–694, Dec. 1970.

Analytical Representation of Root Loci

26. G. A. BENDRIKOV and K. F. TEODORCHIK, "The Analytic Theory of Constructing Root Loci," *Automation and Remote Control*, pp. 340–344, Mar. 1959.

27. K. STEIGLITZ, "An Analytical Approach to Root Loci," *IRE Trans. Automatic Control*, Vol. AC-6, pp. 326–332, Sept. 1961.

28. C. WOJCIK, "Analytical Representation of Root Locus," *Trans. ASME, J. Basic Engineering*, Ser. D, Vol. 86, Mar. 1964.

29. C. S. CHANG, "An Analytical Method for Obtaining the Root Locus with Positive and Negative Gain," *IEEE Trans. Automatic Control*, Vol. AC-10, pp. 92–94, Jan. 1965.

30. B. P. BHATTACHARYYA, "Root Locus Equations of the Fourth Degree," *Internat. J. Control*, Vol. 1, No. 6, pp. 533–556, 1965.

31. J. D. S. MUHLENBERG, "Some Expressions for Root Locus Gain Calibration," *IEEE Trans. Automatic Control*, Vol. AC-12, pp. 796–797, Dec. 1967.

Computer-Aided Plotting of Root Loci

32. D. J. DODA, "The Digital Computer Makes Root Locus Easy," *Control Eng.*, May 1958.

33. Z. KLAGSBRUNN and Y. WALLACH, "On Computer Implementation of Analytic Root-Locus Plotting," *IEEE Trans. Automatic Control*, Vol. AC-13, pp. 744–745, Dec. 1968.

34. R. H. ASH and G. R. ASH, "Numerical Computation of Root Loci Using the Newton–Raphson Technique," *IEEE Trans. Automatic Control*, Vol. AC-13, pp. 576–582, Oct. 1968.

Root Locus Diagram for Design

35. Y. CHU, "Synthesis of Feedback Control Systems by Phase Angle Loci," *Trans. AIEE*, Vol. 71, Part II, 1952.

36. F. M. REZA, "Some Mathematical Properties of Root Loci for Control Systems Design," *Trans. AIEE Commun. Electronics*, Vol. 75, Part I, pp. 103–108, Mar. 1956.

37. J. A. ASELTINE, "Feedback System Synthesis by the Inverse Root-Locus Method," *IRE Natl. Convention Record*, Part 2, pp. 13–17, 1956.

38. G. A. BENDRIKOV and K. F. TEODORCHIK, "The Theory of Derivative Control in Third Order Linear Systems," *Automation and Remote Control*, pp. 516–517, May 1957.

39. R. J. HRUBY, "Design of Optimum Beam Flexural Damping in a Missile by Application of Root-Locus Techniques," *IRE Trans. Automatic Control*, Vol. AC-5, pp. 237–246, Aug. 1960.

40. A. KATAYAMA, "Semi-graphical Design of Servomechanisms Using Inverse Root-Locus," *J. IEE,* Vol. 80, pp. 1140–1149, Aug. 1960.

41. D. A. CALAHAN, "A Note on Root-Locus Synthesis," *IRE Trans. Automatic Control*, Vol. AC-7, p. 84, Jan. 1962.

42. A. KATAYAMA, "Adaptive Control System Design Using Root-Locus," *IRE Trans. Automatic Control*, Vol. AC-7, pp. 81–83, Jan. 1962.

43. J. R. MITCHELL and W. L. McDANIEL, JR., "A Generalized Root Locus Following Technique," *IEEE Trans. Automatic Control*, Vol. AC-15, pp. 483–485, Aug. 1970.

Root Sensitivity

44. J. G. TRUXAL and I. M. HOROWITZ, "Sensitivity Considerations in Active Network Synthesis," *Proc. Second Midwest Symposium on Circuit Theory*, East Lansing, Mich., Dec. 1956.

45. R. Y. HUANG, "The Sensitivity of the Poles of Linear Closed-Loop Systems," *Trans. AIEE Appl. Ind.*, Vol. 77, Part 2, pp. 182–187, Sept. 1958.

46. H. UR, "Root Locus Properties and Sensitivity Relations in Control Systems," *IRE Trans. Automatic Control*, Vol. AC-5, pp. 57–65, Jan. 1960.

Root Locus for Discrete-Data Systems

47. M. MORI, "Root Locus Method of Pulse Transfer Function for Sampled-Data Control Systems," *IRE Trans. Automatic Control*, Vol. AC-3, pp. 13–20, Nov. 1963.

48. B. C. KUO, *Analysis and Synthesis of Sampled-Data Control Systems*, Prentice-Hall, Inc., Englewood Cliffs, N.J., 1963.

49. B. C. KUO, *Discrete-Data Control Systems*, Science-Tech, P.O. Box 2277, Station A, Champaign, Ill. 1974.

Root Locus for Nonlinear Systems

50. M. J. ABZUG, "A Root-Locus Method for the Analysis of Nonlinear Servomechanisms," *IRE Trans. Automatic Control*, Vol. AC-4, No. 3, pp. 38–44, Dec. 1959.

Root Locus for Systems with Time Delays

51. Y. CHU, "Feedback Control System with Dead-Time Lag or Distributed Lag by Root-Locus Method," *Trans. AIEE*, Vol. 70, Part II, p. 291, 1951.

PROBLEMS

8.1. Sketch the root locus diagram for each of the following feedback control systems. In each case determine everything about the locus of roots for $-\infty < K < \infty$ and sketch the root loci. Indicate on each locus the starting point, the ending point, and the direction of increasing value of K. The poles and zeros of $G(s)H(s)$ of the systems are given as follows:

(a) Poles at 0, -2, and -3; zero at -5.

(b) Poles at 0, 0, -2, and -3; zero at -5.

(c) Poles at $-2 + j2$, and $-2 - j2$; zero at -3.

(d) Poles at 0, $-10 + j10$, and $-10 - j10$; zero at -20.

(e) Poles at 0, -20, $-10 + j10$, and $-10 - j10$; no finite zeros.

(f) Poles at -20, $-10 + j10$, and $-10 - j10$; zero at -30.

(g) Poles at 0, 0, -12, and -12; zeros at -4 and -8.

8.2. The open-loop transfer function of a unity-feedback control system is given by

$$G(s) = \frac{K(s + 3)}{s(s^2 + 2s + 2)(s + 5)(s + 6)}$$

(a) Sketch the root locus diagram as a function of $K(-\infty < K < \infty)$.

(b) Determine the value of K that makes the relative damping ratio of the closed-loop complex poles equal to 0.4.

8.3. A unity feedback control system has an open-loop transfer function

$$G(s) = \frac{K(1 + 0.2s)(1 + 0.025s)}{s^3(1 + 0.001s)(1 + 0.005s)}$$

Sketch the complete $(-\infty < K < \infty)$ root locus diagram for the system. Indicate the crossing points of the loci on the $j\omega$ axis, and the corresponding values of K and ω at these points.

8.4. A unity feedback control system has an open-loop transfer function

$$G(s) = \frac{K}{s(1 + 0.02s)(1 + 0.01s)}$$

(a) Sketch the root locus diagram of the system $(0 < K < \infty)$.

(b) Determine the marginal value of K that will cause instability.

(c) Determine the value of K when the system is critically damped.

8.5. The transfer functions of a feedback control system are given as

$$G(s) = \frac{K}{s^2(s + 2)(s + 5)} \quad \text{and} \quad H(s) = 1$$

(a) Sketch the root locus diagram for the system. Indicate the crossing points of the loci on the $j\omega$ axis and the corresponding values of K and ω at these points (positive values of K only).

(b) The transfer function of the feedback loop element is now changed to $H(s) = 1 + 2s$. Determine the stability of the modified system as a function of K. Investigate the effect on the root locus diagram due to this change in $H(s)$.

8.6. The characteristic equation of a feedback control system is given by

$$s^3 + 3s^2 + (K + 2)s + 10K = 0$$

Sketch the root locus diagram (positive K only) for this system.

8.7. For the following loop transfer function, sketch the root locus diagram as a function of T (T varies from 0 to ∞). Determine the value of T so that the damping ratio of the complex roots of the characteristic equation is 0.2.

$$G(s)H(s) = \frac{1000(1 + Ts)}{s(1 + 0.1s)(1 + 0.001s)}$$

8.8. For the following loop transfer function, sketch the root locus diagram as a function of T. Determine the value of T so that the damping ratio of the complex roots of the characteristic equation is 0.2.

$$G(s)H(s) = \frac{30}{s(1 + 0.1s)(1 + 0.2s)(1 + Ts)}$$

8.9. For the bridged-T network of Fig. P8-9,

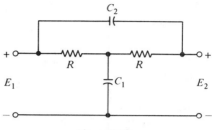

Figure P8-9.

(a) Sketch the root locus diagrams of the zeros and poles of E_2/E_1 as a function of C_1 (C_1 varies from 0 to ∞).

(b) Sketch the root locus diagrams of the zeros and poles of E_2/E_1 as a function of C_2.

8.10. The open-loop transfer function of a control system with positive feedback is given by

$$G(s) = \frac{K}{s(s^2 + 4s + 4)}$$

Sketch the root locus diagram of the system as a function of K ($0 < K < \infty$).

8.11. It is desired that the closed-loop transfer function of a control system be

$$\frac{C(s)}{R(s)} = \frac{1}{(1 + 0.03s)(1 + 0.2s + 0.02s^2)}$$

Determine the open-loop transfer function $G(s)$ of the system. Assume that the system has unity feedback.

8.12. Given the equation

$$s^3 + as^2 + Ks + K = 0$$

It is desired to investigate the root loci of this equation for $-\infty < K < \infty$ and for several values of a.

(a) Sketch the root loci ($-\infty < K < \infty$) for $a = 10$.

(b) Repeat for $a = 3$.

(c) Determine the value of a so that there is only one nonzero breakaway point on the complete root loci. Sketch the loci.

In sketching the root loci you should apply all known rules whenever they are applicable.

8.13. The open-loop transfer function of a control system with unity feedback is given by

$$G(s) = \frac{K(s + a)}{s^2(s + 1)}$$

Determine the values of a so that the root locus diagram will have zero, one, and two breakaway points, respectively, not counting the one at $s = 0$. Sketch the root loci for $-\infty < K < \infty$ for all three cases.

8.14. For the sampled-data control system shown in Fig. P8-14,

$$G(s) = \frac{K}{s(1 + 0.2s)}$$

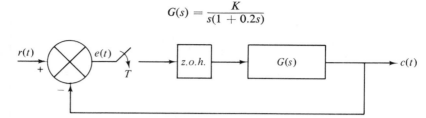

Figure P8-14.

(a) Sketch the root loci for the system ($0 < K < \infty$) without the zero-order hold, for $T = 0.1$ sec and $T = 1$ sec. Determine the marginal value of K for stability in each case.

(b) Repeat part (a) when the system has a zero-order hold.

8.15. The following polynomial in z represents the characteristic equation of a certain discrete-data control system. Sketch the root loci ($-\infty < K < \infty$) for the system. Determine the marginal value of K for stability.

$$z^3 + Kz^2 + 1.5Kz - (K + 1) = 0$$

8.16. Sketch the root loci ($0 \leq K < \infty$) in the z-plane for the discrete-data control system shown in Fig. P8-16.

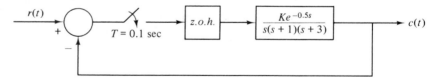

Figure P8-16.

9

Frequency-Domain Analysis

of Control Systems

9.1 Introduction

It was pointed out earlier that in practice the performance of a feedback control system is more preferably measured by its time-domain response characteristics. This is in contrast to the analysis and design of systems in the communication field, where the frequency response is of more importance, since in this case most of the signals to be processed are either sinusoidal or periodic in nature. However, analytically, the time response of a control system is usually difficult to determine, especially in the case of high-order systems. In the design aspects, there are no unified ways of arriving at a designed system given the time-domain specifications, such as peak overshoot, rise time, delay time, and settling time. On the other hand, there is a wealth of graphical methods available in the frequency-domain analysis, all suitable for the analysis and design of linear feedback control systems. Once the analysis and design are carried out in the frequency domain, the time-domain behavior of the system can be interpreted based on the relationships that exist between the time-domain and the frequency-domain properties. Therefore, we may consider that the main purpose of conducting control systems analysis and design in the frequency domain is merely to use the techniques as a convenient vehicle toward the same objectives as with time-domain methods.

The starting point in frequency-domain analysis is the transfer function. We shall first discuss transfer function relations based on, first, the state variable representation and, second, the classical approach.

In Section 3.3 the transfer function of a multivariable closed-loop control system is derived. Referring to Fig. 3-9, the closed-loop transfer function matrix relation is written [Eq. (3-44)]

$$\mathbf{C}(s) = [\mathbf{I} + \mathbf{G}(s)\mathbf{H}(s)]^{-1}\mathbf{G}(s)\mathbf{R}(s) \qquad (9\text{-}1)$$

where $\mathbf{C}(s)$ is a $q \times 1$ vector and $\mathbf{R}(s)$ is a $p \times 1$ vector. The closed-loop transfer function matrix is defined as

$$\mathbf{M}(s) = [\mathbf{I} + \mathbf{G}(s)\mathbf{H}(s)]^{-1}\mathbf{G}(s) \qquad (9\text{-}2)$$

which is a $q \times p$ matrix.

Under the sinusoidal steady state, we set $s = j\omega$; then Eq. (9-2) becomes

$$\mathbf{M}(j\omega) = [\mathbf{I} + \mathbf{G}(j\omega)\mathbf{H}(j\omega)]^{-1}\mathbf{G}(j\omega) \qquad (9\text{-}3)$$

The ijth element of $\mathbf{M}(j\omega)$ is defined as

$$M_{ij}(j\omega) = \frac{C_i(j\omega)}{R_j(j\omega)}\bigg|_{\text{all other inputs}=0} \qquad (9\text{-}4)$$

where i represents the row and j the column of $\mathbf{M}(j\omega)$.

State-Variable Representation

For a system that is represented by state equations,

$$\dot{\mathbf{x}}(t) = \mathbf{A}\mathbf{x}(t) + \mathbf{B}\mathbf{u}(t) \qquad (9\text{-}5)$$

$$\mathbf{c}(t) = \mathbf{D}\mathbf{x}(t) + \mathbf{E}\mathbf{u}(t) \qquad (9\text{-}6)$$

For the present case we assume that the feedback is described by

$$\mathbf{u}(t) = \mathbf{r}(t) - \mathbf{H}\mathbf{c}(t) \qquad (9\text{-}7)$$

where

$$\mathbf{x}(t) = n \times 1 \text{ state vector}$$
$$\mathbf{u}(t) = p \times 1 \text{ control vector}$$
$$\mathbf{c}(t) = q \times 1 \text{ output vector}$$
$$\mathbf{r}(t) = p \times 1 \text{ input vector}$$

$\mathbf{A}$, $\mathbf{B}$, $\mathbf{D}$, and $\mathbf{E}$ are constant matrices of appropriate dimensions, and $\mathbf{H}$ is the $p \times q$ feedback matrix. The transfer function relation of the system is

$$\mathbf{C}(s) = [\mathbf{D}(s\mathbf{I} - \mathbf{A})^{-1}\mathbf{B} + \mathbf{E}]\mathbf{U}(s) \qquad (9\text{-}8)$$

The open-loop transfer function matrix is defined as

$$\mathbf{G}(s) = \mathbf{D}(s\mathbf{I} - \mathbf{A})^{-1}\mathbf{B} + \mathbf{E} \qquad (9\text{-}9)$$

The closed-loop transfer function relation is described by the equation

$$\mathbf{C}(s) = [\mathbf{I} + \mathbf{G}(s)\mathbf{H}]^{-1}\mathbf{G}(s)\mathbf{R}(s) \qquad (9\text{-}10)$$

Thus

$$\mathbf{M}(s) = [\mathbf{I} + \mathbf{G}(s)\mathbf{H}]^{-1}\mathbf{G}(s) \qquad (9\text{-}11)$$

It should be noted that the matrix $\mathbf{H}$ in Eq. (9-11) has only constant elements.

In general, the elements of the transfer function matrices are rational functions of s. In Chapter 4 it is proved that if the system is completely controllable and observable, there will be no pole–zero cancellations in the transfer functions. Under this condition the poles of the transfer function will also be the eigenvalues of the system.

The analysis techniques in the frequency domain discussed in the following sections are conducted with the single-variable notation. Because linear systems satisfy the principle of superposition, these basic techniques can all be applied to multivariable systems.

For a single-loop feedback system, the closed-loop transfer function is written

$$M(s) = \frac{C(s)}{R(s)} = \frac{G(s)}{1 + G(s)H(s)} \tag{9-12}$$

Under the sinusoidal steady state, we set $s = j\omega$; then Eq. (9-12) becomes

$$M(j\omega) = \frac{C(j\omega)}{R(j\omega)} = \frac{G(j\omega)}{1 + G(j\omega)H(j\omega)} \tag{9-13}$$

The sinusoidal steady-state transfer relation $M(j\omega)$, which is a complex function of ω, may be expressed in terms of a real and an imaginary part; that is,

$$M(j\omega) = \text{Re}\,[M(j\omega)] + j\,\text{Im}\,[M(j\omega)] \tag{9-14}$$

Or, $M(j\omega)$ can be expressed in terms of its magnitude and phase as

$$M(j\omega) = M(\omega)\underline{/\phi_m(\omega)} \tag{9-15}$$

where

$$M(\omega) = \left| \frac{G(j\omega)}{1 + G(j\omega)H(j\omega)} \right| \tag{9-16}$$

and

$$\phi_m(\omega) = \underline{/\frac{G(j\omega)}{1 + G(j\omega)H(j\omega)}}$$
$$= \underline{/G(j\omega)} - \underline{/1 + G(j\omega)H(j\omega)} \tag{9-17}$$

Since the analysis is now in the frequency domain, some of the terminology used in communication systems may be applied to the present control system characterization. For instance, $M(\omega)$ of Eq. (9-16) may be regarded as the magnification of the feedback control system. The significance of $M(\omega)$ to a control system is similar to the gain or amplification of an electronic amplifier. In an audio amplifier, for instance, an ideal design criterion is that the amplifier must have a flat gain for all frequencies. Of course, realistically, the design criterion becomes that of having a flat gain in the audio frequency range. In control systems the ideal design criterion is similar. If it is desirable to keep the output $C(j\omega)$ identical to the input $R(j\omega)$ at all frequencies, $M(j\omega)$ must be unity for all frequencies. However, from Eq. (9-13) it is apparent that $M(j\omega)$

can be unity only when $G(j\omega)$ is infinite, while $H(j\omega)$ is finite and nonzero. An infinite magnitude for $G(j\omega)$ is, of course, impossible to achieve in practice, nor would it be desirable, since most control systems become unstable when its loop gain becomes very high. Furthermore, all control systems are subjected to noise. Thus, in addition to responding to the input signal, the system should be able to reject and suppress noise and unwanted signals. This means that the frequency response of a control system should have a cutoff characteristic in general, and sometimes even a band-pass characteristic.

The phase characteristics of the frequency response are also of importance. The ideal situation is that the phase must be a linear function of frequency within the frequency range of interest. Figure 9-1 shows the gain and phase characteristics of an ideal low-pass filter, which is impossible to realize physically. Typical gain and phase characteristics of a feedback control system are shown in Fig. 9-2. The fact is that the great majority of control systems have the characteristics of a low-pass filter, so the gain decreases as the frequency increases.

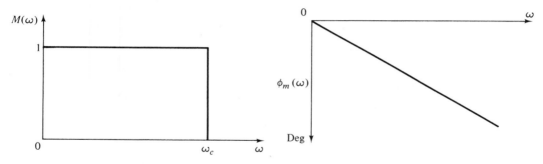

Fig. 9-1. Gain-phase characteristics of an ideal low-pass filter.

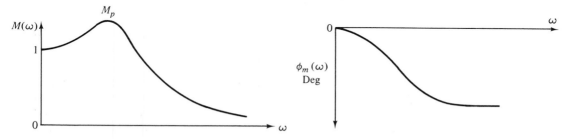

Fig. 9-2. Typical gain and phase characteristics of a feedback control system.

9.2 Frequency-Domain Characteristics

If a control system is to be designed or analyzed using frequency-domain techniques, we need a set of specifications to describe the system performance. The following frequency-domain specifications are often used in practice.

Peak resonance M_p***.*** The peak resonance M_p is defined as the maximum value of $M(\omega)$ that is given in Eq. (9-16). In general, the magnitude of M_p gives an indication of the relative stability of a feedback control system. Normally, a large M_p corresponds to a large peak overshoot in the step response. For most design problems it is generally accepted that an optimum value of M_p should be somewhere between 1.1 and 1.5.

Resonant frequency ω_p***.*** The resonant frequency ω_p is defined as the frequency at which the peak resonance M_p occurs.

Bandwidth. The bandwidth, BW, is defined as the frequency at which the magnitude of $M(j\omega)$, $M(\omega)$, drops to 70.7 per cent of its zero-frequency level, or 3 dB down from the zero-frequency gain. In general, the bandwidth of a control system indicates the noise-filtering characteristics of the system. Also, bandwidth gives a measure of the transient response properties, in that a large bandwidth corresponds to a faster rise time, since higher-frequency signals are passed on to the outputs. Conversely, if the bandwidth is small, only signals of relatively low frequencies are passed, and the time response will generally be slow and sluggish.

Cutoff rate. Often, bandwidth alone is inadequate in the indication of the characteristics of the system in distinguishing signals from noise. Sometimes it may be necessary to specify the cutoff rate of the frequency response at the high frequencies. However, in general, a steep cutoff characteristic may be accompanied by a large M_p, which corresponds to a system with a low stability margin.

The performance criteria defined above for the frequency-domain analysis are illustrated on the closed-loop frequency response, as shown in Fig. 9-3.

There are other criteria that may be used to specify the relative stability and performance of a feedback control system. These are defined in the ensuing sections of this chapter.

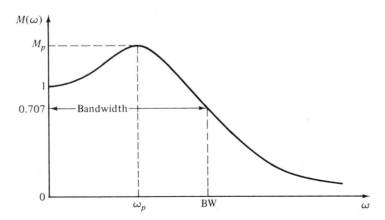

Fig. 9-3. Typical magnification curve of a feedback control system.

9.3 M_p, ω_p, and the Bandwidth of a Second-Order System

For a second-order feedback control system, the peak resonance M_p, the resonant frequency ω_p, and the bandwidth are all uniquely related to the damping ratio ζ and the natural undamped frequency ω_n of the system. Consider the second-order sinusoidal steady-state transfer function of a closed-loop system,

$$M(j\omega) = \frac{C(j\omega)}{R(j\omega)} = \frac{\omega_n^2}{(j\omega)^2 + 2\zeta\omega_n(j\omega) + \omega_n^2}$$
$$= \frac{1}{1 + j2(\omega/\omega_n)\zeta - (\omega/\omega_n)^2} \tag{9-18}$$

We may simplify the last expression by letting $u = \omega/\omega_n$. Then, Eq. (9-18) becomes

$$M(ju) = \frac{1}{1 + j2u\zeta - u^2} \tag{9-19}$$

The magnitude and phase of $M(j\omega)$ are

$$|M(ju)| = M(u) = \frac{1}{[(1 - u^2)^2 + (2\zeta u)^2]^{1/2}} \tag{9-20}$$

and

$$\underline{/M(ju)} = \phi_m(u) = -\tan^{-1}\frac{2\zeta u}{1 - u^2} \tag{9-21}$$

The resonant frequency is determined first by taking the derivative of $M(u)$ with respect to u and setting it equal to zero. Thus

$$\frac{dM(u)}{du} = -\frac{1}{2}[(1 - u^2)^2 + (2\zeta u)^2]^{-3/2}(4u^3 - 4u + 8u\zeta^2) = 0 \tag{9-22}$$

from which

$$4u^3 - 4u + 8u\zeta^2 = 0 \tag{9-23}$$

The roots of Eq. (9-23) are

$$u_p = 0 \tag{9-24}$$

and

$$u_p = \sqrt{1 - 2\zeta^2} \tag{9-25}$$

The solution in Eq. (9-24) merely indicates that the slope of the $M(\omega)$ versus ω curve is zero at $\omega = 0$; it is not a true maximum. The solution of Eq. (9-25) gives the resonant frequency,

$$\omega_p = \omega_n\sqrt{1 - 2\zeta^2} \tag{9-26}$$

Since frequency is a real quantity, Eq. (9-26) is valid only for $1 \geq 2\zeta^2$ or $\zeta \leq 0.707$. This means simply that for all values of ζ greater than 0.707, the solution of $\omega_p = 0$ becomes the valid one, and $M_p = 1$.

Substituting Eq. (9-25) into Eq. (9-20) and simplifying, we get

$$M_p = \frac{1}{2\zeta\sqrt{1-\zeta^2}} \tag{9-27}$$

It is important to note that M_p is a function of ζ only, whereas ω_p is a function of ζ and ω_n.

It should be noted that information on the magnitude and phase of $M(j\omega)$ of Eq. (9-18) may readily be derived from the Bode plot of Eq. (A-51), Figs. A-10 and A-12. In other words, Fig. A-12 is an exact representation of Eq. (9-21). The magnitude of $M(j\omega)$, however, may be represented in decibels versus frequency, which is the Bode plot in Fig. A-10, or it may be plotted in absolute magnitude, $M(\omega)$, versus ω. Figure 9-4 illustrates the plots of $M(u)$ of Eq. (9-20)

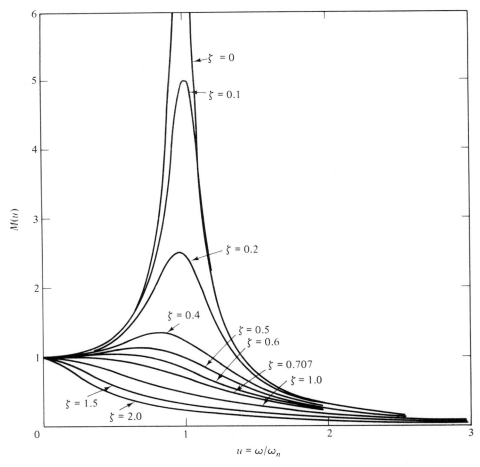

Fig. 9-4. Magnification-versus-normalized frequency of a second-order closed-loop control system.

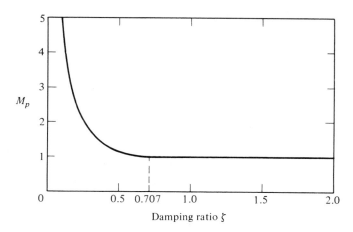

Fig. 9-5. M_p-versus-damping ratio for a second-order system, $M_p = 1/(2\zeta\sqrt{1-\zeta^2})$.

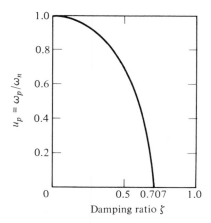

Fig. 9-6. Normalized resonant frequency-versus-damping ratio for a second-order system, $u_p = \sqrt{1-2\zeta^2}$.

versus u for various values of ζ. Notice that if the frequency scale were unnormalized, the value of ω_p would increase when ζ decreases, as indicated by Eq. (9-26). When $\zeta = 0$, ω_p and ω_n become identical. Figures 9-5 and 9-6 illustrate the relationship between M_p and ζ, and $u = \omega_p/\omega_n$ and ζ, respectively.

As defined in Section 9.2, the bandwidth, BW, of a system is the frequency at which $M(\omega)$ drops to 70.7 per cent of its zero-frequency value, or 3 dB down from the zero-frequency gain. Equating Eq. (9-20) to 0.707, we have

$$M(u) = \frac{1}{[(1-u^2)^2 + (2\zeta u)^2]^{1/2}} = 0.707 \qquad (9\text{-}28)$$

Thus

$$[(1-u^2)^2 + (2\zeta u)^2]^{1/2} = \sqrt{2} \qquad (9\text{-}29)$$

This equation leads to

$$u^2 = (1 - 2\zeta^2) \pm \sqrt{4\zeta^4 - 4\zeta^2 + 2} \qquad (9\text{-}30)$$

In the last expression the plus sign should be chosen, since u must be a positive real quantity for any ζ. Therefore, from Eq. (9-30), the bandwidth of the second-order system is determined as

$$\text{BW} = \omega_n[(1 - 2\zeta^2) + \sqrt{4\zeta^4 - 4\zeta^2 + 2}]^{1/2} \qquad (9\text{-}31)$$

Figure 9-7 gives a plot of BW/ω_n as a function of ζ. It is of interest to note that for a fixed ω_n, as the damping ratio ζ decreases from unity, the bandwidth increases and the resonance peak M_p also increases.

For the second-order system under consideration, we easily establish some simple relationships between the time-domain response and the frequency-domain response of the system.

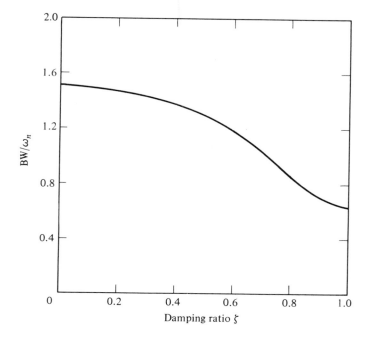

Fig. 9-7. Bandwidth/ω_n-versus-damping ratio for a second-order system, BW $= \omega_n[(1 - 2\zeta^2) + \sqrt{4\zeta^4 - 4\zeta^2 + 2}]^{1/2}$.

1. The maximum overshoot of the unit step response in the time domain depends upon ζ only [Eq. (6-96)].
2. The resonance peak of the closed-loop frequency response M_p depends upon ζ only [Eq. (9-27)].
3. The rise time increases with ζ, and the bandwidth decreases with the increase of ζ, for a fixed ω_n, Eq. (6-102), Eq. (9-31), Fig. 6-14, and Fig. 9-7. Therefore, bandwidth and rise time are inversely proportional to each other.
4. Bandwidth is directly proportional to ω_n.
5. Higher bandwidth corresponds to larger M_p.

9.4 Effects of Adding a Zero to the Open-Loop Transfer Function

The relationships established between the time-domain and the frequency-domain responses arrived at in Section 9.3 are valid only for the second-order closed-loop transfer function of Eq. (9-18). When other transfer functions or more parameters are involved, the relationships between the time-domain and the frequency-domain responses are altered and may be more complex. It is of interest to consider the effects on the frequency-domain characteristics of a feedback control system when poles and zeros are added to the open-loop transfer function. It would be a simpler procedure to study the effects of adding

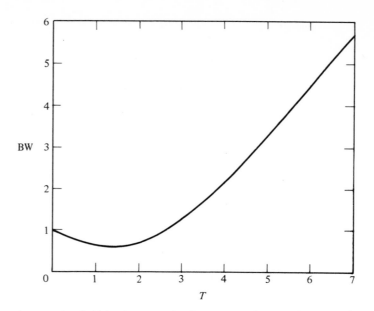

Fig. 9-8. Bandwidth of a second-order system with open-loop transfer function $G(s) = (1 + Ts)/[s(s + 1.414)]$.

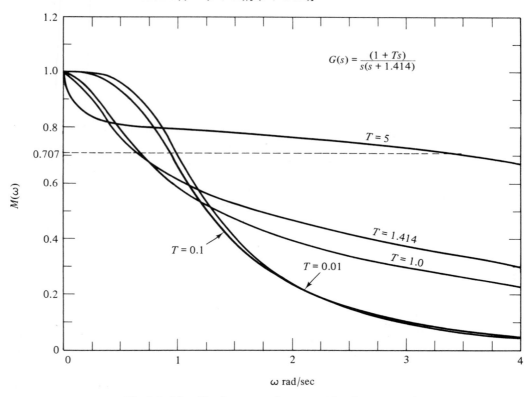

Fig. 9-9. Magnification curves for a second-order system with an open-loop transfer function $G(s)$.

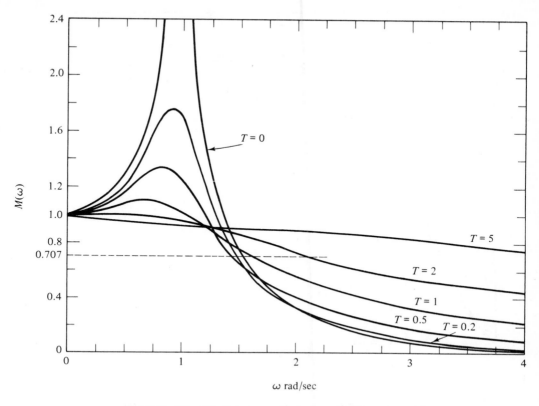

Fig. 9-10. Magnification curves for a second-order system with an open-loop transfer function $G(s) = (1 + Ts)/[s(s + 0.4)]$.

poles and zeros to the closed-loop transfer function. However, it is more realistic to consider modifying the open-loop transfer function directly.

The closed-loop transfer function of Eq. (9-18) may be considered as that of a unity feedback control system with an open-loop transfer function of

$$G(s) = \frac{\omega_n^2}{s(s + 2\zeta\omega_n)} \tag{9-32}$$

Let us add a zero at $s = -1/T$ to the last transfer function so that Eq. (9-32) becomes

$$G(s) = \frac{\omega_n^2(1 + Ts)}{s(s + 2\zeta\omega_n)} \tag{9-33}$$

This corresponds to the second-order system with derivative control studied in Section 6.7. The closed-loop transfer function of the system is given by Eq. (6-156), and is repeated here:

$$M(s) = \frac{C(s)}{R(s)} = \frac{\omega_n^2(1 + Ts)}{s^2 + (2\zeta\omega_n + T\omega_n^2)s + \omega_n^2} \tag{9-34}$$

In principle, M_p, ω_p, and BW of the system can all be derived using the

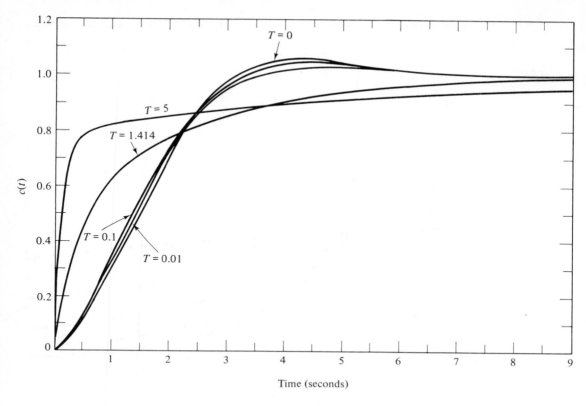

Fig. 9-11. Unit step responses of a second-order system with an open-loop transfer function $G(s) = (1 + Ts)/[s(s + 1.414)]$.

same steps as illustrated in Section 9.3. However, since there are now three parameters in ζ, ω_n, and T, the exact expressions for M_p, ω_p, and BW are difficult to obtain even though the system is still of the second order. For instance, the bandwidth of the system is

$$\text{BW} = \left(-b + \frac{1}{2}\sqrt{b^2 + 4\omega_n^4}\right)^{1/2} \tag{9-35}$$

where

$$b = 4\zeta^2\omega_n^2 + 4\zeta\omega_n^3 T - 2\omega_n^2 - \omega_n^4 T^2 \tag{9-36}$$

It is difficult to see how each of the parameters in Eq. (9-35) affects the bandwidth. Figure 9-8 shows the relationship between BW and T for $\zeta = 0.707$ and $\omega_n = 1$. Notice that the general effect of adding a zero to the open-loop transfer function is to increase the bandwidth of the closed-loop system. However, for small values of T over a certain range the bandwidth is actually decreased. Figures 9-9 and 9-10 give the plots for $M(\omega)$ for the closed-loop system that has the $G(s)$ of Eq. (9-33) as its open-loop transfer function; $\omega_n = 1$, T is given various values, and $\zeta = 0.707$ and $\zeta = 0.2$, respectively. These curves show that for large values of T the bandwidth of the closed-loop system is

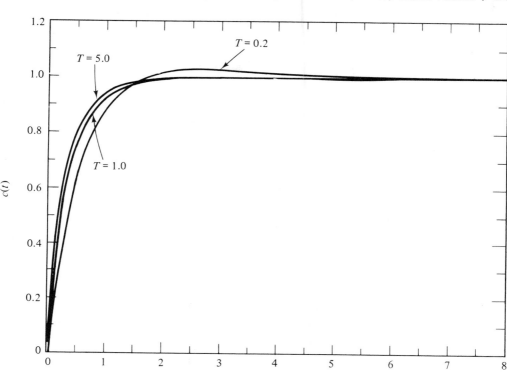

Fig. 9-12. Unit step responses of a second-order system with an open-loop transfer function $G(s) = (1 + Ts)/[s(s + 0.4)]$.

increased, whereas there exists a range of smaller values of T in which the BW is decreased by the addition of the zero to $G(s)$. Figures 9-11 and 9-12 show the corresponding unit step responses of the closed-loop system. The time-domain responses indicate that a high bandwidth corresponds to a faster rise time. However, as T becomes very large, the zero of the closed-loop transfer function, which is at $s = -1/T$, moves very close to the origin, causing the system to have a large time constant. Thus Fig. 9-11 illustrates the situation that the rise time is fast but the large time constant of the zero near the origin of the s-plane causes the time response to drag out in reaching the final steady state.

9.5 Effects of Adding a Pole to the Open-Loop Transfer Function

The addition of a pole to the open-loop transfer function generally has the effect of decreasing the bandwidth of the closed-loop system. The following transfer function is arrived at by adding $(1 + Ts)$ to the denominator of Eq. (9-32):

$$G(s) = \frac{\omega_n^2}{s(s + 2\zeta\omega_n)(1 + Ts)} \qquad (9\text{-}37)$$

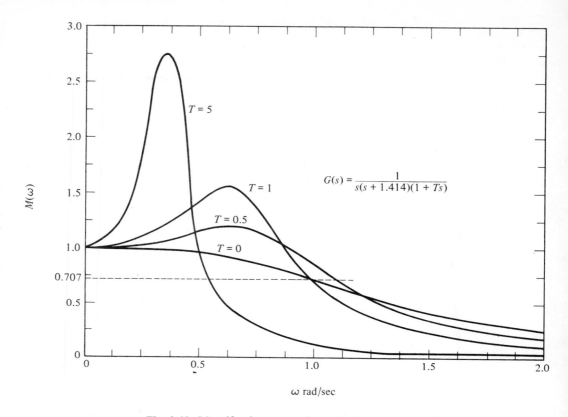

Fig. 9-13. Magnification curves for a third-order system with an open-loop transfer function $G(s) = 1/[s(s + 1.414)(1 + Ts)]$.

The derivation of the bandwidth of the closed-loop system which has the $G(s)$ of Eq. (9-37) as its open-loop transfer function is quite difficult. It can be shown that the BW is the real solution of the following equation:

$$T^2\omega^6 + (1 + 4\zeta^2\omega_n^2 T^2)\omega^4 + (4\zeta^2\omega_n^2 - 2\omega_n^2 - 4\zeta\omega_n^3 T)\omega^2 - \omega_n^4 = 0 \qquad (9\text{-}38)$$

We may obtain a qualitative indication on the bandwidth properties by referring to Fig. 9-13, which shows the plots for $M(\omega)$ for $\omega_n = 1$, $\zeta = 0.707$, and various values of T. Since the system is now of the third order, it can be unstable for a certain set of system parameters. However, it can be easily shown by use of the Routh–Hurwitz criterion that for $\omega_n = 1$ and $\zeta = 0.707$, the system is stable for all positive values of T. The $M(\omega)$ versus ω curves of Fig. 9-13 show that for small values of T, the bandwidth of the system is slightly increased but M_p is also increased. When T becomes large, the pole added to $G(s)$ has the effect of decreasing the bandwidth but increasing M_p. Therefore, we may conclude that, in general, the effect of adding a pole to the open-loop transfer function is to make the closed-loop system less stable. The unit step responses of Fig. 9-14 clearly show that for the larger values of T, $T = 1$ and $T = 5$, the

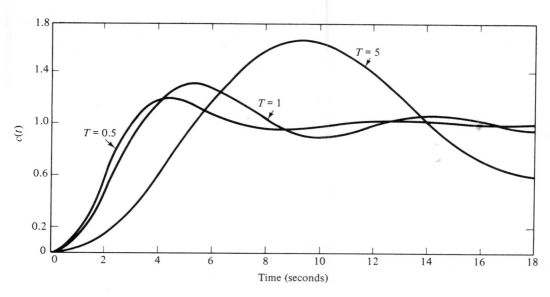

Fig. 9-14. Unit step responses of a third-order system with an open-loop transfer function $G(s) = 1/[s(s + 1.414)(1 + Ts)]$.

rise time increases with the decrease of the bandwidth, and the larger values of M_p also correspond to greater peak overshoots in the step responses. However, it is important to point out that the correlation between M_p and the peak overshoot is meaningful only when the closed-loop system is stable. When the magnitude of $G(j\omega)$ equals unity, $M(\omega)$ is infinite, but if the closed-loop system is unstable with $|G(j\omega)| > 1$ at $\underline{/G(j\omega)} = 180°$, $M(\omega)$ is finite and can assume an arbitrarily small number.

The objective of these last two sections is to demonstrate the simple relationships between the bandwidth and M_p, and the characteristics of the time-domain response. The effects of adding a pole and a zero to the open-loop transfer function are discussed. However, no attempt is made to include all general cases.

9.6 Relative Stability—Gain Margin, Phase Margin, and M_p

We have demonstrated in the last three sections the general relationship between the resonance peak M_p of the frequency response and the peak overshoot of the step response. Comparisons and correlations between frequency-domain and time-domain parameters such as these are useful in the prediction of the performance of a feedback control system. In general, we are interested not only in systems that are stable, but also in systems that have a certain degree of stability. The latter is often termed *relative stability*. In many situations we may use M_p to indicate the relative stability of a feedback control system. Another way of measuring relative stability of a closed-loop system is by means of the Nyquist plot of the loop transfer function, $G(s)H(s)$. The closeness of the $G(j\omega)H(j\omega)$ plot in the polar coordinates to the $(-1, j0)$ point gives an indication of how stable or unstable the closed-loop system is.

To demonstrate the concept of relative stability, the Nyquist plots and the corresponding step responses and frequency responses of a typical third-order

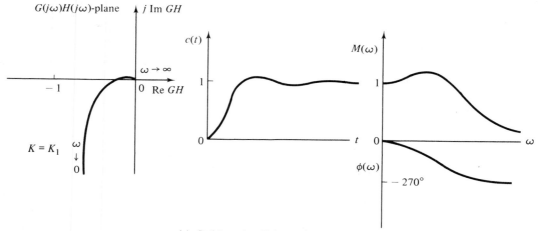

(a) Stable and well-damped system.

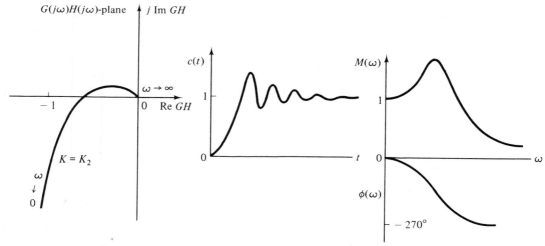

(b) Stable but oscillatory system.

Fig. 9-15. Correlation among Nyquist plots, step responses, and frequency responses.

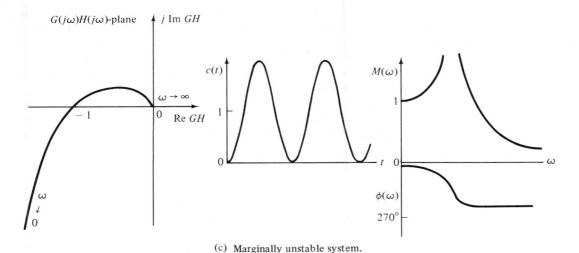

(c) Marginally unstable system.

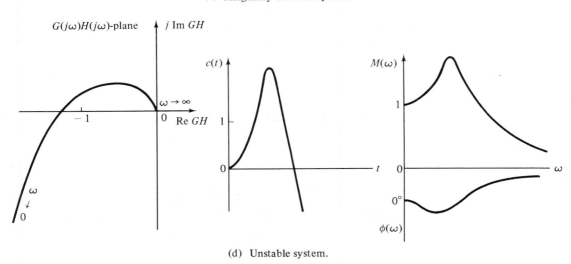

(d) Unstable system.

Fig. 9-15 (Cont.). Correlation among Nyquist plots, step responses, and frequency responses.

system are shown in Fig. 9-15 for four different values of loop gain K. Let us consider the case shown in Fig. 9-15(a), in which the loop gain K is low, so the Nyquist plot of $G(s)H(s)$ intersects the negative real axis at a point (the phase-crossover point) quite far away from the $(-1, j0)$ point. The corresponding step response is shown to be quite well behaved, and M_p is low. As K is increased, Fig. 9-15(b) shows that the phase-crossover point is moved closer to the $(-1, j0)$ point; the system is still stable, but the step response has a higher peak overshoot, and M_p is larger. The phase curve for ϕ_m does not give as good an indication of relative stability as M_p, except that one should note the slope of the ϕ_m curve which gets steeper as the relative stability decreases. The Nyquist plot of Fig. 9-15(c) intersects the $(-1, j0)$ point, and the system is unstable with constant-amplitude oscillation, as shown by the step response; M_p becomes infinite. If K is increased still further, the Nyquist plot will enclose the $(-1, j0)$ point, and the system is unstable with unbounded response, as shown in Fig. 9-15(d). In this case the magnitude curve $M(\omega)$ ceases to have any significance, and the only symptom of instability from the closed-loop frequency response is that the phase curve now has a positive slope at the resonant frequency.

Gain Margin

To give a quantitative way of measuring the relative distance between the $G(s)H(s)$ plot and the $(-1, j0)$ point, we define a quantity that is called the *gain margin*.

Specifically, the gain margin is a measure of the closeness of the phase-crossover point to the $(-1, j0)$ point in the $G(s)H(s)$-plane. With reference to Fig. 9-16, the phase-crossover frequency is denoted by ω_c, and the magnitude of $G(j\omega)H(j\omega)$ at $\omega = \omega_c$ is designated by $|G(j\omega_c)H(j\omega_c)|$. Then, the gain margin

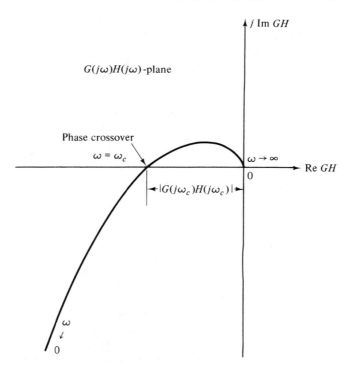

Fig. 9-16. Definition of the gain margin in the polar coordinates.

of the closed-loop system that has $G(s)H(s)$ as its loop transfer function is defined as

$$\text{gain margin} = \text{G.M.} = 20 \log_{10} \frac{1}{|G(j\omega_c)H(j\omega_c)|} \quad d\text{B} \quad (9\text{-}39)$$

On the basis of this definition, it is noticed that in the $G(j\omega)H(j\omega)$ plot of Fig. 9-16, if the loop gain is increased to the extent that the plot passes through the $(-1, j0)$ point, so that $|G(j\omega_c)H(j\omega_c)|$ equals unity, the gain margin becomes 0 dB. On the other hand, if the $G(j\omega)H(j\omega)$ plot of a given system does not intersect the negative real axis, $|G(j\omega_c)H(j\omega_c)|$ equals zero, and the gain margin defined by Eq. (9-39) is infinite in decibels. Based on the above evaluation, the physical significance of gain margin can be stated as follows: *Gain margin is the amount of gain in decibels that can be allowed to increase in the loop before the closed-loop system reaches instability.*

When the $G(j\omega)H(j\omega)$ plot goes through the $(-1, j0)$ point, the gain margin is 0 dB, which implies that the loop gain can no longer be increased as the system is already on the margin of instability. When the $G(j\omega)H(j\omega)$ plot does not intersect the negative real axis at any finite nonzero frequency, and the Nyquist stability criterion indicates that the $(-1, j0)$ point must not be enclosed for system stability, the gain margin is infinite in decibels; this means that, theoretically, the value of the loop gain can be increased to infinity before instability occurs.

When the $(-1, j0)$ point is to the right of the phase-crossover point, the magnitude of $G(j\omega_c)H(j\omega_c)$ is greater than unity, and the gain margin as given by Eq. (9-39) in decibels is negative. In the general case, when the above mentioned condition implies an unstable system, the gain margin is negative in decibels. It was pointed out in Chapter 7 that if $G(s)H(s)$ has poles or zeros in the right half of the s-plane, the $(-1, j0)$ point may have to be encircled by the $G(j\omega)H(j\omega)$ plot in order for the closed-loop system to be stable. Under this condition, a stable system yields a negative gain margin. In practice, we must first determine the stability of the system (i.e., stable or unstable), and then the magnitude of the gain margin is evaluated. Once the stability or instability condition is ascertained, the magnitude of the gain margin simply denotes the margin of stability or instability, and the sign of the gain margin becomes insignificant.

Phase Margin

The gain margin is merely one of many ways of representing the relative stability of a feedback control system. In principle, a system with a large gain margin should be relatively more stable than one that has a smaller gain margin. Unfortunately, gain margin alone does not sufficiently indicate the relative stability of all systems, especially if parameters other than the loop gain are variable. For instance, the two systems represented by the $G(j\omega)H(j\omega)$ plots of Fig. 9-17 apparently have the same gain margin. However, locus A actually

$G(j\omega)H(j\omega)$-plane

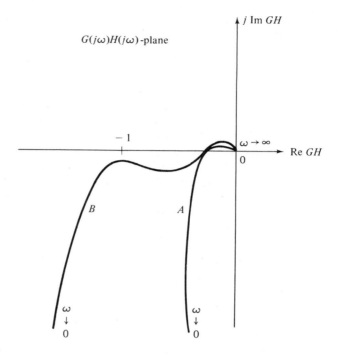

Fig. 9-17. Nyquist plots showing systems with same gain margin but different amount of relative stability.

corresponds to a more stable system than locus B. The reason is that with any change in a system parameter (or parameters) other than the loop gain, it is easier for locus B to pass through or even enclose the $(-1, j0)$ point. Furthermore, system B has a much larger M_p than system A.

In order to strengthen the representation of relative stability of a feedback control system, we define the *phase margin* as a supplement to gain margin. *Phase margin is defined as the angle in degrees through which the $G(j\omega)H(j\omega)$ plot must be rotated about the origin in order that the gain-crossover point on the locus passes through the $(-1, j0)$ point.* Figure 9-18 shows the phase margin as the angle between the phasor that passes through the gain-crossover point and the negative real axis of the $G(j\omega)H(j\omega)$-plane. In contrast to the gain margin, which gives a measure of the effect of the loop gain on the stability of the closed-loop system, the phase margin indicates the effect on stability due to changes of system parameters, which theoretically alter the phase of $G(j\omega)H(j\omega)$ only.

The analytical procedure of computing the phase margin involves first the calculation of the phase of $G(j\omega)H(j\omega)$ at the gain-crossover frequency, and then subtracting $180°$ from this phase; that is,

$$\text{phase margin} = \phi.M. = \underline{/G(j\omega_g)H(j\omega_g)} - 180° \qquad (9\text{-}40)$$

where ω_g denotes the gain-crossover frequency. If the $(-1, j0)$ point is encircled

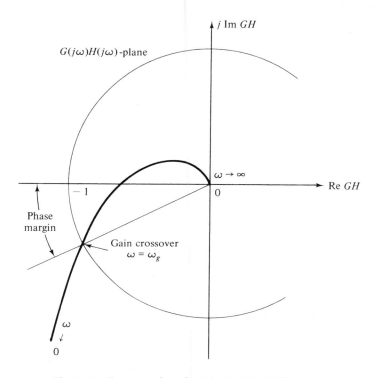

Fig. 9-18. Phase margin defined in the $G(j\omega)H(j\omega)$-plane.

by the $G(j\omega)H(j\omega)$ plot, the gain-crossover point would be found in the second quadrant of the $G(j\omega)H(j\omega)$-plane, and Eq. (9-40) would give a negative phase margin.

Graphic Methods of Determining Gain Margin and Phase Margin

Although the formulas for the gain margin and the phase margin are simple to understand, in practice it is more convenient to evaluate these quantities graphically from the Bode plot or the magnitude-versus-phase plot. As an illustrative example, consider that the open-loop transfer function of a control system with unity feedback is given by

$$G(s) = \frac{10}{s(1 + 0.02s)(1 + 0.2s)} \tag{9-41}$$

The Bode plot of $G(j\omega)$ is shown in Fig. 9-19. Using the asymptotic approximation of $|G(j\omega)|$, the gain-crossover and the phase-crossover points are determined as shown in the figure. The phase-crossover frequency is approximately 16 rad/sec, and the magnitude of $G(j\omega)$ at this frequency is about $-15\,\text{dB}$. This means that if the loop gain of the system is increased by 15 dB, the magnitude curve will cross the 0-dB axis at the phase-crossover frequency. This condition

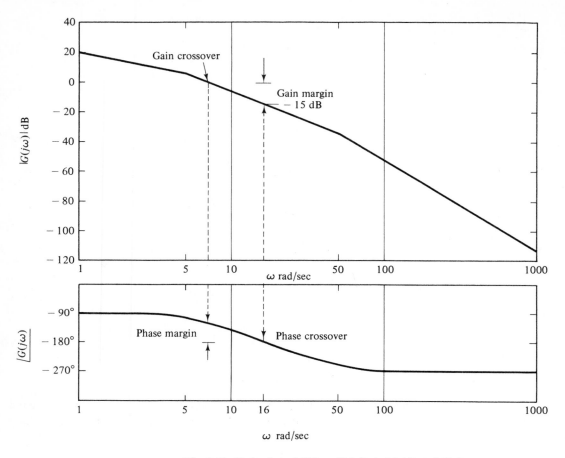

Fig. 9-19. Bode plot of $G(s) = 10/[s(1 + 0.2s)(1 + 0.02s)]$.

corresponds to the Nyquist plot of $G(j\omega)$ passing through the $(-1, j0)$ point, and the system becomes marginally unstable. Therefore, from the definition of the gain margin, the gain margin of the system is 15 dB.

To determine the phase margin, we note that the gain-crossover frequency is at $\omega = 7$ rad/sec. The phase of $G(j\omega)$ at this frequency is approximately $-125°$. The phase margin is the angle the phase curve must be shifted so that it will pass through the $-180°$ axis at the gain-crossover frequency. In this case,

$$\phi.\text{M.} = 180° - 125° = 55° \qquad (9\text{-}42)$$

In general, the procedure of determining the gain margin and the phase margin from the Bode plot may be outlined as follows:

1. The gain margin is measured at the phase-crossover frequency ω_c:

$$\text{G.M.} = -|G(j\omega_c)H(j\omega_c)| \qquad \text{dB} \qquad (9\text{-}43)$$

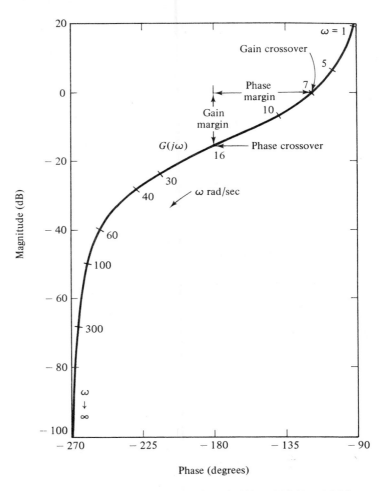

Fig. 9-20. Magnitude-versus-phase plot of $G(s) = 10/[s(1 + 0.2s)(1 + 0.02s)]$.

2. The phase margin is measured at the gain-crossover frequency ω_g:

$$\phi.\text{M.} = 180° + \underline{/G(j\omega_g)H(j\omega_g)} \qquad (9\text{-}44)$$

The gain and phase margins are even better illustrated on the magnitude-versus-phase plot. For the transfer function of Eq. (9-41), the magnitude-versus-phase plot is shown in Fig. 9-20, which is constructed by use of the data from the Bode plot of Fig. 9-19. On the magnitude-versus-phase plot of $G(j\omega)$, the phase crossover point is where the locus intersects the $-180°$ axis, and the gain crossover is where the locus intersects the 0-dB axis. Therefore, the gain margin is simply the distance in decibels measured from the phase crossover to

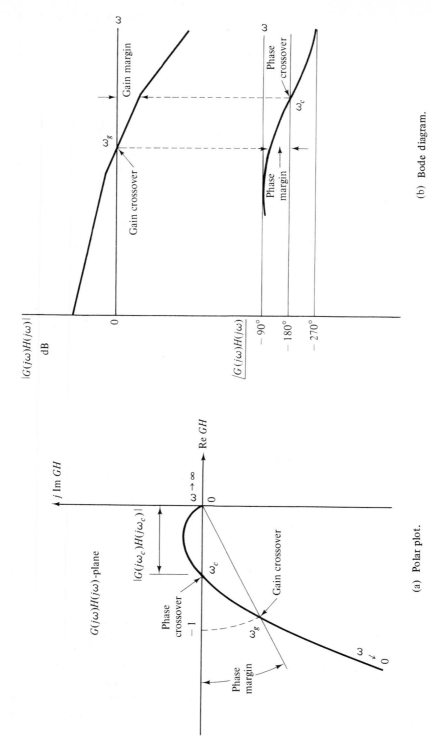

Fig. 9-21. Polar plot, Bode diagram, and magnitude-versus-phase plot, showing gain and phase margins in these domains.

(a) Polar plot.

(b) Bode diagram.

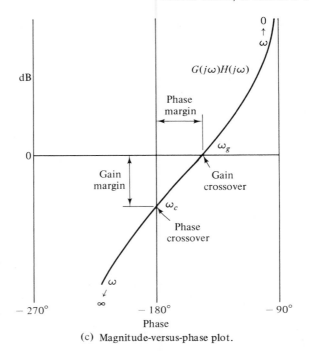

(c) Magnitude-versus-phase plot.

Fig. 9-21 (Cont.). Polar plot, Bode diagram, and magnitude-versus-phase plot, showing gain and phase margins in these domains.

the critical point at 0 dB and $-180°$, and the phase margin is the horizontal distance in degrees measured from the gain crossover to the critical point.

In summarizing, the relations between the measurements of the gain margin and the phase margin in the polar plot, the Bode plot, and the magnitude-versus-phase plot are illustrated in Fig. 9-21.

9.7 Relative Stability As Related to the Slope of the Magnitude Curve of the Bode Plot

In general, a definite relation between the relative stability of a closed-loop system and the slope of the magnitude curve of the Bode plot of $G(j\omega)H(j\omega)$ at the gain crossover can be established. For example, in Fig. 9-19, if the loop gain of the system is decreased from the nominal value, the gain crossover may be moved to the region in which the slope of the magnitude curve is only -20 dB/decade; the corresponding phase margin of the system would be increased. On the other hand, if the loop gain is increased, the relative stability of the system will deteriorate, and if the gain is increased to the extent that the gain crossover occurs in the region where the slope of the magnitude curve is -60 dB/decade, the system will definitely be unstable. The example cited above is a simple one, since the slope of the magnitude curve decreases monotonically as ω increases. Let us consider a conditionally stable system for the purpose of

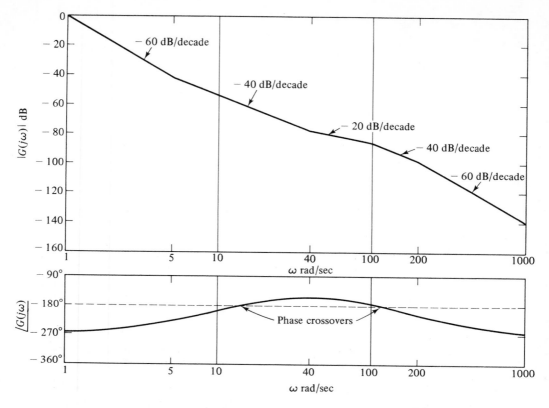

Fig. 9-22. Bode plot of $G(s) = [K(1 + 0.2s)(1 + 0.025s)]/[s^3(1 + 0.01s)(1 + 0.005s)]$.

illustrating relative stability. Consider that a control system with unity feedback has the open-loop transfer function

$$G(s) = \frac{K(1 + 0.2s)(1 + 0.025s)}{s^3(1 + 0.01s)(1 + 0.005s)} \qquad (9\text{-}45)$$

The Bode plot of $G(s)$ is shown in Fig. 9-22 for $K = 1$. The gain-crossover frequency is 1 rad/sec, and the phase margin is negative. The closed-loop system is unstable even for a very small value of K. There are two phase crossover points: one at $\omega = 15$ rad/sec and the other at $\omega = 120$ rad/sec. The phase characteristics between these two frequencies indicate that if the gain crossover lies in this range, the system is stable. From the magnitude curve of the Bode plot, the range of K for stable operation is found to be between 60 and 90 dB. For values of K above and below this range, the phase lag of $G(j\omega)$ exceeds $-180°$ and the system is unstable. This system serves as a good example of the relation between relative stability and the slope of the magnitude curve at the gain crossover. As observed from Fig. 9-22, at both very low and very high frequencies, the slope of the magnitude curve is -60 dB/decade; if the gain crossover falls in either one of these two regions, the phase margin becomes negative and the system is unstable. In the two sections of the magnitude curve that have a slope of -40 dB/decade, the system is stable only if the gain cross-over falls in about half of these regions, but even then the resultant phase margin is small. However, if the gain crossover falls in the region in which the magnitude curve has a slope of -20 dB/decade, the system is stable.

The correlation between the slope of the magnitude curve of the Bode plot at the gain crossover and the relative stability can be used in a qualitative way for design purposes.

9.8 Constant M Loci in the $G(j\omega)$-Plane

In previous sections it was shown that the resonance peak M_p of the closed-loop frequency response is directly related to the maximum overshoot of the transient response. Normally, the magnification curve of $M(\omega)$ versus ω may be constructed by the Bode plot method if the closed-loop transfer function $M(s)$ is given and if its numerator and denominator are in factored form. Unfortunately, this is not the usual case, as the open-loop transfer function $G(s)$ is normally given. For the purpose of analysis we can always obtain the magnification curve by digital computation on a computer. However, our motivation is to be able to predict M_p from the plots of $G(j\omega)$, and eventually to design a system with a specified M_p.

Consider that the closed-loop transfer function of a feedback control system with unity feedback is given by

$$M(s) = \frac{C(s)}{R(s)} = \frac{G(s)}{1 + G(s)} \tag{9-46}$$

For sinusoidal steady state, $G(s) = G(j\omega)$ is written

$$\begin{aligned} G(j\omega) &= \operatorname{Re} G(j\omega) + j \operatorname{Im} G(j\omega) \\ &= x + jy \end{aligned} \tag{9-47}$$

Then

$$\begin{aligned} M(\omega) = |M(j\omega)| &= \left| \frac{G(j\omega)}{1 + G(j\omega)} \right| \\ &= \frac{\sqrt{x^2 + y^2}}{\sqrt{(1 + x)^2 + y^2}} \end{aligned} \tag{9-48}$$

For simplicity, let $M = M(\omega)$; then Eq. (9-48) leads to

$$M\sqrt{(1 + x)^2 + y^2} = \sqrt{x^2 + y^2} \tag{9-49}$$

Squaring both sides of the last equation gives

$$M^2[(1 + x)^2 + y^2] = x^2 + y^2 \tag{9-50}$$

Rearranging this equation yields

$$(1 - M^2)x^2 + (1 - M^2)y^2 - 2M^2x = M^2 \qquad (9\text{-}51)$$

This equation is conditioned by dividing through by $(1 - M^2)$ and adding the term $[M^2/(1 - M^2)]^2$ on both sides. We have

$$x^2 + y^2 - \frac{2M^2}{1 - M^2}x + \left(\frac{M^2}{1 - M^2}\right)^2 = \frac{M^2}{1 - M^2} + \left(\frac{M^2}{1 - M^2}\right)^2 \qquad (9\text{-}52)$$

which is finally simplified to

$$\left(x - \frac{M^2}{1 - M^2}\right)^2 + y^2 = \left(\frac{M}{1 - M^2}\right)^2 \qquad (9\text{-}53)$$

For a given M, Eq. (9-53) represents a circle with the center at $x = M^2/(1 - M^2)$, $y = 0$. The radius of the circle is $r = |M/(1 - M^2)|$. Equation (9-53) is invalid for $M = 1$. For $M = 1$, Eq. (9-50) gives

$$x = -\frac{1}{2} \qquad (9\text{-}54)$$

which is the equation of a straight line parallel to the $j\,\mathrm{Im}\,G(j\omega)$ axis and passing through the $(-\frac{1}{2}, j0)$ point in the $G(j\omega)$-plane.

When M takes on different values, Eq. (9-53) describes in the $G(j\omega)$-plane a family of circles that are called the *constant M loci* or the *constant M circles*. The coordinates of the centers and the radii of the constant M loci for various values of M are given in Table 9-1, and some of the loci are shown in Fig. 9-23.

Table 9-1   Constant M Circles

M	Center $x = \dfrac{M^2}{1 - M^2}$, $y = 0$	Radius $r = \left\|\dfrac{M}{1 - M^2}\right\|$
0.3	0.01	0.33
0.5	0.33	0.67
0.7	0.96	1.37
1.0	∞	∞
1.1	-5.76	5.24
1.2	-3.27	2.73
1.3	-2.45	1.88
1.4	-2.04	1.46
1.5	-1.80	1.20
1.6	-1.64	1.03
1.7	-1.53	0.90
1.8	-1.46	0.80
1.9	-1.38	0.73
2.0	-1.33	0.67
2.5	-1.19	0.48
3.0	-1.13	0.38
4.0	-1.07	0.27
5.0	-1.04	0.21
6.0	-1.03	0.17

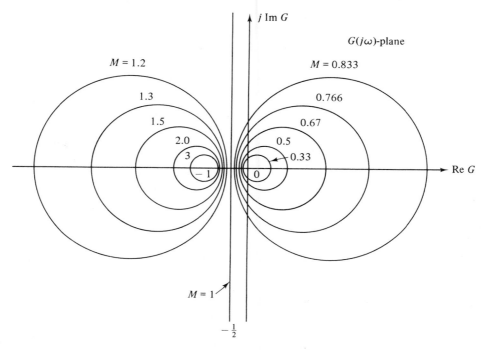

Fig. 9-23. Constant M circles in the polar coordinates.

Note that when M becomes infinite, the circle degenerates into a point at the critical point, $(-1, j0)$. This agrees with the well-known fact that when the Nyquist plot of $G(j\omega)$ passes through the $(-1, j0)$ point, the system is marginally unstable and M_p is infinite. Figure 9-23 shows that the constant M loci in the $G(j\omega)$-plane are symmetrical with respect to the $M = 1$ line and the real axis. The circles to the left of the $M = 1$ locus correspond to values of M greater than 1, and those to the right of the $M = 1$ line are for M less than 1.

Graphically, the intersections of the $G(j\omega)$ plot and the constant M loci give the value of M at the frequency denoted on the $G(j\omega)$ curve. If it is desired to keep the value of M_p less than a certain value, the $G(j\omega)$ curve must not intersect the corresponding M circle at any point, and at the same time must not enclose the $(-1, j0)$ point. The constant M circle with the smallest radius that is tangent to the $G(j\omega)$ curve gives the value of M_p, and the resonant frequency ω_p is read off at the tangent point on the $G(j\omega)$ curve.

Figure 9-24(a) illustrates the Nyquist plot of $G(j\omega)$ for a unity feedback control system, together with several constant M loci. For a given loop gain $K = K_1$, the intersects between the $G(j\omega)$ curve and the constant M loci give the points on the $M(\omega)$-versus-ω curve. The peak resonance M_p is found by locating the smallest circle that is tangent to the $G(j\omega)$ plot. The resonant frequency is found at the point of tangency and is designated as ω_{p1}. If the loop gain is increased to K_2, and if the system is still stable, a constant M circle with a smaller radius that corresponds to a larger M is found tangent to the $G(j\omega)$

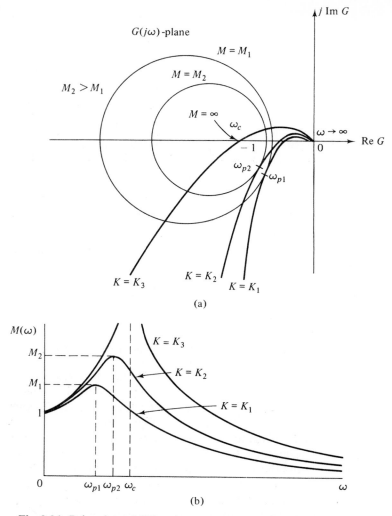

Fig. 9-24. Polar plots of $G(s)$ and constant M loci showing the procedure of determining M_p and the magnification curves.

curve, and thus the peak resonance M_p is larger. The resonant frequency is shown to be ω_{p2}, which is closer to the phase-crossover frequency ω_c than ω_{p1}. When K is increased to K_3 so that the $G(j\omega)$ curve passes through the $(-1, j0)$ point, the system is marginally unstable, M_p is infinite, and $\omega_{p3} = \omega_c$. In all cases the bandwidth of the closed-loop system is found at the intersect of the $G(j\omega)$ curve and the $M = 0.707$ locus. For values of K beyond K_3, the system is unstable, and the magnification curve and M_p no longer have any meaning. When enough points of intersections between the $G(j\omega)$ curve and the constant M loci are obtained, the magnification curves are plotted as shown in Fig. 9-24(b).

9.9 Constant Phase Loci in the $G(j\omega)$-Plane

The loci of constant phase of the closed-loop system may also be determined in the $G(j\omega)$-plane by a method similar to that used to secure the constant M loci. With reference to Eqs. (9-46) and (9-47), the phase of the closed-loop system is written as

$$\phi_m(\omega) = \underline{/M(j\omega)} = \tan^{-1}\left(\frac{y}{x}\right) - \tan^{-1}\left(\frac{y}{1+x}\right) \qquad (9\text{-}55)$$

Taking the tangent on both sides of the last equation and letting $\phi_m = \phi_m(\omega)$, we have

$$\tan\phi_m = \frac{y}{x^2 + x + y^2} \qquad (9\text{-}56)$$

Let $N = \tan\phi_m$; then Eq. (9-56) becomes

$$x^2 + x + y^2 - \frac{y}{N} = 0 \qquad (9\text{-}57)$$

Adding the term $(1/4) + (1/4N^2)$ to both sides of Eq. (9-57) yields

$$x^2 + x + \frac{1}{4} + y^2 - \frac{y}{N} + \frac{1}{4N^2} = \frac{1}{4} + \frac{1}{4N^2} \qquad (9\text{-}58)$$

which is regrouped to give

$$\left(x + \frac{1}{2}\right)^2 + \left(y - \frac{1}{2N}\right)^2 = \frac{1}{4} + \frac{1}{4N^2} \qquad (9\text{-}59)$$

When N assumes various values, this equation represents a family of circles with centers at $(x, y) = (-1/2, 1/2N)$. The radii are given by

$$r = \left(\frac{N^2 + 1}{4N^2}\right)^{1/2} \qquad (9\text{-}60)$$

The centers and the radii of the constant N circles for various values of N are tabulated in Table 9-2, and the loci are shown in Fig. 9-25.

Table 9-2 Constant N Circles

$\phi_m = 180°n$ $n = 0, 1, 2 \ldots$	$N = \tan\phi_m$	Center $x = -\dfrac{1}{2},\ y = \dfrac{1}{2N}$	Radius $r = \sqrt{\dfrac{N^2 + 1}{4N^2}}$
-90	$-\infty$	0	0.500
-60	-1.732	-0.289	0.577
-45	-1.000	-0.500	0.707
-30	-0.577	-0.866	1.000
-15	-0.268	-1.866	1.931
0	0	∞	∞
15	0.268	1.866	1.931
30	0.577	0.866	1.000
45	1.000	0.500	0.707
60	1.732	0.289	0.577
90	∞	0	0.500

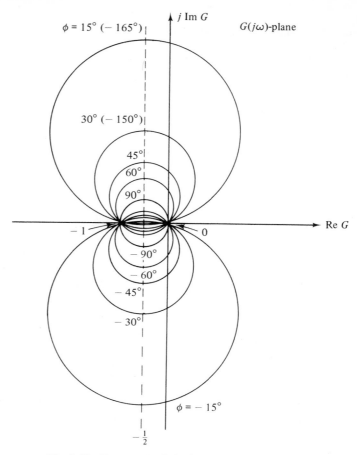

Fig. 9-25. Constant N circles in the polar coordinates.

9.10　Constant M and N Loci in the Magnitude-Versus-Phase Plane—The Nichols Chart

In principle we need both the magnitude and the phase of the closed-loop frequency response to analyze the performance of the system. However, we have shown that the magnitude curve, which includes such information as M_p, ω_p, and BW, normally is more useful for relative stability studies.

A major disadvantage in working with the polar coordinates for the $G(j\omega)$ plot is that the curve no longer retains its original shape when a simple modification such as the change of the loop gain is made to the system. In design problems, frequently not only the loop gain must be altered, but series or feedback controllers are to be added to the original system which require the complete reconstruction of the resulting $G(j\omega)$. For design purposes it is far more convenient to work in the Bode diagram or the magnitude-versus-phase domain. In the Bode diagram the magnitude curve is shifted up and down without distortion when the loop gain is varied; in the magnitude-versus-phase plot, the

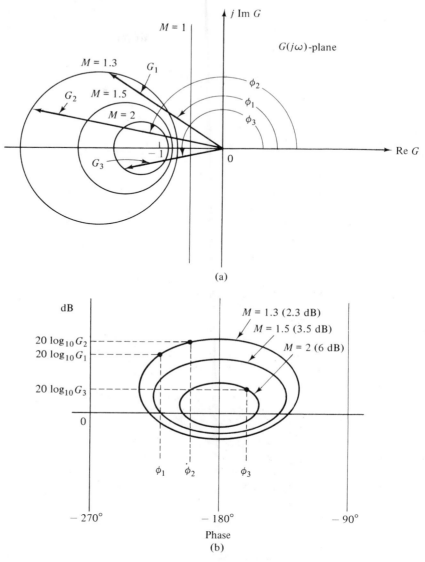

Fig. 9-26. (a) Constant *M* circles in the *G(jω)*-plane. (b) Nichols chart in the magnitude-versus-phase coordinates.

entire *G(jω)* curve is shifted up or down vertically when the gain is altered. In addition, the Bode plot can be easily modified to accommodate any modifications made to *G(jω)* in the form of added poles and zeros.

The constant *M* and constant *N* loci in the polar coordinates may be transferred to the magnitude-versus-phase coordinates without difficulty. Figure 9-26 illustrates how this is done. Given a point on a constant *M* circle in the *G(jω)*-plane, the corresponding point in the magnitude-versus-phase plane may be determined by drawing a vector directly from the origin of the

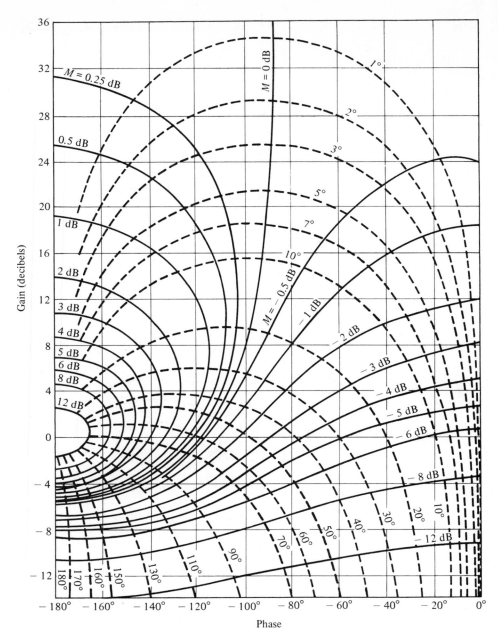

Fig. 9-27. Nichols chart (for phase from $-180°$ to $0°$).

$G(j\omega)$-plane to the particular point on the constant M circle; the length of the vector in decibels and the phase angle in degrees give the corresponding point in the magnitude-versus-phase plane. Figure 9-26 illustrates the process of locating three arbitrary corresponding points on the constant M loci in the magnitude-versus-phase plane. The critical point, $(-1, j0)$, in the $G(j\omega)$-plane corresponds to the point with 0 dB and $-180°$ in the magnitude-versus-phase plane.

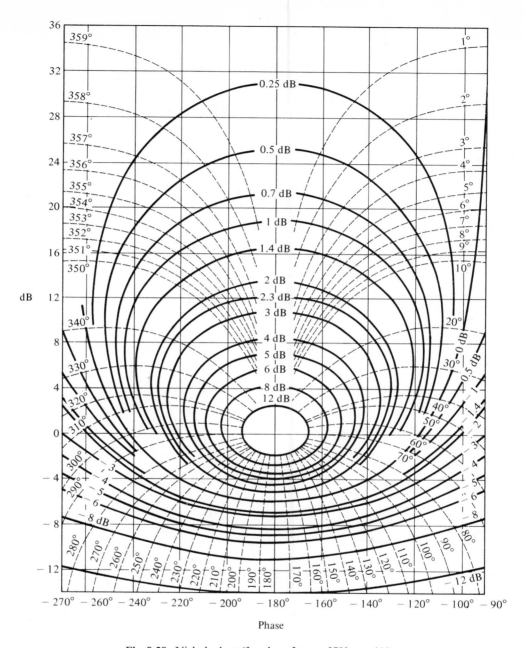

Fig. 9-28. Nichols chart (for phase from −270° to −90°).

Using the same procedure as described above, the constant N loci are also transferred into the magnitude-versus-phase plane. These constant M and N loci in the magnitude-versus-phase coordinates were originated by Nichols[2] and called the *Nichols chart*. A typical Nichols chart is constructed in Fig. 9-27 for the phase angle that extends from −180° to 0°. The chart that corresponds to the phase from −360° to −180° is a mirror image of that in Fig. 9-27 with

493

respect to the $-180°$ axis. In Fig. 9-28 the Nichols chart is shown for $-90°$ to $-270°$, which is the useful region of phase for many practical control systems. The values of M on these constant M loci are given in decibels. To determine bandwidth, the -3-dB locus should be used.

The following example will illustrate the relationships among the analysis methods using the Bode plot, the magnitude-versus-phase plot, and the Nichols chart.

EXAMPLE 9-1   Let us consider the positional control system discussed in Section 6.6. When the inductance of the *dc* motor is 0.1 H, the open-loop transfer function of the system is given by Eq. (6-149) and is repeated here:

$$G(s) = \frac{250A}{s(s^2 + 50.5s + 1725)} \qquad (9\text{-}61)$$

The Bode plot for $G(s)$ is drawn as shown in Fig. 9-29 for $A = 200$. The gain

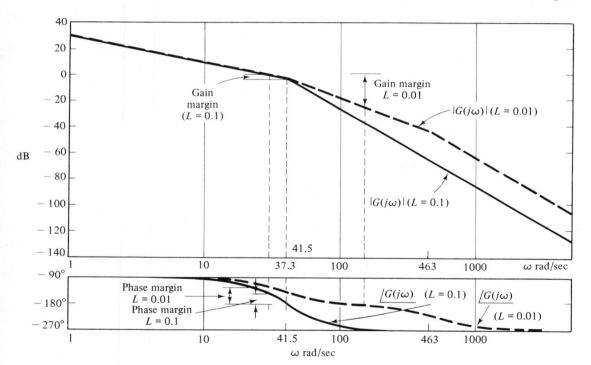

Fig. 9-29. Bode plots of the system in Example 9-1.

margin and the phase margin are determined from the Bode plot to be 5 dB and 40°, respectively. The data on the magnitude and phase of $G(j\omega)$ are transferred to the magnitude-versus-phase plot of Fig. 9-30. From Fig. 9-30, the peak resonance M_p is found to be approximately 5 dB (or 1.78), the resonant frequency is 40 rad/sec, the bandwidth of the system is 48 rad/sec, and the results for the gain and phase margins are as given above.

When the motor inductance is reduced to 0.01 henry, the open-loop transfer

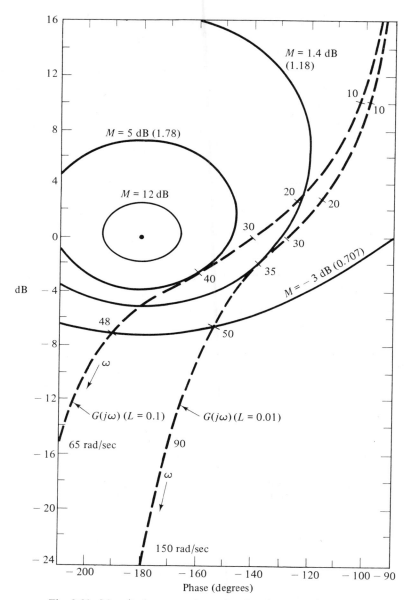

Fig. 9-30. Magnitude-versus-phase plots for the system in Example 9-1.

function of the system becomes

$$G(s) = \frac{2500A}{s(s^2 + 500s + 17270)} \tag{9-62}$$

With $A = 200$, and for the purpose of constructing the Bode plot, $G(s)$ is written

$$G(s) = \frac{29}{s(1 + 0.00216s)(1 + 0.0268s)} \tag{9-63}$$

The Bode plot of $G(s)$ is shown in Fig. 9-29. It is shown that by decreasing the inductance, the gain margin is improved to approximately 24 dB, and the phase margin is about 50°. The magnitude-versus-phase plot in Fig. 9-30 indicates that the bandwidth of the system is not noticeably changed, but M_p is reduced to approximately 1.4 dB, or 1.18. The frequency-domain results obtained in this example for the two values of inductance correlate quite well with the time-domain analysis that resulted in the time responses of Fig. 6-21. Table 9-3 gives a comparison of the time-domain and frequency-domain parameters.

Table 9-3   Comparison of the Time-Domain and Frequency-Domain Performances
of the System in Example 9-1

L (henrys)	Peak Overshoot (%)	Rise Time (sec)	Delay Time (sec)	BW (rad/sec)	M_p	ω_p	Gain Margin (dB)	Phase Margin (deg)
0.1	41	0.4	0.5	48	1.78	40	5	40
0.01	11.5	0.6	0.4	50	1.18	35	24	50

9.11   Closed-Loop Frequency Response Analysis of Nonunity Feedback Systems

The constant M and N loci and the Nichols chart analysis discussed in preceding sections are limited to closed-loop systems with unity feedback, whose transfer function is given by Eq. (9-46). When a system has nonunity feedback, the closed-loop transfer function is

$$M(s) = \frac{C(s)}{R(s)} = \frac{G(s)}{1 + G(s)H(s)} \tag{9-64}$$

the constant M and N loci derived earlier and the Nichols charts of Figs. 9-27 and 9-28 cannot be applied directly. However, we can show that with a slight modification these loci can still be applied to systems with nonunity feedback.

Let us consider the function

$$P(s) = \frac{G(s)H(s)}{1 + G(s)H(s)} \tag{9-65}$$

Comparing Eq. (9-64) with Eq. (9-65), we have

$$P(s) = M(s)H(s) \tag{9-66}$$

Information on the gain margin and the phase margin of the system of Eq. (9-64) can be obtained in the usual fashion by constructing the Bode plot of $G(s)H(s)$. However, the $G(j\omega)H(j\omega)$ curve and the Nichols chart together do not give the magnitude and phase plots for $M(j\omega)$, but for $P(j\omega)$. Since $M(j\omega)$ and $P(j\omega)$ are related through Eq. (9-66), once the plots for $P(\omega)$ versus ω and $\underline{/P(j\omega)}$ versus ω are obtained, the curves for $M(\omega)$ and $\phi_m(\omega)$ versus ω are determined from the following relationships:

$$M(\omega) = \frac{P(\omega)}{H(\omega)} \tag{9-67}$$

$$\phi_m(\omega) = \underline{/P(j\omega)} - \underline{/H(j\omega)} \tag{9-68}$$

9.12 Sensitivity Studies in the Frequency Domain[4]

The frequency-domain study of feedback control systems has an advantage in that the sensitivity of a transfer function with respect to a given parameter can be clearly interpreted. We shall show how the Nyquist plot and the Nichols chart can be utilized for analysis and design of a control system based on sensitivity considerations.

Consider that a control system with unity feedback has the transfer function

$$M(s) = \frac{C(s)}{R(s)} = \frac{G(s)}{1 + G(s)} \tag{9-69}$$

The sensitivity of $M(s)$ with respect to $G(s)$ is defined as

$$S_G^M(s) = \frac{dM(s)/M(s)}{dG(s)/G(s)} \tag{9-70}$$

or

$$S_G^M(s) = \frac{dM(s)}{dG(s)} \frac{G(s)}{M(s)} \tag{9-71}$$

Substituting Eq. (9-69) into Eq. (9-71) and simplifying, we have

$$S_G^M(s) = \frac{1}{1 + G(s)} \tag{9-72}$$

Clearly, the sensitivity function is a function of the complex variable s.

In general, it is desirable to keep the sensitivity to a small magnitude. From a design standpoint, it is possible to formulate a design criterion in the following form:

$$|S_G^M(s)| = \frac{1}{|1 + G(s)|} \leq k \tag{9-73}$$

In the sinusoidal steady state, Eq. (9-73) is easily interpreted in the polar coordinate by a Nyquist plot. Equation (9-73) is written

$$|1 + G(j\omega)| \geq \tfrac{1}{k} \tag{9-74}$$

Figure 9-31 illustrates the Nyquist plot of a stable closed-loop system. The constraint on sensitivity given in Eq. (9-74) is interpreted as the condition that the $G(j\omega)$ locus must not enter the circle with radius k. It is interesting to note that the sensitivity criterion is somewhat similar to the relative stability specifications of gain and phase margins. When the value of k is unity, the $G(j\omega)$ locus must be tangent or outside the circle with a unity radius and centered at the $(-1, j0)$ point. This corresponds to a very stringent stability requirement, since the gain margin is infinite. On the other hand, if the Nyquist plot of $G(j\omega)$ passes through the $(-1, j0)$ point, the system is unstable, and the sensitivity is infinite.

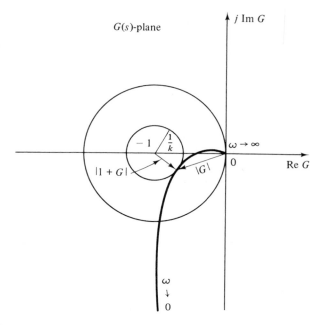

Fig. 9-31. Interpretation of sensitivity criterion with the Nyquist plot.

Equation (9-74) and Fig. 9-31 also indicate clearly that for low sensitivity, the magnitude of $G(j\omega)$ should be high, which reduces the stability margin. This again points to the need of compromise among the criteria in designing control systems.

Although Fig. 9-31 gives a clear interpretation of the sensitivity function in the frequency domain, in general, the Nyquist plot is awkward to use for design purposes. In this case the Nichols chart is again more convenient for the purpose of analysis and design of a feedback control system with a prescribed sensitivity. Equation (9-72) is written

$$S_G^M(j\omega) = \frac{G^{-1}(j\omega)}{1 + G^{-1}(j\omega)} \qquad (9\text{-}75)$$

which clearly indicates that the magnitude and phase of $S_G^M(j\omega)$ can be obtained by plotting $G^{-1}(j\omega)$ in the Nichols chart and making use of the constant M loci for constant sensitivity function. Since the vertical coordinate of the Nichols chart is in decibels, the $G^{-1}(j\omega)$ curve in the magnitude-versus-phase coordinates can be easily obtained if the $G(j\omega)$ is already available, since

$$|G^{-1}(j\omega)|\,\mathrm{dB} = -|G(j\omega)|\,\mathrm{dB} \qquad (9\text{-}76)$$

$$\underline{/G^{-1}(j\omega)} = -\underline{/G(j\omega)} \qquad (9\text{-}77)$$

As an illustrative example, the function $G^{-1}(j\omega)$ for Eq. (9-61), Example 9-1, is plotted in the Nichols chart as shown in Fig. 9-32, using the $G(j\omega)$ plot in Fig. 9-30. The intersects of the $G^{-1}(j\omega)$ curve in the Nichols chart with the constant M loci give the magnitudes of $S_G^M(j\omega)$ at the corresponding frequencies.

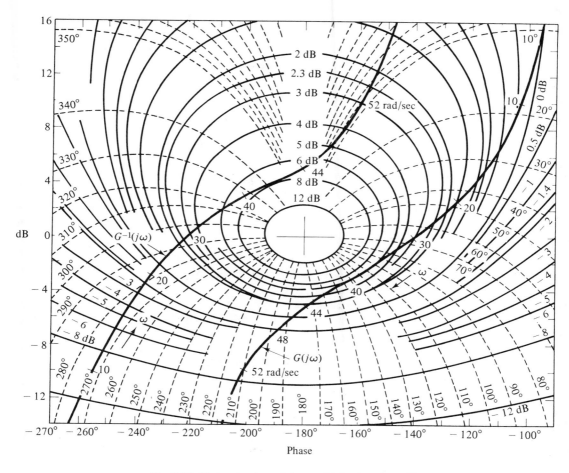

Fig. 9-32. Determination of the sensitivity function S_G^M in the Nichols chart.

Figure 9-32 indicates several interesting points with regard to the sensitivity function of the feedback control system. The sensitivity function S_G^M approaches 0 dB or unity as ω approaches infinity. S_G^M becomes zero as ω approaches zero. A peak value of 8 dB is reached by S_G^M at $\omega = 42$ rad/sec. This means that the closed-loop transfer function is most sensitive to the change of $G(j\omega)$ at this frequency and more generally in this frequency range. This result is not difficult to comprehend, since from the Nichols chart of Fig. 9-30 it is observed that the stability and dynamic behavior of the closed-loop system is more directly governed by the $G(j\omega)$ curve near ω_p, which is 40 rad/sec. Changes made to portions of $G(j\omega)$ at frequencies much higher and much lower than 40 rad/sec are not going to have a great effect on the relative stability of the system directly. When the loop gain of the system increases, the $G(j\omega)$ curve is raised in the Nichols chart domain, and the $G^{-1}(j\omega)$ curve must be lowered. If the $G(j\omega)$ curve passes through the critical point at 0 dB and $-180°$, the system becomes marginally unstable; the $G^{-1}(j\omega)$ also passes through the same point, and the sensitivity is infinite.

In this section we have simply demonstrated the use of the Nichols chart for the analysis of the sensitivity function of a closed-loop system. In a design problem, the objective may be to find a controller such that the sensitivity due to certain system parameters is small.

REFERENCES

1. Y. CHU, "Correlation Between Frequency and Transient Response of Feedback Control Systems," *AIEE Trans. Application and Industry*, Part II, Vol. 72, p. 82, 1953.

2. H. M. JAMES, N. B. NICHOLS, and R. S. PHILLIPS, *Theory of Servomechanisms*, McGraw-Hill Book Company, New York, 1947.

3. B. H. WILLIS and R. W. BROCKETT, "The Frequency Domain Solution of Regulator Problems," *IEEE Trans. Automatic Control*, Vol. AC-10, pp. 262–267, July 1965.

4. A. GELB, "Graphical Evaluation of the Sensitivity Function Using the Nichols Chart," *IRE Trans. Automatic Control*, Vol. AC-7, pp. 57–58, July 1962.

PROBLEMS

9.1. The pole–zero configuration of a closed-loop transfer function is shown in Fig. P9-1(a).
   (a) Compute the bandwidth of the system.
   (b) A zero is added to the closed-loop system, as shown in Fig. P9-1(b); how is the bandwidth affected?

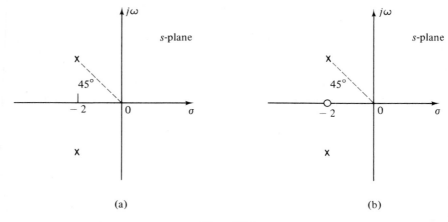

(a)   (b)

Figure P9-1.

(c) Another pole is inserted on the negative real axis in Fig. P9-1(b), but at a distance 10 times farther from the origin than the zero; how is the band-width affected?

9.2. The specification given on a certain second-order feedback control system is that the overshoot of the step response should not exceed 25 per cent.
(a) What are the corresponding limiting values of the damping ratio and peak resonance M_p?
(b) Determine the corresponding values for ω_p and t_{max}.

9.3. Sketch the closed-loop frequency response $|M(j\omega)|$ as a function of frequency for the systems and responses shown in Fig. P9-3.

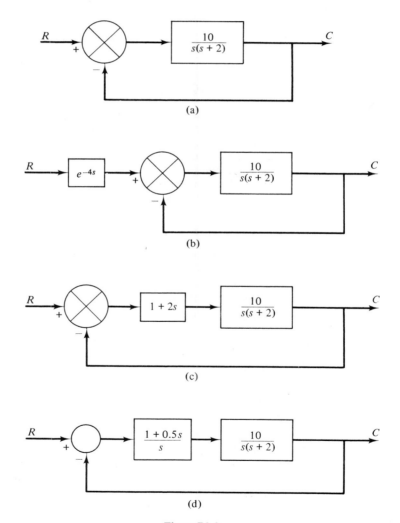

Figure P9-3.

(e) Sketch the unit step response for the system whose $|M|$-versus-ω curve is as shown. Assume that the system is of second order.

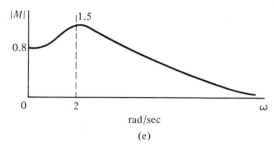

(e)

Figure P9-3 (Cont.).

9.4. The closed-loop transfer function of a feedback control system is given by

$$M(s) = \frac{C(s)}{R(s)} = \frac{1}{(1 + 0.01s)(1 + 0.05s + 0.01s^2)}$$

(a) Plot the frequency response curve for the closed-loop system.

(b) Determine the peak resonance peak M_p and the resonant frequency ω_p of the system.

(c) Determine the damping ratio ζ and the natural undamped frequency ω_n of the second-order system that will produce the same M_p and ω_p determined for the original system.

9.5. The open-loop transfer function of a unity feedback control system is

$$G(s) = \frac{K}{s(1 + 0.1s)(1 + s)}$$

(a) Determine the value of K so that the resonance peak M_p of the system is equal to 1.4.

(b) Determine the value of K so that the gain margin of the system is 20 dB.

(c) Determine the value of K so that the phase margin of the system is 60°.

9.6. The open-loop transfer function of a unity feedback control system is

$$G(s) = \frac{K(1 + Ts)}{s(1 + s)(1 + 0.01s)}$$

Determine the *smallest* possible value of T so that the system has an infinite gain margin.

9.7. The open-loop transfer function of a unity feedback control system is

$$G(s) = \frac{K}{s(1 + 0.1s)(1 + 0.001s)}$$

Determine the value of K if the steady-state error of the output position must be less than or equal to 0.1 per cent for a ramp function input. With this value of K, what are the gain margin and the phase margin of the system? Plot $G(s)$ in the gain-phase plot and determine the resonance peak M_p and the resonant frequency ω_p.

9.8. A random compensation network is added to the forward path of the system in Problem 9.7, so that now the open-loop transfer function reads

$$G(s) = \frac{K(1 + 0.0167s)}{s(1 + 0.00222s)(1 + 0.1s)(1 + 0.001s)}$$

where K is determined in part (a) of Problem 9.7. Plot the gain-phase diagram of $G(s)$. Evaluate M_p, ω_p, the gain margin, the phase margin, and the bandwidth of the compensated system.

9.9. The Bode diagram of the open-loop transfer function $G(s)$ of a unity feedback control system is shown in Fig. P9-9.

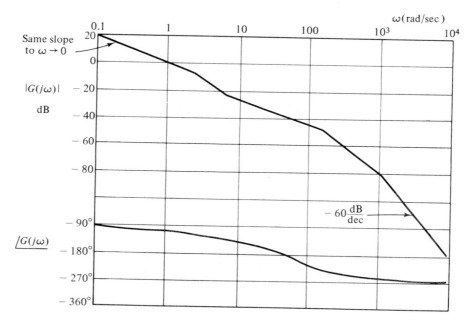

Figure P9-9.

(a) Find the gain margin and the phase margin of the system.
(b) If the open-loop transfer function is changed to $e^{-Ts}G(s)$, find the value of T so that the phase margin of the system is 45°. Then find the value of T so that the gain margin is 20 dB.
(c) What is the velocity error constant of the system in part (a)? in part (b)?

10

Introduction to Control
Systems Design

10.1 Introduction

Design of control systems represents an interesting and complex subject in control systems studies. In a simplified manner the design problem of control systems can be described with the aid of the block diagram of Fig. 10-1. The figure

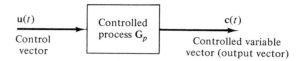

Fig. 10-1. Block diagram of a controlled process.

shows a controlled process whose output vector $\mathbf{c}(t)$ represents q controlled variables, and the control vector $\mathbf{u}(t)$ represents p control signals. The problem is to find a set of "appropriate" signals, $\mathbf{u}(t)$, so that the controlled variable vector $\mathbf{c}(t)$ behaves as desired. The description of the basic design problem is simplified by overlooking the possible existence of external disturbances.

Once the desired control vector $\mathbf{u}(t)$ for satisfactory control is determined, a controller is usually needed to generate this control from the reference inputs and the state vector $\mathbf{x}(t)$ or output $\mathbf{c}(t)$. Figure 10-2 illustrates the block diagram of a control system whose control vector is derived from the input vector and the state vector. This type of system is also referred to as one with state feedback. The block diagram of Fig. 10-2 is intended only for the purpose of illustrating the philosophy of designing control systems, and no attempt is made for it to include all possible configurations.

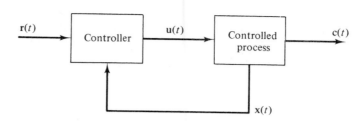

Fig. 10-2. Block diagram of a control system with state feedback.

It is interesting to give a brief review on the history of development of control system design theory, which may help gain perspective in the understanding of the subject.

The early stage of the theoretical development of the design of control systems was characterized by the works of Nyquist, Hall, Nichols,[2] and Bode,[1] who developed such classical methods as the Nyquist plot, Bode diagram, and Nichols chart. A unique feature of these methods is that they are all graphical techniques which are conducted in the frequency domain. As was pointed out earlier that in the design of control systems, it is the time response that is of importance, rather than the frequency response. The use of the frequency-domain techniques is simply due to the fact that the graphical techniques are convenient to apply.

The classical design methods are characterized by first fixing the configuration of the system to be designed. In other words, the designer must first choose a suitable system configuration. For instance, Fig. 10-3(a) shows the block diagram of a system with the controller located in the forward path of the system. This is a very common practice because of the versatility of the scheme, and the system is said to have a cascade or series compensation. Figure 10-3(b) shows another scheme of compensation by having a controller in the feedback path, and this is often referred to as the feedback compensation. In general, other configurations, such as having controllers in both the forward path and the feedback path, may be used if desired. In practice, the controllers or compensators used in control systems may assume a great variety of forms. In the simple cases, the controller may be passive networks in the form of low-pass, high-pass, or band-pass filters, or networks with active elements. In elaborate cases, the controller may even be a mini-computer.

The proper selection of the system configuration as well as the contents of the controller depend to a great extent on the experience and ingenuity on the part of the designer. In the frequency-domain design, the design specifications usually are given in terms of such criteria as gain margin, phase margin, peak resonance, and bandwidth. These criteria, however, should be related to the time-domain specifications, such as rise time, overshoot, and settling time, which are more direct measurements of the system's performance.

The classical design of control systems is very much a trial-and-error proposition. This is a distinct disadvantage of the method, since it does not indicate whether a solution even actually exists for the design problem at the outset. It

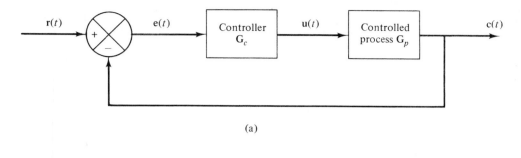

(a)

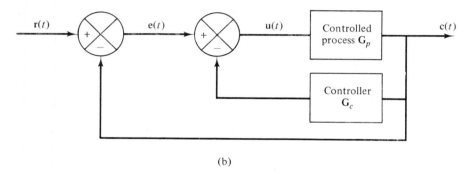

(b)

Fig. 10-3. Block diagrams of control systems with two different schemes of compensation. (a) Series compensation. (b) Feedback compensation.

is entirely possible that the design requirements are so stringent or may even be contradictory so that they cannot be satisfied by any system configuration or controllers that are physically realizable. Even when a solution does exist, the classical design yields a system that is very seldom the best by any standards. For instance, gain margin and phase margin are measures of the relative stability of a control system. A system having a gain margin of, say, 20 dB or a phase margin of 45° does not imply that it is optimal in any sense.

The introduction of the root locus technique by Evans[4] in 1950 made possible the design of control systems to be carried out in the s-plane. The main advantage of the root locus method is that information on frequency-domain as well as time-domain characteristics can be derived directly from the pole–zero configuration in the s-plane. With the knowledge of the closed-loop transfer function poles and zeros, the time-domain response is determined readily by means of inverse Laplace transform, and the frequency response is obtained from the Bode plot. However, the root locus design is still basically a trial-and-error procedure, and it relies on the reshaping of the root loci to obtain a satisfactory pole–zero configuration for the closed-loop transfer function.

The work by Norbert Wiener[3] in the late 1940s opened a new horizon to the design of control systems. Wiener introduced not only the statistical considerations of control systems but also the idea of the performance index. For the first time, the design engineer was able to start from a set of design criteria and carry out the design by means of a completely analytical procedure. He is able to design a control system that is optimum or the best possible with respect to a given performance criterion.

In many practical applications of control systems, the actual signals and disturbances subjected by a control system may be random in nature. Unlike the deterministic signals, such as the step function and the sinusoidal function considered in the preceding chapters, random signals can be adequately described only by their statistical properties. For instance, in the problem of controlling the antenna of a radar system, the wind force acting on the antenna is best described by some probabilistic function rather than by a sine wave or any other deterministic signals. The main difference between a deterministic signal and a random signal is that the magnitude of the latter can only be described as what is the probability that it will lie in a certain range at a given time.

The principle of Wiener's optimization technique is demonstrated by the block diagram shown in Fig. 10-4. The design objective is to determine the closed-loop transfer function $C(s)/R(s)$ of the system such that the error between the desired output and the actual output is minimized. In Wiener's statistical design technique, the mean-square value of the error $e(t)$ is used as the performance index, J; that is,

$$J = \lim_{T \to \infty} \frac{1}{T} \int_0^T e^2(t)\, dt \tag{10-1}$$

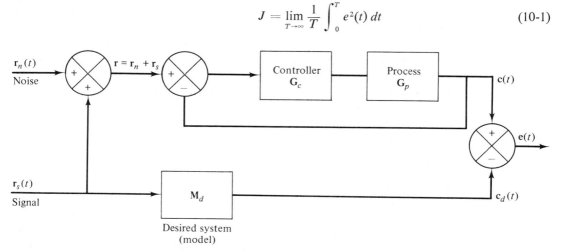

Fig. 10-4. Block diagram of control systems designed by Wiener's optimization technique.

The reason for using the mean-square error as the performance index is that the minimization of this particular performance index induces an analytical design procedure which makes use of the mathematical functions already defined in the theory of probability and statistics. However, in practice, a system that is optimum in the sense of minimum mean-square error may be ideal only for certain situations but not for others. In fact, it is not difficult to see from Eq. (10-1) that the mean-square-error criterion places heavier emphasis on large errors than on smaller ones.

In reality, the configuration shown in Fig. 10-4 can be used for the analytical design of systems with random inputs as well as systems with deterministic inputs. When the input signal is considered to be deterministic, other performance indices such as those listed below can be treated mathematically:

$$\int_0^\infty e^2(t)\, dt \qquad \int_0^\infty |e(t)|\, dt \qquad \int_0^\infty t\,|e(t)|\, dt$$

In fact, the first of these criteria is known as the integral-square error (ISE) and is the most popular one used for the analytical design of control systems. The reason for this popularity is due to the fact that the integral is directly related to the Laplace transform domain through Parseval's theorem.

The importance of Wiener's work and the analytical design is not so much because the techniques have found significant applications in control systems practice, but because these represent a revolution in design principle from the conventional trial-and-error methods.

At approximately the same time the analytical design principle and techniques were being developed, Truxal[5] proposed a synthesis procedure through pole–zero configuration in the s-plane. The synthesis still makes use of the conventional design specifications, such as the relative damping ratio, error constants, bandwidth, and rise time, and the input signals are deterministic. Based on the design specifications, the closed-loop transfer function of the control system is first determined, and then the corresponding open-loop transfer function is found. The advantage of this synthesis method over the classical frequency-domain design is that the designer is able to determine if the given set of specifications are consistent at the beginning of the design, so the amount of guesswork and trial and error are cut to a minimum. Furthermore, Truxal's synthesis starts with the closed-loop transfer function and then works toward the transfer function of the controller, whereas the frequency-domain design starts out with the controller and then works toward the closed-loop transfer function to see if the design specifications are satisfied.

It is very difficult to pinpoint exactly when the modern control systems theory was inaugurated. In fact, the mathematical foundation of certain aspects of modern control theory can be traced far back to works that were completed some seventy years ago. For instance, the state-variable approach to linear systems is well known to mathematicians as the theory of solutions of first-order differential equations. Liapunov's method on stability was based on his Ph.D thesis, which was completed in 1892. The linear programming technique, which has significant impact on modern control theory and practice, was developed about 1939. These significant contributions, among many others, did not become widely known until recent years because they were much too far ahead of their time.

The launch of the space age has placed a challenge to the control engineer to find new methods of design of more complex control systems and to meet more rigid requirements. The control engineer soon discovered that the conventional design was no longer adequate and rigorous enough to handle the complicated problems encountered in modern fire control systems, autopilot systems, missile guidance systems, spacecraft rendezvous control systems, and many others. Modern societal systems are modeled by complex models with multiple inputs and outputs which are difficult to handle even with optimal control theory. Consequently, not only new design principles have been developed, but many of the mathematical contributions that have long been neglected were rediscovered and made applicable to the practical control problems. In a sense, the classical control system design is truly an engineering endeavor, and the modern control design contains, first, the development and formulation of the mathematical theory, and second, the application of the mathematical principles to practical design problems. Indeed, at present, many areas of the advanced control theory are still at the theoretical stage, and it is generally recognized that a gap still exists between theory and practice in the design of control systems.

The objective of modern control design can be described by two words: *optimal control*. In other words, a system is to be designed so that it is optimum in a prescribed sense. For instance, with reference to the block diagram of the multivariable controlled process shown in Fig. 10-2, one of the common problems in optimal control is to determine the control vector $\mathbf{u}(t)$ over the time interval $t_0 \leq t < t_f$ so that the state vector $\mathbf{x}(t)$ is brought from the initial state $\mathbf{x}(t_0)$ to the desired final state $\mathbf{x}(t_f)$ in the shortest possible time, subject to the given controlled process and possibly other constraints. The problem is usually referred to as the *minimal-time problem* or *time-optimal problem*.

In general, the performance index J can be represented by the following integral:

$$J = \int_{t_0}^{t_f} F[\mathbf{x}(t), \mathbf{u}(t), t]\, dt \tag{10-2}$$

where F is a scalar function. For the minimal-time problem, F is set at unity, so

$$J = \int_{t_0}^{t_f} dt = t_f - t_0 \tag{10-3}$$

Minimizing J is therefore equivalent to minimizing the time interval between the initial and the final times.

As another example, if it is desired to drive the state vector $\mathbf{x}(t)$ as close as possible to the desired state $\mathbf{x}_d$ over the time interval (t_0, t_f), while keeping the magnitudes of the controls within reason, the following performance index may be used:

$$J = \int_{t_0}^{t_f} \{[\mathbf{x}(t) - \mathbf{x}_d]'\mathbf{Q}[\mathbf{x}(t) - \mathbf{x}_d] + \mathbf{u}'(t)\mathbf{R}\mathbf{u}(t)\}\, dt \tag{10-4}$$

where $\mathbf{Q}$ and $\mathbf{R}$ are symmetric matrices.

The extensive use of applied mathematics in modern control theory has made it difficult for one to make a quick transition from the classical design to the modern. The classical design is characterized by such terminology and tools as transfer function, poles and zeros, frequency response, root loci, Bode plot, and Nyquist plot. In optimal control studies we shall find a set of new terms such as state variables, state equations, state transition matrix, maximum or minimum principle, Liapunov's methods, gradient technique, linear programming, dynamic programming, controllability, and observability.

In this section we have given a brief discussion of the historical development of control systems theory. The discussions and reference made are by no means exhaustive. It is hoped that these introductory remarks will give the reader a general idea of the basic problems involved in the design of control systems before entering the details on the subject of design. The remaining part of this chapter will contain subjects on the classical design methods. The analytical design method and certain aspects of optimal control are covered in Chapter 11.

10.2 Classical Design of Control Systems

In this section the classical design of control systems will be carried out in the frequency domain and the s-domain. The designs are mainly affected by means of Bode plot, Nichols chart, and the root locus.

To illustrate the basic principle of the classical design, let us consider the following example. Let us begin by considering the transfer function of a controlled process

$$G_p(s) = \frac{K}{s(1 + s)(1 + 0.0125s)} \tag{10-5}$$

The closed-loop system is considered to have a unity feedback. It is required that when a unit ramp input is applied to the closed-loop system, the steady-state error of the system does not exceed 1 per cent of the amplitude of the input ramp, which is unity. Thus when we use the approach discussed in Chapter 6 on steady-state error, we can find the minimum value of K in order to fulfill this error requirement:

$$\text{steady-state error} = e_{ss} = \lim_{s \to 0} \frac{1}{sG_p(s)} = \frac{1}{K} \le 0.01 \tag{10-6}$$

Therefore, K must be greater than 100. However, applying the Routh–Hurwitz criterion to the characteristic equation of the closed-loop system it is easy to show that the system is unstable for all values of K greater than 81. This means that some kind of compensation scheme or controller should be applied to the system so that the steady-state error and the relative stability requirements can be satisfied simultaneously. It is apparent that this controller must be able to keep the zero-frequency gain of s times the open-loop transfer function of the compensated system effectively at 100 while maintaining a prescribed degree of relative stability. The principle of the design in the frequency domain is best illustrated by the Nyquist plot of $G_p(s)$ shown in Fig. 10-5. In practice, we seldom use the Nyquist plot but rather the Bode plot for design, because the latter is easier to construct. When $K = 100$, the system is unstable, and the Nyquist plot of $G_p(s)$ is shown to enclose the $(-1, j0)$ point. Let us assume that we desire to realize a resonance peak of $M_p = 1.25$. This means that the Nyquist plot of $G_p(s)$ must be tangent to the constant-M circle for $M = 1.25$ from below. If K is the only parameter that we can adjust to achieve the objective of $M_p = 1.25$, Fig. 10-5 shows that the desired value of K is 1. However, with this value of K,

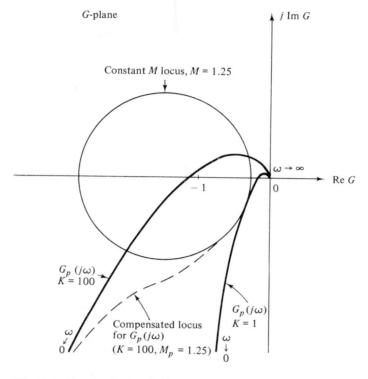

Fig. 10-5. Nyquist plot for $G_p(s) = K/[s(1 + s)(1 + 0.0125s)]$.

the velocity error constant is only 1 sec^{-1}, and the steady-state error requirement is not satisfied. Since the steady-state performance of the system is governed by the characteristics of the transfer function at the low frequency, Fig. 10-5 shows that in order to simultaneously satisfy the transient and the steady-state requirements, the Nyquist plot of $G_p(s)$ has to be reshaped so that the high-frequency portion of the locus follows the $K = 1$ plot and the low-frequency portion follows the $K = 100$ locus. The significance of this locus reshaping is that the compensated locus shown in Fig. 10-5 will be tangent to the $M = 1.25$ circle at a relatively high frequency, while the zero-frequency gain is maintained at 100 to satisfy the steady-state requirement. When we inspect the loci of Fig. 10-5, it is clear that there are two alternative approaches in arriving at the compensated locus:

1. Starting from the $K = 100$ locus and reshaping the locus in the region near the resonant frequency ω_p, while keeping the low-frequency region of $G_p(s)$ relatively unaltered.
2. Starting from the $K = 1$ locus and reshaping the low-frequency portion of $G_p(s)$ to obtain a velocity error constant of $K_v = 100$ while keeping the locus near $\omega = \omega_p$ relatively unchanged.

In the first approach, the high-frequency portion of $G_p(s)$ is pushed in the counterclockwise direction, which means that more phase is added to the system in the positive direction in the proper frequency range. This scheme is basically referred to as *phase-lead* compensation, and controllers used for this purpose are often of the high-pass-filter type. The second approach apparently involves the shifting of the low-frequency part of the $K = 1$ trajectory in the clockwise direction, or alternatively, reducing the magnitude of $G_p(s)$ with $K = 100$ at the high-frequency range. This scheme is often referred to as phase-lag compensation, since more phase lag is introduced to the system in the low-frequency range. The type of network that is used for phase-lag compensation is often referred to as low-pass filters.

Figures 10-6 and 10-7 further illustrate the philosophy of design in the frequency domain using the Bode diagram. In this case the relative stability of the system is more conveniently represented by the gain margin and the phase margin. In Fig. 10-6 the Bode plots of $G_p(j\omega)$ show that when $K = 100$, the gain and phase margins are both negative, and the system is unstable. When $K = 1$, the gain and phase margins are both positive, and the system has quite a comfortable safety margin. Using the first approach, the phase-lead compensation, as described earlier, we add more phase lead to $G_p(j\omega)$ so as to improve the phase margin. However, in attempting to reshape the phase curve by use of a high-pass filter, the magnitude curve of $G_p(j\omega)$ is unavoidably altered as shown in Fig. 10-6. If the design is carried out properly, it is possible to obtain a net gain in relative stability using this approach. The Bode diagram of Fig. 10-7 serves

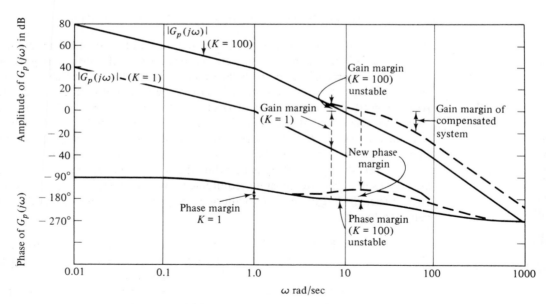

Fig. 10-6. Bode plot of $G_p(s) = K/[s(1 + s)(1 + 0.0125s)]$ with phase-lead compensation.

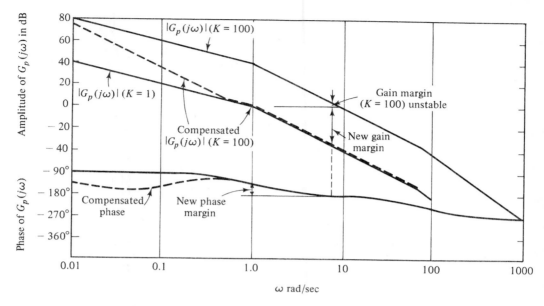

Fig. 10-7. Bode plot of $G_p(s) = K/[s(1 + s)(1 + 0.0125s)]$ with phase-lag compensation.

to illustrate the principle of phase-lag compensation. If, instead of adding more positive phase to $G_p(j\omega)$ at the high-frequency range as in Fig. 10-6, we attenuate the amplitude of $G_p(j\omega)$ at the low frequency by means of a low-pass filter, a similar stabilization effect can be achieved. The Bode diagram of Fig. 10-7 shows that if the attenuation is affected at a sufficiently low frequency range, the effect on the phase shift due to the phase-lag compensation is negligible at the phase-crossover frequency. Thus the net effect of the compensating scheme is the improvement on the relative stability of the system.

It will be shown later that the transfer function of a simple passive network controller can be represented in the form of

$$G_c(s) = \frac{s + z_1}{s + p_1} \tag{10-7}$$

where for a high-pass filter, $p_1 > z_1$, and for a low-pass filter, $p_1 < z_1$. Using this transfer function, it is now possible to explain the two basic approaches of compensation using the root locus diagram.

The root locus diagram of the closed-loop control system which has the $G_p(s)$ of Eq. (10-5) as its open-loop transfer function is shown in Fig. 10-8. The root loci clearly indicate the instability condition when $K = 100$. Using the phase-lead compensation, with the transfer function of Eq. (10-7), $p_1 > z_1$, the resulting root loci are shown in Fig. 10-8. If the values of z_1 and p_1 are chosen appropriately, the complex roots of the closed-loop system with $K = 100$ may be moved into the left half of the s-plane.

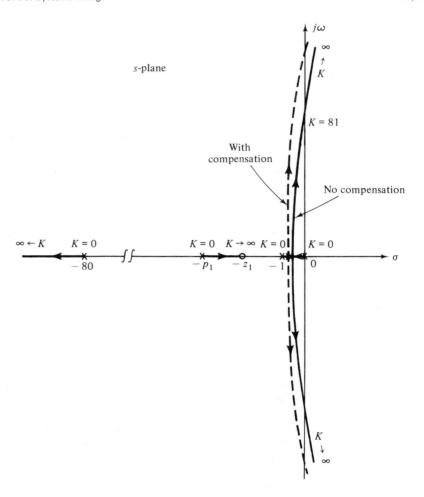

Fig. 10-8. Root locus diagrams of $G_p(s) = K/[s(1 + s)(1 + 0.0125s)]$ and $G_c(s)G_p(s) = [K(s + z_1)]/[s(s + p_1)(1 + s)(1 + 0.0125s)]$ $(p_1 > z_1)$.

The root locus diagram of Fig. 10-9 illustrates the principle of phase-lag compensation in the s-domain. In this case p_1 is chosen to be less than z_1, but for effective control, *the two values are chosen to be very close to each other and are relatively small.* Because of the closeness of z_1 and p_1, the portion of the root loci that represents the dominant eigenvalues of the system are not much affected by the compensation. However, the stability of the compensated system is improved since the points at which $K = 100$ are shifted into the left half of the s-plane. Details of the design using the above mentioned techniques are given in the following sections. One should not be misled that any given control system can always be compensated satisfactorily by either of the two schemes discussed here. It will be shown that, for systems with certain characteristics, satisfactory

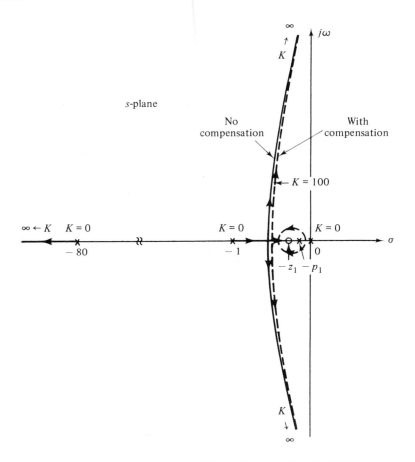

Fig. 10-9. Root locus diagrams of $G_p(s) = K/[s(1 + s)(1 + 0.0125s)]$ and
$G_c(s)G_p(s) = [K(s + z_1)]/[s(s + p_1)(1 + s)(1 + 0.0125s)]$ $(p_1 < z_1)$.

compensation cannot be accomplished by phase-lead networks alone. However, this does not mean that proper compensation may then be achieved by using phase-lag networks, for it is quite common that neither scheme is feasible, and some combination of lead and lag characteristics is needed.

10.3 Phase-Lead Compensation

In Chapter 6 a simple phase-lead compensation scheme using the transfer function $(1 + Ts)$ was described under the name "derivative control." Although this transfer function represents the simplest configuration for a phase-lead scheme, in practice, $(1 + Ts)$ cannot be realized as a transfer function by any passive network. A practical and simple phase-lead network is shown in Fig. 10-10. Although the network may be simplified still further by eliminating R_1, such a

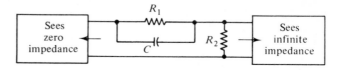

Fig. 10-10. Passive phase-lead network.

network would block dc signals completely and cannot be used as a series compensator for a control system.

The transfer function of the network is derived as follows by assuming that the source impedance which the lead network sees is zero, and the output load impedance is infinite. This assumption is necessary in the derivation of the transfer function of any four-terminal network.

$$\frac{E_2(s)}{E_1(s)} = \frac{R_2 + R_1R_2Cs}{R_1 + R_2 + R_1R_2Cs} \tag{10-8}$$

or

$$\frac{E_2(s)}{E_1(s)} = \frac{R_2}{R_1 + R_2} \frac{1 + R_1Cs}{1 + \dfrac{R_1R_2}{R_1 + R_2}Cs} \tag{10-9}$$

Let

$$a = \frac{R_1 + R_2}{R_2} \qquad a > 1 \tag{10-10}$$

and

$$T = \frac{R_1R_2}{R_1 + R_2}C \tag{10-11}$$

then Eq. (10-9) becomes

$$\frac{E_2(s)}{E_1(s)} = \frac{1}{a}\frac{1 + aTs}{1 + Ts} \qquad a > 1 \tag{10-12}$$

In the following sections, the characteristics of the RC phase-lead network and various design considerations are considered.

Characteristics of the RC Phase-Lead Network

Pole–zero configuration of the RC phase-lead network. As seen from Eq. (10-12), the transfer function of the phase-lead network has a real zero at $s = -1/aT$ and a real pole at $s = -1/T$. These are represented in the s-plane as shown in Fig. 10-11. By varying the values of a and T, the pole and zero may be located at any point on the negative real axis in the s-plane. Since $a > 1$, the zero is always located to the right of the pole, and the distance between them is determined by the constant a.

Polar plot of the RC phase-lead network. When using the RC phase-lead network of Fig. 10-10 as a compensator for control systems, the attenuation, $1/a$, of Eq. (10-12) is overcome by the amplifier gain of the system, so it is neces-

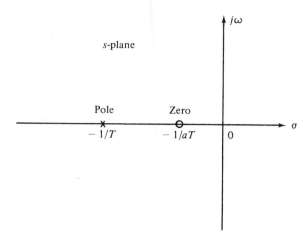

Fig. 10-11. Pole–zero configuration of a passive phase-lead network. $E_2(s)/E_1(s) = (1/a)(1 + aTs)/(1 + Ts)$.

sary only to investigate the transfer function

$$a\frac{E_2(s)}{E_1(s)} = \frac{1 + aTs}{1 + Ts} \tag{10-13}$$

The polar plot of Eq. (10-13) is shown in Fig. 10-12 for several different values of a. For any particular value of a, the angle between the tangent line drawn from the origin to the semicircle and the real axis gives the maximum phase lead ϕ_m which the network can provide. The frequency at the tangent point, ω_m, represents the frequency at which ϕ_m occurs. It is seen that, as a increases, the maximum phase lead, ϕ_m, also increases, approaching a limit of 90° as a approaches infinity. The frequency ω_m decreases with the increase in a.

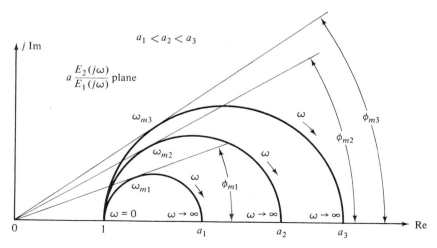

Fig. 10-12. Polar plot of $a[E_2(s)/E_1(s)] = (1 + aTs)/(1 + Ts)$.

Bode plot of the RC phase-lead network. In terms of the Bode plot, the RC network of Fig. 10-10 has two corner frequencies: a positive corner frequency at $\omega = 1/aT$ and a negative corner frequency at $\omega = 1/T$. The Bode diagram of $aE_2(j\omega)/E_1(j\omega)$ is shown in Fig. 10-13.

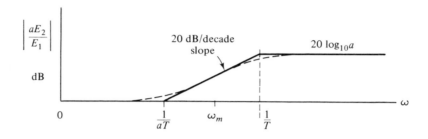

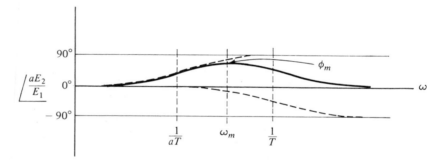

Fig. 10-13. Bode plot of the phase-lead network $aE_2(s)/E_1(s) = (1 + aTs)/(1 + Ts)$ $(a > 1)$.

Analytically, ϕ_m and ω_m may be related to the parameters a and T. Since ω_m is the geometric mean of the two corner frequencies, we can write

$$\log_{10} \omega_m = \frac{1}{2}\left(\log_{10}\frac{1}{aT} + \log_{10}\frac{1}{T}\right) \tag{10-14}$$

Thus

$$\omega_m = \frac{1}{\sqrt{a}\,T} \tag{10-15}$$

To determine the maximum phase lead, ϕ_m, we write the phase of $aE_2(j\omega)/E_1(j\omega)$ as

$$\phi = \text{Arg}\left[a\frac{E_2(j\omega)}{E_1(j\omega)}\right] = \tan^{-1} aT\omega - \tan^{-1} T\omega \tag{10-16}$$

from which we have

$$\tan\phi = \frac{aT\omega - T\omega}{1 + (aT\omega)(T\omega)} \tag{10-17}$$

When $\phi = \phi_m$,

$$\omega = \omega_m = \frac{1}{\sqrt{a}\,T} \tag{10-18}$$

Therefore, Eq. (10-17) gives

$$\tan \phi_m = \frac{(a-1)(1/\sqrt{a})}{1+1} = \frac{a-1}{2\sqrt{a}} \tag{10-19}$$

or

$$\sin \phi_m = \frac{a-1}{a+1} \tag{10-20}$$

This last expression is a very useful relationship in the proper selection of the value of a for compensation design.

Design of Phase-Lead Compensation by the Bode Plot Method

Design of linear control systems in the frequency domain is more preferably carried out with the aid of the Bode plot. The reason is simply because the effect of the compensation network is easily obtained by adding its magnitude and phase curves, respectively, to that of the original process. For the phase-lead compensation employing the RC network of Fig. 10-10, the general outline of the design procedure is as follows. It is assumed that the design specifications simply include a steady-state error and phase margin-versus-gain margin requirements.

1. The magnitude and phase-versus-frequency curves for the uncompensated process $G_p(s)$ are plotted with the gain constant K set according to the steady-state error requirement.
2. The phase margin and the gain margin of the original system are read from the Bode plot, and the additional amount of phase lead needed to provide the required degree of relative stability is determined. From the additional phase lead required, the desired value of ϕ_m is estimated accordingly, and a is calculated from Eq. (10-20).
3. Once a is determined, it is necessary only to obtain the proper value of T, and the design is in principle completed. The important step is to place the corner frequencies of the phase-lead network, $1/aT$, and $1/T$ such that ϕ_m is located at the new gain-crossover frequency.
4. The Bode plot of the compensated system is investigated to check that all performance specifications are met; if not, a new value of ϕ_m must be chosen and the steps repeated.
5. If the specifications are all satisfied, the transfer function of the phase-lead network is established from the values of a and T.

The following numerical example will illustrate the steps involved in the phase-lead design.

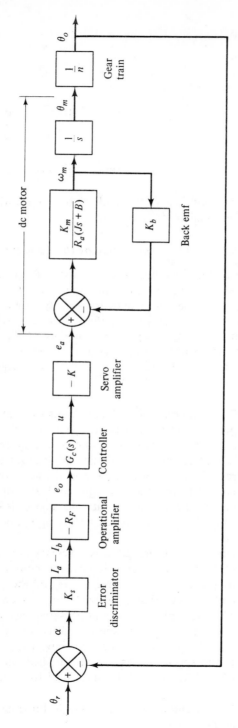

Fig. 10-14. Block diagram of a sun-seeker control system.

EXAMPLE 10-1   Consider the sun-seeker control system described in Section 5.13. The block diagram of the system is shown in Fig. 10-14. It is assumed that for small signal, the error discriminator can be approximated by a linear gain; that is,

$$\frac{I_a - I_b}{\alpha} = K_s = \text{constant} \tag{10-21}$$

The tachometer in the original system has also been eliminated for the present design problem. A block representing the controller, in case it is needed, is inserted between the two amplifiers. The parameters of the system are given as

$$R_F = 10,000 \ \Omega$$

$$K_b = 0.0125 \ \text{volt/rad/sec}$$

$$K_m = 0.0125 \ \text{newton-m/amp}$$

$$R_a = 6.25 \ \Omega$$

$$J = 10^{-6} \ \text{kg-m}^2$$

$$K_s = 0.1 \ \text{amp/rad}$$

$$K = \text{variable}$$

$$B = 0$$

$$n = 800$$

The open-loop transfer function of the uncompensated system is written

$$\frac{\theta_0(s)}{\alpha(s)} = \frac{K_s R_F K K_m / n}{R_a J s^2 + K_m K_b s} \tag{10-22}$$

Substituting the numerical values of the system parameters, Eq. (10-22) gives

$$\frac{\theta_0(s)}{\alpha(s)} = \frac{2500K}{s(s + 25)} \tag{10-23}$$

The specifications of the system are given as

1. The phase margin of the system should be greater than 45°.
2. The steady-state error of $\alpha(t)$ due to a unit ramp function input should be less than or equal to 0.01 rad per rad/sec of the final steady-state output velocity. In other words, the steady-state error due to a ramp input should be less than or equal to 1 per cent.

The following steps are carried out in the design of the phase-lead compensation:

1. Applying the final-value theorem, we have

$$\lim_{t \to \infty} \alpha(t) = \lim_{s \to 0} s\alpha(s) = \lim_{s \to 0} s\frac{\theta_r(s)}{1 + [\theta_0(s)/\alpha(s)]} \tag{10-24}$$

Since $\theta_r(s) = 1/s^2$, using Eq. (10-23), Eq. (10-24) becomes

$$\lim_{t \to \infty} \alpha(t) = 0.01/K \tag{10-25}$$

Thus we see that if $K = 1$, we have the steady-state error equal to 0.01. However, for this amplifier gain, the damping ratio of the closed-loop system is merely 25 per cent, which corresponds to an overshoot of over

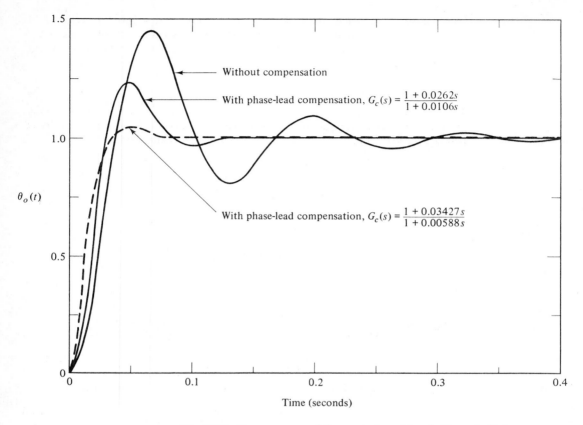

Fig. 10-15. Step responses of the sun-seeker system in Example 10-1.

44.4 per cent. Figure 10-15 shows the unit step response of the closed-loop system with $K = 1$. It is seen that the step response is quite oscillatory.

2. The Bode plot of $\theta_0(s)/\alpha(s)$ of the uncompensated system, with $K = 1$, is sketched as shown in Fig. 10-16.

3. The phase margin of the uncompensated system, read at the gain-crossover frequency, $\omega_c = 50$ rad/sec, is $25°$. Since the phase margin is less than the desired value of $45°$, more phase lead should be added to the open-loop system.

4. Let us choose to use the phase-lead network of Fig. 10-10 as the controller. Then, the transfer function for the controller of Fig. 10-14 is

$$\frac{U(s)}{E_0(s)} = G_c(s) = \frac{1 + aTs}{1 + Ts} \tag{10-26}$$

As mentioned earlier, the attenuation, $1/a$, accompanied by the phase-lead network is assumed to be absorbed by the other amplifier gains.

Since the desired phase margin is $45°$ and the uncompensated system has a phase margin of $25°$, the phase-lead network must provide the additional $20°$ in the vicinity of the gain-crossover frequency. However, by inserting the phase-lead network, the magnitude curve of the Bode plot is also affected in

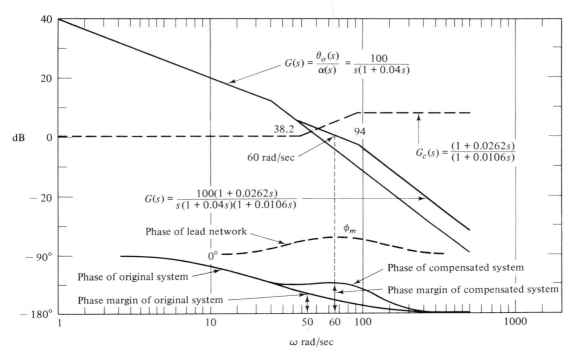

Fig. 10-16. Bode plots of compensated and uncompensated systems in Example 10-1.

such a way that the gain-crossover frequency is shifted to a higher frequency. Although it is a simple matter to adjust the corner frequencies, $1/aT$ and $1/T$, so that the maximum phase of the network, ϕ_m, falls exactly at the new gain-crossover frequency, the original phase curve at this point is no longer 25°, and could be considerably less. This represents one of the main difficulties in the phase-lead design. In fact, if the phase of the uncompensated system decreases rapidly with increasing frequency near the gain-crossover frequency, phase-lead compensation may become ineffective.

In view of the above mentioned difficulty, when estimating the necessary amount of phase lead, it is essential to include some safety margin to account for the inevitable phase dropoff. Therefore, in the present design, instead of selecting a ϕ_m of 20°, we let ϕ_m be 25°. Using Eq. (10-20), we have

$$\sin \phi_m = \sin 25° = 0.422 = \frac{a-1}{a+1} \qquad (10\text{-}27)$$

from which we get

$$a = 2.46 \qquad (10\text{-}28)$$

5. To determine the proper location of the two corner frequencies, $1/aT$ and $1/T$, it is known from Eq. (10-18) that the maximum phase lead ϕ_m occurs at the geometrical mean of the corners. To achieve the maximum phase margin with the value of a already determined, ϕ_m should occur at the new gain-crossover frequency ω'_c, which is not known. Thus the problem now

is to locate the two corner frequencies so that ϕ_m occurs at ω_c'. This may be accomplished graphically as follows:

(a) The zero-frequency attenuation of the phase-lead network is calculated:

$$20 \log_{10} a = 20 \log_{10} 2.46 = 7.82 \text{ dB} \qquad (10\text{-}29)$$

(b) The geometric mean ω_m of the two corner frequencies $1/aT$ and $1/T$ should be located at the frequency at which the magnitude of the uncompensated transfer function $\theta_0(j\omega)/\alpha(j\omega)$ in decibels is equal to the negative value in decibels of one half of this attenuation. This way, the magnitude plot of the compensated transfer function will pass through the 0-dB axis at $\omega = \omega_m$. Thus ω_m should be located at the frequency where

$$\left| \frac{\theta_0(j\omega)}{\alpha(j\omega)} \right| = \frac{-7.82}{2} = -3.91 \text{ dB} \qquad (10\text{-}30)$$

From Fig. 10-16, this frequency is found to be $\omega_m = 60$ rad/sec. Now using Eq. (10-15), we have

$$\frac{1}{T} = \sqrt{a} \, \omega_m = \sqrt{2.46} \times 60 = 94 \text{ rad/sec} \qquad (10\text{-}31)$$

Then

$$\frac{1}{aT} = 38.2 \text{ rad/sec} \qquad (10\text{-}32)$$

The parameters of the phase-lead controller are now determined. Figure 10-16 shows that the phase margin of the compensated system is approximately 48°. The transfer function of the phase-lead network is

$$\frac{U(s)}{E_0(s)} = \frac{1}{a} \frac{1 + aTs}{1 + Ts} = \frac{1}{2.46} \frac{1 + 0.0262s}{1 + 0.0106s} \qquad (10\text{-}33)$$

Since it is assumed that the amplifier gains are increased by a factor of 2.46, the open-loop transfer function of the compensated sun-seeker system becomes

$$\frac{\theta_0(s)}{\alpha(s)} = \frac{6150(s + 38.2)}{s(s + 25)(s + 94)} \qquad (10\text{-}34)$$

In Fig. 10-17 the magnitude and phase of the original and the compensated systems are plotted on the Nichols chart. These plots are obtained by taking the data directly from the Bode plot of Fig. 10-16. From the Nichols chart, the resonance peak, M_p, of the uncompensated system is found to be 1.88, or 5.5 dB. The value of M_p with compensation is 1.175, or 1.4 dB. One more important point is that the resonant frequency of the system is decreased from 50 rad/sec to approximately 37 rad/sec, but the bandwidth is increased from 70 rad/sec to 110 rad/sec.

The unit step response of the compensated system is shown in Fig. 10-15. Note that the response of the system with the phase-lead controller is far less oscillatory than that of the original system. The overshoot is reduced from 44.4 per cent to 24.5 per cent, and the rise time is also reduced. The reduction of the rise time is due to the increase of the bandwidth by the phase-lead controller. On the other hand, excessive bandwidth may be objectionable in certain systems where noise and disturbance signals may be critical.

In the present design problem, we notice that a specification of 45° for the phase margin yields an overshoot of 24.5 per cent in the step response. To demonstrate the capability of the phase-lead controller, we select a to be 5.828. The resulting controller

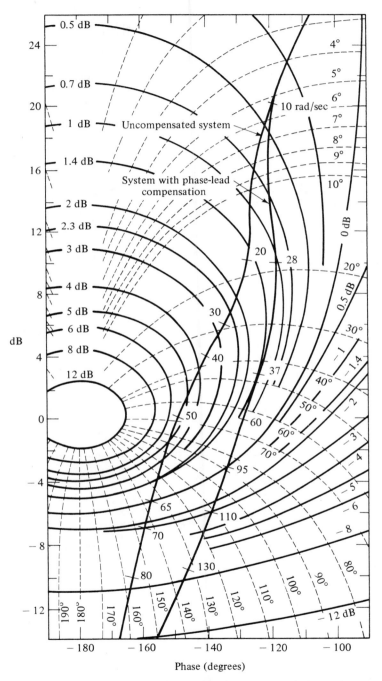

Fig. 10-17. Plots of $G(s)$ in Nichols chart for the system in Example 10-1.

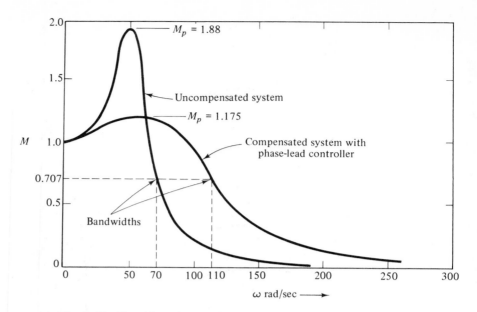

Fig. 10-18. Closed-loop frequency responses of the sun-seeker system in Example 10-1.

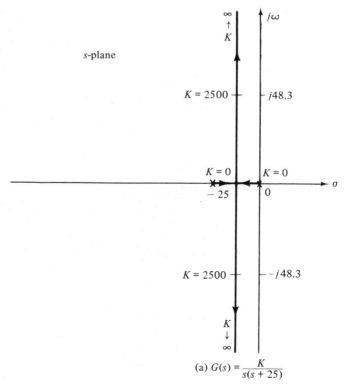

(a) $G(s) = \dfrac{K}{s(s + 25)}$

Fig. 10-19. Root locus diagrams of the sun-seeker system in Example 10-1. (a) Uncompensated system.

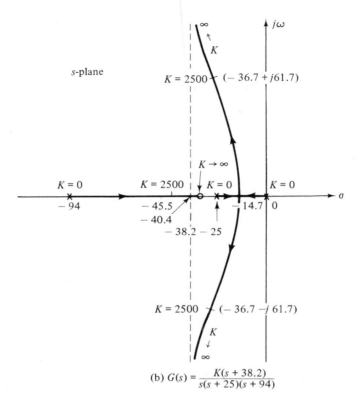

(b) $G(s) = \dfrac{K(s + 38.2)}{s(s + 25)(s + 94)}$

Fig. 10-19 (Cont.). (b) Compensated system with phase-lead controller.

has the transfer function

$$G_c(s) = \frac{1}{5.828} \frac{1 + 0.03427s}{1 + 0.00588s} \tag{10-35}$$

The unit step response of the compensated system is plotted as shown in Fig. 10-15. In this case the rise time is shorter still, and the peak overshoot is reduced to 7.7 per cent.

Using the magnitude-versus-phase plots and the Nichols chart of Fig. 10-17, the closed-loop frequency responses for the sun-seeker system, before and after compensation, are plotted as shown in Fig. 10-18.

To show the effects of the phase-lead compensation, the root locus diagrams of the system, before and after compensation, are shown in Fig. 10-19. It is clear from these diagrams that the phase-lead compensation has the effect of bending the complex root loci toward the left, thus improving the stability of the system. The eigenvalues of the compensated closed-loop system are at $s = -45.5$, $-36.7 + j61.7$, and $-36.7 - j61.7$. Therefore, the damping ratio of the complex eigenvalues of the compensated system is approximately 50 per cent.

In general, it is difficult to visualize how the phase-lead design can be carried out with the root locus method independently. The root loci of Fig. 10-19 are drawn based on the results of the Bode plot design. In the following section, we shall illustrate a procedure of designing phase-lead compensation using the root contour method.

Design of Phase-Lead Compensation by the Root Locus Method

The root contour method can be used for the design of control systems. Let us use the sun-seeker system studied in Example 10-1 to illustrate the design procedure.

The open-loop transfer function of the uncompensated system is

$$\frac{\theta_0(s)}{\alpha(s)} = G(s) = \frac{2500}{s(s + 25)} \tag{10-36}$$

The characteristic equation of the closed-loop system is

$$s^2 + 25s + 2500 = 0 \tag{10-37}$$

and the eigenvalues are $s_1 = -12.5 + j48.4$ and $s_2 = -12.5 - j48.4$.

The open-loop transfer function of the system with phase-lead compensation is written

$$G(s) = \frac{2500(1 + aTs)}{s(s + 25)(1 + Ts)} \tag{10-38}$$

The problem is to determine the values of a and T so that the system will perform as desired. In this case, rather than using the frequency-domain specifications, such as the phase margin and gain margin, it is more convenient to specify the relative positions of the complex eigenvalues.

To begin with the root contour design, we first set a to zero in Eq. (10-38). Then, the characteristic equation of the compensated system becomes

$$s(s + 25)(1 + Ts) + 2500 = 0 \tag{10-39}$$

Since T is the variable parameter, we divide both sides of Eq. (10-39) by the terms that do not contain T. We have

$$1 + \frac{Ts^2(s + 25)}{s^2 + 25s + 2500} = 0 \tag{10-40}$$

This equation is of the form of $1 + G_1(s) = 0$, where $G_1(s)$ is an equivalent transfer function that can be used to study the root loci of Eq. (10-39). The root contour of Eq. (10-39) is drawn as shown in Fig. 10-20, starting with the poles and zeros of $G_1(s)$. Of significance is that the poles of $G_1(s)$ are the eigenvalues of the system when $a = 0$ and $T = 0$. As can be seen from the figure, the factor $1 + Ts$ in the denominator of Eq. (10-38) alone would not improve the system performance at all. In fact, the eigenvalues of the system are moved toward the right half of the s-plane, and the system becomes unstable when T is greater than 0.0133. To achieve the full effect of the phase-lead compensation, we must restore the value of a in Eq. (10-38). The characteristic equation of the compensated system now becomes

$$s(s + 25)(1 + Ts) + 2500(1 + aTs) = 0 \tag{10-41}$$

Now we must consider the effect of varying a while keeping T constant. This is accomplished by dividing both sides of Eq. (10-41) by the terms that do not contain a. We have

$$1 + \frac{2500aTs}{s(s + 25)(1 + Ts) + 2500} = 0 \tag{10-42}$$

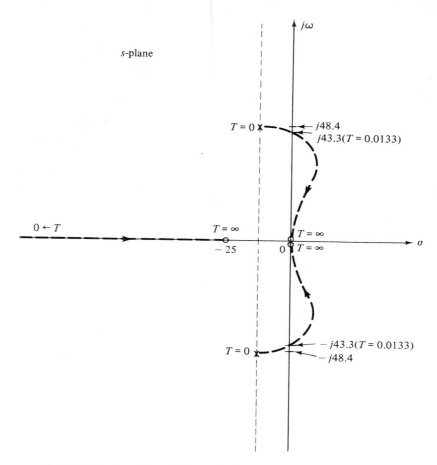

Fig. 10-20. Root contours of the sun-seeker system of Example 10-1 with $a = 0$, and T varies from 0 to ∞.

This equation is again of the form of $1 + G_2(s) = 0$, and for a given T, the root loci of Eq. (10-41) can be obtained based upon the poles and zeros of

$$G_2(s) = \frac{2500aTs}{s(s + 25)(1 + Ts) + 2500} \tag{10-43}$$

Notice also that the denominator of $G_2(s)$ is identical to the left side of Eq. (10-39), which means that the poles of $G_2(s)$ must lie on the root contours of Fig. 10-20, for a given T. In other words, the root contours of Eq. (10-41) as a varies must start from the points on the trajectories of Fig. 10-20. These root contours end at $s = 0$, ∞, ∞, which are the zeros of $G_2(s)$. The complete root contours of the system with phase-lead compensation are now sketched in Fig. 10-21.

From the root contours of Fig. 10-21, we can see that for effective phase-lead compensation, the value of T should be small. For large values of T, the bandwidth of the system increases very rapidly as a increases, while very little

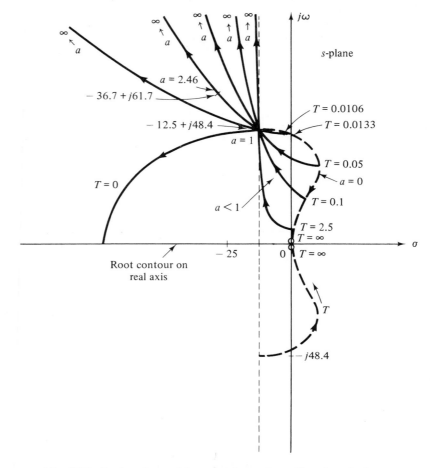

Fig. 10-21. Root contours of the sun-seeker system with a phase-lead controller, $G_c(s) = (1 + aTs)/(1 + Ts)$, Example 10-1.

improvement is made on the damping of the system. We must remember that these remarks are made with respect to the phase-lead design that corresponds to values of a greater than unity.

Effects and Limitations of Phase-Lead Compensation

From the results of the last illustrative example, we may summarize the general effects of phase-lead compensation on the performance of control systems as follows:

1. The phase of the open-loop transfer function in the vicinity of the gain-crossover frequency is increased. Thus the phase margin is usually improved.
2. The slope of the magnitude curve representing the magnitude of the open-loop transfer function is reduced at the gain-crossover fre-

quency. This usually corresponds to an improvement in the relative stability of the system. In other words, the phase and gain margins are improved.

3. The bandwidths of the open-loop system and the closed-loop system are increased.
4. The overshoot of the step response is reduced.
5. The steady-state error of the system is not affected.

It was mentioned earlier that certain types of systems cannot be effectively compensated satisfactorily by phase-lead compensation. The sun-seeker system studied in Example 10-1 happens to be one for which phase-lead control is effective and practical. In general, successful application of phase-lead compensation is hinged upon the following considerations:

1. Bandwidth considerations: If the original system is unstable, the additional phase lead required to obtain a certain desirable phase margin may be excessive. This requires a relatively large value of a in Eq. (10-12), which, as a result, will give rise to a large bandwidth for the compensated system, and the transmission of noise may become objectionable. Also, if the value of a becomes too large, the values of the network elements of the phase-lead network may become disproportionate, such as a very large capacitor. Therefore, in practice, the value of a is seldom chosen greater than 15. If a larger value of a is justified, sometimes two or more phase-lead controllers are connected in cascade to achieve the large phase lead.
2. If the original system is unstable or has low stability margin, the phase plot of the open-loop transfer function has a steep negative slope near the gain-crossover frequency. In other words, the phase decreases rapidly near the gain crossover. Under this condition, phase-lead compensation usually becomes ineffective because the additional phase lead at the new gain crossover is added to a much smaller phase angle than that at the old gain crossover. The desired phase margin may be realized only by using a very large value of a. However, the resulting system may still be unsatisfactory because a portion of the phase curve may still be below the 180° axis, which corresponds to a conditionally stable system.

In general, the following situations may also cause the phase to change rapidly near the gain-crossover frequency:

1. The open-loop transfer function has two or more poles that are close to each other and are close to the gain-crossover frequency.
2. The open-loop transfer function has one or more pairs of complex conjugate poles near the gain-crossover frequency.

The following example will illustrate a typical situation under which phase-lead compensation is ineffective.

EXAMPLE 10-2　　Let the open-loop transfer function of a control system with unity feedback be

$$G_p(s) = \frac{K}{s(1 + 0.1s)(1 + 0.2s)} \qquad (10\text{-}44)$$

It is desired that the system satisfies the following performance specifications:

1. $K_v = 100$; or the steady-state error of the system due to a unit ramp function input is 0.01 in magnitude.
2. Phase margin $\geq 40°$.

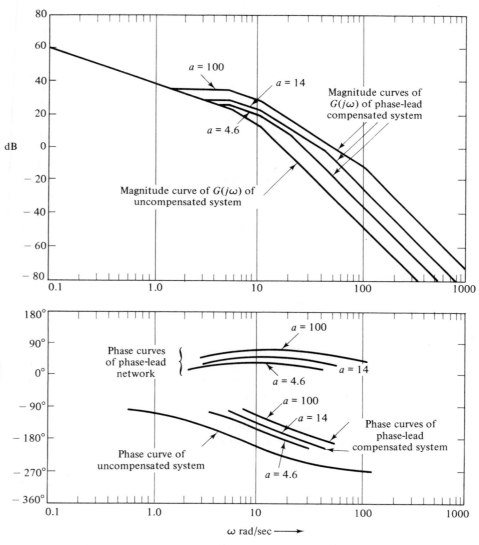

Fig. 10-22. Bode plots of $G_p(s) = 100/[s(1 + 0.1s)(1 + 0.2s)]$ and the effects of using phase-lead compensation.

From the steady-state error requirement, we set $K = 100$. The Bode plot of $G_p(s)$ when $K = 100$ is shown in Fig. 10-22. As observed from this Bode plot, the phase margin of the system is approximately $-40°$, which means that the system is unstable. In fact, the system is unstable for all values of K greater than 15. The rapid decrease of phase at the gain-crossover frequency, $\omega_c = 17$ rad/sec, implies that the phase-lead compensation may be ineffective for this case. To illustrate the point, the phase-lead network of Fig. 10-10 and Eq. (10-12) with $a = 4.6$, 14, and 100, respectively, is used to compensate the system. Figure 10-22 illustrates the effects of phase-lead compensation when the values of T are chosen according to the procedure described in Example 10-1.

It is clearly shown in Fig. 10-22 that as more phase lead is being added, the gain-crossover frequency is also being pushed to a higher value. Therefore, for this case, in which the uncompensated system is very unstable at the outset, it may be impossible to realize a phase margin of 40° by the phase-lead network of Fig. 10-10.

In a similar fashion, we may use the root contour method to illustrate why the phase-lead compensation is ineffective in this case of a highly unstable system. Figure 10-23 shows the root locus diagram of the uncompensated system with the process described by the transfer function of Eq. (10-44). When $K = 100$, the two complex eigenvalues of the closed-loop system are at $s = 3.8 + j14.4$ and $3.8 - j14.4$.

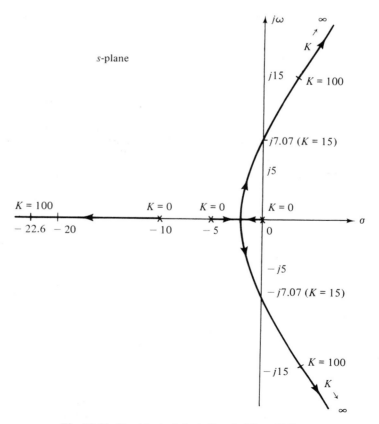

Fig. 10-23. Root loci of $s(s + 5)(s + 10) + 50K = 0$.

Let the controller be represented by

$$G_c(s) = \frac{1 + aTs}{1 + Ts} \qquad a > 1 \tag{10-45}$$

Then, the open-loop transfer function becomes

$$G_c(s)G_p(s) = \frac{5000(1 + aTs)}{s(s + 5)(s + 10)(1 + Ts)} \tag{10-46}$$

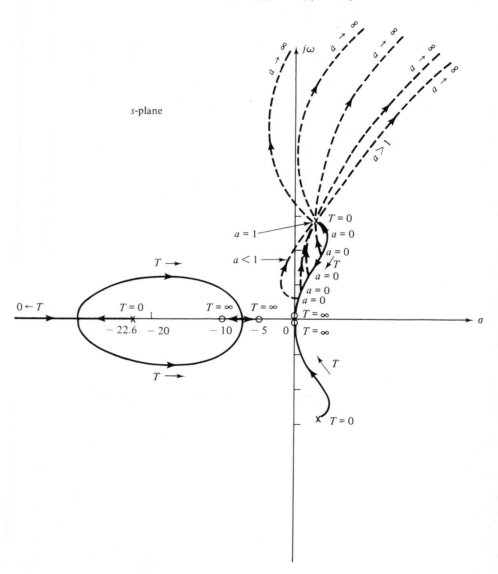

Fig. 10-24. Root contours of $s(s + 5)(s + 10)(1 + Ts) + 5000 = 0$, and of $s(s + 5)(s + 10)(1 + Ts) + 5000(1 + aTs) = 0$.

First, we set $a = 0$ while we vary T from zero to infinity. The characteristic equation becomes

$$s(s + 5)(s + 10)(1 + Ts) + 5000 = 0 \qquad (10\text{-}47)$$

The root contours of Eq. (10-47) are constructed from the pole–zero configuration of

$$G_1(s) = \frac{Ts^2(s + 5)(s + 10)}{s(s + 5)(s + 10) + 5000} \qquad (10\text{-}48)$$

Thus, in Fig. 10-24 the poles of $G_1(s)$ are labeled as $T = 0$ points, and the zeros of $G_1(s)$ are points at which $T = \infty$.

Next, we restore the value of a in Eq. (10-46), and the root contours of Eq. (10-47) become the trajectories on which $a = 0$. In other words, the characteristic equation of the overall system is written

$$s(s + 5)(s + 10)(1 + Ts) + 5000(1 + aTs) = 0 \qquad (10\text{-}49)$$

When a is considered the variable parameter, we divide both sides of Eq. (10-49) by the terms that do not contain a; we have

$$1 + \frac{5000aTs}{s(s + 5)(s + 10)(1 + Ts) + 5000} = 0 \qquad (10\text{-}50)$$

Thus, as we have stated, the root contours with a varying, start ($a = 0$) at the poles of

$$G_2(s) = \frac{5000aTs}{s(s + 5)(s + 10)(1 + Ts) + 5000} \qquad (10\text{-}51)$$

The dominant part of the root loci of Eq. (10-49) is sketched in Fig. 10-24. Notice that, since for phase-lead compensation the value of a is limited to greater than 1, the root contours that correspond to this range are mostly in the right-half plane. It is apparent from this root contour plot that the ineffectiveness of phase-lead compensation, in this case, may be attributed to the eigenvalues of the uncompensated system being in the right-half plane.

We shall show in the following that for the system considered in this example, it is far more effective to compensate it by a phase-lag network, or by means of an auxiliary feedback loop.

10.4 Phase-Lag Compensation

In contrast to using a high-pass filter as a compensator for control systems, we may use a low-pass filter for similar purposes. Figure 10-25 shows a simple RC network that may be used for the low-pass or phase-lag compensation of control systems. If we assume that the input impedance of the network is zero and that the output impedance which the network sees is infinite, the transfer function of the network is

$$\frac{E_2(s)}{E_1(s)} = \frac{1 + R_2Cs}{1 + (R_1 + R_2)Cs} \qquad (10\text{-}52)$$

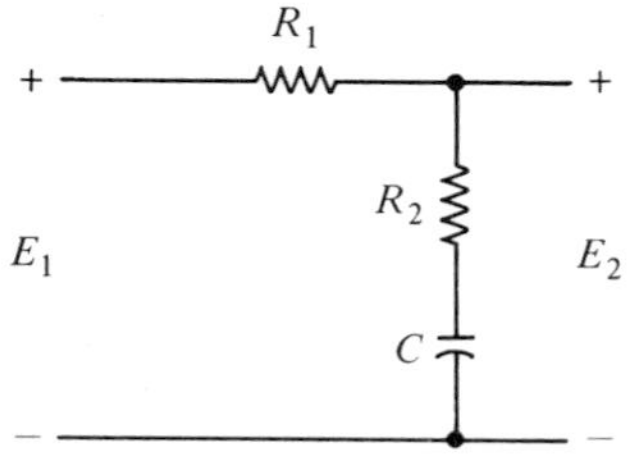

Let

$$aT = R_2C \qquad (10\text{-}53)$$

and

$$a = \frac{R_2}{R_1 + R_2} \qquad a < 1 \qquad (10\text{-}54)$$

Fig. 10-25. RC phase-lag network.

Equation (10-52) becomes

$$\frac{E_2(s)}{E_1(s)} = \frac{1 + aTs}{1 + Ts} \qquad a < 1 \qquad (10\text{-}55)$$

Characteristics of the RC Phase-Lag Network

Pole–zero configuration of the RC phase-lag network. The transfer function of the phase-lag network of Fig. 10-25 has a real zero at $s = -1/aT$ and a real pole at $s = -1/T$. As shown in Fig. 10-26, since a is less than unity, the pole is always located to the right of the zero, and the distance between them is determined by a.

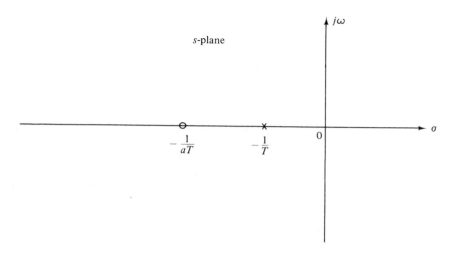

Fig. 10-26. Pole–zero configuration of the transfer function $(1 + aTs)/(1 + Ts)$, $a < 1$, of a phase-lag network.

Polar plot of the RC phase-lag network. As seen from Eq. (10-52), the transfer function of the phase-lag network does not have any attenuation at zero frequency, contrary to the case of the phase-lead network.

When we let $s = j\omega$, Eq. (10-55) becomes

$$\frac{E_2(j\omega)}{E_1(j\omega)} = \frac{1 + j\omega aT}{1 + j\omega T} \qquad a < 1 \qquad (10\text{-}56)$$

The polar plot of this transfer function is shown in Fig. 10-27 for three values of a, $1 > a_1 > a_2 > a_3$. Just as in the case of the phase-lead network, for any value of a ($a < 1$), the angle between the tangent line drawn from the origin to the semicircle and the real axis gives the maximum phase lag ϕ_m ($\phi_m < 0°$) of the network. As the value of a decreases, the maximum phase lag ϕ_m becomes more negative, approaching the limit of $-90°$ as a approaches zero. As the value of a decreases, the frequency at which ϕ_m occurs, ω_m, increases; that is, in Fig. 10-26, $\omega_{m3} > \omega_{m2} > \omega_{m1}$.

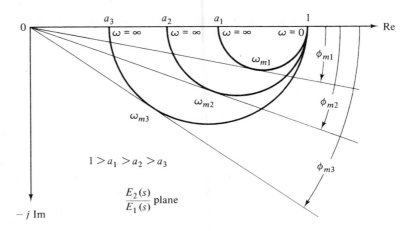

Fig. 10-27. Polar plots of $E_2(s)/E_1(s) = (1 + aTs)/(1 + Ts)$ $(a < 1)$.

Bode plot of the RC phase-lag network. The Bode plot of the transfer function of Eq. (10-56) is shown in Fig. 10-28. The magnitude plot has a positive corner frequency at $\omega = 1/aT$, and a negative corner frequency at $\omega = 1/T$. Since the transfer functions of the phase-lead and phase-lag networks are identical in form except for the zero-frequency attenuation and the value of a, it can

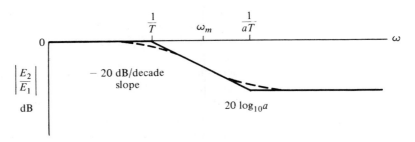

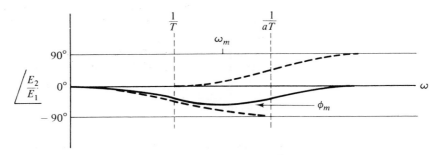

Fig. 10-28. Bode plot of the phase-lag network $E_2(s)/E_1(s) = (1 + aTs)/(1 + Ts)$ $(a < 1)$.

readily be shown that the maximum phase lag ϕ_m of the phase curve of Fig. 10-28 satisfies the following relation:

$$\sin \phi_m = \frac{a-1}{a+1} \tag{10-57}$$

Design of Phase-Lag Compensation by Bode Plot

Unlike the design of phase-lead compensation, which utilizes the maximum phase lead of the network, the design of phase-lag compensation utilizes the attenuation of the network at the high frequencies. It was pointed out earlier that, for phase-lead compensation, the function of the network is to increase the phase in the vicinity of the gain-crossover frequency while keeping the magnitude curve of the Bode plot relatively unchanged near that frequency. However, usually in phase-lead design, the gain crossover frequency is increased because of the phase-lead network, and the design is essentially the finding of a compromise between the increase in bandwidth and the desired amount of relative stability (phase margin or gain margin). In phase-lag compensation, however, the objective is to move the gain-crossover frequency to a lower frequency while keeping the phase curve of the Bode plot relatively unchanged at the gain-crossover frequency.

The design procedure for phase-lag compensation using the Bode plot is outlined as follows:

1. The Bode plot of the open-loop transfer function of the uncompensated system is made. The open-loop gain of the system is set according to the steady-state error requirement.

2. The phase margin and the gain margin of the uncompensated system are determined from the Bode plot. For a certain specified phase margin, the frequency corresponding to this phase margin is found on the Bode plot. The magnitude plot must pass through the 0-dB axis at this frequency in order to realize the desired phase margin. In other words, the gain-crossover frequency of the compensated system must be located at the point where the specified phase margin is found.

3. To bring the magnitude curve down to 0 dB at the new prescribed gain-crossover frequency, ω_c', the phase-lag network must provide the amount of attenuation equal to the gain of the magnitude curve at the new gain-crossover frequency. In other words, let the open-loop transfer function of the uncompensated system be $G_p(s)$; then

$$|G_p(j\omega_c')| = -20 \log_{10} a \quad \text{dB} \quad a < 1 \tag{10-58}$$

from which

$$a = 10^{-|G_p(j\omega_c')|/20} \quad a < 1 \tag{10-59}$$

4. Once the value of a is determined, it is necessary only to select the proper value of T to complete the design. Up to this point, we have assumed that although the gain-crossover frequency is altered by attenuating the gain at ω_c, the original phase curve is not affected.

This is not possible, however, since any modification of the magnitude curve will bring change to the phase curve, and vice versa. With reference to the phase characteristics of the phase-lag network shown in Fig. 10-28, it is apparent that if the positive corner frequency, $1/aT$, is placed far below the new gain-crossover frequency, ω_c', the phase characteristics of the compensated system will not be appreciably affected near ω_c' by phase-lag compensation. On the other hand, the value of $1/aT$ should not be too much less than ω_c', or the bandwidth of the system will be too low, causing the system to be too sluggish. Usually, as a general guideline, it is recommended that the corner frequency $1/aT$ be placed at a frequency that is approximately 1 decade below the new gain-crossover frequency, ω_c'; that is,

$$\frac{1}{aT} = \frac{\omega_c'}{10} \qquad \text{rad/sec} \tag{10-60}$$

Therefore,

$$\frac{1}{T} = \frac{\omega_c'}{10} a \qquad \text{rad/sec} \tag{10-61}$$

5. The Bode plot of the phase-lag compensated system is investigated to see if the performance specifications are met.
6. If all the design specifications are met, the values of a and T are substituted in Eq. (10-55) to give the desired transfer function of the phase-lag network.

EXAMPLE 10-3   In this example we shall design a phase-lag controller for the sun-seeker system considered in Example 10-1. The open-loop transfer function of the sun-seeker system is given by Eq. (10-23),

$$\frac{\theta_0(s)}{\alpha(s)} = \frac{2500K}{s(s+25)} \tag{10-62}$$

The specifications of the system are repeated as follows:

1. The phase margin of the system should be greater than 45°.
2. The steady-state error of $\alpha(t)$ due to a unit ramp function input should be less than or equal to 0.01 rad/rad/sec of the final steady-state output velocity. This is translated into the requirement of $K \geq 1$.

The Bode plot of $\theta_0(s)/\alpha(s)$ with $K = 1$ is shown in Fig. 10-29. As seen, the phase margin is only 25°.

For a phase-lag compensation, let us choose the network of Fig. 10-25, whose transfer function is

$$\frac{U(s)}{E_0(s)} = G_c(s) = \frac{1 + aTs}{1 + Ts} \qquad a < 1 \tag{10-63}$$

From Fig. 10-29 it is observed that the desired 45° phase margin can be obtained if the gain-crossover frequency ω_c' is at 25 rad/sec. This means that the phase-lag controller must reduce the magnitude of $\theta_0(j\omega)/\alpha(j\omega)$ to 0 dB at $\omega = 25$ rad/sec, while it does not appreciably affect the phase curve in the vicinity of this frequency. Since, actually, a small negative phase shift is still accompanied by the phase-lag network at the new

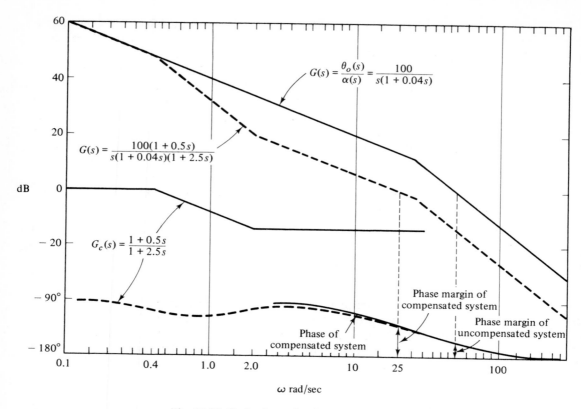

Fig. 10-29. Bode plots of compensated and uncompensated systems in Example 10-3.

gain-crossover frequency, it is a safe measure to choose this new gain-crossover frequency somewhat less than 25 rad/sec, say, 20 rad/sec.

From the magnitude plot of $\theta_0(j\omega)/\alpha(j\omega)$, the value of $|\theta_0(j\omega)/\alpha(j\omega)|$ at $\omega'_c = 20$ rad/sec is 14 dB. This means that the phase-lag controller must provide an attenuation of 14 dB at this frequency, in order to bring the magnitude curve down to 0 dB at $\omega'_c = 20$ rad/sec. Thus, using Eq. (10-59), we have

$$a = 10^{-|\theta_0(j\omega_c')/\alpha(j\omega_c')|/20}$$
$$= 10^{-0.7} = 0.2 \tag{10-64}$$

The value of a indicates the required distance between the two corner frequencies of the phase-lag controller, in order that the attenuation of 14 dB is realized.

In order that the phase lag of the controller does not appreciably affect the phase at the new gain-crossover frequency, we choose the corner frequency $1/aT$ to be at 1 decade below $\omega'_c = 20$ rad/sec. Thus

$$\frac{1}{aT} = \frac{\omega'_c}{10} = \frac{20}{10} = 2 \text{ rad/sec} \tag{10-65}$$

which gives

$$\frac{1}{T} = 0.4 \text{ rad/sec} \tag{10-66}$$

The transfer function of the phase-lag controller is

$$\frac{U(s)}{E_0(s)} = \frac{1 + 0.5s}{1 + 2.5s} \tag{10-67}$$

and the open-loop transfer function of the compensated system is

$$\frac{\theta_0(s)}{\alpha(s)} = \frac{500(s + 2)}{s(s + 0.4)(s + 25)} \tag{10-68}$$

The Bode plot of the open-loop transfer function of the compensated system is shown in Fig. 10-29. We see that the magnitude curve beyond $\omega = 0.4$ rad/sec is attenuated by the phase-lag controller while the low-frequency portion is not affected. In the meantime, the phase curve is not much affected by the phase-lag characteristic near the new gain-crossover frequency, which is at 25 rad/sec. The phase margin of the compensated system is determined from Fig. 10-29 to be about 50°.

The magnitude-versus-phase curves of the uncompensated and the compensated systems are plotted on the Nichols chart, as shown in Fig. 10-30. It is seen that the resonant peak, M_p, of the compensated system is approximately 1.4 dB. The bandwidth of the system is reduced from 70 rad/sec to 32 rad/sec.

The unit step responses of the uncompensated and the compensated systems are shown in Fig. 10-31. The effects of the phase-lag controller are that the overshoot is reduced from 44.4 per cent to 22 per cent, but the rise time is increased considerably. The latter effect is apparently due to the reduction of the bandwidth by the phase-lag controller. Figure 10-31 also gives the step responses of the system when the value of T of the phase-lag controller is changed to 5 and then to 10. It is seen that larger values of T give only slight improvements on the overshoot of the step response. Earlier it was pointed out that the value of T is not critical; when $T = 5$, it is equivalent to setting $1/aT$ at 20 times below the gain-crossover frequency of $\omega_c' = 20$ rad/sec. Similarly, $T = 10$ corresponds to placing $1/aT$ at 40 times below ω_c'.

Phase-Lag Compensation by the Root Locus Method

The design of phase-lag compensation is best illustrated by the root locus diagram of the system considered in Example 10-3, which is shown in Fig. 10-32. In this figure the root loci of the uncompensated system and the compensated system are shown. It is important to note that since the simple pole and zero of the phase-lag controller are placed close together and are very near the origin, they have very little effect on the shape of the original root loci, especially near the region where the eigenvalues should be located to achieve the desired performance. However, the values of K that correspond to similar points on the two root loci are different. For example, when $K = 1$, which gives the desired steady-state response, the eigenvalues of the uncompensated system are $s = -12.5 + j48.4$ and $s = -12.5 - j48.4$, which correspond to a damping ratio of 25 per cent. The comparable points to these eigenvalues on the compensated root loci correspond to $K = 5$, which is 5 times greater. In fact, when $K = 1$ on the root loci of the compensated system, the eigenvalues are at $s = -11.6 + j18$ and $s = -11.6 - j18$, which correspond to a damping ratio (for these complex roots) of 56.2 per cent.

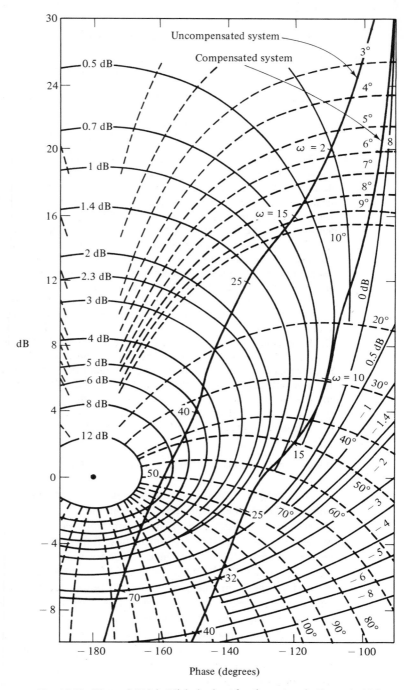

Fig. 10-30. Plots of $G(s)$ in Nichols chart for the system in Example 10-3.

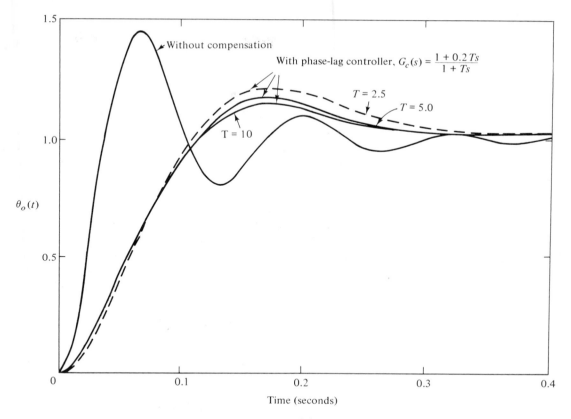

Fig. 10-31. Step responses of the sun-seeker system in Example 10-3.

It is simple to show that the values of K on the root loci of the compensated system at points relatively far away from the origin are 5 times greater than those values of K at similar points on the root loci of the uncompensated system. For instance, at the root $s = -11.6 + j18$ on the compensated root loci of Fig. 10-32, the value of K is 1; the comparable point on the uncompensated root loci is $s = -12.5 + j18.4$, and the value of K at that point is 0.2. The Bode diagram result of Fig. 10-29 already reflects this relation on the values of K, since the phase-lag network hardly altered the phase plot near the gain-crossover frequency, but reduced the gain by 14 dB, which is a factor of 5. Another explanation of the relation between the values of K of the uncompensated loci and those of the compensated may be obtained by referring to the open-loop transfer functions of the two systems. From Eq. (10-62), the value of K at a point s_1 on the root loci for the uncompensated system is written

$$|K| = \frac{|s_1||s_1 + 25|}{2500} \qquad (10\text{-}69)$$

Assuming that a point s_1 is on the compensated root loci and is far from the pole and zero of the phase-lag network at $s = -0.4$ and $s = -2$, respectively, the value of K at s_1 is given by

$$|K| = \frac{|s_1||s_1 + 0.4||s_1 + 25|}{500|s_1 + 2|} \tag{10-70}$$

or

$$|K| \cong \frac{|s_1||s_1 + 25|}{500}$$

since the distance from s_1 to -0.4 will be approximately the same as that from s_1 to -2. This argument also points to the fact that the exact location of the pole and the zero of the phase-lag network is not significant as long as they are close to the origin, and that the distance between the pole and the zero is a fixed desired quantity. In the case of the last example, the ratio between $1/aT$ and $1/T$ is 5.

Based on the discussions given above, we may outline a root locus design procedure for the phase-lag design of control systems as follows:

Since the design will be carried out in the s-plane, the specifications on the transient response or the relative stability should be given in terms of the damping ratio of the dominant roots, and other quantities such as rise time, bandwidth, and maximum overshoot, which can be correlated with the location of the eigenvalues.

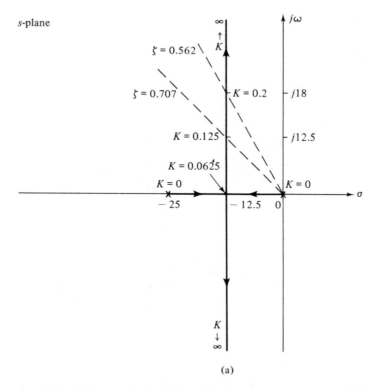

(a)

Fig. 10-32. (a) Root loci of $G(s) = 2500K/[s(s + 25)]$ of the sun-seeker system.

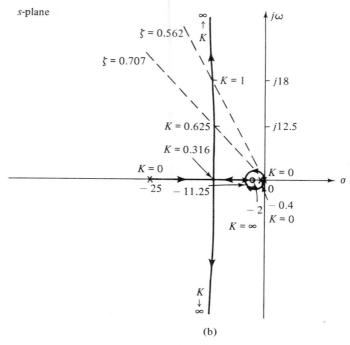

(b)

Fig. 10-32 (Cont.). (b) Root loci of $G(s) = 500(s + 2)/[s(s + 25)(s + 0.4)]$ of the sun-seeker system.

1. Sketch the root loci of the characteristic equation of the uncompensated system.
2. Determine on these root loci where the desired eigenvalues should be located to achieve the desired relative stability of the system. Find the value of K that corresponds to these eigenvalues.
3. Compare the value of K required for steady-state performance and the K found in the last step. The ratio of these two Ks is a ($a < 1$), which is the desired ratio between the pole and the zero of the phase-lag controller.
4. The exact value of T is not critical as long as it is relatively large. We may choose the value of $1/aT$ to be many orders of magnitudes smaller than the smallest pole of the process transfer function.

Let us repeat the design of the system in Example 10-3 by means of the root locus method just outlined. Instead of using the phase-margin specification, we require that the damping ratio of the dominant eigenvalues of the closed-loop system be approximately 56.2 per cent. This value of the damping ratio is chosen so that we can compare the result of the root locus design with that which was obtained independently by the Bode diagram method. The root locus diagram of the original uncompensated system is drawn as shown in Fig. 10-32(a), based on the pole–zero configuration of Eq. (10-62). The steady-state performance specification requires that the value of K should be equal to 1 or greater. From the root loci of Fig. 10-32(a), it is found that a damping ratio of 56.2 per cent may be attained by setting K of the original system equal to 0.2. Therefore, the ratio of the two values of K is

$$a = \frac{0.2}{1} = \frac{1}{5} \tag{10-71}$$

The desired relative stability is attained by setting K to 0.2 in the open-loop transfer function of Eq. (10-62) to give

$$\frac{\theta_0(s)}{\alpha(s)} = \frac{500}{s(s + 25)} \tag{10-72}$$

The open-loop transfer function of the compensated system is given by

$$\frac{\theta_0(s)}{\alpha(s)} = \frac{2500K(1 + aTs)}{s(s + 25)(1 + Ts)}$$

$$= \frac{2500aK(s + 1/aT)}{s(s + 25)(s + 1/T)} \tag{10-73}$$

If the values of aT and T are chosen to be large, Eq. (10-73) is approximately

$$\frac{\theta_0(s)}{\alpha(s)} \simeq \frac{2500aK}{s(s + 25)} \tag{10-74}$$

from the transient standpoint. Since K is necessarily equal to unity, to have the right sides of Eqs. (10-72) and (10-74) equal to each other, $a = 1/5$, as is already determined in Eq. (10-71). Theoretically, the value of T can be arbitrarily large. However, in order that the bandwidth of the closed-loop system is not too small, we must not make T too large. By setting $T = 2.5$, the root loci of the compensated system will be identical to that of Fig. 10-32(b).

As an alternative, the root locus design can also be carried out by means of the root contour method. The root contour method was applied to the sunseeker system earlier in the phase-lead design. The design carried out by Eqs. (10-38) through (10-43) and Figs. 10-20 and 10-21 is still valid for phase-lag control, except that in the present case $a < 1$. Thus in Fig. 10-21 only the portions of the root contours that correspond to $a < 1$ are applicable for phase-lag compensation. These root contours show that for effective phase-lag control, the value of T should be relatively large. In Fig. 10-33 we illustrate further that the complex eigenvalues of the closed-loop system are rather insensitive to the value of T when the latter is relatively large.

EXAMPLE 10-4   Consider the system given in Example 10-2 for which the phase-lead compensation is ineffective. The open-loop transfer function of the original system and the performance specifications are repeated as follows:

$$G_p(s) = \frac{K}{s(1 + 0.1s)(1 + 0.2s)} \tag{10-75}$$

The performance specifications are

    1. $K_v = 100 \text{ sec}^{-1}$.
    2. Phase margin $\geq 40°$.

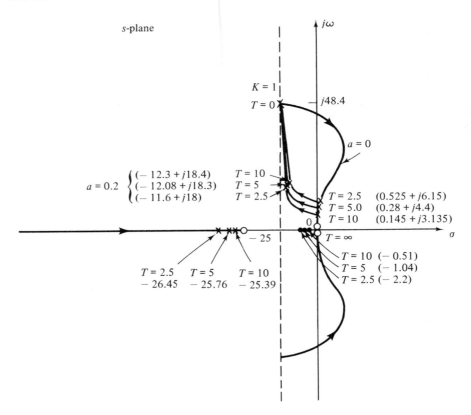

Fig. 10-33. Root contours of $s(s + 25)(1 + Ts) + 2500K(1 + aTs) = 0$; $K = 1$.

The phase-lag design procedure is as follows:

1. The Bode plot of $G_p(s)$ is made as shown in Fig. 10-34 for $K = 100$.
2. The phase margin at the gain-crossover frequency, $\omega_c = 17$ rad/sec, is approximately $-45°$, and the closed-loop system is unstable.
3. The desired phase margin is $40°$; and from Fig. 10-34 this can be realized if the gain-crossover frequency is moved to approximately 4 rad/sec. This means that the phase-lag controller must reduce the magnitude of $G_p(j\omega)$ to 0 dB, while it does not affect the phase curve at this new gain crossover frequency, ω_c'. Since actually a small negative phase is still introduced by the phase-lag controller at ω_c', it is a safe measure to choose the new gain-crossover frequency somewhat less then 4 rad/sec, say at 3.5 rad/sec. As an alternative, we may select a larger phase margin of $45°$.
4. From the Bode plot, the magnitude of $G_p(j\omega)$ at $\omega_c' = 3.5$ rad/sec is 30 dB, which means that the controller must introduce 30 dB of attenuation at this frequency, in order to bring down the magnitude curve of $G_p(j\omega)$ to 0 dB. Thus, from Eq. (10-59),

$$a = 10^{-|G_p(j\omega_c')|/20} = 10^{-1.5} = 0.032 \qquad (10-76)$$

This equation implies that the two corners of the phase-lag controller

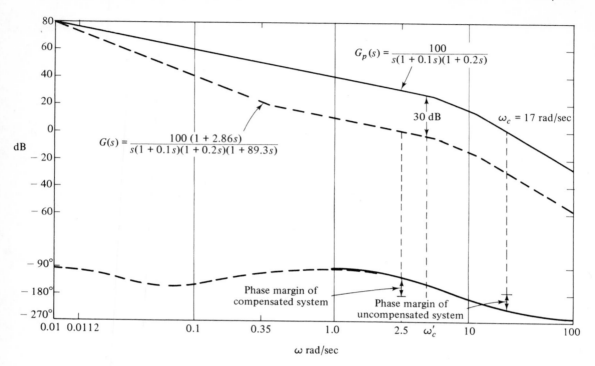

Fig. 10-34. Bode plots of compensated and uncompensated systems in Example 10-4.

must be placed 1.5 decades apart, in order to produce the required 30 dB of attenuation.

5. Using the guideline that the upper corner frequency of the controller, $1/aT$, is placed at the frequency of 1 decade below the new gain-crossover frequency, we have

$$\frac{1}{aT} = \frac{\omega'_c}{10} = \frac{3.5}{10} = 0.35 \text{ rad/sec} \tag{10-77}$$

which gives

$$T = 89.3$$

6. The Bode plot of the compensated system, with the phase-lag controller transfer function given by

$$G_c(s) = \frac{1 + aTs}{1 + Ts} = \frac{1 + 2.86s}{1 + 89.3s} \tag{10-78}$$

is sketched in Fig. 10-34. It is seen that the phase margin of the compensated system is approximately 40°.

7. The open-loop transfer function of the compensated system is

$$G(s) = G_c(s)G_p(s) = \frac{100(1 + 2.86s)}{s(1 + 0.1s)(1 + 0.2s)(1 + 89.3s)} \tag{10-79}$$

The magnitude-versus-phase curves of the compensated and the uncompensated systems are plotted on the Nichols chart, as shown in Fig. 10-35. These curves show

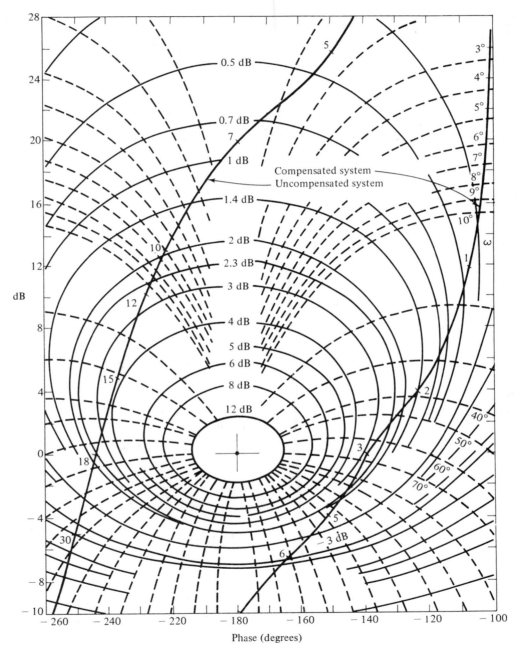

Fig. 10-35. Plots of $G(s)$ of the compensated and the uncompensated systems in the Nichols chart for Example 10-4.

that the uncompensated system is unstable, but the compensated system has the following performance data as measured by the frequency-domain criteria:

resonant peak $M_p = 3$ dB (1.41)

phase margin $= 40$ deg

gain margin $= 10$ dB

bandwidth $= 6$ rad/sec

When $K = 100$, the open-loop transfer function of the uncompensated system and the compensated system may be written

$$G(s) = \frac{5000}{s(s+5)(s+10)} \tag{10-80}$$

$$G(s) = \frac{160(s+0.35)}{s(s+5)(s+10)(s+0.0112)} \tag{10-81}$$

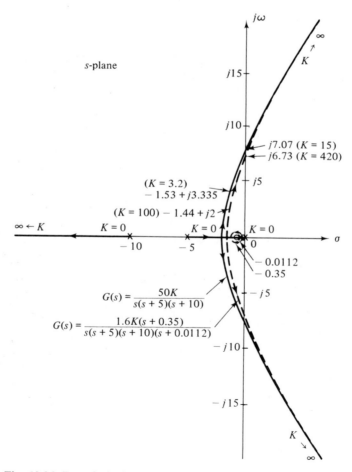

Fig. 10-36. Root loci of compensated and uncompensated systems in Example 10-5.

respectively. These transfer functions give indication of the improvements made on the system's performance from the root locus point of view. The gain constant of $G(s)$ in Eq. (10-80) is 5000, whereas that of $G(s)$ in Eq. (10-81) is only 160; the ratio of 160 to 5000 is the constant, a, which is determined earlier to be 0.032. This means that since the pole and zero of the phase-lag controller are very close to the origin, as compared with the poles at $s = -5$ and -10, the controller effectively reduced the loop gain of the system by a factor of 0.032. Figure 10-36 shows the root loci of the uncompensated and the compensated systems. Again, the loci of the complex roots are very close to each other for the two cases. The critical value of K for the uncompensated system is 15, whereas for the compensated system it is 420. When $K = 100$, the compensated system has eigenvalues at $s = -11.33$, $s = -0.8$, $s = -1.436 + j2$, and $s = -1.436 - j2$. The unit step response of the system with the phase-lag controller is shown in Fig. 10-37. The peak overshoot of the system is approximately 35 per cent.

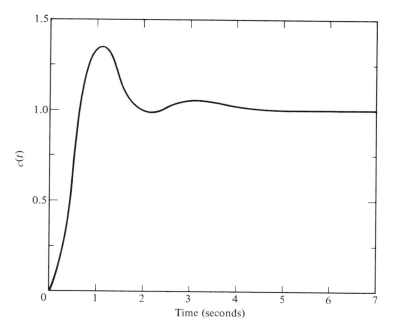

Fig. 10-37. Unit step response of the system with phase-lag compensation in Example 10-4.

Effects and Limitations of Phase-Lag Compensation

From the results of the preceding illustrative examples, the effects and limitations of phase-lag compensation on the performance of a control system may be summarized as follows:

1. For a given relative stability, the velocity error constant is increased.
2. The gain-crossover frequency is decreased; thus the bandwidth of the closed-loop system is decreased.
3. For a given loop gain, K, the magnitude of the open-loop transfer

function is attenuated at the gain-crossover frequency, thus allowing improvement in the phase margin, gain margin, and resonance peak of the system.

4. The rise time of the system with phase-lag compensation is usually slower, since the bandwidth is usually decreased. For additional reduction in the overshoot, the rise time is further increased.

10.5 Lag–Lead Compensation

We have learned from the preceding sections that phase-lead compensation usually improves the rise time and the overshoot but increases the bandwidth of a feedback control system, and it is effective for systems without too severe a stability problem. On the other hand, phase-lag compensation improves the overshoot and relative stability but often results in longer rise time because of reduced bandwidth. Therefore, we can say that each of these two types of control has its advantages and disadvantages. However, there are many systems that cannot be satisfactorily improved by the use of either scheme. Therefore, it is natural to consider the use of a combination of the lead and the lag controllers, so that the advantages of both schemes may be utilized; at the same time, some of the undesirable features in each are eliminated in the combined structure.

The transfer function of such a lag–lead (or lead–lag) controller may be written

$$G_c(s) = \left(\frac{1 + aT_1s}{1 + T_1s}\right)\left(\frac{1 + bT_2s}{1 + T_2s}\right) \qquad (10\text{-}82)$$

$$| \leftarrow \text{lead} \rightarrow || \leftarrow \text{lag} \rightarrow |$$

where $a > 1$ and $b < 1$, and the attenuation factor $1/a$ is not shown in the equation if we assume that adequate loop gain is available in the system to compensate for this loss.

Usually it is not necessary to cascade the lead and the lag networks of Figs. 10-10 and 10-25 for the realization of Eq. (10-82) if a and b need not be specified independently. A network that has lag–lead characteristics, but with fewer number of elements, is shown in Fig. 10-38. The transfer function of the network is

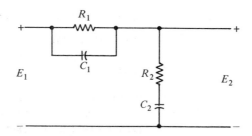

Fig. 10-38. Lag–lead network.

$$G_c(s) = \frac{E_2(s)}{E_1(s)} = \frac{(1 + R_1C_1s)(1 + R_2C_2s)}{1 + (R_1C_1 + R_1C_2 + R_2C_2)s + R_1R_2C_1C_2s^2} \quad (10\text{-}83)$$

Comparing Eq. (10-82) with Eq. (10-83), we have

$$aT_1 = R_1C_1 \quad (10\text{-}84)$$

$$bT_2 = R_2C_2 \quad (10\text{-}85)$$

$$T_1T_2 = R_1R_2C_1C_2 \quad (10\text{-}86)$$

From Eqs. (10-84) and (10-85), we have

$$abT_1T_2 = R_1R_2C_1C_2 \quad (10\text{-}87)$$

Thus

$$ab = 1 \quad (10\text{-}88)$$

which means that a and b cannot be specified independently.

EXAMPLE 10-5   In this example we shall design a lag–lead controller for the control system considered in Examples 10-2 and 10-4. The open-loop transfer function of the original system is repeated as

$$G_p(s) = \frac{K}{s(1 + 0.1s)(1 + 0.2s)} \quad (10\text{-}89)$$

The performance specifications are

1. $K_v = 100$ sec^{-1}.
2. Phase margin $\geq 40°$.

These requirements have been satisfied by the phase-lag controller designed in Example 10-4. However, it is noted that the phase-lag controller yielded a step response that has a relatively large rise time. In this example we shall design a lag–lead controller so that the rise time is reduced.

Let the series controller be represented by the transfer function

$$G_c(s) = \frac{(1 + aT_1s)(1 + bT_2s)}{(1 + T_1s)(1 + T_2s)} \quad (10\text{-}90)$$

For the first part we consider that the lag–lead controller is realized by the network of Fig. 10-38, so that the coefficients a and b are related through Eq. (10-88).

In general, there is no fixed procedure available for the design of the lag–lead controller. Usually, a trial-and-error procedure, using the design techniques outlined for the phase-lag and the phase-lead controllers, may provide a satisfactory design arrangement.

Let us first determine the phase-lag portion of the compensation by selecting the proper values of T_2 and b of Eq. (10-90). The Bode plot of $G_p(s)$ of Eq. (10-89) is sketched in Fig. 10-39 for $K = 100$. We arbitrarily choose to move the gain-crossover frequency of $G_p(j\omega)$ from 17 rad/sec to 6 rad/sec. Recall that in Example 10-4, the desired new gain-crossover frequency is 3.5 rad/sec when phase-lag compensation alone is used. With the gain-crossover frequency at 6 rad/sec, the phase margin of the system should be improved to approximately 10°. Using the phase-lag design technique, we notice that the attenuation needed to bring the magnitude of $G_p(j\omega)$ down to 0 dB at $\omega_c' = 6$ rad/sec is -22 dB. Thus using Eq. (10-59), we have

$$b = 10^{-22/20} = 10^{-1.1} = 0.08 \quad (10\text{-}91)$$

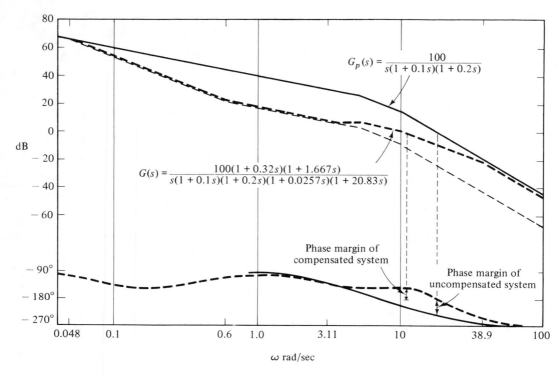

Fig. 10-39. Bode plots of uncompensated system and compensated system with lag-lead controller in Example 10-5.

Placing $1/bT_2$ at 1 decade of frequency below $\omega_c' = 6$ rad/sec, we have

$$\frac{1}{bT_2} = \frac{6}{10} = 0.6 \text{ rad/sec} \tag{10-92}$$

Thus

$$T_2 = 20.83 \tag{10-93}$$

and

$$\frac{1}{T_2} = 0.048 \tag{10-94}$$

The phase-lag portion of the controller is described by

$$\frac{1 + 1.667s}{1 + 20.83s}$$

Now we turn to the phase-lead part of the controller. Since a is equal to the inverse of b, the value of a is found to be 12.5. From Eq. (10-20), the maximum phase lead that corresponds to this value of a is

$$\sin \phi_m = \frac{a - 1}{a + 1} = 0.8518 \tag{10-95}$$

or

$$\phi_m = 58.3 \text{ deg}$$

The zero-frequency attenuation of the phase-lead controller is

$$20 \log_{10} a = 20 \log_{10} 12.5 = 21.9 \text{ dB} \tag{10-96}$$

Using the procedure outlined earlier, the new gain-crossover frequency is found to be at $\omega_m = 11$ rad/sec. Then, using Eq. (10-15), we get

$$\frac{1}{T_1} = \sqrt{a} \, \omega_m = 38.88 \tag{10-97}$$

and

$$\frac{1}{aT_1} = 3.11$$

The transfer function of the lag-lead controller is determined as

$$G_c(s) = \frac{(1 + 0.32s)(1 + 1.667s)}{(1 + 0.0257s)(1 + 20.83s)} \tag{10-98}$$

where as a usual practice, the attenuation factor in the phase-lead portion has been dropped.

Figure 10-39 shows that the phase margin of the compensated system is approximately $40°$. The open-loop transfer function of the compensated system with $K = 100$ is

$$G(s) = G_c(s)G_p(s) = \frac{4981(s + 3.11)(s + 0.6)}{s(s + 5)(s + 10)(s + 38.88)\,(s + 0.048)} \tag{10-99}$$

The unit step response of the compensated system is shown in Fig. 10-40. It is apparent

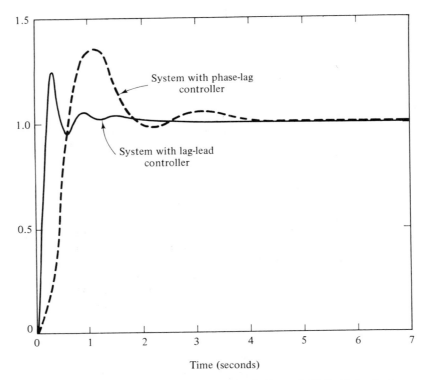

Time (seconds)

Fig. 10-40. Unit step response of the system in Example 10-5 with lag-lead controller.

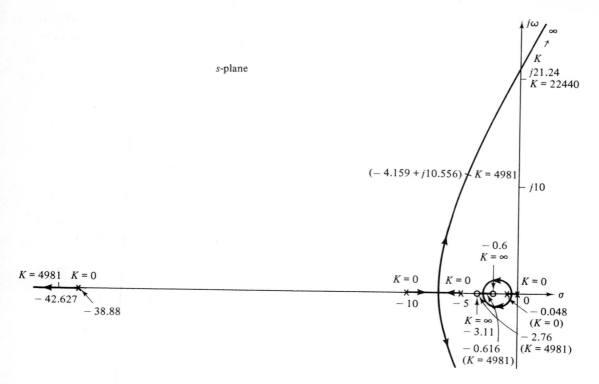

Fig. 10-41. Root locus diagram of the system in Example 10-5 with lag-lead controller.

that the step response of the system with the lag–lead controller is much improved over that of the phase-lag compensated system. Not only the overshoot is smaller, but the rise time is also greatly reduced. We may attribute these improvements to the addition of the phase-lead portion of the controller. It can be easily verified that the bandwidth of the closed-loop system is now approximately 18 rad/sec, which is 3 times that of the system with the phase-lag controller alone.

The root locus diagram of Fig. 10-41 shows that the dominant complex roots of the lag–lead compensated system correspond to a much larger natural undamped frequency than that of the system with only the phase-lag controller designed in Example 10-4, Fig. 10-36. This accounts for the fact that the lag–lead compensation gives a wider bandwidth and thus a shorter rise time. The root locus diagram of Fig. 10-41 also shows that for all practical purposes, the transfer function of the closed-loop system with the lag–lead controller can be approximated by

$$\frac{C(s)}{R(s)} = \frac{5487}{(s + 42.627)(s + 4.159 + j10.556)(s + 4.159 - j10.556)} \quad (10\text{-}100)$$

where the closed-loop poles at -0.616 and -2.76 are effectively canceled by the zeros at -0.6 and -3.11, respectively.

It should be pointed out that we have been fortunate to have found a solution to the design problem on the first trial. In general, any trial-and-error design scheme may not be as straightforward, especially since there is a constraint between a and b of Eq. (10-90). In other words, once we selected a value for b, a is also determined. It is entirely possible that the combination does not satisfy the design specifications.

10.6 Bridged-T Network Compensation

Many controlled processes have transfer functions that contain one or more pairs of complex-conjugate poles. One of the distinct features of a pair of complex poles is that the gradient of the phase with respect to frequency near the gain-crossover frequency is large. This is similar to the situation when two or more simple poles of the transfer function are placed close to each other. With reference to the system considered in Examples 10-2 and 10-4, the reason the phase-lead controller is ineffective in improving the relative stability of the system is because the phase of the process has a steep negative slope near the gain-crossover frequency (see Fig. 10-22). When a process has complex poles, especially if the poles are close to the imaginary axis, the design problem may become more acute. We may suggest the use of a controller that has a transfer function with zeros so selected as to cancel the undesired complex poles of the original process, and the poles of the controller are placed at the desired locations in the s-plane. For instance, if the transfer function of a process is

$$G_p(s) = \frac{K}{s(s^2 + s + 10)} \tag{10-101}$$

in which the complex-conjugate poles are causing stability problems in the closed-loop system, the suggested series controller may have the transfer function

$$G_c(s) = \frac{s^2 + s + 10}{s^2 + 2\zeta\omega_n s + \omega_n^2} \tag{10-102}$$

The constants ζ and ω_n are determined according to the performance specifications of the system. Although the transfer function of Eq. (10-102) can be realized by various types of passive networks, the bridged-T networks have the advantage of containing only RC elements. Figure 10-42 illustrates two basic types of the bridged-T RC networks. In the following discussion, the network shown in Fig. 10-42(a) is referred to as the bridged-T type 1, and that of Fig. 10-42(b) is referred to as type 2.

With the assumption of zero input source impedance and infinite output impedance, the transfer function of the bridged-T type 1 network is given by

$$\frac{E_2(s)}{E_1(s)} = \frac{1 + 2RC_2 s + R^2 C_1 C_2 s^2}{1 + R(C_1 + 2C_2)s + R^2 C_1 C_2 s^2} \tag{10-103}$$

and that of the bridged-T type 2 network is

$$\frac{E_2(s)}{E_1(s)} = \frac{1 + 2R_1 C s + C^2 R_1 R_2 s^2}{1 + C(R_2 + 2R_1)s + C^2 R_1 R_2 s^2} \tag{10-104}$$

When these two equations are compared, it is apparent that the two net-

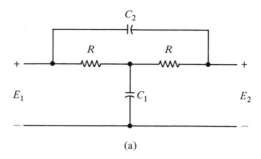

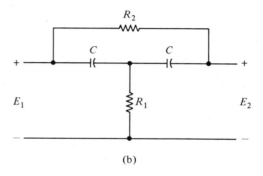

(a)

(b)

Fig. 10-42. Two basic types of bridged-T network. (a) Type 1 network. (b) Type 2 network.

works have similar transfer characteristics. In fact, if R, C_1, and C_2 in Eq. (10-103) are replaced by C, R_2, and R_1, respectively, Eq. (10-103) becomes the transfer function of the type 2 network given by Eq. (10-104).

It is useful to study the behavior of the poles and zeros of the transfer functions of the bridged-T networks of Eqs. (10-103) and (10-104) when the network parameters are varied. Owing to the similarity of the two networks, only type 1 will be studied here.

Equation (10-103) is written as

$$\frac{E_2(s)}{E_1(s)} = \frac{s^2 + \dfrac{2}{RC_1}s + \dfrac{1}{R^2C_1C_2}}{s^2 + \dfrac{C_1 + 2C_2}{RC_1C_2}s + \dfrac{1}{R^2C_1C_2}} \tag{10-105}$$

If both the numerator and the denominator polynomials of Eq. (10-105) are written in the standard form

$$s^2 + 2\zeta\omega_n s + \omega_n^2 = 0 \tag{10-106}$$

we have, for the numerator,

$$\omega_{nz} = \pm\frac{1}{R\sqrt{C_1C_2}} \tag{10-107}$$

$$\zeta_z = \sqrt{\frac{C_2}{C_1}} \tag{10-108}$$

and for the denominator,

$$\omega_{np} = \pm \frac{1}{R\sqrt{C_1 C_2}} = \omega_{nz} \qquad (10\text{-}109)$$

$$\zeta_p = \frac{C_1 + 2C_2}{2\sqrt{C_1 C_2}} = \frac{1 + 2C_2/C_1}{2\sqrt{C_2/C_1}} = \frac{1 + 2\zeta_z^2}{2\zeta_z} \qquad (10\text{-}110)$$

The loci of the poles and zeros of $E_2(s)/E_1(s)$ of Eq. (10-105) when C_1, C_2, and R vary individually are sketched in Fig. 10-43. When R varies, the numerator and the denominator of Eq. (10-105) contain R in the form of R^2, and the root locus method cannot be applied directly to this nonlinear problem. Fortunately,

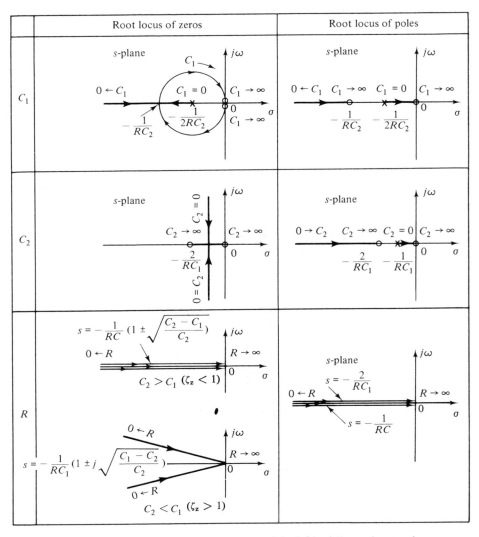

Fig. 10-43. Loci of poles and zeros of the bridged-T type 1 network.

in this case, the equations that are of the form of Eq. (10-106) are of the second order and can be solved easily. Therefore, the pole and zero loci of Fig. 10-43 show that the two zeros of the bridged-T network type 1 can be either real or complex; for complex zeros, C_2 must be less than C_1. The poles of the transfer function always lie on the negative real axis.

The ζ and ω_n parameters of the denominator and the numerator of the transfer function of the type 2 network may be obtained by replacing R, C_1, and C_2 in Eqs. (10-107) through (10-110) by C, R_2, and R_1, respectively. Thus

$$\omega_{nz} = \pm \frac{1}{C\sqrt{R_1 R_2}} \qquad (10\text{-}111)$$

$$\zeta_z = \sqrt{\frac{R_1}{R_2}} \qquad (10\text{-}112)$$

$$\omega_{np} = \omega_{nz} \qquad (10\text{-}113)$$

$$\zeta_p = \frac{R_2 + 2R_1}{2\sqrt{R_1 R_2}} \qquad (10\text{-}114)$$

The root loci shown in Fig. 10-43 can still be used for the type 2 bridged-T network if the corresponding symbols are altered.

EXAMPLE 10-6   The control system shown in Fig. 10-44 is selected to demonstrate the use of the bridged-T network as a controller and the principle of improving system performance by pole–zero cancellation.

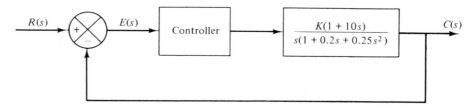

Fig. 10-44. Feedback control system for Example 10-6.

The controlled process has the transfer function

$$G_p(s) = \frac{K(1 + 10s)}{s(1 + 0.2s + 0.25s^2)} \qquad (10\text{-}115)$$

The root locus diagram of the uncompensated closed-loop system is shown in Fig. 10-45. It is shown that although the closed-loop system is always stable, the damping of the system is very low, and the step response of the system will be quite oscillatory for any positive K. Figure 10-47 illustrates the unit step response of the system when $K = 1$. Notice that the zero at $s = -0.1$ of the open-loop transfer function causes the closed-loop system to have an eigenvalue at $s = -0.091$, which corresponds to a time constant of 11 sec. Thus this small eigenvalue causes the step response of Fig. 10-47 to oscillate about a level that is below unity, and it would take a long time for the response to reach the desired steady state.

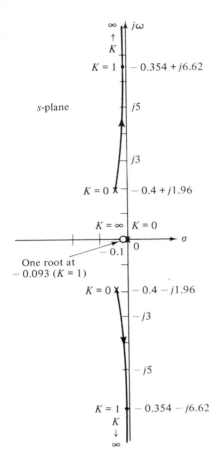

Fig. 10-45. Root loci of the feedback control system in Example 10-6 with $G_p(s) = [K(1 + 10s)]/[s(1 + 0.2s + 0.25s^2)]$.

Let us select the type 1 bridged-T network as series compensation to improve the relative stability of the system. The complex-conjugate zeros of the network should be so placed that they will cancel the undesirable poles of the controlled process. Therefore, the transfer function of the bridged-T network should be

$$G_c(s) = \frac{s^2 + 0.8s + 4}{s^2 + 2\zeta_p\omega_{np}s + \omega_{np}^2} \qquad (10\text{-}116)$$

Using Eqs. (10-107) and (10-108), we have

$$\omega_{nz} = 2 = \frac{1}{R\sqrt{C_1 C_2}} \qquad (10\text{-}117)$$

$$\zeta_z = \sqrt{\frac{C_2}{C_1}} = 0.2 \qquad (10\text{-}118)$$

From Eqs. (10-109) and (10-110),

$$\omega_{np} = \omega_{nz} = 2 \qquad (10\text{-}119)$$

and

$$\zeta_p = \frac{1 + 2\zeta_z^2}{2\zeta_z} = 2.7 \qquad (10\text{-}120)$$

The transfer function of the type 1 bridged-T network is

$$G_c(s) = \frac{s^2 + 0.8s + 4}{(s + 0.384)(s + 10.42)} \qquad (10\text{-}121)$$

The open-loop transfer function of the compensated system is

$$G(s) = G_c(s)G_p(s) = \frac{40K(s + 0.1)}{s(s + 0.384)(s + 10.42)} \qquad (10\text{-}122)$$

The root loci of the compensated system are sketched as shown in Fig. 10-46. The root loci show that for K greater than 0.64, two of the eigenvalues of the system are complex. However, the dynamic response of the system is still dominated by the eigenvalue that lies near the origin of the s-plane. Figure 10-47 illustrates the unit step response of the compensated system when $K = 1$. In this case the complex eigenvalues of the compensated system cause the response to oscillate only slightly, but the eigenvalue at $s = -0.093$ causes the system response to take a very long time in reaching its steady state of unity. It is apparent that the relative stability of the system is greatly improved by the bridged-T controller through pole–zero cancellation.

Although the preceding design is carried out on the basis that the undesirable poles of the controlled process are exactly canceled by the zeros of the controller, in practice this ideal situation is difficult to achieve. One of the difficulties lies in the fact that we cannot design the controller so that its poles and zeros will fall exactly at the specified locations. Another problem is lack of knowledge as to exactly where the poles of the process are. Usually, in the determination of the transfer function of a process, assumptions and approximations have to be made, so that the mathematical description of the system is never precise. In

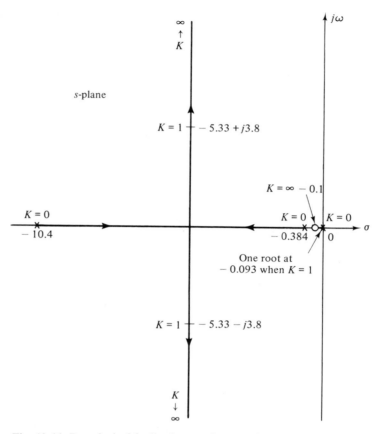

Fig. 10-46. Root loci of feedback control system in Example 10-6 with bridged-T controller.

view of these practical considerations, in practice, a more realistic situation in the attempt of the pole–zero cancellation design is at best a "near cancellation."

Let us consider that the design of the bridged-T compensation of the system in Example 10-6 resulted in an inexact pole–zero cancellation, so that the transfer function of the controller is

$$G_c(s) = \frac{s^2 + 0.76s + 3.8}{(s + 0.366)(s + 10.4)} \qquad (10\text{-}123)$$

Since the complex poles of the controlled process are not canceled by the zeros of $G_c(s)$, the open-loop transfer function of the system becomes

$$G(s) = \frac{40K(s + 0.1)(s^2 + 0.76s + 3.8)}{s(s + 0.366)(s + 10.4)(s^2 + 0.8s + 4)} \qquad (10\text{-}124)$$

For all practical purposes, the complex poles and zeros of $G(s)$ of Eq. (10-124) can still be considered as close enough for cancellation. For $K = 1$, the closed-loop transfer function is

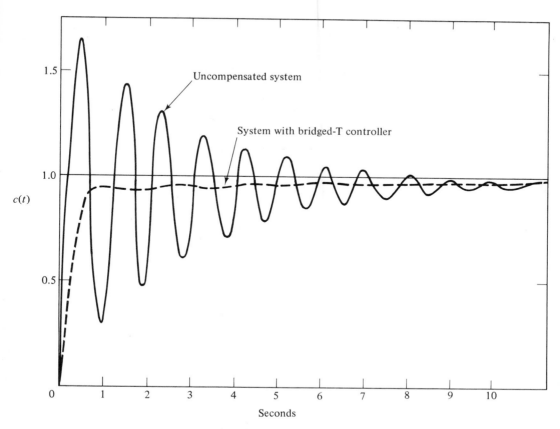

Fig. 10-47. Unit step responses of the compensated and the uncompensated systems in Example 10-6.

$$\frac{C(s)}{R(s)} = \frac{40(s + 0.1)(s^2 + 0.76s + 3.8)}{(s + 0.093)(s^2 + 0.81s + 3.805)(s^2 + 10.66s + 42.88)} \quad (10\text{-}125)$$

Since the zeros of the open-loop transfer function are retained as zeros of the closed-loop transfer function, Eq. (10-125) verifies that near cancellation of poles and zeros in the open-loop transfer function will result in the same situation for the closed-loop transfer function. Therefore, we may conclude that from the transient response standpoint, the effectiveness of the cancellation compensation is not diminished even if the pole–zero cancellation is not exact. Alternatively, we can show that if the partial-fraction expansion of Eq. (10-125) is carried out, the coefficients that correspond to the roots of $s^2 + 0.81s + 3.805 = 0$ will be very small, so that the contribution to the time response from these roots will be negligible.

If the undesirable poles of the open-loop transfer function are very close to or right on the imaginary axis of the s-plane, inexact cancellation may result in a conditionally stable or an unstable system. Figure 10-48(a) illustrates a situation in which the relative positions of the poles and zeros intended for cancellation

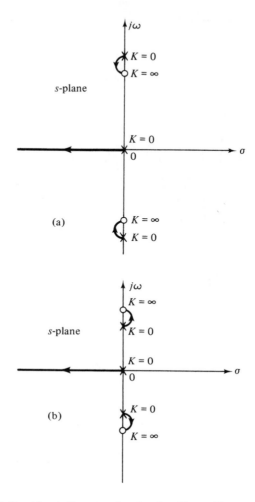

Fig. 10-48. Root locus diagrams showing the effects of inexact pole–zero cancellations.

result in a stable system, whereas in Fig. 10-48(b), the inexact cancellation is unacceptable, although the relative distance between the poles and zeros intended for cancellation is small. Normally, the two situations may be distinguished simply by the use of the angles of departure properties of the root loci.

REFERENCES

1. H. W. Bode, *Network Analysis and Feedback Amplifier Design*, D. Van Nostrand Reinhold Company, New York, 1945.

2. H. M. James, N. B. Nichols, and R. S. Phillips, *Theory of Servomechanisms*, McGraw-Hill Book Company, New York, 1947.

3. N. WIENER, *Extrapolation, Interpolation, and Smoothing of Stationary Time Series*, Technology Press, MIT, Cambridge, Mass., 1949.

4. W. R. EVANS, "Control System Synthesis by Root Locus Method," *Trans. AIEE*, Vol. 69, pp. 66–69, 1950.

5. J. G. TRUXAL, *Automatic Feedback Control System Synthesis*, McGraw-Hill Book Company, New York, 1955.

PROBLEMS

10.1. The open-loop transfer function of a feedback control system is given by

$$G(s) = \frac{K}{s(1 + 0.2s)(1 + 0.5s)}$$

The feedback is unity. The output of the system is to follow a reference ramp input to yield a velocity of 2 rpm with a maximum steady-state error of 2°.

(a) Determine the smallest value of K that will satisfy the specification given above. With this value of K, analyze the system performance by evaluating the system gain margin, phase margin, M_p, and bandwidth.

(b) A lead compensation with the transfer function $(1 + 0.4s)(1 + 0.08s)$ is inserted in the forward path of the system. Evaluate the values of the gain margin, phase margin, M_p, and bandwidth of the compensated system. Comment on the effects of the lead compensation of the system performance.

(c) Sketch the root loci of the compensated and the uncompensated systems.

10.2. A type-2 control system is shown in Fig. P10-2. The system must meet the following performance specifications:

(a) Acceleration constant $K_a = 5$ sec^{-2}.

(b) The resonance peak $M_p \leq 1.5$.

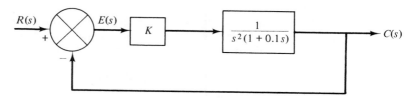

Figure P10-2.

Design a series phase-lead controller to satisfy these requirements. Sketch the root loci for the uncompensated and compensated systems. What are the values of the damping ratio (of the complex roots) and the bandwidth of the compensated system?

10.3. Figure P10-3 illustrates the block diagram of a positioning control system. The system may be used for the positioning of a shaft by a command from a remote location. The parameters of the system are given as follows:

$$\tau_e = 0.01 \text{ sec}$$

$$J = 0.05 \text{ oz-in-sec}^2$$

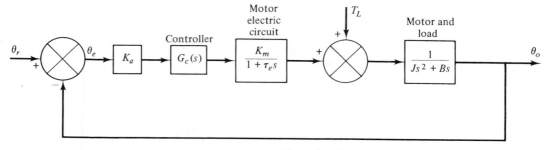

Figure P10-3.

$$B = 0.5 \text{ oz-in-sec}$$

$$K_m = 10 \text{ oz-in/volt}$$

$$T_L = \text{disturbance torque}$$

(a) Determine the minimum value of the error sensor and amplifier gain K_a so that the steady-state error of θ_0 due to a unit step torque disturbance is less than or equal to 0.01 (1 per cent).

(b) Determine the stability of the system when K_a is set at the value determined in part (a).

Design a phase-lead controller in the form

$$G_c(s) = \frac{1 + aTs}{1 + Ts} \qquad a > 1$$

so that the closed-loop system has a phase margin of approximately 40°. Determine the bandwidth of the compensated system. Plot the output response $\theta_0(t)$ when the input $\theta_r(t)$ is a unit step function. ($T_L = 0$.)

10.4. Repeat the design problem in Problem 10-3 with a phase-lag controller

$$G_c(s) = \frac{1 + aTs}{1 + Ts} \qquad a < 1$$

10.5. Human beings breathe in order to provide the means for gas exchange for the entire body. A respiratory control system is needed to ensure that the body's needs for this gas exchange are adequately met. The criterion of control is adequate ventilation, which ensures satisfactory levels of both oxygen and carbon dioxide in the arterial blood. Respiration is controlled by neural impulses that originate within the lower brain and are transmitted to the chest cavity and diaphragm to govern the rate and tidal volume. One source of the signals is the chemoreceptors located near the respiratory center, which are sensitive to carbon dioxide and oxygen concentrations. Figure P10-5 shows the block diagram of a simplified model of the human respiratory control system. The objective is to control the effective ventilation of the lungs so that satisfactory balance of concentrations of carbon dioxide and oxygen is maintained in the blood circulated at the chemoreceptor.

(a) A normal value of the chemoreceptor gain K_f is 0.1. Determine the gain margin and the phase margin of the system.

(b) Assume that the chemoreceptor is defective so that its gain K_f is increased to 1. Design a controller in the forward path (to be inserted in front of the block representing the lungs) so that a phase margin of 45° is maintained.

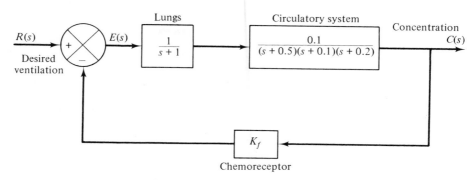

Figure P10-5.

10.6. This problem deals with the cable-reel unwind process described in Problem 5-16 and shown in Fig. P5-16. The inertia of the reel is

$$J_R = 18R^4 - 200 \text{ ft-lb-sec}^2$$

where R is the effective radius of the cable reel.
(a) Assume that R and J_R are constant between layers of the cable. Determine the maximum value of the amplifier gain K so that the entire unwinding process is stable from beginning until end.
(b) Let $K = 10$. Design a series controller so that the system has a phase margin of $45°$ at the end of the unwind process ($R = 2$ ft). With this controller, what are the phase and gain margins of the system at the beginning of the unwind process? Sketch the root loci of the compensated process with $K = 10$ and indicate the variation of the roots on the loci as the unwind process progresses. It should be noted that the treatment of this problem by a transfer function is only an approximation; strictly, the process is nonlinear as well as time varying.

10.7. Figure P10-7 shows the block diagram of the speed control system for an electrical power generating system. The speed governor valve controls the steam flow input to the turbine. The turbine drives the generator, which puts out electric power at a frequency proportional to the generator speed ω_g. The

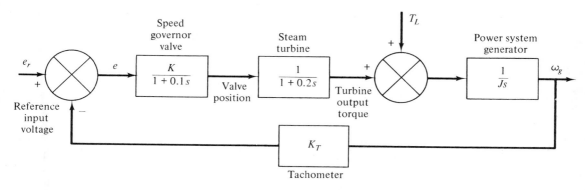

Figure P10-7.

desired steady-state speed of the generator is 1200 rpm, at which the generated output voltage is 60 Hz. $J = 100$.

(a) Let the speed governor valve gain K be set at 10 rad/volt. Determine the tachometer gain so that the complex eigenvalues of the closed-loop system correspond to a damping ratio of 0.707. Sketch the root loci as a function of K_T and indicate the location of the roots with the desired damping.

(b) Determine the desired reference input voltage, with the value of K_T set at the value determined in (a), so that the generator speed is 1200 rpm $(T_L = 0)$.

(c) In Fig. P10-7, T_L denotes a load change. Determine the per cent change in the steady-state speed due to a constant load change when K_T is as determined in (a).

(d) It is desired to keep the frequency variation due to load change within ± 0.1 per cent. At the same time, the relative damping of the complex eigenvalues of the overall system must be approximately 0.707. Can both requirements be satisfied by only changing the values of K and K_T? If not, design a series controller in the forward path for this purpose. Sketch the root locus for the compensated system with $K = 10$, with K_T as the variable parameter.

10.8. The phase-lock loop technique is a popular method for *dc* motor speed control. A basic phase-lock loop motor speed control system is shown in Fig. P10-8.

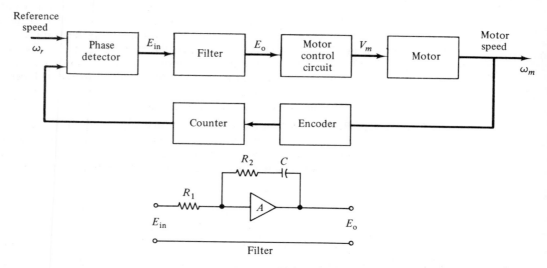

Figure P10-8.

The encoder produces digital pulses that represent motor speed. The pulse train from the encoder is compared with the reference frequency by a phase comparator or detector. The output of the phase detector is a voltage that is proportional to the phase difference between the reference speed and the actual motor speed. This error voltage upon filtering is used as the control signal for the motor. The system parameters and transfer functions are as follows:

Phase detector	$K_p = 0.0609$ volt/rad
Motor control circuit gain	$K_a = 1$
Motor transfer function	$\dfrac{\omega_m(s)}{V_m(s)} = \dfrac{K_m}{s(1 + T_m s)}$ $(K_m = 10, T_m = 0.04)$
Encoder gain	$K_e = 5.73$ pulses/rad
Filter transfer function	$\dfrac{E_o}{E_{in}} = \dfrac{R_2 C s + 1}{R_1 C s}$
	$R_1 = 1.745 \times 10^6 \ \Omega, \ C_1 = 1 \ \mu\text{F}$
Counter	$\dfrac{1}{N} = 1$

Determine the value of R_2 so that the complex roots of the closed-loop system have a damping ratio of approximately 0.707. If the problem has more than one solution, use the one that corresponds to the smallest system bandwidth. Sketch the root locus diagram for the characteristic equation roots as R_2 varies.

10.9. Tachometer feedback is employed frequently in feedback control systems for the purpose of stabilization. Figure P10-9 shows the block diagram of a system with tachometer feedback. Choose the tachometer gain constant K_t so that the relative damping ratio of the system is 50 per cent. How does the tachometer feedback affect the bandwidth of the system?

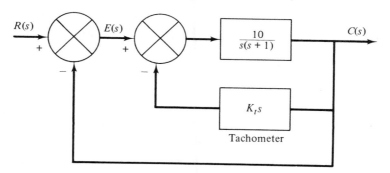

Figure P10-9.

10.10. The block diagram of a control system is shown in Fig. P10-10. By means of the root contour method, show the effect of variation in the value of T on the location of the closed-loop poles of the system.

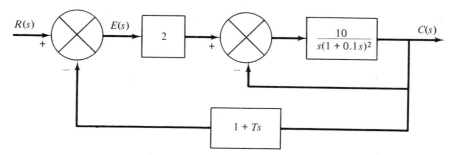

Figure P10-10.

10.11. A computer-tape-drive system utilizing a permanent-magnet dc motor is shown in Fig. P10-11(a). The system is modeled by the diagram shown in Fig. P10-11(b). The constant K_L represents the spring constant of the elastic tape, and B_L denotes the viscous frictional coefficient between the tape and the capstans. The system parameters are as follows:

$$\frac{K_a}{R_a} = 36 \text{ oz-in/volt} \qquad J_m = 0.023 \text{ oz-in-sec}^2$$

$$K_e = 6.92 \text{ oz-in/rad/sec} \qquad K_L = 2857.6 \text{ oz-in/rad}$$

$$B_L = 10 \text{ oz-in-sec} \qquad J_L = 7.24 \text{ oz-in-sec}^2$$

where K_b = back emf constant, $K_e = K_b K_a/R_a + B_m$, K_a = torque constant in oz-in/amp, and B_m = motor viscous friction constant.
(a) Write the state equations of the system shown in Fig. P10-11(b) with θ_L, ω_L, θ_m, and ω_m as the state variables in the prescribed order. Derive the transfer functions

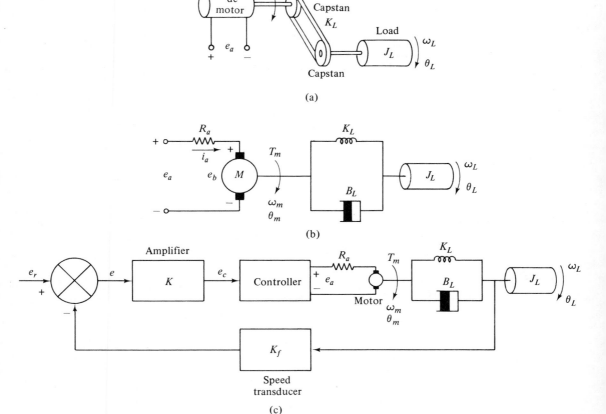

(a)

(b)

(c)

Figure P10-11.

$$\frac{\omega_m(s)}{E_a(s)} \quad \text{and} \quad \frac{\omega_L(s)}{E_a(s)}$$

(b) The objective of the system is to control the speed of the load, ω_L, accurately. Figure P10-11(c) shows a closed-loop system in which the load speed is fed back through a speed transducer and compared with the reference input, with $K_f = 0.01$. Design a controller and select the amplifier gain so that the following specifications are satisfied: (1) no steady-state speed error when the input e_r is a step function; (2) the dominant roots of the characteristic equation correspond to a damping ratio of approximately 0.707; and (3) what should be the value of the input e_r if the steady-state speed is to be 100 rpm? Sketch the root loci of the designed system with K as the variable parameter.

10.12. The computer-tape-drive system described in Problem 10-11 has the following system parameters:

$$\frac{K_a}{R_a} = 36 \text{ oz-in/volt} \qquad\qquad J_m = 0.023 \text{ oz-in-sec}^2$$

$$K_e = \frac{K_b K_a}{R_a} + B_m = 6.92 \text{ oz-in/rad/sec} \qquad B_L = 0$$

$$K_L = 28{,}576 \text{ oz-in/rad} \qquad\qquad K_f = 0.01$$

$$J_L = 7.24 \text{ oz-in-sec}^2$$

(a) Show that the closed-loop system without compensation has an oscillatory response in ω_L for any positive K.

(b) In order to reduce the speed oscillations, a bridged-T network is proposed to cancel the undesirable poles of the open-loop transfer function. Figure P10-12 shows the block diagram of the overall system. The phase-lag controller preceding the bridged-T controller is for the purpose of ensuring zero steady-state speed error when a step input is applied. Determine the transfer function of the bridged-T network and the amplifier gain so that the dominant roots of the characteristic equation correspond to a damping ratio of approximately 70.7 per cent.

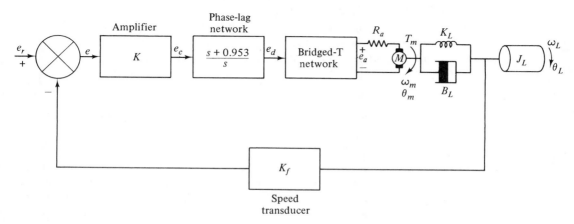

Figure P10-12.

11

Introduction to Optimal
Control

11.1 Introduction

From the discussions given in Chapter 10 we realize that the conventional design
of feedback control systems has many limitations. The most important disadvantage of the conventional design approach is that these methods are not
rigorous and rely heavily on trial and error. For a system with multiple inputs
and outputs and a high degree of complexity, the trial-and-error approach may
not lead to a satisfactory design.

The optimal control design is aimed at obtaining a system that is the best
possible with respect to a certain performance index or design criterion. The
state-variable formulation is often used for the representation of the dynamic
process. Since the optimal control problem is more of a mathematical development, the problem is stated as:

Find the control u(t) that minimizes the performance index,

$$J = \int_{t_0}^{t_f} F[\mathbf{x}(t), \mathbf{u}(t), t] \, dt \qquad (11\text{-}1)$$

*given the initial state $\mathbf{x}(t_0) = \mathbf{x}_0$ and subject to the constraint that the process is
described by*

$$\dot{\mathbf{x}}(t) = \mathbf{f}[\mathbf{x}(t), \mathbf{u}(t), t] \qquad (11\text{-}2)$$

The performance index is a scalar quantity that consists of an integral of
a scalar function of the state vector and the control vector. Notice that the
state equations which describe the dynamics of the system to be controlled are
interpreted as *constraints* to the design problem. In general, additional constraints on the state variables and/or the control are often used.

In the optimal control design, the performance index J replaces the conventional design criteria such as peak overshoot, damping ratio, gain margin, and phase margin. Of course, the designer must be able to select the performance index properly so that the resulting system from the design will perform satisfactorily according to physical standards which are more easily interpreted by conventional performance criteria.

It must be noted that the control $\mathbf{u}(t)$ is usually not what has been referred to as the reference input in conventional design terminology. Since the state equation of Eq. (11-2) describes the controlled process only, the statement of the optimal control problem may give one the misconception that the problem deals with the design of an open-loop system. However, when the design is accomplished, usually, the optimal control will be of such a nature that it depends on the output or the state variables, so that a closed-loop control system is naturally formed.

To illustrate the various possible formulations of the performance index J, the following examples are given.

Minimum-Time Problem

One of the most frequently encountered problems in optimal control is the design of a system that will be able to drive its states from some initial condition to the final state in minimum time. In terms of the performance index of Eq. (11-1), the minimum-time problem may be stated so as to minimize

$$J = t_f - t_0 = \int_{t_0}^{t_f} dt \tag{11-3}$$

Thus, in Eq. (11-1), $F[\mathbf{x}(t), \mathbf{u}(t), t] = 1$.

Linear Regulator Problem

Given a linear system, the design objective may be to keep the states at the equilibrium state, and the system should be able to return to the equilibrium state from any initial state. The performance index for the linear regulator problem is

$$J = \tfrac{1}{2} \int_{t_0}^{t_f} \mathbf{x}' \mathbf{Q} \mathbf{x} \, dt \tag{11-4}$$

where $\mathbf{Q}$ is a symmetric positive-definite matrix. Or,

$$J = \tfrac{1}{2} \int_{t_0}^{t_f} [\mathbf{x}' \mathbf{Q} \mathbf{x} + \mathbf{u}' \mathbf{R} \mathbf{u}] \, dt \tag{11-5}$$

where a constraint is placed on the control vector $\mathbf{u}$. The matrices $\mathbf{Q}$ and $\mathbf{R}$ are called the *weighting matrices*, as their components put various weights on the elements of $\mathbf{x}$ and $\mathbf{u}$ in the optimization process.

Tracking Problem

If it is desired that the state $\mathbf{x}$ of a given system track, or be as close as possible to a certain target trajectory $\mathbf{x}_d$, the problem may be formulated by minimizing

$$J = \int_{t_0}^{t_f} (\mathbf{x} - \mathbf{x}_d)' \mathbf{Q} (\mathbf{x} - \mathbf{x}_d) \, dt \tag{11-6}$$

In addition to these examples, there are the minimum-energy criterion, minimum-fuel criterion, and many others.

11.2 Analytical Design[1,2]

One of the first design methods that eliminates the trial-and-error uncertainties is analytical design, sometimes known as *minimum integral-square-error design*. The design is significant in that it forms a bridge between the conventional design philosophy and the modern optimal control principle. For the first time, the designer can simply set up the performance index and then carry out the complete design in a mathematical and analytical way. The only problems the designer has to be concerned with are: (1) if the design has a solution, and (2) if the solution can be physically implemented. The similarity of analytical design to the conventional approach is that the design is carried out in the frequency domain or, more appropriately stated, the *s*-domain. However, none of the frequency-domain graphical techniques is necessary.

We first begin with the system configuration of Fig. 11-1. The assumptions are that the system is linear, time invariant, and that there is only one input and one output. Systems with multivariables may be designed by the analytical method, but the procedure is much more involved. Therefore, in the current discussion we are restricted to single-variable systems.

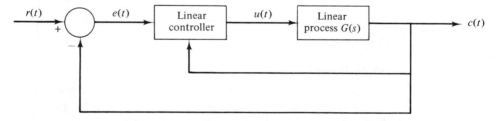

Fig. 11-1. Block diagram of a linear system with a controller.

The linear controlled process of the system in Fig. 11-1 is described by the transfer function $G(s)$. The control signal is $u(t)$, and the output is $c(t)$. Notice that the block diagram of Fig. 11-1 is not an algebraic one in that the controller is portrayed in such a way that its specific input–output configuration is not yet defined. This allows the flexibility of selecting the final controller configuration, such as a series controller or a feedback controller, once the design is completed.

The performance index J is composed of the following two integrals:

$$J_e = \int_0^\infty e^2(t)\, dt \tag{11-7}$$

$$J_u = \int_0^\infty u^2(t)\, dt \tag{11-8}$$

where $e(t)$ is the error signal and $u(t)$ is the control signal. J_e is referred to as the *integral square error* (ISE) and J_u is the *integral square control* (ISC).

The significance of using J_e is that this performance index gives a flexible restriction on such criteria as peak overshoot, rise time, and relative stability. If the error between the output and the reference input is large, the value of J_e will be large. The ISC index J_u may be regarded as a measure of the energy required by the control. Since $u^2(t)$ is proportional to the power consumed for control, the time integral of $u^2(t)$ is a measure of the energy consumed by the system. Ordinarily, the analytical design problem may be formulated as one that desires the minimization of J_e while keeping J_u within a certain limit, or minimizing J_u while keeping J_e not greater than a certain value. Using the calculus-of-variation method, the problem of minimizing one function while constraining another function to a constant value can be solved by adjoining the constraint to the function to be minimized (see Appendix C). For instance, if the optimization problem is:

$$\text{minimize } J_e$$

and

$$J_u = K = \text{constant}$$

the problem is equivalent to

$$\text{minimize } J = J_e + k^2 J_u \tag{11-9}$$

where k^2 is a parameter yet to be determined and is called the *Lagrange multiplier*. On the other hand, if we wish to

$$\text{minimize } J_u$$

and

$$J_e = K = \text{constant}$$

then the problem is equivalent to minimizing

$$J = J_u + k^2 J_e \tag{11-10}$$

The major development of the analytical design is the use of an equivalent frequency-domain expression for Eq. (11-9) or Eq. (11-10) in terms of the system transfer function through the use of Parseval's theorem. The version of the Parseval's theorem that is applied to the present problem is

$$\int_0^\infty e^2(t)\, dt = \frac{1}{2\pi j} \int_{-j\infty}^{j\infty} E(s) E(-s)\, ds \tag{11-11}$$

where $E(s)$ is the Laplace transform of $e(t)$.

Let us formulate the problem as being that of finding the optimal control for the system of Fig. 11-1 so that

$$J = J_e + k^2 J_u = \int_0^\infty [e^2(t) + k^2 u^2(t)]\, dt \tag{11-12}$$

is minimized.

Applying Parseval's theorem to Eq. (11-12), we have

$$J = \frac{1}{2\pi j} \int_{-j\infty}^{j\infty} [E(s) E(-s) + k^2 U(s) U(-s)]\, ds \tag{11-13}$$

With reference to Fig. 11-1,

$$E(s) = R(s) - G(s)U(s) \tag{11-14}$$

Substituting Eq. (11-14) into Eq. (11-13), we get

$$
\begin{aligned}
J = \frac{1}{2\pi j} \int_{-j\infty}^{j\infty} [R(s) - G(s)U(s)][R(-s) - G(-s)U(-s)]\,ds \\
+ \frac{k^2}{2\pi j} \int_{-j\infty}^{j\infty} U(s)U(-s)\,ds
\end{aligned}
\tag{11-15}
$$

Let $U_{ok}(s)$ be the optimal control, subject to the determination of k^2. We can write any arbitrary $U(s)$ as

$$U(s) = U_{ok}(s) + \lambda U_1(s) \tag{11-16}$$

where λ is a constant and $U_1(s)$ is any arbitrary function that has only poles in the left half of the s-plane. The last condition is due to the requirement that $u(t)$ must be bounded. Substitution of Eq. (11-16) into Eq. (11-15) yields

$$J = J_e + k^2 J_u = J_1 + \lambda(J_2 + J_3) + \lambda^2 J_4 \tag{11-17}$$

where

$$J_1 = \frac{1}{2\pi j} \int_{-j\infty}^{j\infty} (R\bar{R} - R\bar{G}\bar{U}_{ok} - \bar{R}GU_{ok} + k^2 U_{ok}\bar{U}_{ok} + G\bar{G}U_{ok}\bar{U}_{ok})\,ds \tag{11-18}$$

$$J_2 = \frac{1}{2\pi j} \int_{-j\infty}^{j\infty} (k^2\bar{U}_{ok} - G\bar{R} + G\bar{G}\bar{U}_{ok})U_1\,ds \tag{11-19}$$

$$J_3 = \frac{1}{2\pi j} \int_{-j\infty}^{j\infty} (k^2 U_{ok} - \bar{G}R + G\bar{G}U_{ok})\bar{U}_1\,ds \tag{11-20}$$

$$J_4 = \frac{1}{2\pi j} \int_{-j\infty}^{j\infty} (k^2 + G\bar{G})U_1\bar{U}_1\,ds \tag{11-21}$$

For simplicity, we have used G for $G(s)$, U_{ok} for $U_{ok}(s)$, $\bar{U}_{ok}$ for $U_{ok}(-s)$, and so on.

In Eq. (11-17), J_1 is the optimum performance index, that is, the minimum J. Equations (11-19) and (11-20) indicate that $J_2 = J_3$, since the integrand of J_2 is identical to that of J_3 with s replaced by $-s$. Also, J_4 is always positive, since its integrand is symmetrical with respect to the imaginary axis of the s-plane.

The necessary and sufficient conditions for J in Eq. (11-17) to be a minimum are

$$\frac{\partial J}{\partial \lambda}\bigg|_{\lambda=0} = 0 \tag{11-22}$$

and

$$\frac{\partial^2 J}{\partial \lambda^2}\bigg|_{\lambda=0} > 0 \tag{11-23}$$

Applying Eq. (11-22) to Eq. (11-17) leads to

$$J_2 + J_3 = 2J_3 = 0 \tag{11-24}$$

and Eq. (11-23) gives

$$J_4 > 0 \tag{11-25}$$

which is always satisfied. Therefore, the necessary and sufficient conditions for minimum J are obtained from Eqs. (11-24) and (11-20),

$$J_3 = \frac{1}{2\pi j} \int_{-j\infty}^{j\infty} (k^2 U_{ok} - \bar{G}R + G\bar{G}U_{ok})\bar{U}_1 \, ds = 0 \qquad (11\text{-}26)$$

Since we have assumed that all the poles of U_1 are inside the left half of the s-plane, the poles of $\bar{U}_1$ must all lie in the right half of the s-plane. One solution of Eq. (11-26) is that all the poles of the integrand are in the right half of the s-plane, and the line integral is evaluated as a contour integral around the left half of the s-plane. Since the contour does not enclose any singularity of the integrand, the contour integration is zero. Thus J_3 of Eq. (11-26) is identically zero if all the poles of

$$X(s) = G\bar{G}U_{ok} - \bar{G}R + k^2 U_{ok} \qquad (11\text{-}27)$$

are located in the right half of the s-plane. Equation (11-27) is written

$$X(s) = (k^2 + G\bar{G})U_{ok} - \bar{G}R \qquad (11\text{-}28)$$

The function $(k^2 + G\bar{G})$ is symmetrical with respect to the imaginary axis in the s-plane. We let

$$Y(s)Y(-s) = Y\bar{Y} = k^2 + G\bar{G} \qquad (11\text{-}29)$$

where Y is a function that has only poles and zeros in the left half of the s-plane, and $\bar{Y}$ is Y with s replaced by $-s$ and has only right-half-plane poles and zeros. The following notation is defined:

$$Y = \{k^2 + G\bar{G}\}^+ \qquad (11\text{-}30)$$

$$\bar{Y} = \{k^2 + G\bar{G}\}^- \qquad (11\text{-}31)$$

The process of finding Y and $\bar{Y}$ is called *spectral factorization*; this is relatively straightforward for scalar functions but is quite complex for multivariable systems since matrix functions are involved.

Equation (11-28) is now written

$$X = Y\bar{Y}U_{ok} - \bar{G}R \qquad (11\text{-}32)$$

We must now separate the right side of Eq. (11-32) into two parts, one with poles only in the left half of the s-plane, and one with poles only in the right-half plane. This is accomplished by dividing both sides of the equation by $\bar{Y}$, yielding

$$\frac{X}{\bar{Y}} = YU_{ok} - \frac{\bar{G}R}{\bar{Y}} \qquad (11\text{-}33)$$

For convenience, we assume that R has no poles in the right-half plane. Performing partial fraction of $\bar{G}R/\bar{Y}$ in Eq. (11-33), the result is written

$$\frac{\bar{G}R}{\bar{Y}} = \left[\frac{\bar{G}R}{\bar{Y}}\right]_+ + \left[\frac{\bar{G}R}{\bar{Y}}\right]_- \qquad (11\text{-}34)$$

where

$$\left[\frac{\bar{G}R}{\bar{Y}}\right]_+ = \text{expansion with poles in left-half plane} \qquad (11\text{-}35)$$

$$\left[\frac{\bar{G}R}{\bar{Y}}\right]_- = \text{expansion with poles in right-half plane} \qquad (11\text{-}36)$$

Then, Eq. (11-33) becomes

$$\frac{X}{Y} + \left[\frac{\bar{G}R}{\bar{Y}}\right]_{-} = YU_{ok} - \left[\frac{\bar{G}R}{\bar{Y}}\right]_{+} \tag{11-37}$$

In this equation, the left-side terms have only poles in the right-half plane, and the right-side terms have only poles in the left-half plane. Therefore, the left-side terms must be independent of the right-side terms, or

$$U_{ok} = \frac{1}{Y}\left[\frac{\bar{G}R}{\bar{Y}}\right]_{+} \tag{11-38}$$

Note that U_{ok} is still a function of k^2 and it is the optimal control that minimizes the performance index of Eq. (11-12) for a given k^2. To find the absolute optimum value for U_{ok}, which is U_o, we must use the constraint condition

$$J_u = \frac{1}{2\pi j}\int_{-j\infty}^{j\infty} U(s)U(-s)\,ds = K \tag{11-39}$$

to determine k^2. The integral of Eq. (11-39) may be evaluated with the aid of Table 11-1.

Table 11-1   Tabulation of the Definite Integral

$$J_n = \frac{1}{2\pi j}\int_{-j\infty}^{j\infty} \frac{N(s)N(-s)}{D(s)D(-s)}\,ds$$

$$N(s) = N_{n-1}s^{n-1} + N_{n-2}s^{n-2} + \ldots + N_1 s + N_0$$

$$D(s) = D_n s^n + D_{n-1}s^{n-1} + \ldots + D_1 s + D_0$$

$$J_1 = \frac{N_0^2}{2D_0 D_1}$$

$$J_2 = \frac{N_1^2 D_0 + N_0^2 D_2}{2D_0 D_1 D_2}$$

$$J_3 = \frac{N_2^2 D_0 D_1 + (N_1^2 - 2N_0 N_2)D_0 D_3 + N_0^2 D_2 D_3}{2D_0 D_3(-D_0 D_3 + D_1 D_2)}$$

$$J_4 = \frac{N_3^2(-D_0^2 D_3 + D_0 D_1 D_2) + (N_2^2 - 2N_1 N_3)D_0 D_1 D_4 + (N_1^2 - 2N_0 N_2)D_0 D_3 D_4 + N_0^2(-D_1 D_4^2 + D_2 D_3 D_4)}{2D_0 D_4(-D_0 D_3^2 - D_1^2 D_4 + D_1 D_2 D_3)}$$

Once $U_{ok}(s)$ is determined, if desired, the optimal closed-loop transfer function can be easily determined. Let

$$M_{ok}(s) = \frac{C(s)}{R(s)} \tag{11-40}$$

which is the optimal closed-loop transfer function subject to finding k^2. Since

$$C(s) = G(s)U_{ok}(s) \tag{11-41}$$

$$M_{ok}(s) = \frac{G(s)}{R(s)}U_{ok}(s) \tag{11-42}$$

or

$$M_{ok} = \frac{G}{RY}\left[\frac{\bar{G}R}{\bar{Y}}\right]_{+} \tag{11-43}$$

Equations (11-38) and (11-43) are made more useful by defining

$$G(s) = \frac{N(s)}{D(s)} \tag{11-44}$$

where $N(s)$ and $D(s)$ are polynomials of s.
Then,

$$
\begin{aligned}
Y &= \{k^2 + G\bar{G}\}^+ \\
&= \left\{ \frac{k^2 D\bar{D} + N\bar{N}}{D\bar{D}} \right\}^+ = \frac{\{k^2 D\bar{D} + N\bar{N}\}^+}{D}
\end{aligned} \tag{11-45}
$$

and

$$\bar{Y} = \frac{\{k^2 D\bar{D} + N\bar{N}\}^-}{\bar{D}} \tag{11-46}$$

Equation (11-38) now gives

$$U_{ok} = \frac{D}{\{k^2 D\bar{D} + N\bar{N}\}^+} \left[\frac{\bar{N}R}{\{k^2 D\bar{D} + N\bar{N}\}^-} \right]_+ \tag{11-47}$$

Similarly, Eq. (11-43) leads to

$$M_{ok} = \frac{N}{R\{k^2 D\bar{D} + N\bar{N}\}^+} \left[\frac{\bar{N}R}{\{k^2 D\bar{D} + N\bar{N}\}^-} \right]_+ \tag{11-48}$$

which is the optimal transfer function of the closed-loop system, subject to finding k^2.

EXAMPLE 11-1   Referring to the block diagram of Fig. 11-1, consider that

$$G(s) = \frac{10}{s^2} \tag{11-49}$$

and the input transform is

$$R(s) = \frac{0.5}{s} \tag{11-50}$$

Determine the optimal control $U_0(s)$ and the optimal closed-loop transfer function $M_0(s)$ so that

$$(1) \quad J_e = \int_0^\infty e^2(t)\, dt = \text{minimum} \tag{11-51}$$

$$(2) \quad J_u = \int_0^\infty u^2(t)\, dt \leq 2.5 \tag{11-52}$$

Since $N = 10$ and $D = s^2$, we have

$$\{k^2 D\bar{D} + N\bar{N}\}^+ = \{k^2 s^4 + 100\}^+ \tag{11-53}$$

To carry out the spectral factorization of the last equation, we write

$$
\begin{aligned}
k^2 s^4 + 100 &= (a_2 s^2 + a_1 s + a_0)(a_2 s^2 - a_1 s + a_0) \\
&= a_2^2 s^4 - (a_1^2 - 2a_2 a_0)s^2 + a_0^2
\end{aligned} \tag{11-54}
$$

Equating like coefficients on both sides of Eq. (11-54),

$$a_2^2 = k^2 \tag{11-55}$$

or

$$a_2 = k \tag{11-56}$$

$$a_1^2 - 2a_2a_0 = 0 \tag{11-57}$$

$$a_0^2 = 100 \tag{11-58}$$

which gives

$$a_0 = 10 \tag{11-59}$$

Substituting the results of a_0 and a_2 into Eq. (11-57), we get

$$a_1 = \sqrt{20k} \tag{11-60}$$

Therefore,

$$\{k^2 D\bar{D} + N\bar{N}\}^+ = (a_2 s^2 + a_1 s + a_0)$$
$$= ks^2 + \sqrt{20k}s + 10 \tag{11-61}$$

and

$$\{k^2 D\bar{D} + N\bar{N}\}^- = ks^2 - \sqrt{20k}s + 10 \tag{11-62}$$

The second term on the right side of Eq. (11-47) is

$$\left[\frac{\bar{N}R}{\{k^2 D\bar{D} + N\bar{N}\}^-} \right]_+ = \left[\frac{10(0.5/s)}{ks^2 - \sqrt{20k}s + 10} \right]_+$$
$$= \frac{0.5}{s} \tag{11-63}$$

where we have considered that the pole at the origin is slightly over in the left-half plane. Equation (11-47) now gives

$$U_{ok} = \frac{s^2}{ks^2 + \sqrt{20k}s + 10} \frac{0.5}{s} = \frac{0.5s}{ks^2 + \sqrt{20k}s + 10} \tag{11-64}$$

Substituting U_{ok} into Eq. (11-39) for U, we have

$$J_u = \frac{1}{2\pi j} \int_{-j\infty}^{j\infty} \frac{0.5s}{ks^2 + \sqrt{20k}s + 10} \frac{-0.5s}{ks^2 - \sqrt{20k}s + 10} ds \tag{11-65}$$

This integral is evaluated by use of the formula in Table 11-1. From Table 11-1,

$$J_u = \frac{N_1^2 D_0 + N_0^2 D_2}{2 D_0 D_1 D_2} \tag{11-66}$$

where $N_1 = 0.5$, $N_0 = 0$, $D_0 = 10$, $D_1 = \sqrt{20k}$, $D_2 = k$. Thus

$$J_u = \frac{0.125}{k\sqrt{20k}} = 2.5 \tag{11-67}$$

from which

$$k = 0.05$$

The optimal control is given by

$$U_0(s) = \frac{0.5s}{0.05s^2 + s + 10} \tag{11-68}$$

and the optimal closed-loop transfer function is

$$M_0(s) = \frac{G(s)}{R(s)} U_0(s) = \frac{10}{0.05s^2 + s + 10}$$
$$= \frac{200}{s^2 + 20s + 200} \tag{11-69}$$

The following features of the design should be noted:

1. The performance index is the integral square error with constraint on the control signal in the form of integral square control. No specifications are given in the form of overshoot, bandwidth, rise time, gain margin, and so on.

2. The optimal control and the optimal closed-loop transfer function are determined without specifying the final system configuration.

It is interesting to note that the optimal system represented by the transfer function of Eq. (11-69) has a damping ratio of 0.707 and a bandwidth of 14.14 rad/sec.

There is a great deal of flexibility in arriving at the final system configuration and the transfer function of the controller. Figure 11-2(a) shows the system with a series controller. The transfer function of the series controller is determined as

$$G_c(s) = \frac{1}{G(s)} \frac{M_0(s)}{1 - M_0(s)}$$
$$= \frac{20s}{s + 20}$$

(11-70)

As an alternative, Fig. 11-2(b) shows the system with a feedback controller which is a tachometer, and series controller, with

$$G_c(s) = 20$$

and

$$H(s) = 0.1s$$

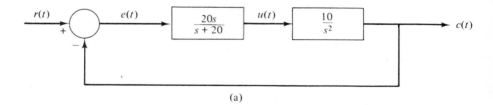

(a)

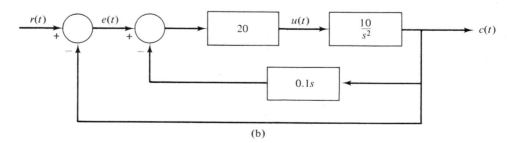

(b)

Fig. 11-2. (a) Analytically designed system in Example 11-1 with a series controller. (b) Analytically designed system in Example 11-1 with a feedback controller.

EXAMPLE 11-2   Referring to the system of Fig. 11-1, consider that

$$G(s) = \frac{s-1}{s^2} \qquad (11\text{-}71)$$

The input is a unit step function. It is desired to determine the optimal closed-loop transfer function $M_0(s)$ so that $J_e = $ minimum and $J_u \leq 1$.

Using the notation of Eq. (11-44), $N = s - 1$, $D = s^2$. Then, $\bar{N} = -s - 1$ and $\bar{D} = s^2$.

$$\{k^2 D\bar{D} + N\bar{N}\}^+ = \{k^2 s^4 - s^2 + 1\}^+ \qquad (11\text{-}72)$$

The spectral factorization of Eq. (11-72) is carried out by writing

$$
\begin{aligned}
k^2 s^4 - s^2 + 1 &= (a_2 s^2 + a_1 s + a_0)(a_2 s^2 - a_1 s + a_0) \\
&= a_2^2 s^4 - (a_1^2 - 2a_2 a_0)s^2 + a_0^2
\end{aligned}
\qquad (11\text{-}73)
$$

Equating like terms on both sides of Eq. (11-73) yields

$$a_2 = k \qquad (11\text{-}74)$$

$$a_1 = \sqrt{2k+1} \qquad (11\text{-}75)$$

$$a_0 = 1 \qquad (11\text{-}76)$$

Thus

$$\{k^2 D\bar{D} + N\bar{N}\}^+ = ks^2 + \sqrt{2k+1}\,s + 1 \qquad (11\text{-}77)$$

and

$$\{k^2 D\bar{D} + N\bar{N}\}^- = ks^2 - \sqrt{2k+1}\,s + 1 \qquad (11\text{-}78)$$

The second term on the right side of Eq. (11-47) is

$$\left[\frac{\bar{N}R}{\{k^2 D\bar{D} + N\bar{N}\}^-}\right]_+ = \left[\frac{-(s+1)}{s(ks^2 - \sqrt{2k+1}\,s + 1)}\right]_+ = -\frac{1}{s} \qquad (11\text{-}79)$$

where we have considered that the pole at the origin is slightly to the left of the imaginary axis. The optimal control is now found from Eq. (11-47),

$$U_{ok} = \frac{-s}{ks^2 + \sqrt{2k+1}\,s + 1} \qquad (11\text{-}80)$$

subject to the determination of k.

Using Eq. (11-39), and evaluating the integral with the help of Table 11-1, we have

$$k = 0.378$$

The optimal closed-loop transfer function is found by use of Eq. (11-48),

$$M_0(s) = \frac{-2.645(s-1)}{s^2 + 3.5s + 2.645} \qquad (11\text{-}81)$$

It is worth noting that the closed-loop transfer function contains the right-half-plane zero of $G(s)$. This is a necessary condition for $u(t)$ to be bounded. In other words, if $M_0(s)$ does not contain the term $(s - 1)$ in its numerator in this problem, it would not be possible to satisfy the $J_u \leq 1$ requirement. However, the analytical design formulation derived in this section naturally leads to the proper solution for this case.

It should be pointed out that the analytical design can yield a physically realizable solution only if J_e and J_u are finite. Given an arbitrary input and process transfer function $G(s)$, a minimum integral-square-error solution may not exist. For instance, it is well known that the steady-state error of a type-0 system is nonzero for a step input. Thus J_e would be infinite if $G(s)$ is of the form

$$G(s) = \frac{K}{(s + a)(s + b)(s + c)} \qquad a, b, c \neq 0 \tag{11-82}$$

11.3 Parameter Optimization

The analytical design method introduced in Section 11.2 is restricted to single-input systems only. For multivariable systems, the spectral factorization problem causes the solution to be a great deal more complex. The advantage with the analytical design is that the configuration of the system to be designed is semi-fixed, so that the end result of the design is the optimal closed-loop transfer function.

As a variation to the analytical design, we may consider a fixed-configuration system but with free parameters that are to be optimized so that the following specifications are satisfied:

$$J_e = \int_0^\infty e^2(t)\, dt \tag{11-83}$$

$$J_u = \int_0^\infty u^2(t)\, dt \leq K \tag{11-84}$$

As in the analytical design, we consider the case when $J_u = K$, and the problem becomes that of minimizing the following performance index:

$$J = J_e + k^2 J_u \tag{11-85}$$

where k^2 is the Lagrange multiplier whose value is to be determined. The system is optimized with respect to a given reference input.

The advantage with the parameter optimization method is that it can be applied to systems with multiple inputs. The disadvantage is that the actual implementation of the design is quite tedious for complex systems without the use of a digital computer.

Consider that the control system to be designed has a fixed configuration for the process and the controllers that contain n free parameters, $K_1, K_2, \ldots, K_n$. These parameters are to be determined so that J of Eq. (11-85) is minimized, subject to the constraint

$$J_u = K \tag{11-86}$$

Applying Parseval's theorem and Table 11-1, the performance index J is expressed as a function of the n free parameters and k^2; that is,

$$J = J(K_1, K_2, \ldots, K_n, k^2) \tag{11-87}$$

The necessary condition for J to be a minimum is

$$\frac{\partial J}{\partial K_i} = 0 \qquad i = 1, 2, \ldots, n \tag{11-88}$$

and the sufficient condition is that the following Hessian matrix is positive definite:

$$\nabla^2 J = \begin{bmatrix} \dfrac{\partial^2 J}{\partial K_1^2} & \dfrac{\partial^2 J}{\partial K_1 \partial K_2} & \cdots & \dfrac{\partial^2 J}{\partial K_1 \partial K_n} \\[2ex] \dfrac{\partial^2 J}{\partial K_2 \partial K_1} & \dfrac{\partial^2 J}{\partial K_2^2} & \cdots & \dfrac{\partial^2 J}{\partial K_2 \partial K_n} \\[1ex] \cdots\cdots\cdots\cdots\cdots\cdots\cdots\cdots\cdots\cdots \\[1ex] \dfrac{\partial^2 J}{\partial K_n \partial K_1} & \dfrac{\partial^2 J}{\partial K_n \partial K_2} & \cdots & \dfrac{\partial^2 J}{\partial K_n^2} \end{bmatrix} \tag{11-89}$$

Since $\partial^2 J/\partial K_i\,\partial K_j = \partial^2 J/\partial K_j\,\partial K_i$, the Hessian matrix is always symmetric. In addition to the necessary and sufficient conditions of Eqs. (11-88) and (11-89), the equality constraint of Eq. (11-86) must be satisfied.

The following examples illustrate how the parameter optimization design is carried out.

EXAMPLE 11-3   Consider the control system of Example 11-1. Let us use the cascade controller configuration of Fig. 11-2, with the controller parameters set as K_1 and K_2. The block diagram of the system is shown in Fig. 11-3. The problem is to determine the optimal values of K_1 and K_2 such that

$$J_e = \int_0^\infty e^2(t)\,dt = \text{minimum} \tag{11-90}$$

$$J_u = \int_0^\infty u^2(t)\,dt = 2.5 \tag{11-91}$$

The reference input to the system is a step function with an amplitude of 0.5.

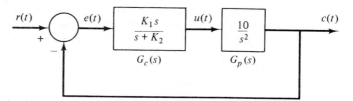

Fig. 11-3. Control system for parameter optimization.

The problem is equivalent to that of minimizing

$$J = J_e + k^2 J_u = \frac{1}{2\pi j} \int_{-j\infty}^{j\infty} [E(s)E(-s) + k^2 U(s)U(-s)]\,ds \tag{11-92}$$

From Fig. 11-3 we have

$$E(s) = \frac{R(s)}{1 + G_c(s)G_p(s)} = \frac{0.5(s + K_2)}{s^2 + K_2 s + 10K_1} \tag{11-93}$$

and

$$U(s) = E(s)G_c(s) = \frac{0.5K_1 s}{s^2 + K_2 s + 10K_1} \tag{11-94}$$

Substituting Eqs. (11-93) and (11-94) in Eq. (11-92), and using Parseval's theorem and Table 11-1, we have

$$J = \frac{0.25(10K_1 + K_2^2)}{20K_1K_2} + k^2\frac{2.5K_1^3}{20K_1K_2} \tag{11-95}$$

Applying the necessary condition for J to be a minimum, we get from setting the necessary conditions and the equality constraint,

$$\frac{\partial J}{\partial K_1} = 0 \qquad -0.25K_2^2 + 5k^2K_1^3 = 0 \tag{11-96}$$

$$\frac{\partial J}{\partial K_2} = 0 \qquad K_2^2 - 10K_1 - 10k^2K_1^3 = 0 \tag{11-97}$$

$$J_u = 2.5 \qquad K_1^2 = 20K_2 \tag{11-98}$$

The last three simultaneous equations give the solutions

$$K_1 = 20 \qquad K_2 = 20 \qquad k^2 = 0.05 \tag{11-99}$$

which are identical to the results found in Example 11-1. It can be shown that the Hessian matrix $\nabla^2 J$ is positive definite, so that the sufficient condition is also satisfied.

The fact that K_1 equals K_2 in this problem is purely coincidental. We could have formulated a less flexible problem by assigning $K_1 = K_2$. Then there would be only one free parameter to optimize. The solution for such a situation would be quite simple for the design criteria given in Eqs. (11-90) and (11-91), since the optimal value of the parameter can be obtained by using only the constraint requirement of Eq. (11-91).

11.4 Design of System with Specific Eigenvalues—An Application of Controllability[3-9]

An interesting and useful application of state controllability is the design of a linear feedback control system with specific eigenvalues. The system under consideration may be represented by the block diagram of Fig. 11-4. The linear process is described by the state equation

$$\dot{\mathbf{x}} = \mathbf{A}\mathbf{x} + \mathbf{B}u \tag{11-100}$$

where the scalar control is

$$u = -\mathbf{G}\mathbf{x} + r \tag{11-101}$$

The matrix $\mathbf{G}$ is an $1 \times n$ feedback matrix. Combining Eqs. (11-100) and (11-101), the closed-loop system is represented by the state equation

$$\dot{\mathbf{x}} = (\mathbf{A} - \mathbf{BG})\mathbf{x} + \mathbf{B}r \tag{11-102}$$

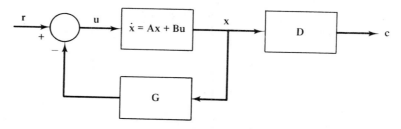

Fig. 11-4. Linear system with state feedback.

It can be shown that if $[\mathbf{A}, \mathbf{B}]$ is a controllable pair, then a matrix $\mathbf{G}$ exists that will give an arbitrary set of eigenvalues of $(\mathbf{A} - \mathbf{BG})$. In other words, the roots of the characteristic equation

$$|\lambda\mathbf{I} - (\mathbf{A} - \mathbf{BG})| = 0 \qquad (11\text{-}103)$$

can be arbitrarily placed.

It has been shown in Section 4.7 that if a system is state controllable, it can always be represented in the phase-variable canonical form; that is, in Eq. (11-100),

$$\mathbf{A} = \begin{bmatrix} 0 & 1 & 0 & \ldots & 0 \\ 0 & 0 & 1 & \ldots & 0 \\ 0 & 0 & 0 & \ldots & 0 \\ \multicolumn{5}{c}{\dotfill} \\ 0 & 0 & 0 & & 1 \\ -a_n & -a_{n-1} & -a_{n-2} & \ldots & -a_1 \end{bmatrix} \qquad \mathbf{B} = \begin{bmatrix} 0 \\ 0 \\ 0 \\ \cdot \\ \cdot \\ \cdot \\ 1 \end{bmatrix}$$

The reverse is also true in that if the system is represented in the phase-variable canonical form, it is always state controllable. To show this, we form the following matrices:

$$\mathbf{AB} = \begin{bmatrix} 0 \\ 0 \\ 0 \\ \cdot \\ \cdot \\ \cdot \\ 1 \\ -a_1 \end{bmatrix} \qquad \mathbf{A^2B} = \begin{bmatrix} 0 \\ 0 \\ 0 \\ \cdot \\ \cdot \\ \cdot \\ 1 \\ -a_1 \\ a_1^2 - a_2 \end{bmatrix} \qquad \mathbf{A^3B} = \begin{bmatrix} 0 \\ 0 \\ 0 \\ \cdot \\ \cdot \\ \cdot \\ 1 \\ -a_1 \\ a_1^2 - a_2 \\ -a_1^3 + a_2 a_1 - a_3 \end{bmatrix}$$

Continuing with the matrix product through $\mathbf{A^{n-1}B}$, it will become apparent that regardless of what the values of $a_1, a_2, \ldots, a_n$, are, the determinant of $\mathbf{S} = [\mathbf{B} \quad \mathbf{AB} \quad \mathbf{A^2B} \ldots \mathbf{A^{n-1}B}]$ will always be equal to -1, since $\mathbf{S}$ is a triangular matrix with 1s on the main diagonal. Therefore, we have proved that if the system is represented by the phase-variable canonical form, it is always state controllable.

The feedback matrix $\mathbf{G}$ can be written

$$\mathbf{G} = [g_1 \quad g_2 \quad \ldots \quad g_n] \qquad (11\text{-}104)$$

Then

$$\mathbf{A} - \mathbf{BG} = \begin{bmatrix} 0 & 1 & 0 & 0 \\ 0 & 0 & 1 & \ldots & 0 \\ 0 & 0 & 0 & \ldots & 0 \\ \multicolumn{5}{c}{\dotfill} \\ 0 & 0 & 0 & 1 \\ -a_n - g_1 & -a_{n-1} - g_2 & \ldots & -a_1 - g_n \end{bmatrix} \qquad (11\text{-}105)$$

The eigenvalues of $\mathbf{A} - \mathbf{BG}$ are then found from the characteristic equation

$$|\lambda\mathbf{I} - (\mathbf{A} - \mathbf{BG})| = \lambda^n + (a_1 + g_n)\lambda^{n-1} + (a_2 + g_{n-1})\lambda^{n-2} + \ldots + (a_n + g_1)$$
$$= 0 \qquad (11\text{-}106)$$

Clearly, the eigenvalues can be arbitrarily chosen by the choice of $g_1, g_2, \ldots, g_n$.

EXAMPLE 11-4   Consider that a linear process is described by the transfer function

$$\frac{C(s)}{U(s)} = \frac{10}{s(s+1)(s+2)} \qquad (11\text{-}107)$$

It is desired to design a feedback controller with state feedback so that the eigenvalues of the closed-loop system are at -2, $-1 + j1$, and $-1 - j1$. Since the transfer function does not have common poles and zeros, the process is completely state controllable.

By direct decomposition, the state equations for Eq. (11-107) are written

$$\begin{bmatrix} \dot{x}_1 \\ \dot{x}_2 \\ \dot{x}_3 \end{bmatrix} = \begin{bmatrix} 0 & 1 & 0 \\ 0 & 0 & 1 \\ 0 & -2 & -3 \end{bmatrix}\begin{bmatrix} x_1 \\ x_2 \\ x_3 \end{bmatrix} + \begin{bmatrix} 0 \\ 0 \\ 1 \end{bmatrix}u \qquad (11\text{-}108)$$

Since this is a third-order system, the feedback matrix $\mathbf{G}$ is of the form

$$\mathbf{G} = [g_1 \quad g_2 \quad g_3] \qquad (11\text{-}109)$$

The characteristic equation of the closed-loop system is

$$\lambda^3 + (3 + g_3)\lambda^2 + (2 + g_2)\lambda + g_1 = 0 \qquad (11\text{-}110)$$

and with the prescribed eigenvalues it is

$$\lambda^3 + 4\lambda^2 + 6\lambda + 4 = 0 \qquad (11\text{-}111)$$

Equating the like coefficients of the last two equations, we have

$$g_1 = 4 \qquad g_2 = 4 \qquad g_3 = 1$$

or

$$\mathbf{G} = [4 \quad 4 \quad 1] \qquad (11\text{-}112)$$

The state diagram of the overall system is shown in Fig. 11-5.

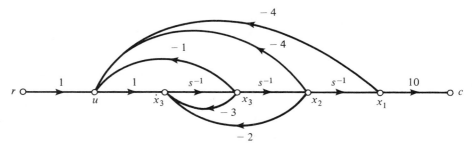

Fig. 11-5. Linear control system with state feedback for Example 11-4.

Stabilizability[8]

We have shown that if a system is completely controllable, its eigenvalues can be arbitrarily located by complete state feedback. This also implies that any unstable system can be stabilized by complete state feedback if all the states are controllable. On the other hand, if the system is not completely controllable, as long as the uncontrollable states are stable, the entire system is still stabilizable.

11.5 Design of State Observers[10-13]

Another area of design of control systems concerned with the concepts of controllability and observability is the design of *state observers*. Suppose that in the design of the state feedback controller discussed in Section 11.4, we determined the feedback matrix **G** so that

$$u(t) = -\mathbf{G}\mathbf{x}(t) + \dot{r}(t) \tag{11-113}$$

However, in order to implement the state feedback, we still need to feed back all the state variables. Unfortunately, in practice, not all the state variables are accessible, and we can assume that only the outputs and the inputs are measurable. The subsystem that performs the observation of the state variables based on information received from the measurements of the input $u(t)$ and the output $\mathbf{c}(t)$ is called an *observer*.

Figure 11-6 shows the overall system structure including the observer. The scalar control is ordinarily given by

$$u(t) = E r(t) - \mathbf{G}\mathbf{x}(t) \tag{11-114}$$

where E and $\mathbf{G}$ are matrices of appropriate dimensions. Figure 11-6 shows that the inputs to the observer are the output $\mathbf{c}(t)$ and the control u. The output of the observer is the observed state vector $\mathbf{x}_e(t)$. Therefore, the actual control is

$$u(t) = E r(t) - \mathbf{G}\mathbf{x}_e(t) \tag{11-115}$$

We must establish the condition under which an observer exists. The following theorem indicates that the design of the state observer is closely related to the condition of observability.

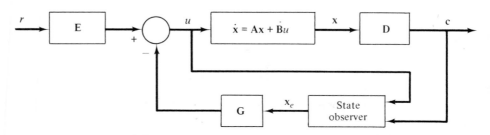

Fig. 11-6. Linear system with state feedback and state observer.

Theorem 11-1. Given an nth-order linear time-invariant system that is described by the dynamic equations,

$$\dot{\mathbf{x}}(t) = \mathbf{A}\mathbf{x}(t) + \mathbf{B}u(t) \tag{11-116}$$

$$\mathbf{c}(t) = \mathbf{D}\mathbf{x}(t) \tag{11-117}$$

*where **x** is an n-vector, u is the scalar control, and **c** is a p-vector. The state vector **x*** *may be constructed from linear combinations of the output* **c**, *input u, and derivatives of these variables if the system is completely observable.*

Proof: Assuming that the system is completely observable, the matrix

$$\mathbf{T} = [\mathbf{D}' \quad \mathbf{A}'\mathbf{D}' \quad (\mathbf{A}')^2\mathbf{D}' \ldots (\mathbf{A}')^{n-1}\mathbf{D}'] \tag{11-118}$$

is of rank n. We shall show that the states can be expressed as linear combinations of the output $\mathbf{c}(t)$, the input $u(t)$, and their derivatives. Starting with Eq. (11-117), we take the derivative on both sides with respect to t. We have

$$\dot{\mathbf{c}}(t) = \mathbf{D}\dot{\mathbf{x}}(t) = \mathbf{D}\mathbf{A}\mathbf{x}(t) + \mathbf{D}\mathbf{B}u(t) \tag{11-119}$$

or

$$\dot{\mathbf{c}}(t) - \mathbf{D}\mathbf{B}u(t) = \mathbf{D}\mathbf{A}\mathbf{x}(t) \tag{11-120}$$

Repeating the time derivative process to Eq. (11-120), we get

$$\ddot{\mathbf{c}}(t) - \mathbf{D}\mathbf{B}\dot{u}(t) = \mathbf{D}\mathbf{A}\dot{\mathbf{x}}(t) = \mathbf{D}\mathbf{A}^2\mathbf{x}(t) + \mathbf{D}\mathbf{A}\mathbf{B}u(t) \tag{11-121}$$

Rearranging the last equation yields

$$\ddot{\mathbf{c}}(t) - \mathbf{D}\mathbf{B}\dot{u}(t) - \mathbf{D}\mathbf{A}\mathbf{B}u(t) = \mathbf{D}\mathbf{A}^2\mathbf{x}(t) \tag{11-122}$$

Continuing the process, we have, after taking n derivatives,

$$\mathbf{c}^{(n)}(t) = \mathbf{D}\mathbf{B}u^{(n-1)}(t) - \mathbf{D}\mathbf{A}\mathbf{B}u^{(n-2)}(t) - \ldots - \mathbf{D}\mathbf{A}^{n-1}\mathbf{B}u(t) = \mathbf{D}\mathbf{A}^{n-1}\mathbf{x}(t) \tag{11-123}$$

In matrix form, Eqs. (11-117) and (11-120) through (11-123) become

$$\begin{bmatrix} \mathbf{c} \\ \dot{\mathbf{c}} - \mathbf{D}\mathbf{B}u \\ \ddot{\mathbf{c}} - \mathbf{D}\mathbf{B}\dot{u} - \mathbf{D}\mathbf{A}\mathbf{B}u \\ \cdots\cdots\cdots\cdots \\ \cdots\cdots\cdots\cdots \\ \cdots\cdots\cdots\cdots \\ \mathbf{c}^{(n)} - \mathbf{D}\mathbf{B}u^{(n-1)} - \mathbf{D}\mathbf{A}\mathbf{B}u^{(n-2)} - \ldots - \mathbf{D}\mathbf{A}^{n-1}\mathbf{B}u \end{bmatrix} = \begin{bmatrix} \mathbf{D} \\ \mathbf{D}\mathbf{A} \\ \mathbf{D}\mathbf{A}^2 \\ \cdot \\ \cdot \\ \cdot \\ \mathbf{D}\mathbf{A}^{n-1} \end{bmatrix} \mathbf{x} \tag{11-124}$$

Therefore, in order to express the state vector $\mathbf{x}$ in terms of $\mathbf{c}$, u, and their derivatives, the matrix

$$\mathbf{T} = [\mathbf{D} \quad \mathbf{D}\mathbf{A} \quad \mathbf{D}\mathbf{A}^2 \ldots \mathbf{D}\mathbf{A}^{n-1}]'$$
$$= [\mathbf{D}' \quad \mathbf{A}'\mathbf{D}' \quad (\mathbf{A}')^2\mathbf{D}' \ldots (\mathbf{A}')^{n-1}\mathbf{D}']$$

must be of rank n.

There are many ways of designing a state observer, and there is more than one way of judging the closeness of $\mathbf{x}_e(t)$ to $\mathbf{x}(t)$. Intuitively, the observer should have the same state equations as the original system. However, the observer should have $u(t)$ and $\mathbf{c}(t)$ as inputs and should have the capability of minimizing the error between $\mathbf{x}(t)$ and $\mathbf{x}_e(t)$. Since we cannot measure $\mathbf{x}(t)$ directly, an alternative is to compare $\mathbf{c}(t)$ and $\mathbf{c}_e(t)$, where

$$\mathbf{c}_e(t) = \mathbf{D}\mathbf{x}_e(t) \tag{11-125}$$

Based on the above arguments, a logical arrangement for the state observer is shown in Fig. 11-7. The state observer design is formulated as a feedback

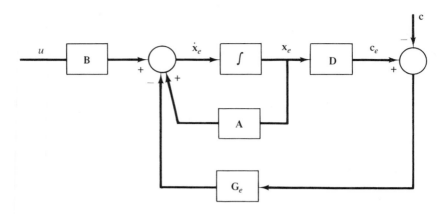

Fig. 11-7. Block diagram of an observer.

control problem with $\mathbf{G}_e$ as the feedback matrix. The design objective is to select the feedback matrix $\mathbf{G}_e$ such that $\mathbf{c}_e(t)$ will approach $\mathbf{c}(t)$ as fast as possible. When $\mathbf{c}_e(t)$ equals $\mathbf{c}(t)$, the dynamics of the state observer are described by

$$\dot{\mathbf{x}}_e(t) = \mathbf{A}\mathbf{x}_e(t) + \mathbf{B}u(t) \tag{11-126}$$

which is identical to the state equation of the system to be observed. In general, with u and $\mathbf{c}(t)$ as inputs to the oberver, the dynamics of the observer are represented by

$$\dot{\mathbf{x}}_e(t) = (\mathbf{A} - \mathbf{G}_e\mathbf{D})\mathbf{x}_e(t) + \mathbf{B}u(t) + \mathbf{G}_e\mathbf{c}(t) \tag{11-127}$$

The block diagram of the overall system with the observer is shown in Fig. 11-8. Since $\mathbf{c}(t)$ and $\mathbf{x}(t)$ are related through Eq. (11-117), the last equation is also written as

$$\dot{\mathbf{x}}_e(t) = \mathbf{A}\mathbf{x}_e(t) + \mathbf{B}u(t) + \mathbf{G}_e\mathbf{D}[\mathbf{x}(t) - \mathbf{x}_e(t)] \tag{11-128}$$

The significance of this expression is that if the initial values of $\mathbf{x}(t)$ and $\mathbf{x}_e(t)$ are identical, the equation reverts to that of Eq. (11-126), and the response of the observer will be identical to that of the original system. Therefore, the design of the feedback matrix $\mathbf{G}_e$ for the observer is significant only if the initial conditions to $\mathbf{x}(t)$ and $\mathbf{x}_e(t)$ are different.

There are many possible ways of carrying out the design of the feedback matrix $\mathbf{G}_e$. One way is to utilize the eigenvalue assignment method discussed in Section 11.4. If we subtract Eq. (11-128) from Eq. (11-116), we have

$$[\dot{\mathbf{x}}(t) - \dot{\mathbf{x}}_e(t)] = (\mathbf{A} - \mathbf{G}_e\mathbf{D})[\mathbf{x}(t) - \mathbf{x}_e(t)] \tag{11-129}$$

which may be regarded as the homogeneous state equation of a linear system with coefficient matrix $\mathbf{A} - \mathbf{G}_e\mathbf{D}$. The characteristic equation of $\mathbf{A} - \mathbf{G}_e\mathbf{D}$ and of the state observer is then

$$|\lambda\mathbf{I} - (\mathbf{A} - \mathbf{G}_e\mathbf{D})| = 0 \tag{11-130}$$

Since we are interested in driving $\mathbf{x}_e(t)$ as close to $\mathbf{x}(t)$ as possible, the objective of the observer design may be stated as to select the elements of $\mathbf{G}_e$ so that the natural response of Eq. (11-129) decays to zero as quickly as possible. In other

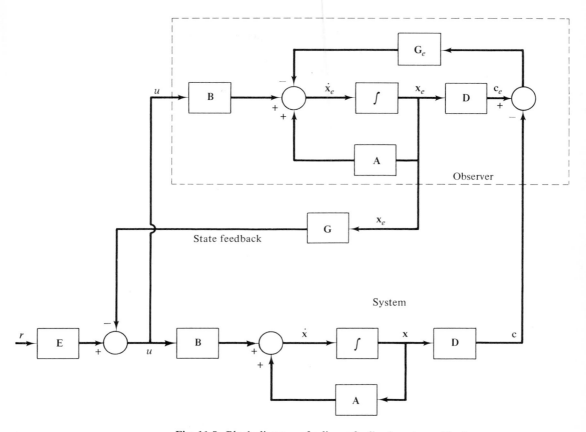

Fig. 11-8. Block diagram of a linear feedback system with observer.

words, the eigenvalues of $\mathbf{A} - \mathbf{G}_e\mathbf{D}$ should be selected so that $\mathbf{x}_e(t)$ approaches $\mathbf{x}(t)$ rapidly. However, it must be kept in mind that the approach of assigning the eigenvalues of $\mathbf{A} - \mathbf{G}_e\mathbf{D}$ may not always be satisfactory for the purpose of matching all the observed states to the real states, since the eigenvalues control only the denominator polynomial of the transfer relation, while the numerator polynomial is not controlled. In Section 11.4 the condition for the arbitrary assignment of the eigenvalues of $(\mathbf{A} - \mathbf{BG})$ given $\mathbf{A}$ and $\mathbf{B}$ has been established. We must first condition Eq. (11-129) so that the similarity between the two cases may be established. It is easy to see that the eigenvalues of $(\mathbf{A} - \mathbf{G}_e\mathbf{D})$ are identical to those of $(\mathbf{A} - \mathbf{G}_e\mathbf{D})'$. The latter matrix is written

$$(\mathbf{A} - \mathbf{G}_e\mathbf{D})' = (\mathbf{A}' - \mathbf{D}'\mathbf{G}_e') \tag{11-131}$$

Thus, invoking the results of Section 11.4, the condition of arbitrary assignment of the eigenvalues of $(\mathbf{A} - \mathbf{G}_e\mathbf{D})$ or of $(\mathbf{A}' - \mathbf{D}'\mathbf{G}_e')$ is that the pair $[\mathbf{A}', \mathbf{D}']$ be completely controllable. This is equivalent to requiring that the pair $[\mathbf{A}, \mathbf{D}]$ be completely observable. *Therefore, the observability of* $[\mathbf{A}, \mathbf{D}]$ *ensures not only that a state observer can be constructed from linear combinations of the output* $\mathbf{c}$, *input u, and the derivatives of these, but also that the eigenvalues of the observer can be arbitrarily assigned by choosing the feedback matrix* $\mathbf{G}_e$.

At this point an illustrative example may be deemed necessary to show the procedure of the observer design.

EXAMPLE 11-5   Consider that a linear system is described by the following transfer function and dynamic equations:

$$\frac{C(s)}{U(s)} = \frac{2}{(s+1)(s+2)} \tag{11-132}$$

$$\dot{\mathbf{x}}(t) = \mathbf{A}\mathbf{x}(t) + \mathbf{B}u(t) \tag{11-133}$$

$$\mathbf{c}(t) = \mathbf{D}\mathbf{x}(t) \tag{11-134}$$

where

$$\mathbf{A} = \begin{bmatrix} 0 & 1 \\ -2 & -3 \end{bmatrix} \quad \mathbf{B} = \begin{bmatrix} 0 \\ 1 \end{bmatrix}$$

$$\mathbf{D} = [2 \quad 0]$$

It is desired to design a state observer so that the eigenvalues of $|\lambda \mathbf{I} - (\mathbf{A} - \mathbf{G}_e\mathbf{D})| = 0$ are at $\lambda = -10, -10$.

Let the feedback matrix be designated as

$$\mathbf{G}_e = \begin{bmatrix} g_{e1} \\ g_{e2} \end{bmatrix} \tag{11-135}$$

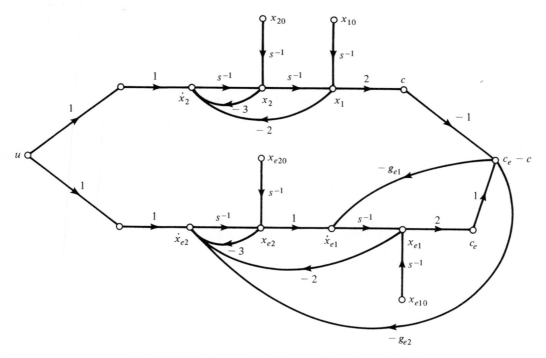

Fig. 11-9. State diagram of the observer and the original system for Example 11-5.

Then, the characteristic equation of the state observer is

$$|\lambda \mathbf{I} - (\mathbf{A} - \mathbf{G}_e \mathbf{D})| = \begin{vmatrix} \lambda + 2g_{e1} & -1 \\ 2 + 2g_{e2} & \lambda + 3 \end{vmatrix}$$

$$= \lambda^2 + (2g_{e1} + 3)\lambda + (6g_{e1} + 2 + 2g_{e2}) = 0 \qquad (11\text{-}136)$$

For the eigenvalues to be at $\lambda = -10, -10$, the characteristic equation should be

$$\lambda^2 + 20\lambda + 100 = 0 \qquad (11\text{-}137)$$

Equating like terms of Eqs. (11-136) and (11-137) gives

$$g_{e1} = 8.5$$

$$g_{e2} = 23.5$$

The state diagram for the observer together with the original system is shown in Fig. 11-9.

As was mentioned earlier, if the initial values of $\mathbf{x}(t)$ and $\mathbf{x}_e(t)$ are identical, the responses of the observer will be identical to those of the original system, and the feedback matrix $\mathbf{G}_e$ will have no effect on the responses of the observer whatsoever. Figure 11-10 illustrates the unit step responses of $x_1(t)$ and $x_{e1}(t)$ for the following

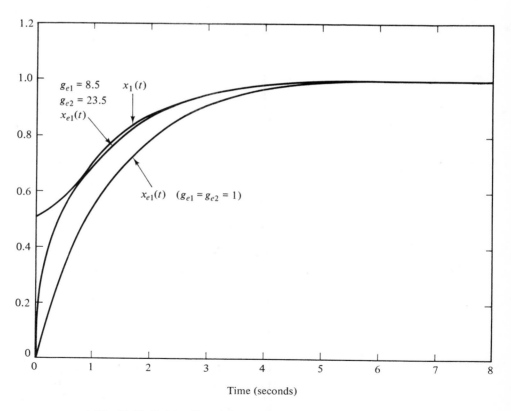

Fig. **11-10.** State $x_1(t)$ and the observed states of the observer in Example 11-5.

initial states:

$$\mathbf{x}(0) = \begin{bmatrix} 0.5 \\ 0 \end{bmatrix} \qquad \mathbf{x}_e(0) = \begin{bmatrix} 0 \\ 0 \end{bmatrix}$$

Shown in the same figure is the response of $x_{e1}(t)$ when the state observer is designed for eigenvalues at $\lambda = -2.5, -2.5$, in which case $g_{e1} = 1$ and $g_{e2} = 1$. Since the state observer dynamics are now slower, it is seen from Fig. 11-10 that the $x_{e1}(t)$ response deviates more from the state $x_1(t)$.

Figure 11-11 illustrates the responses $x_2(t)$ and $x_{e2}(t)$ for the two cases of observer design. The reason that both $x_{e2}(t)$ responses are not very close to the actual state $x_2(t)$ is due to the fact that the elements of the feedback matrix, g_{e1} and g_{e2}, appear in the numerator of the transfer function relation between $X_2(s) - X_{e2}(s)$ and $x_2(0) - x_{e2}(0)$, but not in the relation between $X_1(s) - X_{e1}(s)$ and $x_1(0) - x_{e1}(0)$. Taking the Laplace transform on both sides of Eq. (11-129) and rearranging, we have

$$\mathbf{X}(s) - \mathbf{X}_e(s) = [s\mathbf{I} - (\mathbf{A} - \mathbf{G}_e\mathbf{D})]^{-1}[\mathbf{x}(0) - \mathbf{x}_e(0)] \qquad (11\text{-}138)$$

where

$$[s\mathbf{I} - (\mathbf{A} - \mathbf{G}_e\mathbf{D})]^{-1} = \frac{\begin{bmatrix} s+3 & 1 \\ -2 - 2g_{e2} & s + 2g_{e1} \end{bmatrix}}{|s - (\mathbf{A} - \mathbf{G}_e\mathbf{D})|} \qquad (11\text{-}139)$$

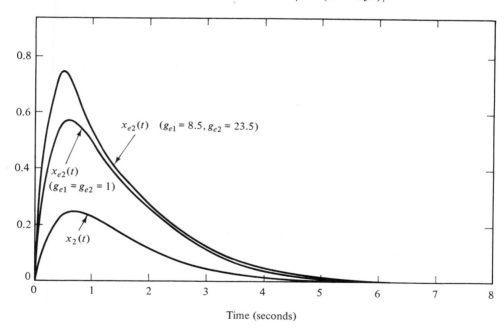

Fig. 11-11. State $x_2(t)$ and the observed states of the observer in Example 11-5.

Thus, we notice from Eq. (11-139) that selecting larger values for g_{e1} and g_{e2} to give faster transient response for the observer will not affect the numerator terms of the $X_{e1}(s)$ response. However, a larger value of g_{e2} will increase the gain factor between $X_2(s) - X_{e2}(s)$ and $x_2(0) - x_{e2}(0)$. This explains the fact that in Fig. 11-11 the $x_{e2}(t)$ response for $g_{e2} = 1$ is a better observation of $x_2(t)$ than the one for $g_{e2} = 23.5$.

A more versatile design method for the observer may be effected by the use of a performance index on the error between $\mathbf{x}$ and $\mathbf{x}_e$. A popular performance index is the quadratic form,

$$J = \tfrac{1}{2} \int_0^\infty (\mathbf{x} - \mathbf{x}_e)'\mathbf{Q}(\mathbf{x} - \mathbf{x}_e)\,dt \qquad (11\text{-}140)$$

The matrix $\mathbf{Q}$ is symmetric and positive semidefinite. The elements of $\mathbf{Q}$ may be selected to give various weightings on the observation of the states.

Closed-Loop Control with Observer

We are now ready to investigate the use of an observer in a system with state-variable feedback. The block diagram of such a system is shown in Fig. 11-8.

From Fig. 11-8 the state equations of the entire system are written

$$\dot{\mathbf{x}} = \mathbf{A}\mathbf{x} - \mathbf{B}\mathbf{G}\mathbf{x}_e + \mathbf{B}\mathbf{E}r \qquad (11\text{-}141)$$

$$\dot{\mathbf{x}}_e = (\mathbf{A} - \mathbf{D}\mathbf{G}_e - \mathbf{B}\mathbf{G})\mathbf{x}_e + \mathbf{D}\mathbf{G}_e\mathbf{x} + \mathbf{B}\mathbf{E}r \qquad (11\text{-}142)$$

A state diagram for the system is drawn as shown in Fig. 11-12 using Eqs. (11-141) and (11-142). The initial states are included on the state diagram, and the branch gains are matrix quantities. It can be easily shown that the system with the observer is uncontrollable (state) but observable. However, the lack of controllability is in all the states $\mathbf{x}$ and $\mathbf{x}_e$. The reason for this is simple; the main objective is to control $\mathbf{x}$ while keeping $\mathbf{x}_e$ close to $\mathbf{x}$. It is not necessary to control $\mathbf{x}$ and $\mathbf{x}_e$ independently.

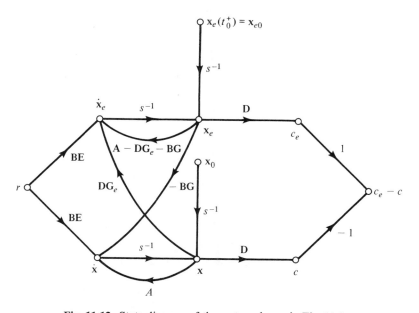

Fig. 11-12. State diagram of the system shown in Fig. 11-8.

From Fig. 11-12 the transfer relations of the state vector and the observed state vector are derived. These are

$$\mathbf{X}(s) = [s\mathbf{I} - (\mathbf{A} - \mathbf{BG})]^{-1}\mathbf{B}E R(s) - \mathbf{BG}\boldsymbol{\Delta}^{-1}\mathbf{x}_{e0} + [s\mathbf{I} - (\mathbf{A} - \mathbf{G}_e\mathbf{D} - \mathbf{BG})]\boldsymbol{\Delta}^{-1}\mathbf{x}_0 \tag{11-143}$$

$$\mathbf{X}_e(s) = [s\mathbf{I} - (\mathbf{A} - \mathbf{BG})]^{-1}\mathbf{B}E R(s) + (s\mathbf{I} - \mathbf{A})\boldsymbol{\Delta}^{-1}\mathbf{x}_{e0} + \mathbf{G}_e\mathbf{D}\boldsymbol{\Delta}^{-1}\mathbf{x}_0 \tag{11-144}$$

where

$$\boldsymbol{\Delta} = s^2\mathbf{I} - (2\mathbf{A} - \mathbf{G}_e\mathbf{D} - \mathbf{BG})s + [\mathbf{G}_e\mathbf{DBG} + \mathbf{A}(\mathbf{A} - \mathbf{G}_e\mathbf{D} - \mathbf{BG})] \tag{11-145}$$

When $\mathbf{x}_0 = \mathbf{x}_{e0}$, it can be shown that

$$\mathbf{X}(s) = \mathbf{X}_e(s) = [s\mathbf{I} - (\mathbf{A} - \mathbf{BG})]^{-1}\mathbf{B}E R(s) + [s\mathbf{I} - (\mathbf{A} - \mathbf{BG})]^{-1}\mathbf{x}_0 \tag{11-146}$$

Therefore, when the observer and the system have the same initial conditions, the two systems will have identical state and output responses. More important is the fact that *the transfer relation in Eq. (11-146) is identical to that when the observer is absent; that is, the true state variables are fed back for control purposes.* This means that when the initial states of the original system and the observer are identical, the responses of the system will not be effected by the feedback matrix $\mathbf{G}_e$. Since, in general, $\mathbf{x}_{e0} \neq \mathbf{x}_0$, the design of the observer, and subsequently of $\mathbf{G}_e$, will affect the system response through the initial-condition term due to $\mathbf{x}_0$, as indicated by Eq. (11-143). However, the steady-state response is due to the first term on the right side of Eq. (11-143), and the eigenvalues of $\mathbf{A} - \mathbf{BG}$ are not functions of $\mathbf{G}_e$.

EXAMPLE 11-6   As an example of the design of a closed-loop system with observed state feedback, consider the linear process that has the transfer function

$$\frac{C(s)}{U(s)} = \frac{100}{s(s + 5)} \tag{11-147}$$

The dynamic equations of the process are written

$$\dot{\mathbf{x}}(t) = \mathbf{A}\mathbf{x}(t) + \mathbf{B}u(t) \tag{11-148}$$

$$c(t) = \mathbf{D}\mathbf{x}(t) \tag{11-149}$$

where

$$\mathbf{A} = \begin{bmatrix} 0 & 1 \\ 0 & -5 \end{bmatrix} \quad \mathbf{B} = \begin{bmatrix} 0 \\ 100 \end{bmatrix}$$

$$\mathbf{D} = [1 \quad 0]$$

It is desired to design a state feedback control that is given by

$$u = r - \mathbf{Gx} \tag{11-150}$$

where

$$\mathbf{G} = [g_1 \quad g_2] \tag{11-151}$$

The eigenvalues of the closed-loop system should be located at $\lambda = -7.07 \pm j7.07$. The corresponding damping ratio is 0.707, and the natural undamped frequency is 10 rad/sec.

Assuming that the states x_1 and x_2 are inaccessible so that a state observer is to be designed, the overall system block diagram is identical to that of Fig. 11-8, with $E = 1$.

First, the pair [**A, B**] is completely controllable, so the eigenvalues of **A** − **BG** can be arbitrarily assigned. Also, it is simple to show that the pair [**A, D**] is completely observable, so that an observer may be constructed from c and u.

The characteristic equation of the closed-loop system with state feedback is

$$|\lambda\mathbf{I} - (\mathbf{A} - \mathbf{BG})| = 0 \tag{11-152}$$

or

$$\lambda^2 + (5 + 100g_2)\lambda + 100g_1 = 0 \tag{11-153}$$

To realize the desired eigenvalues stated earlier, the coefficients of Eq. (11-153) must satisfy the following conditions:

$$5 + 100g_2 = 14.14$$

$$100g_1 = 100$$

Thus $g_1 = 1$ and $g_2 = 0.0914$.

The state diagram of the overall system, which includes the observer, is shown in Fig. 11-13.

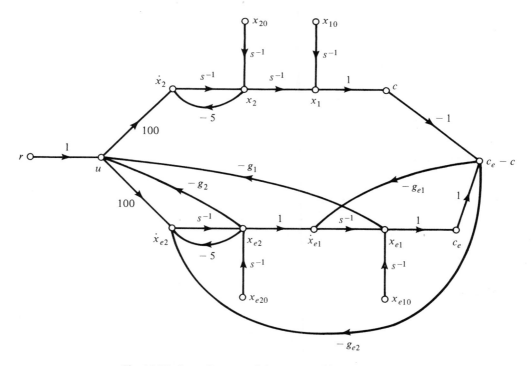

Fig. 11-13. State diagram of the system with observer in Example 11-6.

The design of the observer may be carried out in the usual manner. The characteristic equation of the observer is

$$|\lambda\mathbf{I} - (\mathbf{A} - \mathbf{G}_e\mathbf{D})| = 0 \tag{11-154}$$

where

$$\mathbf{G}_e = \begin{bmatrix} g_{e1} \\ g_{e2} \end{bmatrix} \tag{11-155}$$

Let us assume that the eigenvalues of the observer must be at $\lambda = -50, -50$. These eigenvalues are of much greater magnitude than the real parts of the eigenvalues of the system. Therefore, the transient of the observer due to the difference between the initial states $\mathbf{x}_0$ and $\mathbf{x}_{e0}$ should decay very rapidly to zero. The corresponding values of g_{e1} and g_{e2} are found to be

$$g_{e1} = 95$$
$$g_{e2} = 2025$$
(11-156)

The system of Fig. 11-13 is simulated on a digital computer with $\mathbf{G}$ and $\mathbf{G}_e$ as indicated above. The responses of $x_1(t)$ when $r(t)$ is a unit step function are computed and plotted as shown in Fig. 11-14 for $\mathbf{x}_{e0} = \mathbf{x}_0 = \mathbf{0}$, and $\mathbf{x}_0 = \mathbf{0}$, $\mathbf{x}_e = [0.5 \quad 0]'$.

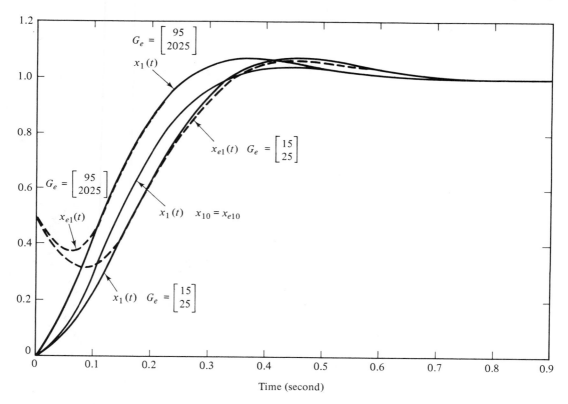

Fig. 11-14. Unit step responses of $x_1(t)$ and $x_{e1}(t)$ of the system in Example 11-6.

When $\mathbf{x}_{e0} = \mathbf{x}_0$, the response is that with the real state-variable feedback. When $\mathbf{x}_0 \neq \mathbf{x}_{e0}$, and the elements of $\mathbf{G}_e$ are as given in Eq. (11-156), the step response of $x_1(t)$ is faster than the ideal $x_1(t)$, and the overshoot is greater. Also shown in Fig. 11-14 is the step response of $x_1(t)$ when the observer is designed to have its eigenvalues at $\lambda = -10$ and -10, which correspond to $g_{e1} = 15$ and $g_{e2} = 25$. The initial states of the observer are again given by $\mathbf{x}_{e0} = [0.5 \quad 0]'$. Note that since the observer dynamics are much slower in this case, the corresponding step response of $x_1(t)$ is also slower than the true response, but the overshoot is still greater. The $x_{e1}(t)$ responses are shown to follow their corresponding $x_1(t)$ responses very closely, starting from the initial state of $x_{e1}(0) = 0.5$.

It is interesting to investigate the effect of having various differences in the initial values between $x_1(t)$ and $x_{e1}(t)$. In practice, the initial value of $x_1(t)$ (as well as that of the other state variables) is not fixed and could be of any value. The initial values of the observed states must be set so that they are as close to that of the real states as possible. For illustrative purposes, we set $\mathbf{x}(0) = \mathbf{x}_0 = \mathbf{0}$ and $x_{e2}(0) = 0$ but vary $x_{e1}(0)$. Figure 11-15 shows the unit step responses of $x_1(t)$ when $x_{e1}(0) = 0$, 0.25, 0.5, and 0.8, with $\mathbf{G}_e = [15 \quad 25]'$. The case with $x_{e1}(0) = x_1(0) = 0$ is the ideal response. These curves illustrate the importance of making $\mathbf{x}_{e0}$ as close to $\mathbf{x}_0$ as possible.

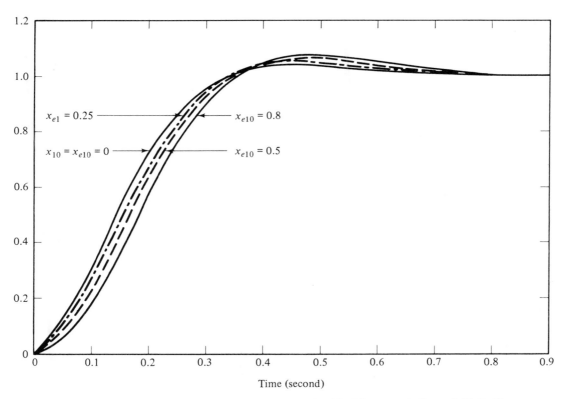

Fig. 11-15. Unit step responses of $x_1(t)$ of the system in Example 11-6 with various values of x_{e10}.

11.6 Optimal Linear Regulator Design[13-16]

One of the modern optimal control design methods that has found practical applications is *linear regulator design theory*. The design is based on the minimization of a quadratic performance index, and the result is the optimal control through state feedback. The basic advantages of optimal linear regulator design are as follows:

1. The solution is carried out without cut-and-try; a mathematical algorithm leads to a unique solution.
2. Can be used to design linear time-varying or time-invariant systems with multivariables.
3. Guarantees stability.

Consider that an nth-order linear system is described by the state equation

$$\dot{\mathbf{x}}(t) = \mathbf{A}\mathbf{x}(t) + \mathbf{B}\mathbf{u}(t) \qquad (11\text{-}157)$$

where $\mathbf{x}(t)$ is the $n \times 1$ state vector and $\mathbf{u}(t)$ is the $r \times 1$ control vector; $\mathbf{A}$ and $\mathbf{B}$ are coefficient matrices of appropriate dimensions. In general, $\mathbf{A}$ and $\mathbf{B}$ may contain time-varying elements. The design objective is that given the initial state $\mathbf{x}(t_0)$, determine the optimal control $\mathbf{u}°(t)$ so that the performance index

$$J = \tfrac{1}{2}\mathbf{x}'(t_f)\mathbf{P}\mathbf{x}(t_f) + \tfrac{1}{2}\int_{t_0}^{t_f}[\mathbf{x}'(t)\mathbf{Q}\mathbf{x}(t) + \mathbf{u}'(t)\mathbf{R}\mathbf{u}(t)]\,dt \qquad (11\text{-}158)$$

is minimized. The performance index J is known as a quadratic performance index, since $\mathbf{P}$, $\mathbf{Q}$, and $\mathbf{R}$ are required to be symmetric matrices. In addition, it is required that $\mathbf{P}$ and $\mathbf{Q}$ be positive semidefinite and $\mathbf{R}$ be positive definite.

The first term on the right side of Eq. (11-158) is the terminal cost, which is a constraint of the final state, $\mathbf{x}(t_f)$. The inclusion of the terminal cost in the performance index is desirable since in the finite-time problem, t_f is finite and $\mathbf{x}(t_f)$ may not be specified.

The first term under the integral sign of J represents a constraint on the state variables. The simplest form of $\mathbf{Q}$ that one can use is a diagonal matrix, unless no constraint is placed on the states, in which case $\mathbf{Q} = \mathbf{0}$. For an nth-order system, a diagonal $\mathbf{Q}$ may be defined as

$$\mathbf{Q} = \begin{bmatrix} q_1 & 0 & 0 & \cdots & 0 \\ 0 & q_2 & 0 & \cdots & 0 \\ 0 & 0 & q_3 & \cdots & 0 \\ \cdots\cdots\cdots\cdots\cdots\cdots \\ 0 & 0 & 0 & \cdots & q_n \end{bmatrix} \qquad (11\text{-}159)$$

The ith entry of $\mathbf{Q}$ represents the amount of weight the designer places on the constraint on the state variable $x_i(t)$. The larger the value of q_i, relative to the other qs, the more limiting is placed on the state $x_i(t)$. The second term under the integral of Eq. (11-158) places a constraint on the control vector $\mathbf{u}(t)$. Therefore, the weighting matrix $\mathbf{R}$ has the same significance for $\mathbf{u}(t)$ as $\mathbf{Q}$ has for $\mathbf{x}(t)$. Physically, the time integral of $\mathbf{u}'(t)\mathbf{R}\mathbf{u}(t)$ has the dimension of energy. Thus the minimization of the performance index of Eq. (11-158) has the physical interpretation of keeping the states near the equilibrium state $\mathbf{x} = \mathbf{0}$, while holding the energy consumption to a reasonable level. For this design, the reference input is not considered and is set to zero, although the designed system may still be subject to nonzero inputs. The term "regulator" simply refers to the condition that the primary purpose of the design is that the system is capable of damping out any initial state $\mathbf{x}(t_0)$ quickly without excessive overshoot and oscillations in the state responses.

The derivation of the necessary condition of the optimal control may be carried out in a number of ways. In this section we shall use the *principle of optimality*[16] and the *Hamilton–Jacobi equation* as development tools.

Principle of Optimality

Many algorithms and solutions in optimal control theory follow basic laws of nature that do not need complex mathematical development to explain their validity. One of these algorithms is the principle of optimality: *An optimal control strategy has the property that whatever the initial state and the control law of the initial stages, the remaining control must form an optimal control with respect to the state resulting from the control of the initial stages.*

Figure 11-16 gives a simple explanation of the principle in the form of a state trajectory of a second-order system. Let us consider that the control objective is to drive the system from $\mathbf{x}(t_0)$ to $\mathbf{x}(t_f)$, with $\mathbf{x}(t_0)$ given, in such a way that the performance index

$$J = \int_{t_0}^{t_f} F(\mathbf{x}, \mathbf{u})\, dt \tag{11-160}$$

is minimized. The state equations of the system may be that given in Eq. (11-157). Let t_1 be some intermediate time between t_0 and t_f ($t_0 < t_1 < t_f$), and $\mathbf{x}(t_1)$ is the corresponding state, as illustrated in Fig. 11-16. We can break up the performance index in Eq. (11-160) so that for the time interval $t_1 \leq t \leq t_f$, it is expressed as

$$J_1 = \int_{t_1}^{t_f} F(\mathbf{x}, \mathbf{u})\, dt \tag{11-161}$$

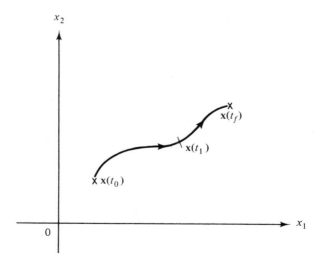

Fig. 11-16. Principle of optimality.

The principle of optimality states that regardless of how the state reaches $\mathbf{x}(t_1)$ at $t = t_1$, once $\mathbf{x}(t_1)$ is known, then using $\mathbf{x}(t_1)$ as the initial state for the last portion of the state transition, in order for the entire control process to be optimal in the sense of minimum J, the control for the time interval $t_1 \leq t \leq t_f$ must be so chosen that J_1 is a minimum. In other words, the last section of an optimal trajectory is also an optimal trajectory.

Using another more readily understandable example, we may consider that a track star desires to run a 440-yard distance in the shortest possible time. If he plans his strategy based on dividing the total distance into four parts, the optimal strategy for this minimum-time control problem, according to the principle of optimality, is

No matter what the control laws of the first n parts, n = 1, 2, 3, are, the remaining 4 − n parts must be run in the shortest possible time, based on the state of the physical condition at the time of the decision.

Hamilton–Jacobi Equation

In this section we shall use the principle of optimality to derive the Hamilton–Jacobi equation, which is then used to develop the linear regulator design.

To show that the theory is generally applicable to time-varying and non-linear systems, we shall use the following state equation representation:

$$\dot{\mathbf{x}}(t) = \mathbf{f}(\mathbf{x}, \mathbf{u}, t) \tag{11-162}$$

where $\mathbf{x}(t)$ is an n-vector and $\mathbf{u}(t)$ is an r-vector.

The problem is to determine the control $\mathbf{u}(t)$ so that the performance index

$$J = \int_{t_0}^{t_f} \mathbf{F}(\mathbf{x}, \mathbf{u}, \tau) \, d\tau \tag{11-163}$$

is a minimum.

Let us denote the performance index over the time interval $t \leq \tau \leq t_f$ by

$$\hat{J} = \int_{t}^{t_f} F(\mathbf{x}, \mathbf{u}, \tau) \, d\tau \tag{11-164}$$

and its minimum value by

$$S[\mathbf{x}(t), t] = \underset{\mathbf{u}}{\text{Min}} \, \hat{J} = \underset{\mathbf{u}}{\text{Min}} \int_{t}^{t_f} F(\mathbf{x}, \mathbf{u}, \tau) \, d\tau \tag{11-165}$$

Then

$$S[\mathbf{x}(t_0), t_0] = \underset{\mathbf{u}}{\text{Min}} \, J = \underset{\mathbf{u}}{\text{Min}} \int_{t_0}^{t_f} F(\mathbf{x}, \mathbf{u}, \tau) \, d\tau \tag{11-166}$$

When $t = t_f$,

$$S[\mathbf{x}(t_f), t_f] = \underset{\mathbf{u}}{\text{Min}} \int_{t_f}^{t_f} F(\mathbf{x}, \mathbf{u}, \tau) \, d\tau = 0 \tag{11-167}$$

Now let us break up the time interval from t to t_f into two intervals: $t \leq \tau \leq t + \Delta$ and $t + \Delta \leq \tau \leq t_f$. Then Eq. (11-165) is written

$$S[\mathbf{x}(t), t] = \underset{\mathbf{u}}{\text{Min}} \left(\int_{t}^{t+\Delta} F(\mathbf{x}, \mathbf{u}, \tau) \, d\tau + \int_{t+\Delta}^{t_f} F(\mathbf{x}, \mathbf{u}, \tau) \, d\tau \right) \tag{11-168}$$

Applying the principle of optimality we have

$$S[\mathbf{x}(t), t] = \underset{\mathbf{u}}{\text{Min}} \left(\int^{t+\Delta} F(\mathbf{x}, \mathbf{u}, \tau) \, d\tau + S[\mathbf{x}(t + \Delta), t + \Delta] \right) \quad (11\text{-}169)$$

where, according to Eq. (11-165),

$$S[\mathbf{x}(t + \Delta), t + \Delta] = \underset{\mathbf{u}}{\text{Min}} \int_{t+\Delta}^{t_f} F(\mathbf{x}, \mathbf{u}, \tau) \, d\tau \quad (11\text{-}170)$$

Let Δ be a very small time interval; then Eq. (11-168) can be written

$$S[\mathbf{x}(t), t] = \underset{\mathbf{u}}{\text{Min}} \left(F(\mathbf{x}, \mathbf{u}, t)\Delta + S[\mathbf{x}(t + \Delta), t + \Delta] \right) + \epsilon_0(\Delta) \quad (11\text{-}171)$$

where $\epsilon_0(\Delta)$ is the error function.

The next step involves the expansion of $S[\mathbf{x}(t + \Delta), t + \Delta]$ into a Taylor series. This would require the expansion of $\mathbf{x}(t + \Delta)$ into a Taylor series about $\Delta = 0$ first. Thus

$$\mathbf{x}(t + \Delta) = \mathbf{x}(t) + \frac{d\mathbf{x}(t)}{dt}\Delta + \cdots \quad (11\text{-}172)$$

Thus Eq. (11-171) is written

$$S[\mathbf{x}(t), t] = \underset{\mathbf{u}}{\text{Min}}\left(F(\mathbf{x}, \mathbf{u}, t)\Delta + S[\mathbf{x}(t) + \dot{\mathbf{x}}(t)\Delta + \cdots, t + \Delta]\right) + \epsilon_0(\Delta) \quad (11\text{-}173)$$

The second term inside the minimization operation in the last equation is of the form $S[\mathbf{x}(t) + \mathbf{h}(t), t + \Delta]$, which can be expanded into a Taylor series as follows:

$$S[\mathbf{x}(t) + \mathbf{h}(t), t + \Delta] = S[\mathbf{x}(t), t] + \frac{\partial S[\mathbf{x}(t), t]}{\partial \mathbf{x}(t)}\mathbf{h}(t) + \frac{\partial S[\mathbf{x}(t), t]}{\partial t}\Delta$$
$$+ \text{ higher-order terms} \quad (11\text{-}174)$$

Therefore,

$$S[\mathbf{x}(t + \Delta), t + \Delta] = S[\mathbf{x}(t) + \mathbf{f}(\mathbf{x}, \mathbf{u}, t)\Delta + \cdots, t + \Delta]$$
$$= S[\mathbf{x}(t), t] + \sum_{i=1}^{n} \frac{\partial S[\mathbf{x}(t), t]}{\partial x_i(t)} f_i(\mathbf{x}, \mathbf{u}, t)\Delta + \frac{\partial S[\mathbf{x}(t), t]}{\partial t}\Delta$$
$$+ \text{ higher-order terms} \quad (11\text{-}175)$$

where we have replaced $\dot{\mathbf{x}}$ by $\mathbf{f}(\mathbf{x}, \mathbf{u}, t)$

Substituting Eq. (11-175) into Eq. (11-173), we have

$$S[\mathbf{x}(t), t] = \underset{\mathbf{u}}{\text{Min}}\left(F(\mathbf{x}, \mathbf{u}, t)\Delta + S[\mathbf{x}(t), t] + \sum_{i=1}^{n} \frac{\partial S[\mathbf{x}(t), t]}{\partial x_i(t)} f_i(\mathbf{x}, \mathbf{u}, t)\Delta \right.$$
$$\left. + \frac{\partial S[\mathbf{x}(t), t]}{\partial t}\Delta \right) + \epsilon_1(\Delta) \quad (11\text{-}176)$$

where $\epsilon_1(\Delta)$ is the error function, which includes the higher-order terms in Eq. (11-175) and $\epsilon_0(\Delta)$.

Since $S[\mathbf{x}(t), t]$ is not a function of the control $\mathbf{u}(t)$, Eq. (11-176) is simplified to the following form after cancellation and dropping the common factor Δ:

$$-\frac{\partial S[\mathbf{x}(t), t]}{\partial t} = \underset{\mathbf{u}}{\text{Min}}\left(F(\mathbf{x}, \mathbf{u}, t) + \sum_{i=1}^{n} \frac{\partial S[\mathbf{x}(t), t]}{\partial x_i(t)} f_i(\mathbf{x}, \mathbf{u}, t) \right) + \epsilon_2(\Delta) \quad (11\text{-}177)$$

As Δ approaches zero, Eq. (11-177) becomes

$$-\frac{\partial S[\mathbf{x}(t), t]}{\partial t} = \underset{\mathbf{u}}{\text{Min}}\left(F(\mathbf{x}, \mathbf{u}, t) + \sum_{i=1}^{n} \frac{\partial S[\mathbf{x}(t), t]}{\partial x_i(t)} f_i(\mathbf{x}, \mathbf{u}, t)\right) \qquad (11\text{-}178)$$

which is known as *Bellman's equation*.

The necessary condition for the last equation to be a minimum is determined by taking the partial derivative on both sides of the equation with respect to $\mathbf{u}$, and setting the result to zero. We have

$$\frac{\partial F(\mathbf{x}, \mathbf{u}, t)}{\partial \mathbf{u}(t)} + \sum_{i=1}^{n} \frac{\partial S[\mathbf{x}(t), t]}{\partial x_i(t)} \frac{\partial f_i(\mathbf{x}, \mathbf{u}, t)}{\partial \mathbf{u}(t)} = \mathbf{0} \qquad (11\text{-}179)$$

or in matrix equation form,

$$\frac{\partial F(\mathbf{x}, \mathbf{u}, t)}{\partial \mathbf{u}(t)} + \frac{\partial \mathbf{f}(\mathbf{x}, \mathbf{u}, t)}{\partial \mathbf{u}(t)} \frac{\partial S[\mathbf{x}(t), t]}{\partial \mathbf{x}(t)} = \mathbf{0} \qquad (11\text{-}180)$$

where $\partial \mathbf{f}(\mathbf{x}, \mathbf{u}, t)/\partial \mathbf{u}(t)$ denotes the matrix transpose of the Jacobian of $\mathbf{f}(\mathbf{x}, \mathbf{u}, t)$; that is,

$$\frac{\partial \mathbf{f}(\mathbf{x}, \mathbf{u}, t)}{\partial \mathbf{u}(t)} = \begin{bmatrix} \dfrac{\partial f_1}{\partial u_1} & \dfrac{\partial f_2}{\partial u_1} & \cdots & \dfrac{\partial f_n}{\partial u_1} \\[2mm] \dfrac{\partial f_1}{\partial u_2} & \dfrac{\partial f_2}{\partial u_2} & \cdots & \dfrac{\partial f_n}{\partial u_2} \\[2mm] \cdots & \cdots & \cdots & \cdots \\[2mm] \dfrac{\partial f_1}{\partial u_r} & \dfrac{\partial f_2}{\partial u_r} & \cdots & \dfrac{\partial f_n}{\partial u_r} \end{bmatrix} = (\text{Jacobian of } \mathbf{f})' \qquad (11\text{-}181)$$

Once the optimal control $\mathbf{u}^\circ(t)$ is determined or known, Eq. (11-178) becomes

$$\frac{\partial S[\mathbf{x}^\circ(t), t]}{\partial t} + F(\mathbf{x}^\circ, \mathbf{u}^\circ, t) + \sum_{i=1}^{n} \frac{\partial S[\mathbf{x}^\circ(t), t]}{\partial x_i^\circ(t)} f_i(\mathbf{x}^\circ, \mathbf{u}^\circ, t) = 0 \qquad (11\text{-}182)$$

where all the variables with superscript zeros indicate optimal quantities. The sum of the last two terms in the last equation is also known as the Hamiltonian $H(\mathbf{x}^\circ, \mathbf{u}^\circ, t)$. Thus Eq. (11-182) is written

$$\frac{\partial S[\mathbf{x}^\circ(t), t]}{\partial t} + H(\mathbf{x}^\circ, \mathbf{u}^\circ, t) = 0 \qquad (11\text{-}183)$$

which is known as the *Hamilton–Jacobi equation* and is the necessary condition for the control to be optimal in the sense of minimum J.

Derivation of the Differential Riccati Equation (Finite-Time Problem)

Now we are ready to use the Hamilton–Jacobi equation to derive the solution of the linear regulator problem. Consider the system

$$\dot{\mathbf{x}}(t) = \mathbf{A}\mathbf{x}(t) + \mathbf{B}\mathbf{u}(t) \qquad (11\text{-}184)$$

with the initial condition, $\mathbf{x}(t_0) = \mathbf{x}_0$. The design problem is to find the optimal control $\mathbf{u}^\circ(t)$ such that the performance index

$$J = \tfrac{1}{2}\mathbf{x}'(t_f)\mathbf{P}\mathbf{x}(t_f) + \tfrac{1}{2}\int_{t_0}^{t_f} [\mathbf{x}'(t)\mathbf{Q}\mathbf{x}(t) + \mathbf{u}'(t)\mathbf{R}\mathbf{u}(t)] \, dt \qquad (11\text{-}185)$$

is a minimum. The matrices $\mathbf{P}$ and $\mathbf{Q}$ are symmetric and positive semidefinite, and $\mathbf{R}$ is symmetric and positive definite.

Let the minimum of J be represented by $S[\mathbf{x}(t), t]$. Then, from the development of the last section we know that the necessary condition of optimal control is that the Hamilton–Jacobi equation must be satisfied. For the present case, Eq. (11-178) is written

$$-\frac{\partial S[\mathbf{x}(t), t]}{\partial t} = \operatorname*{Min}_{\mathbf{u}}\left(\frac{1}{2}\mathbf{x}'(t)\mathbf{Q}\mathbf{x}(t) + \frac{1}{2}\mathbf{u}'(t)\mathbf{R}\mathbf{u}(t) + [\mathbf{A}\mathbf{x}(t) + \mathbf{B}\mathbf{u}(t)]'\frac{\partial S[\mathbf{x}(t), t]}{\partial \mathbf{x}}\right)$$

(11-186)

To carry out the minimization process, we differentiate both sides of Eq. (11-186) with respect to $\mathbf{u}(t)$, and we have the optimal control after setting the result to zero:

$$\mathbf{u}^\circ(t) = -\mathbf{R}^{-1}\mathbf{B}'\frac{\partial S[\mathbf{x}^\circ(t), t]}{\partial \mathbf{x}^\circ(t)}$$

(11-187)

It remains to solve for $\partial S[\mathbf{x}^\circ(t), t]/\partial \mathbf{x}^\circ(t)$ in order to complete the expression of the optimal control. Substituting Eq. (11-187) into Eq. (11-186), we have

$$-\frac{\partial S[\mathbf{x}^\circ(t), t]}{\partial t} = \frac{1}{2}\mathbf{x}^{\circ\prime}(t)\mathbf{Q}\mathbf{x}^\circ(t) + \frac{1}{2}\left(\frac{\partial S[\mathbf{x}^\circ(t), t]}{\partial \mathbf{x}^\circ(t)}\right)'\mathbf{B}\mathbf{R}^{-1}\mathbf{B}'\left(\frac{\partial S[\mathbf{x}^\circ(t), t]}{\partial \mathbf{x}^\circ(t)}\right)$$

$$+ \mathbf{x}^{\circ\prime}(t)\mathbf{A}'\frac{\partial S[\mathbf{x}^\circ(t), t]}{\partial \mathbf{x}^\circ(t)} - \left(\frac{\partial S[\mathbf{x}^\circ(t), t]}{\partial \mathbf{x}^\circ(t)}\right)'\mathbf{B}\mathbf{R}^{-1}\mathbf{B}'\left(\frac{\partial S[\mathbf{x}^\circ(t), t]}{\partial \mathbf{x}^\circ(t)}\right)$$

(11-188)

The last equation is simplified to

$$-\frac{\partial S[\mathbf{x}^\circ(t), t]}{\partial t} = \frac{1}{2}\mathbf{x}^{\circ\prime}(t)\mathbf{Q}\mathbf{x}^\circ(t) - \frac{1}{2}\left(\frac{\partial S[\mathbf{x}^\circ(t), t]}{\partial \mathbf{x}^\circ(t)}\right)'\mathbf{B}\mathbf{R}^{-1}\mathbf{B}'\frac{\partial S[\mathbf{x}^\circ(t), t]}{\partial \mathbf{x}^\circ(t)}$$

$$+ \mathbf{x}^{\circ\prime}(t)\mathbf{A}'\frac{\partial S[\mathbf{x}^\circ(t), t]}{\partial \mathbf{x}^\circ(t)}$$

(11-189)

Since $\mathbf{Q}$ is a symmetric matrix, it can be shown that $S[\mathbf{x}^\circ(t), t]$ is a quadratic function of $\mathbf{x}^\circ(t)$. Thus we let

$$S[\mathbf{x}^\circ(t), t] = \tfrac{1}{2}\mathbf{x}^{\circ\prime}(t)\mathbf{K}(t)\mathbf{x}^\circ(t)$$

(11-190)

where $\mathbf{K}(t)$ is an $n \times n$ symmetric matrix.

The optimal control is determined in terms of $K(t)$ by substituting Eq. (11-190) in Eq. (11-187). We have

$$\mathbf{u}^\circ(t) = -\mathbf{R}^{-1}\mathbf{B}'\mathbf{K}(t)\mathbf{x}^\circ(t)$$

(11-191)

Substitution of Eq. (11-190) in Eq. (11-189) yields

$$-\frac{1}{2}\mathbf{x}^{\circ\prime}(t)\frac{d\mathbf{K}(t)}{dt}\mathbf{x}^\circ(t) = \frac{1}{2}\mathbf{x}^{\circ\prime}(t)\mathbf{Q}\mathbf{x}^\circ(t) - \frac{1}{2}\mathbf{x}^{\circ\prime}(t)\mathbf{K}(t)\mathbf{B}\mathbf{R}^{-1}\mathbf{B}'\mathbf{K}(t)\mathbf{x}^\circ(t)$$

$$+ \mathbf{x}^{\circ\prime}(t)\mathbf{A}'\mathbf{K}(t)\mathbf{x}^\circ(t)$$

(11-192)

The last term in Eq. (11-192) can be written

$$\mathbf{x}^{\circ\prime}(t)\mathbf{A}'\mathbf{K}(t)\mathbf{x}^\circ(t) = \tfrac{1}{2}\mathbf{x}^{\circ\prime}(t)[\mathbf{A}'\mathbf{K}(t) + \mathbf{K}(t)\mathbf{A}]\mathbf{x}^\circ(t)$$

(11-193)

since both sides of the equation represent scalar quantities. Thus, using Eq. (11-193) and comparing both sides of Eq. (11-192), we have

$$\frac{d\mathbf{K}(t)}{dt} = \mathbf{K}(t)\mathbf{B}\mathbf{R}^{-1}\mathbf{B}'\mathbf{K}(t) - \mathbf{K}(t)\mathbf{A} - \mathbf{A}'\mathbf{K}(t) - \mathbf{Q} \qquad (11\text{-}194)$$

which is known as the *Riccati equation*. The matrix $\mathbf{K}(t)$ is sometimes called the *Riccati gain matrix*. The boundary condition needed to solve the Riccati equation is obtained from Eq. (11-190) by setting $t = t_f$. Then

$$\mathbf{K}(t_f) = \mathbf{P} \qquad (11\text{-}195)$$

Another important property of $\mathbf{K}(t)$ is that it is positive definite, since otherwise, $S[\mathbf{x}°(t), t]$ would be negative.

In summarizing the solution of the optimal linear regulator problem, the optimal control is given by Eq. (11-191) in the form of state feedback; that is,

$$\mathbf{u}°(t) = -\mathbf{G}(t)\mathbf{x}°(t) \qquad (11\text{-}196)$$

where

$$\mathbf{G}(t) = \mathbf{R}^{-1}\mathbf{B}'\mathbf{K}(t) \qquad (11\text{-}197)$$

is the time-varying feedback gain matrix. The Riccati gain matrix $\mathbf{K}(t)$ is solved from the Riccati equation of Eq. (11-194), with the boundary condition of $\mathbf{K}(t_f) = \mathbf{P}$. The state equation of the optimal closed-loop system is given by

$$\dot{\mathbf{x}}(t) = [\mathbf{A} - \mathbf{B}\mathbf{R}^{-1}\mathbf{B}'\mathbf{K}(t)]\mathbf{x}(t) \qquad (11\text{-}198)$$

The block diagram of the feedback system is shown in Fig. 11-17.

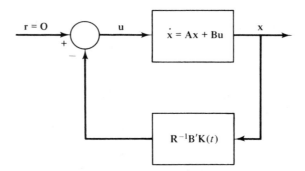

Fig. 11-17. Implementation of the optimal linear regulator design.

A few remarks may be made concerning the solution of the linear regulator problem. It should be noted that the present consideration has a finite terminal time t_f, and nothing is required on the stability, the controllability, or the observability of the system of Eq. (11-184). Indeed, the finite-time linear regulator problem does not require that the process be observable, controllable, or stable.

The Riccati equation of Eq. (11-194) is a set of n^2 nonlinear differential equations. However, there are only $n(n-1)$ unknowns in $\mathbf{K}(t)$, since the matrix is symmetric.

We shall postpone the discussion on the solution of the differential Riccati equation, Eq. (11-194), until we have investigated the infinite-time problem; that is, $t_f = \infty$.

Infinite-Time Linear Regulator Problem

Now let us consider the infinite-time regulator problem for which the performance index becomes

$$J = \tfrac{1}{2} \int_{t_0}^{\infty} [\mathbf{x}'(t)\mathbf{Q}\mathbf{x}(t) + \mathbf{u}'(t)\mathbf{R}\mathbf{u}(t)] \, dt \qquad (11\text{-}199)$$

Notice that the terminal cost is eliminated from J, since as t approaches infinity, the final state $\mathbf{x}(t_f)$ should approach the equilibrium state $\mathbf{0}$, so that the terminal constraint is no longer necessary. The same conditions of positive semidefiniteness and positive definiteness are placed on $\mathbf{Q}$ and $\mathbf{R}$, respectively, and all the derivations of the last section are still valid. However, the infinite-time linear regulator problem imposes additional conditions on the system of Eq. (11-184):

1. The pair $[\mathbf{A}, \mathbf{B}]$ must be completely controllable.
2. The pair $[\mathbf{A}, \mathbf{D}]$ must be completely observable, where $\mathbf{D}$ is any $n \times n$ matrix such that $\mathbf{D}\mathbf{D}' = \mathbf{Q}$.

The solution of the linear regulator problem does not require that the process is stable; in the finite-time case, controllability is not a problem, since for finite t_f, J can still be finite even if an uncontrollable state is unstable. On the other hand, an unstable state that is uncontrollable will yield an unbounded performance index if t_f is infinite. The observability condition is required because the feedback system of Eq. (11-198) must be asymptotically stable. The requirement is again tied in with the controllability of the system in ensuring that the state vector $\mathbf{x}(t)$ approaches the equilibrium state as t approaches infinity, since observability of $[\mathbf{A}, \mathbf{D}]$ is equivalent to the controllability of $[\mathbf{A}', \mathbf{D}']$. A complete proof of these requirements may be found in the literature.[13-16]

For the infinite-time case, the Riccati equation in Eq. (11-194) becomes a set of nonlinear algebraic equations,

$$\mathbf{K}\mathbf{B}\mathbf{R}^{-1}\mathbf{B}'\mathbf{K} - \mathbf{K}\mathbf{A} - \mathbf{A}'\mathbf{K} - \mathbf{Q} = 0 \qquad (11\text{-}200)$$

since as t_f approaches infinity, $d\mathbf{K}(t)/dt = \mathbf{0}$, the Riccati gain matrix becomes a constant matrix. The solution of the Riccati equation, $\mathbf{K}$, is symmetric and positive definite.

Solution of the Differential Riccati Equation

A great deal of work has been done on the solution of the Riccati equation, both for the differential form and for the algebraic form. In general, the algebraic Riccati equation of Eq. (11-200) is more difficult to solve than the differential equation of Eq. (11-194).

The differential Riccati equation is generally solved by one of the following methods:

1. Direct numerical integration.
2. Matrix iterative solution.
3. Negative exponential solution

The direct numerical integration method simply involves the solution of the nonlinear differential equations backward in time on a digital computer, knowing $\mathbf{K}(t_f)$. The matrix iterative method is effected by writing $d\mathbf{K}(t)/dt$ as

$$\frac{d\mathbf{K}(t)}{dt} = \lim_{\Delta \to 0} \frac{\mathbf{K}(t + \Delta) - \mathbf{K}(t)}{\Delta} \tag{11-201}$$

Equation (11-194) can be approximated by a set of difference equations which are then solved iteratively backward in time.

Negative Exponential Solution

Let us define the vector $\partial S(\mathbf{x}, t)/\partial \mathbf{x}(t)$ as the costate vector $\mathbf{y}(t)$; that is,

$$\frac{\partial S(\mathbf{x}, t)}{\partial \mathbf{x}(t)} = \mathbf{y}(t) \qquad (n \times 1) \tag{11-202}$$

From Eqs. (11-187), (11-196), and (11-197), $\mathbf{y}(t)$ is related to the Riccati gain $\mathbf{K}(t)$ and the state vector $\mathbf{x}(t)$ through the following relation:

$$\mathbf{y}(t) = \mathbf{K}(t)\mathbf{x}(t) \tag{11-203}$$

Taking the derivative on both sides of Eq. (11-203), we have

$$\dot{\mathbf{y}}(t) = \dot{\mathbf{K}}(t)\mathbf{x}(t) + \mathbf{K}(t)\dot{\mathbf{x}}(t) \tag{11-204}$$

Substituting Eqs. (11-194) and (11-184) in Eq. (11-204) and simplifying, we have

$$\dot{\mathbf{y}}(t) = -\mathbf{Q}\mathbf{x}(t) - \mathbf{A}'\mathbf{y}(t) \tag{11-205}$$

which is referred to as the *costate equation*.

Equations (11-198) and (11-205) are combined to form the *canonical state equations*,

$$\begin{bmatrix} \dot{\mathbf{x}}(t) \\ \dot{\mathbf{y}}(t) \end{bmatrix} = \begin{bmatrix} \mathbf{A} & -\mathbf{B}\mathbf{R}^{-1}\mathbf{B}' \\ -\mathbf{Q} & -\mathbf{A}' \end{bmatrix} \begin{bmatrix} \mathbf{x}(t) \\ \mathbf{y}(t) \end{bmatrix} \tag{11-206}$$

which represent $2n$ linear homogeneous differential equations. The $2n$ boundary conditions needed to solve Eq. (11-206) come from the n initial states, $\mathbf{x}(t_0)$, and

$$\mathbf{y}(t_f) = \mathbf{K}(t_f)\mathbf{x}(t_f) = \mathbf{P}\mathbf{x}(t_f) \tag{11-207}$$

Let the coefficient matrix of the canonical state equations in Eq. (11-206) be represented by

$$\mathbf{M} = \begin{bmatrix} \mathbf{A} & -\mathbf{B}\mathbf{R}^{-1}\mathbf{B}' \\ -\mathbf{Q} & -\mathbf{A}' \end{bmatrix} \tag{11-208}$$

The matrix M has the following properties:

1. **M** has no purely imaginary eigenvalues.
2. If λ_i is an eigenvalue of **M**, then the negative complex conjugate of λ_i is also an eigenvalue; that is,

$$\lambda_i = -\lambda_i^* \tag{11-209}$$

Let **W** be the $2n \times 2n$ similarity transformation matrix which transforms **M** into a diagonal matrix **T**. Then

$$\mathbf{T} = \mathbf{W}^{-1}\mathbf{MW} = \begin{bmatrix} -\mathbf{\Lambda} & 0 \\ 0 & \mathbf{\Lambda} \end{bmatrix} \tag{11-210}$$

where $\mathbf{\Lambda}$ is an $n \times n$ diagonal matrix that contains the eigenvalues of **M** with *positive* real parts (in the right-half plane).

Let **W** be partitioned as follows:

$$\mathbf{W} = \begin{bmatrix} \mathbf{W}_{11} & \mathbf{W}_{12} \\ \mathbf{W}_{21} & \mathbf{W}_{22} \end{bmatrix} \tag{11-211}$$

where $\mathbf{W}_{11}, \mathbf{W}_{12}, \mathbf{W}_{21},$ and $\mathbf{W}_{22}$ are $n \times n$ matrices. The transformed state and costate variables are denoted by $\hat{\mathbf{x}}(t)$ and $\hat{\mathbf{y}}(t)$, respectively, and are related to the original variables through the following relation:

$$\begin{bmatrix} \mathbf{x}(t) \\ \mathbf{y}(t) \end{bmatrix} = \begin{bmatrix} \mathbf{W}_{11} & \mathbf{W}_{12} \\ \mathbf{W}_{21} & \mathbf{W}_{22} \end{bmatrix} \begin{bmatrix} \hat{\mathbf{x}}(t) \\ \hat{\mathbf{y}}(t) \end{bmatrix} \tag{11-212}$$

Equation (11-207) leads to the following boundary condition for the transformed state equations

$$\hat{\mathbf{y}}(t_f) = -(\mathbf{W}_{22} - \mathbf{PW}_{12})^{-1}(\mathbf{W}_{21} - \mathbf{PW}_{11})\hat{\mathbf{x}}(t_f) \tag{11-213}$$

The transformed canonical state equations are written

$$\begin{bmatrix} \dot{\hat{\mathbf{x}}}(t) \\ \dot{\hat{\mathbf{y}}}(t) \end{bmatrix} = \begin{bmatrix} -\mathbf{\Lambda} & 0 \\ 0 & \mathbf{\Lambda} \end{bmatrix} \begin{bmatrix} \hat{\mathbf{x}}(t) \\ \hat{\mathbf{y}}(t) \end{bmatrix} \tag{11-214}$$

The solutions of the last equation are

$$\hat{\mathbf{x}}(t) = e^{-\mathbf{\Lambda}(t-t_f)}\hat{\mathbf{x}}(t_f) \tag{11-215}$$

$$\hat{\mathbf{y}}(t) = e^{\mathbf{\Lambda}(t-t_f)}\hat{\mathbf{y}}(t_f) \tag{11-216}$$

Since it is awkward to work in backward time, let us introduce a time variable, τ, which represents the time from t until the final time t_f:

$$\tau = t_f - t$$

Let

$$\hat{\mathbf{x}}(t) = \hat{\mathbf{X}}(\tau) = \hat{\mathbf{X}}(t_f - t) \tag{11-217}$$

$$\hat{\mathbf{y}}(t) = \hat{\mathbf{Y}}(\tau) = \hat{\mathbf{Y}}(t_f - t) \tag{11-218}$$

Equations (11-215) and (11-216) become

$$\hat{\mathbf{X}}(\tau) = e^{\mathbf{\Lambda}\tau}\hat{\mathbf{X}}(0) \tag{11-219}$$

$$\hat{\mathbf{Y}}(\tau) = e^{-\mathbf{\Lambda}\tau}\hat{\mathbf{Y}}(0) \tag{11-220}$$

Since $\mathbf{\Lambda}$ represents a diagonal matrix with eigenvalues that have positive real parts, the solutions of Eq. (11-219) will increase with τ. Thus we rearrange Eq. (11-219), and when it is combined with Eq. (11-220), we get

$$\begin{bmatrix} \hat{\mathbf{X}}(0) \\ \hat{\mathbf{Y}}(\tau) \end{bmatrix} = \begin{bmatrix} e^{-\mathbf{\Lambda}\tau} & \mathbf{0} \\ 0 & e^{-\mathbf{\Lambda}\tau} \end{bmatrix} \begin{bmatrix} \hat{\mathbf{X}}(\tau) \\ \hat{\mathbf{Y}}(0) \end{bmatrix} \tag{11-221}$$

Now Eq. (11-221) represents solutions that have only negative exponentials. Using the boundary condition of Eq. (11-213), the relation between $\hat{\mathbf{Y}}(0)$ and $\hat{\mathbf{X}}(0)$ can be written

$$\hat{\mathbf{Y}}(0) = -(\mathbf{W}_{22} - \mathbf{P}\mathbf{W}_{12})^{-1}(\mathbf{W}_{21} - \mathbf{P}\mathbf{W}_{11})\hat{\mathbf{X}}(0) \tag{11-222}$$

Substitution of Eq. (11-222) into Eq. (11-221), the relation between $\hat{\mathbf{Y}}(\tau)$ and $\hat{\mathbf{X}}(\tau)$ is established:

$$\hat{\mathbf{Y}}(\tau) = -e^{-\mathbf{\Lambda}\tau}(\mathbf{W}_{22} - \mathbf{P}\mathbf{W}_{12})^{-1}(\mathbf{W}_{21} - \mathbf{P}\mathbf{W}_{11})e^{-\mathbf{\Lambda}\tau}\hat{\mathbf{X}}(\tau) \tag{11-223}$$

Let

$$\mathbf{H}(\tau) = -e^{-\mathbf{\Lambda}\tau}(\mathbf{W}_{22} - \mathbf{P}\mathbf{W}_{12})^{-1}(\mathbf{W}_{21} - \mathbf{P}\mathbf{W}_{11})e^{-\mathbf{\Lambda}\tau} \tag{11-224}$$

Equation (11-223) becomes

$$\hat{\mathbf{Y}}(\tau) = \mathbf{H}(\tau)\hat{\mathbf{X}}(\tau) \tag{11-225}$$

In order to return to the variables $\mathbf{x}(t)$ and $\mathbf{y}(t)$, we substitute Eq. (11-225) in Eq. (11-212) and using Eqs. (11-217) and (11-218), we have the relationship between $\mathbf{y}(t)$ and $\mathbf{x}(t)$,

$$\mathbf{y}(t) = [\mathbf{W}_{21} + \mathbf{W}_{22}\mathbf{H}(\tau)][\mathbf{W}_{11} + \mathbf{W}_{12}\mathbf{H}(\tau)]^{-1}\mathbf{x}(t) \tag{11-226}$$

Comparing the last equation with Eq. (11-203), the Riccati gain matrix is written

$$\mathbf{K}(t) = [\mathbf{W}_{21} + \mathbf{W}_{22}\mathbf{H}(t_f - t)][\mathbf{W}_{11} + \mathbf{W}_{12}\mathbf{H}(t_f - t)]^{-1} \tag{11-227}$$

or

$$\mathbf{K}(t_f - \tau) = [\mathbf{W}_{21} + \mathbf{W}_{22}\mathbf{H}(\tau)][\mathbf{W}_{11} + \mathbf{W}_{12}\mathbf{H}(\tau)]^{-1} \tag{11-228}$$

Although we have a noniterative solution for $\mathbf{K}(t)$, because it is time dependent, in general, for high-order systems, even a computer solution will be quite complex.

Solution of the Algebraic Riccati Equation

In the infinite-time case, the Riccati equation is a matrix algebraic equation given by Eq. (11-200). Based on the negative exponential solution for the finite-time problem obtained in Section 11.5, we can obtain the solution of the algebraic Riccati equation by simply taking the limit as τ approaches infinity, since

$$\tau = t_f - t = \infty - t = \infty \tag{11-229}$$

Thus, from Eq. (11-224),

$$\lim_{\tau \to \infty} \mathbf{H}(\tau) = \lim_{t_f \to \infty} \mathbf{H}(t_f - t) = \mathbf{0} \tag{11-230}$$

and Eq. (11-228) gives

$$\mathbf{K} = \lim_{\tau \to \infty} \mathbf{K}(t_f - \tau) = \mathbf{W}_{21}\mathbf{W}_{11}^{-1} \tag{11-231}$$

Although the above solution is of a very simple form, in general, to determine the matrices $\mathbf{W}_{21}$ and $\mathbf{W}_{11}$ requires the solution of the eigenvalues and the eigenvectors of the $2n \times 2n$ matrix $\mathbf{M}$. For high-order systems, a computer solution becomes necessary.

In order to illustrate the solution of the Riccati equation, we shall first consider the optimal linear regulator design of a first-order system and carry out the design by pencil and paper. Then a higher-order system will be designed with the aid of a digital computer.

EXAMPLE 11-7   Consider that a linear time-invariant system is described by the state equation

$$\dot{x}(t) = -2x(t) + u(t) \tag{11-232}$$

with the initial condition given as $x(t_0)$. It is desired to determine the optimal control $u^\circ(t)$ such that

$$J = \tfrac{1}{2}x^2(t_f) + \tfrac{1}{2}\int_{t_0}^{t_f} [x^2(t) + u^2(t)]\, dt = \min. \tag{11-233}$$

We shall first consider that the final time t_f is finite. From Eqs. (11-232) and (11-233), we identify that $A = -2$, $B = 1$, $P = 1$, $Q = 1$, and $R = 1$. Using the negative exponential method, we form the matrix $\mathbf{M}$ in Eq. (11-208),

$$\mathbf{M} = \begin{bmatrix} A & -BR^{-1}B' \\ -Q & -A' \end{bmatrix} = \begin{bmatrix} -2 & -1 \\ -1 & 2 \end{bmatrix} \tag{11-234}$$

The eigenvalues of $\mathbf{M}$ are $\lambda_1 = -\sqrt{5}$ and $\lambda_2 = \sqrt{5}$. The eigenvectors of $\mathbf{M}$ are found to be

$$\mathbf{p}_1 = \begin{bmatrix} 1 \\ 0.235 \end{bmatrix} \qquad \mathbf{p}_2 = \begin{bmatrix} 1 \\ -4.235 \end{bmatrix} \tag{11-235}$$

Then

$$\mathbf{W} = \begin{bmatrix} W_{11} & W_{12} \\ W_{21} & W_{22} \end{bmatrix} = \begin{bmatrix} 1 & 1 \\ 0.235 & -4.235 \end{bmatrix} \tag{11-236}$$

It is important to note that the eigenvectors should be so arranged in $\mathbf{W}$ that they correspond to the requirement of the matrix $\mathbf{T}$ in Eq. (11-210). Substituting the elements of $\mathbf{W}$, $P = 1$, and $\Lambda = \lambda_2 = \sqrt{5}$ in Eq. (11-224), we get

$$H(\tau) = -0.146e^{-2\sqrt{5}\tau} \tag{11-237}$$

The time-varying Riccati gain is obtained by substituting $H(\tau)$ in Eq. (11-228). Thus

$$K(t_f - \tau) = \frac{0.235 + 0.6189e^{-2\sqrt{5}\tau}}{1 - 0.146e^{-2\sqrt{5}\tau}} \tag{11-238}$$

We note that when $t = t_f$, $\tau = 0$, $K(t_f) = P = 1$, which agrees with the boundary condition on $K(t_f)$.

Figure 11-18 gives a plot of $K(t)$ as a function of t. The steady-state solution of $K(t)$ is obtained from Eq. (11-238) by setting $t \longrightarrow -\infty$.

For the infinite-time problem, $t_f = \infty$. Since the system in Eq. (11-232) is completely controllable, the optimal solution exists, and the constant Riccati gain K is solved from Eq. (11-238) by setting $\tau = \infty$. Thus

$$K = 0.235 \tag{11-239}$$

which is also the steady-state value of $K(t)$.

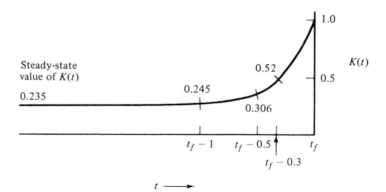

Fig. 11-18. Time-varying Riccati gain.

The optimal control for the finite-time problem is

$$u°(t) = -R^{-1}B'K(t)x(t)$$
$$= -K(t)x(t) \qquad t \le t_f \tag{11-240}$$

and that of the infinite-time problem is

$$u°(t) = -0.235x(t) \tag{11-241}$$

The state diagram for the feedback system designed for the infinite final time is shown in Fig. 11-19.

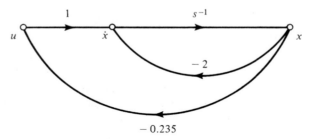

Fig. 11-19. Optimal linear regulator system in Example 11-7.

EXAMPLE 11-8   In this example we consider the linear regulator design of the control of a simplified one-axis model of the Large Space Telescope (LST).

The LST, which will be launched into orbit in 1980, is an optical space telescope aimed at observing stars that cannot be observed by ground-stationed telescopes. One of the major control problems in the design of the LST control system is the pointing stability of the vehicle. In addition to the accuracy requirement, the LST vehicle should be maintained at its equilibrium attitude position during operation. This is a typical regulator design problem.

The block diagram of the simplified one-axis model of the system is shown in Fig. 11-20. An equivalent state diagram of the system is shown in Fig. 11-21, from which are written the state equations

$$\dot{\mathbf{x}}(t) = \mathbf{A}\mathbf{x}(t) + \mathbf{B}u(t) \tag{11-242}$$

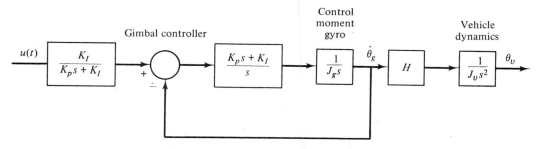

Fig. 11-20. Block diagram of a simplified one-axis model of the LST.

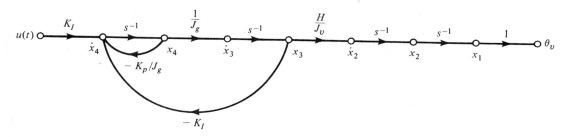

Fig. 11-21. State diagram of the simplified LST system.

where

$$\mathbf{A} = \begin{bmatrix} 0 & 1 & 0 & 0 \\ 0 & 0 & \dfrac{H}{J_v} & 0 \\ 0 & 0 & 0 & \dfrac{1}{J_g} \\ 0 & 0 & -K_I & \dfrac{-K_p}{J_g} \end{bmatrix}$$

$$\mathbf{B} = \begin{bmatrix} 0 \\ 0 \\ 0 \\ K_I \end{bmatrix}$$

The system parameters are defined as:

$H = 600$ ft-lb-sec   control moment gyro angular momentum

$J_g = 2.1$ ft-lb-sec^2 gimbal inertia

$K_I = 9700$ ft-lb/rad   gimbal-rate loop integral gain

$K_p = 216$ ft-lb/rad/sec   gimbal-rate loop proportional gain

$J_v = 10^5$ ft-lb/rad/sec   vehicle inertia

The design objective is to find the optimal control that minimizes the following performance index:

$$J = \int_0^\infty [\mathbf{x}'(t)\mathbf{Q}\mathbf{x}(t) + R u^2(t)]\, dt \qquad (11\text{-}243)$$

where

$$\mathbf{Q} = \begin{bmatrix} q_1 & 0 & 0 & 0 \\ 0 & q_2 & 0 & 0 \\ 0 & 0 & q_3 & 0 \\ 0 & 0 & 0 & q_4 \end{bmatrix} \tag{11-244}$$

Since the states x_1 and x_2 represent the vehicle displacement and velocity, respectively, and are of primary interest, more weights are placed on these variables in the specification of the elements of $\mathbf{Q}$. We let $q_1 = 5 \times 10^7, q_2 = 5000, q_3 = q_4 = 1$, and $R = 1$.

The optimal control is given by

$$u^\circ(t) = -R^{-1}\mathbf{B}'\mathbf{K}\mathbf{x}(t) \tag{11-245}$$

where $\mathbf{K}$ is the positive definite solution of the algebraic Riccati equation,

$$-\mathbf{K}\mathbf{A} - \mathbf{A}'\mathbf{K} + \mathbf{K}\mathbf{B}R^{-1}\mathbf{B}'\mathbf{K} - \mathbf{Q} = \mathbf{0} \tag{11-246}$$

The solutions of the Riccati equation and the optimal control are obtained on a digital computer. The eigenvalues of the feedback system are

$$-2.75 \quad -9700 \quad -1.375 + j2.33 \quad -1.375 - j2.33$$

The feedback matrix is

$$\mathbf{G} = [7071 \quad 5220 \quad 10.6 \quad 0.99] \tag{11-247}$$

The state diagram of the feedback system with state feedback is shown in Fig. 11-22. An input is added to the system through a gain of 7071. The latter is included so that the output of the system will follow a step function input without steady-state error. Figure 11-23 shows the responses of x_1 (θ_v) and x_2 $(\dot{\theta}_v)$ when the input is a step function with a magnitude of 5×10^{-8} rad. The initial conditions of the system are set to zero.

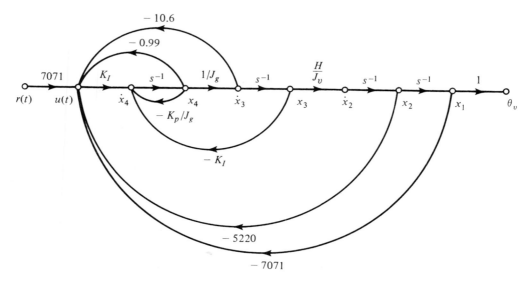

Fig. 11-22. State diagram of the LST system with state feedback from linear regulator design.

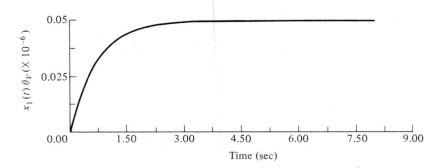

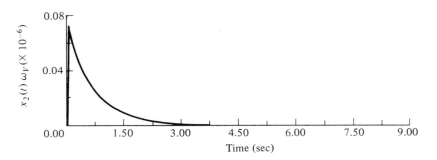

Fig. 11-23. Time responses of $x_1(t)$ and $x_2(t)$ of the LST system when the input is a step function.

11.7 Design with Partial State Feedback

It is demonstrated in Section 11.5 that an observer can be designed for an observable system to observe the state variables from the output variable. However, in general, the complexity of the observer is comparable to that of the system. It is natural to seek a design procedure that relies on the use of feedback from only the accessible state variables through constant feedback gains. It should be pointed out that control is achieved by feedback through constant gains from the accessible states, and no dynamics are involved in the controller. Of course, the design problem would only have a meaningful solution if the system is stabilizable with feedback from the accessible state variables.

Given the linear time-invariant system

$$\dot{\mathbf{x}}(t) = \mathbf{A}\mathbf{x}(t) + \mathbf{B}u(t) \qquad (11\text{-}248)$$

where

$$\mathbf{x}(t) = n \times 1 \text{ state vector}$$

$$u(t) = \text{scalar control}$$

it is assumed that the system is completely controllable.

Let the accessible state variables be denoted by the vector $\mathbf{y}(t)$ ($m \times 1$, $m \leq n$), and

$$\mathbf{y}(t) = \mathbf{C}\mathbf{x}(t) \qquad (11\text{-}249)$$

where $\mathbf{C}$ is an $m \times n$ matrix with constant coefficients. It is assumed that $\mathbf{C}$ has full rank; that is, the rank of $\mathbf{C}$ should be at least equal to m.

The design problem is to find the partial state feedback control

$$u(t) = -\mathbf{F}\mathbf{y}(t) \tag{11-250}$$

where $\mathbf{F}$ is an $1 \times m$ feedback matrix such that the designed system satisfies a certain performance index. The performance index chosen for this design is

$$J = \int_{t_0}^{\infty} [\mathbf{x}'(t)\mathbf{Q}\mathbf{x}(t) + Ru^2(t)]\, dt \tag{11-251}$$

where $\mathbf{Q}$ is a symmetric positive semidefinite matrix and R is a positive constant. The design objective is to find the optimal control $u(t)$ such that J is minimized. The reason for the choice of this performance index is that the design can be associated with the linear regulator design.

Substitution of Eq. (11-249) in Eq. (11-250) yields

$$u(t) = -\mathbf{F}\mathbf{C}\mathbf{x}(t) \tag{11-252}$$

or

$$u(t) = -\mathbf{G}\mathbf{x}(t) \tag{11-253}$$

where

$$\mathbf{G} = \mathbf{F}\mathbf{C} = [g_1 \quad g_2 \quad \ldots \quad g_n] \tag{11-254}$$

It has been shown[13] that the control in Eq. (11-253) is optimal in the sense of minimizing the J in Eq. (11-251) for some $\mathbf{Q}$ and $\mathbf{R}$ if and only if

$$|1 + \mathbf{G}(j\omega\mathbf{I} - \mathbf{A})^{-1}\mathbf{B}| \geq 1 \tag{11-255}$$

for all real ω.

It is interesting to note that the transfer relation between $\mathbf{x}(t)$ and $u(t)$ is

$$\overset{\bullet}{\mathbf{X}}(s) = (s\mathbf{I} - \mathbf{A})^{-1}\mathbf{B}U(s) \tag{11-256}$$

The loop transfer function of the system is $\mathbf{G}(s\mathbf{I} - \mathbf{A})^{-1}\mathbf{B}$. Thus $1 + \mathbf{G}(s\mathbf{I} - \mathbf{A})^{-1}\mathbf{B}$ is equivalent to the function $1 + G(s)$ in the classical control notation, and the condition of Eq. (11-255) is equivalent to the sensitivity relation of Eq. (9-74). In other words, the optimal system designed in the sense of minimizing J in Eq. (11-251) is that *the Nyquist locus of* $\mathbf{G}(j\omega\mathbf{I} - \mathbf{A})^{-1}\mathbf{B}$ *must not intersect or enclose the unit circle centered at* $(-1, j0)$ *in the complex function plane.* Therefore, the system designed based on the optimal linear regulator theory must all satisfy Eq. (11-255).

The design strategy here is to utilize the condition of Eq. (11-255) to select a weighting matrix $\mathbf{Q}$ that is positive semidefinite. The resulting $\mathbf{G}$ is then considered as the optimal feedback gain corresponding to that $\mathbf{Q}$. The approach follows that of the *inverse problem*[13] of the linear regulator design, which involves the determination of the weighting matrices once the feedback matrix is known.

For convenience, we let $R = 1$, since it is relative to the magnitudes of the elements of $\mathbf{Q}$. Let $p(s)$ represent the characteristic equation of the closed-loop system with partial state feedback; that is,

$$p(s) = |s\mathbf{I} - (\mathbf{A} - \mathbf{BG})| \qquad (11\text{-}257)$$

Since the system is controllable, it can be represented in phase-variable canonical form. The matrices $\mathbf{A}$ and $\mathbf{B}$ are written

$$\mathbf{A} = \begin{bmatrix} 0 & 1 & 0 & 0 & \ldots & 0 \\ 0 & 0 & 1 & 0 & \ldots & 0 \\ 0 & 0 & 0 & 1 & \ldots & 0 \\ \multicolumn{6}{c}{\dotfill} \\ 0 & 0 & 0 & 0 & \ldots & 1 \\ -a_1 & -a_2 & -a_3 & -a_4 & \ldots & -a_n \end{bmatrix} \qquad \mathbf{B} = \begin{bmatrix} 0 \\ 0 \\ 0 \\ . \\ . \\ 1 \end{bmatrix} \qquad (11\text{-}258)$$

The characteristic equation of $\mathbf{A}$ is denoted by

$$q(s) = |s\mathbf{I} - \mathbf{A}| = s^n + a_n s^{n-1} + a_{n-1} s^{n-2} + \ldots + a_1 \qquad (11\text{-}259)$$

The characteristic equation of the closed-loop system is written

$$p(s) = s^n + (a_n + g_n)s^{n-1} + \ldots + (a_1 + g_1) \qquad (11\text{-}260)$$

Then

$$(s\mathbf{I} - \mathbf{A})^{-1}\mathbf{B} = \frac{1}{q(s)}\begin{bmatrix} 1 \\ s \\ s^2 \\ . \\ . \\ . \\ s^{n-1} \end{bmatrix} \qquad (11\text{-}261)$$

and

$$\mathbf{G}(s\mathbf{I} - \mathbf{A})^{-1}\mathbf{B} = \frac{g_n s^{n-1} + g_{n-1} s^{n-2} + \ldots + g_1}{q(s)} \qquad (11\text{-}262)$$

Thus

$$[1 + \mathbf{G}(s\mathbf{I} - \mathbf{A})^{-1}\mathbf{B}] = \frac{p(s)}{q(s)} \qquad (11\text{-}263)$$

In view of Eq. (11-255), we can write

$$|1 + \mathbf{G}(j\omega\mathbf{I} - \mathbf{A})^{-1}\mathbf{B}|^2 = [1 + \mathbf{G}(j\omega\mathbf{I} - \mathbf{A})^{-1}\mathbf{B}][1 + \mathbf{G}(-j\omega\mathbf{I} - \mathbf{A})^{-1}\mathbf{B}]$$

$$= \frac{p(j\omega)p(-j\omega)}{q(j\omega)q(-j\omega)} \geq 1 \qquad (11\text{-}264)$$

Subtracting 1 from both sides of the last equation, we have

$$\frac{p(j\omega)p(-j\omega) - q(j\omega)q(-j\omega)}{q(j\omega)q(-j\omega)} \geq 0 \qquad (11\text{-}265)$$

Let the numerator of the last equation be factored as

$$p(j\omega)p(-j\omega) - q(j\omega)q(-j\omega) = d(j\omega)d(-j\omega) \qquad (11\text{-}266)$$

where

$$d(s) = d_n s^{n-1} + d_{n-1} s^{n-2} + \ldots + d_1 \qquad (11\text{-}267)$$

Equation (11-267) is of the $(n - 1)$st order since the s^n terms in $p(s)$ and $q(s)$ are canceled in Eq. (11-266). Let

$$\mathbf{D} = \begin{bmatrix} d_1 \\ d_2 \\ \cdot \\ \cdot \\ \cdot \\ d_n \end{bmatrix} \tag{11-268}$$

Then

$$\mathbf{D}(s\mathbf{I} - \mathbf{A})^{-1}\mathbf{B} = \frac{d(s)}{q(s)} \tag{11-269}$$

Equation (11-264) is written as

$$[1 + \mathbf{G}(j\omega\mathbf{I} - \mathbf{A})^{-1}\mathbf{B}][1 + \mathbf{G}(-j\omega\mathbf{I} - \mathbf{A})^{-1}\mathbf{B}] = 1 + \frac{d(j\omega)\,d(-j\omega)}{q(j\omega)q(-j\omega)} \tag{11-270}$$
$$= 1 + \mathbf{B}'(-j\omega\mathbf{I} - \mathbf{A}')^{-1}\mathbf{D}'\mathbf{D}(j\omega\mathbf{I} - \mathbf{A})^{-1}\mathbf{B}$$

It has been shown by Anderson and Moore[12] that for $\mathbf{G}$ to be the optimal feedback matrix, it is necessary that

$$[1 + \mathbf{G}(j\omega\mathbf{I} - \mathbf{A})^{-1}\mathbf{B}][1 + \mathbf{G}(-j\omega\mathbf{I} - \mathbf{A})^{-1}\mathbf{B}]$$
$$= 1 + \mathbf{B}'(-j\omega\mathbf{I} - \mathbf{A}')^{-1}\mathbf{Q}(j\omega\mathbf{I} - \mathbf{A})^{-1}\mathbf{B} \tag{11-271}$$

Comparing Eqs. (11-270) and (11-271), we have

$$\mathbf{Q} = \mathbf{D}'\mathbf{D} \tag{11-272}$$

Equations (11-266), (11-272), and the condition that $\mathbf{Q}$ is positive semi-definite form a design procedure for the partial state feedback. It should be kept in mind that since the system may not be stabilizable, or the inverse problem does not exist, an optimal solution may not always be possible.

EXAMPLE 11-9   Consider the dynamic equations of a linear time-invariant system,

$$\dot{\mathbf{x}}(t) = \mathbf{A}\mathbf{x}(t) + \mathbf{B}u(t) \tag{11-273}$$
$$y(t) = x_2(t) \tag{11-274}$$

where

$$\mathbf{A} = \begin{bmatrix} 0 & 1 \\ -1 & 0 \end{bmatrix} \qquad \mathbf{B} = \begin{bmatrix} 0 \\ 1 \end{bmatrix}$$

It is desired to design a closed-loop system by feeding back only $x_2(t)$ through a constant gain, so that the performance index

$$J = \int_{t_0}^{\infty} [\mathbf{x}'(t)\mathbf{Q}\mathbf{x}(t) + u^2(t)]\,dt \tag{11-275}$$

is minimized. The control is $u(t) = -\mathbf{G}\mathbf{x}(t)$.

The characteristic equation of $\mathbf{A}$ is

$$q(s) = |s\mathbf{I} - \mathbf{A}| = s^2 + 1 \tag{11-276}$$

and the characteristic equation of the closed-loop system with feedback from x_2 is

$$p(s) = |s\mathbf{I} - (\mathbf{A} - \mathbf{BG})| = s^2 + g_2 s + 1 \tag{11-277}$$

where

$$\mathbf{G} = [0 \quad g_2] \tag{11-278}$$

For the second-order system,

$$d(s) = d_2 s + d_1 \tag{11-279}$$

and

$$\mathbf{D} = [d_1 \quad d_2] \tag{11-280}$$

Then the weighting matrix $\mathbf{Q}$ is of the form

$$\mathbf{Q} = \mathbf{D'D} = \begin{bmatrix} d_1^2 & d_1 d_2 \\ d_1 d_2 & d_2^2 \end{bmatrix} \tag{11-281}$$

From Eq. (11-266) we have

$$\begin{aligned} p(s)p(-s) &= s^4 + (2 - g_2^2)s^2 + 1 \\ &= q(s)q(-s) + d(s)d(-s) = s^4 + (2 - d_2^2)s^2 + d_1^2 + 1 \end{aligned} \tag{11-282}$$

Therefore,

$$d_1 = 0 \qquad d_2^2 = g_2^2$$

and

$$\mathbf{Q} = \begin{bmatrix} 0 & 0 \\ 0 & d_2^2 \end{bmatrix} = \begin{bmatrix} 0 & 0 \\ 0 & g_2^2 \end{bmatrix} \tag{11-283}$$

This result shows that $\mathbf{Q}$ must be of the form defined in Eq. (11-283), or the optimal solution does not exist for the problem given here. However, it does not mean that given any other positive semidefinite $\mathbf{Q}$ an optimal linear regulator solution does not exist.

The nonzero element of $\mathbf{Q}$, d_2^2, represents a constraint on the state x_2 which is the derivative of x_1. It is reasoned that if we place more constraint on x_2, the overshoot of the system will be reduced. This is verified by assigning different values to d_2, and observing the effect on the roots of the characteristic equation of the feedback system.

$$\begin{array}{lll} d_2^2 = 1: & p(s) = s^2 + s + 1 & \text{damping ratio } \zeta = 0.5 \\ d_2^2 = 2: & p(s) = s^2 + \sqrt{2}\,s + 1 & \text{damping ratio } \zeta = 0.707 \\ d_2^2 = 4: & p(s) = s^2 + 2s + 1 & \text{damping ratio } \zeta = 1.0 \end{array}$$

In more complex systems, Eq. (11-266) leads to a set of nonlinear equations that must be solved to obtain the elements of $\mathbf{D}$ and $\mathbf{G}$. The problem is simplified by observing that the elements in the last row of the matrix $\mathbf{A}$ in Eq. (11-258) already represent feedbacks from the state variables. Therefore, it is logical to absorb these coefficients in the feedback matrix $\mathbf{G}$. Then the modified matrix $\mathbf{A}$ would be

$$\mathbf{A}^* = \begin{bmatrix} 0 & 1 & 0 & \cdots & 0 \\ 0 & 0 & 1 & \cdots & 0 \\ & & \cdots\cdots\cdots & & \\ 0 & 0 & 0 & \cdots & 1 \\ 0 & 0 & 0 & \cdots & 0 \end{bmatrix} \tag{11-284}$$

The state equations of the system become

$$\dot{\mathbf{x}}(t) = \mathbf{A}^* \mathbf{x}(t) + \mathbf{B}u(t) \tag{11-285}$$

and the state feedback control is

$$u(t) = -\mathbf{G}^* \mathbf{x}(t) \tag{11-286}$$

where

$$\mathbf{G^*} = \mathbf{G} - [0 \quad 0 \quad \ldots \quad 0 \quad 1]\mathbf{A} \tag{11-287}$$

$$= [g_1 - a_1 \quad g_2 - a_2 \ldots g_n - a_n]$$

It is simple to see that the characteristic equation of $\mathbf{A^*}$ will always be

$$q(s) = s^n \tag{11-288}$$

The following example will illustrate this simplified procedure.

EXAMPLE 11-10   Consider the linear time-invariant system

$$\dot{\mathbf{x}}(t) = \mathbf{A}\mathbf{x}(t) + \mathbf{B}u(t)$$

where

$$\mathbf{A} = \begin{bmatrix} 0 & 1 & 0 \\ 0 & 0 & 1 \\ 0 & -2 & -3 \end{bmatrix} \quad \mathbf{B} = \begin{bmatrix} 0 \\ 0 \\ 1 \end{bmatrix}$$

We wish to find the feedback gain matrix $\mathbf{G} = [g_1 \quad 0 \quad 0]$ such that the performance index

$$J = \int_{t_0}^{\infty} [\mathbf{x}'(t)\mathbf{Q}\mathbf{x}(t) + u^2(t)] \, dt$$

is minimized. The form of $\mathbf{G}$ given indicates that only the state x_1 is fed back for control. Using Eq. (11-287), $\mathbf{G^*} = [g_1 \quad 2 \quad 3]$.

From Eq. (11-284), the characteristic equation of the closed-loop system is

$$p(s) = |s\mathbf{I} - (\mathbf{A} - \mathbf{BG})| = |s\mathbf{I} - (\mathbf{A^*} - \mathbf{BG^*})|$$

$$= s^3 + 3s^2 + 2s + g_1 \tag{11-289}$$

Let

$$d(s) = d_3 s^2 + d_2 s + d_1 \tag{11-290}$$

From Eq. (11-266), we have

$$p(s)p(-s) - q(s)q(-s) = 5s^4 + (6g_1 - 4)s^2 + g_1^2$$

$$= d_3^2 s^4 + (2d_1 d_3 - d_2^2)s^2 + d_1^2 \tag{11-291}$$

Thus equating the coefficients of corresponding terms in the last equation, we have

$$d_1^2 = g_1^2 \tag{11-292}$$

$$d_3^2 = 5 \tag{11-293}$$

$$2d_1 d_3 - d_2^2 = 6g_1 - 4 \tag{11-294}$$

Substitution of Eqs. (11-292) and (11-293) into Eq. (11-294) yields

$$d_2^2 = 4 - 1.5278d_1 \tag{11-295}$$

Since d_2^2 must be positive, Eq. (11-295) leads to

$$d_1 \leq \frac{4}{1.5278} = 2.618 \tag{11-296}$$

From Eq. (11-292),

$$g_1 \leq 2.618 \tag{11-297}$$

It is interesting to note that applying the Routh–Hurwitz criterion to the characteristic equation of Eq. (11-289) shows that the closed-loop system with feedback from x_1 is stable for $g_1 < 6$. However, for the optimal partial state feedback control, g_1

must be less than or equal to 2.618. The difference between the two values of g_1 is the margin of relative stability.

The weighting matrix is given by

$$\mathbf{Q} = \mathbf{D}'\mathbf{D} = \begin{bmatrix} d_1^2 & d_1 d_2 & d_1 d_3 \\ d_1 d_2 & d_2^2 & d_2 d_3 \\ d_1 d_3 & d_2 d_3 & d_3^2 \end{bmatrix} \tag{11-298}$$

whose elements have to satisfy Eqs. (11-292) through (11-294).

REFERENCES

Analytical Design

1. G. C. NEWTON, JR., L. A. GOULD, and J. F. KAISER, *Analytical Design of Linear Feedback Controls*, John Wiley & Sons, Inc., New York, 1957.

2. S. S. L. CHANG, *Synthesis of Optimum Control Systems*, McGraw-Hill Book Company, New York, 1961.

Design by Pole Assignment

3. W. M. WONHAM, "On Pole Assignment in Multi-input Controllable Linear Systems," *IEEE Trans. Automatic Control*, Vol. AC-12, pp. 660–665, Dec. 1967.

4. F. M. BRASCH and J. B. PEARSON, "Pole Placement Using Dynamic Compensators," *IEEE Trans. Automatic Control*, Vol. AC-15, pp. 34–43, Feb. 1970.

5. E. J. DAVISON, "On Pole Assignment in Linear Systems with Incomplete State Feedback," *IEEE Trans. Automatic Control*, Vol. AC-15, pp. 348–351, June 1970.

6. E. J. DAVISON and R. CHATTERJEE, "A Note on Pole Assignment in Linear Systems with Incomplete State Feedback," *IEEE Trans. Automatic Control*, Vol. AC-16, pp. 98–99, Feb. 1971.

7. J. C. WILLEMS and S. K. MITTER, "Controllability, Observability, Pole Allocation, and State Reconstruction," *IEEE Trans. Automatic Control*, Vol. AC-16, pp. 582–593, Dec. 1971.

8. B. SRIDHER and D. P. LINDORFF, "A Note on Pole Assignment," *IEEE Trans. Automatic Control*, Vol. AC-17, pp. 822–823, Dec. 1972.

Design of Observers

9. D. G. LUENBERGER, "Observing the State of a Linear System," *IEEE Trans. Military Electronics*, Vol. MIL-8, pp. 74–80, Apr. 1964.

10. D. G. LUENBERGER, "Observers for Multivariable Systems," *IEEE Trans. Automatic Control*, Vol. AC-11, pp. 190–197, Apr. 1966.

11. D. G. LUENBERGER, "An Introduction to Observers," *IEEE Trans. Automatic Control*, Vol. AC-16, pp. 596–602, Dec. 1971.

12. B. D. O. ANDERSON and J. B. MOORE, *Linear Optimal Control*, Prentice-Hall, Inc., Englewood Cliffs, N.J., 1971.

Linear Regulator Design

13. R. E. KALMAN, "When Is a Linear Control System Optimal?" *Trans. ASME, J. Basic Eng.*, Ser. D, Vol. 86, pp. 51–60, Mar. 1964.

14. R. E. KALMAN, "Contributions to the Theory of Optimal Control," *Bol. Soc. Mat. Mex.*, Vol. 5, pp. 102–119, 1960.

15. M. ATHANS and P. L. FALB, *Optimal Control*, McGraw-Hill Book Company, New York, 1966.

16. A. P. SAGE, *Optimum System Control*, Prentice-Hall, Inc., Englewood Cliffs, N.J., 1968.

PROBLEMS

11.1. Prove that the "integral of $t^2 e^2(t)$ criterion,"

$$J = \int_0^\infty t^2 e^2(t)\, dt$$

can be expressed as

$$J = -\frac{1}{2\pi j} \int_{-j\infty}^{j\infty} F(s)F(-s)\, ds$$

where $F(s) = dE(s)/ds$.

11.2. A unity-feedback control system is shown in Fig. P11-2.

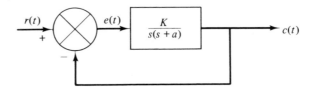

Figure P11-2.

(a) Determine the integral square error (ISE),

$$J = \int_0^\infty e^2(t)\, dt$$

when $K = 10$, $r(t) = u_s(t)$ (unit step input), and $a = 2$.

(b) Find the value of a as a function of K so that the ISE is minimized, when $r(t) = u_s(t)$. Determine the damping ratio of the optimal system.

(c) If K is variable but a is held constant, find the optimal value of K so that the ISE is a minimum.

(d) Repeat part (a) when the input is a unit ramp function, $r(t) = tu_s(t)$.

11.3. The transfer function of a linear process is

$$G(s) = \frac{C(s)}{U(s)} = \frac{1}{s^2}$$

Determine the optimal closed-loop transfer function $C(s)/R(s)$ such that

$$J_e = \int_0^\infty [r(t) - c(t)]^2 \, dt = \text{minimum}$$

$$J_d = \int_0^\infty \left[\frac{du(t)}{dt}\right]^2 \, dt \le 1$$

The reference input $r(t)$ is a unit step function,

11.4. A linear process is described by

$$\dot{x}_1 = x_2$$

$$\dot{x}_2 = u$$

where x_1 and x_2 are state variables and u is the control. It is desired to minimize the quadratic performance index

$$J = \tfrac{1}{2} \int_0^\infty [(x_1 - x_d)^2 + u^2] \, dt$$

where $x_d = $ constant. Determine the optimal control as a function of x_1, x_2, and x_d, by means of the integral-square-error design. Draw a block diagram for the completely designed system.

11.5. For the control system shown in Fig. P11-5, the input $r(t)$ is a unit step function.

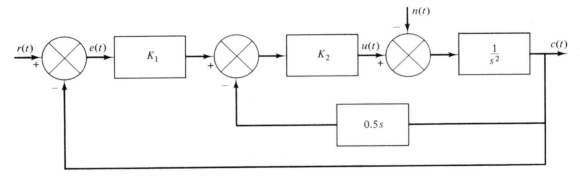

Figure P11-5.

(a) Determine K_1 and K_2 so that the integral square error J_e is minimized, subject to $J_u \le 1$;

$$J_e = \int_0^\infty e^2(t) \, dt \qquad J_u = \int_0^\infty d^2(t) \, dt$$

The noise signal $n(t)$ is assumed to be zero.

(b) If $r(t) = 0$, and the values of K_1 and K_2 are as determined in part (a), find the maximum strength of the impulse disturbance $n(t) = N\delta(t)$ that can be applied to the system such that $J_u \le 1$.

(c) Repeat part (a) with the following design criteria:

$$J_e \le 1 \qquad J_u = \text{minimum}$$

(d) Repeat part (b), and find the maximum strength of the impulse disturbance N such that J_u is less than or equal to the minimum value found in (c).

11.6. The control system shown in Fig. P11-6 has two free parameters in K_1 and K_2. Determine the values of K_1 and K_2 by parameter optimization such that

$$J_e = \int_0^\infty e^2(t)\, dt = \text{minimum}$$

$$J_u = \int_0^\infty u^2(t)\, dt \le 2.5$$

The reference input is a step function with an amplitude of 0.5.

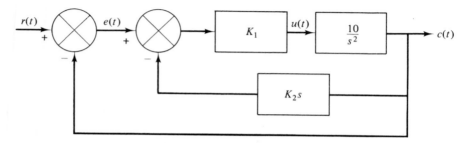

Figure P11-6.

11.7. Consider the linear process

$$\dot{\mathbf{x}}(t) = \mathbf{A}\mathbf{x}(t) + \mathbf{B}u(t)$$

$$c(t) = \mathbf{D}\mathbf{x}(t)$$

where

$$\mathbf{A} = \begin{bmatrix} 1 & 0 \\ 0 & 0 \end{bmatrix} \qquad \mathbf{B} = \begin{bmatrix} 1 \\ 1 \end{bmatrix} \qquad \mathbf{D} = \begin{bmatrix} 2 & -1 \end{bmatrix}$$

Design a state observer so that $\mathbf{x}(t) - \mathbf{x}_e(t)$ will decay as fast as e^{-10t}. Find the characteristic equation of the observer and the feedback matrix $\mathbf{G}_e$. Write the state equation of the observer in matrix form.

11.8. A linear time-invariant process is described by the state equation

$$\dot{x}(t) = -0.5x(t) + u(t)$$

with the initial state $x(t_0) = x_0$. Find the optimal control $u^\circ(t)$ that minimizes the performance index

$$J = \tfrac{1}{2} \int_{t_0}^{t_f} [2x^2(t) + u^2(t)]\, dt$$

where $t_0 = 0$ and $t_f = 1$ sec.

11.9. Solve Problem 11.8 with $t_f = \infty$.

11.10. Consider the linear process

$$\dot{x}(t) = 0.5x(t) + u(t)$$

It is desired to minimize the performance index

$$J = \int_0^T (\tfrac{1}{4}e^{-t}x^2 + e^{-t}u^2)\, dt$$

Show that the optimal control is given by

$$u^\circ(t) = -\frac{1}{2}\frac{1 - e^t e^{-T}}{1 + e^t e^{-T}}x(t)$$

11.11. A second-order process is described by the state equations

$$\dot{x}_1(t) = x_2(t)$$
$$\dot{x}_2(t) = u(t)$$

Find the optimal state-feedback control that minimizes

$$J = \frac{1}{2}\int_0^\infty (x_1^2 + 2x_1x_2 + 4x_2^2 + u^2)\,dt$$

Draw a block diagram for the closed-loop system. Find the damping ratio of the system.

11.12. Consider the linear time-invariant process

$$\dot{\mathbf{x}}(t) = \mathbf{A}\mathbf{x}(t) + \mathbf{B}u(t)$$

where

$$\mathbf{A} = \begin{bmatrix} 0 & 1 & 0 \\ 0 & 0 & 1 \\ 4 & -4 & 1 \end{bmatrix} \quad \mathbf{B} = \begin{bmatrix} 0 \\ 0 \\ 1 \end{bmatrix}$$

It is desired to find the feedback gain matrix $\mathbf{G} = [g_1 \quad 0 \quad g_3]$ such that the performance index

$$J = \int_{t_0}^\infty [\mathbf{x}'(t)\mathbf{Q}\mathbf{x}(t) + u^2(t)]\,dt = \text{minimum}$$

Formulate the problem using the method described by Eqs. (11-284) through (11-288). Find the bounds on g_1 and g_3 such that solutions to the optimal linear regulator problem with the specific partial state feedback exist.

APPENDIX A

Frequency-Domain Plots

Consider that $G(s)H(s)$ is the loop transfer function of a feedback control system. The sinusoidal steady-state transfer function is obtained by setting $s = j\omega$ in $G(s)H(s)$. In control systems studies, frequency-domain plots of the open-loop transfer function $G(j\omega)H(j\omega)$ are made for the purpose of analysis of the performance of the closed-loop control system.

The function $G(j\omega)$ is generally a complex function of the frequency ω and can be written

$$G(j\omega) = |G(j\omega)| \underline{/G(j\omega)} \tag{A-1}$$

where $|G(j\omega)|$ denotes the magnitude of $G(j\omega)$ and $\underline{/G(j\omega)}$ is the phase of $G(j\omega)$.

The following forms of frequency-domain plots of $G(j\omega)$ [or of $G(j\omega)H(j\omega)$] versus ω are most useful in the analysis and design of feedback control systems in the frequency domain.

1. Polar plot: a plot of the magnitude versus phase in the polar coordinates as ω is varied from zero to infinity.
2. Bode plot (corner plot): a plot of the magnitude in decibels versus ω (or $\log_{10} \omega$) in the semilog (or rectangular) coordinates.
3. Magnitude-versus-phase plot: a plot of the magnitude in decibels versus the phase on rectangular coordinates with ω as a variable parameter on the curve.

These various plots are described in the following sections.

A.1   Polar Plots of Transfer Functions

The polar plot of a transfer function $G(s)$ is a plot of the magnitude of $G(j\omega)$ versus the phase of $G(j\omega)$ on the polar coordinates, as ω is varied from zero to infinity. From a mathematical viewpoint, the process may be regarded as a mapping of the positive half of the imaginary axis of the s-plane onto the plane of the function $G(j\omega)$. A simple example of this mapping is shown in Fig. A-1. For any frequency $\omega = \omega_1$, the magnitude and phase of $G(j\omega_1)$ are represented by a phasor that has the corresponding magnitude and phase angle in the $G(j\omega)$-

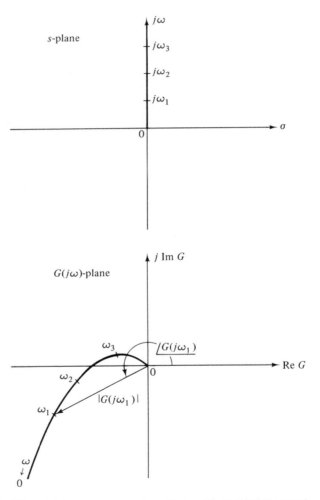

Fig. A-1. Polar plot shown as a mapping of the positive half of the $j\omega$ axis in the s-plane onto the $G(j\omega)$-plane.

plane. In measuring the phase, counterclockwise is referred to as positive, and clockwise as negative.

To illustrate the construction of the polar plot of a transfer function, consider the function

$$G(s) = \frac{1}{1 + Ts} \tag{A-2}$$

where T is a positive constant.

Putting $s = j\omega$, we have

$$G(j\omega) = \frac{1}{1 + j\omega T} \tag{A-3}$$

In terms of magnitude and phase, the last expression is written

$$G(j\omega) = \frac{1}{\sqrt{1 + \omega^2 T^2}} \underline{/-\tan^{-1} \omega T} \tag{A-4}$$

When ω is zero, the magnitude of $G(j\omega)$ is unity, and the phase of $G(j\omega)$ is at $0°$. Thus, at $\omega = 0$, $G(j\omega)$ is represented by a phasor of unit length directed in the $0°$ direction. As ω increases, the magnitude of $G(j\omega)$ decreases, and the phase becomes more negative. As ω increases, the length of the phasor in the polar coordinates decreases, and the phasor rotates in the clockwise (negative) direction. When ω approaches infinity, the magnitude of $G(j\omega)$ becomes zero, and the phase reaches $-90°$. This is often represented by a phasor with an infinitesimally small length directed along the $-90°$ axis in the $G(j\omega)$-plane. By substituting other finite values of ω into Eq. (A-4), the exact plot of $G(j\omega)$ turns out to be a semicircle, as shown in Fig. A-2.

As a second illustrative example, consider the transfer function

$$G(j\omega) = \frac{1 + j\omega T_2}{1 + j\omega T_1} \tag{A-5}$$

where T_1 and T_2 are positive constants. Equation (A-5) is written

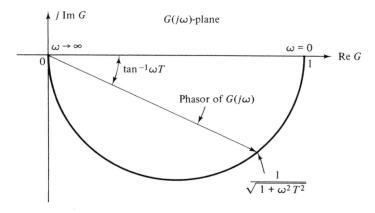

Fig. A-2. Polar plot of $G(j\omega) = 1/(1 + j\omega T)$.

$$G(j\omega) = \sqrt{\frac{1 + \omega^2 T_2^2}{1 + \omega^2 T_1^2}} \underline{/\tan^{-1} \omega T_2 - \tan^{-1} \omega T_1} \qquad (\text{A-6})$$

The polar plot of $G(j\omega)$, in this case, depends upon the relative magnitudes of T_2 and T_1. If T_2 is greater than T_1, the magnitude of $G(j\omega)$ is always greater than unity as ω is varied from zero to infinity, and the phase of $G(j\omega)$ is always positive. If T_2 is less than T_1, the magnitude of $G(j\omega)$ is always less than unity, and the phase is always negative. The polar plots of $G(j\omega)$ of Eq. (A-6) that correspond to the two above-mentioned conditions are shown in Fig. A-3.

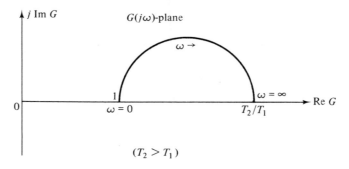

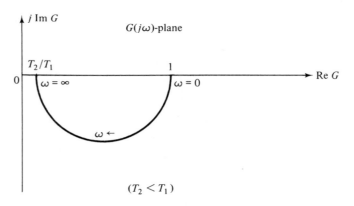

Fig. A-3. Polar plots of $G(j\omega) = (1 + j\omega T_2)/(1 + j\omega T_1)$.

It is apparent that the accurate plotting of the polar plot of a transfer function is generally a tedious process, especially if the transfer function is of high order. In practice, a digital computer can be used to generate the data, or even the final figure of the polar plot, for a wide class of transfer functions. However, from the analytical standpoint, it is essential that the engineer be completely familiar with the properties of the polar plot, so that the computer data may be properly interpreted. In some cases, such as in the Nyquist stability study, only the general shape of the polar plot of $G(j\omega)H(j\omega)$ is needed, and often a rough sketch of the polar plot is quite adequate for the specific objective. In general, the sketching of the polar plot is facilitated by the following information:

1. The behavior of the magnitude and the phase at $\omega = 0$ and at $\omega = \infty$.

2. The points of intersections of the polar plot with the real and imaginary axes, and the values of ω at these intersections.

The general shape of the polar plot may be determined once we have information on the two items listed above.

EXAMPLE A-1   Consider that it is desired to make a rough sketch of the polar plot of the transfer function

$$G(s) = \frac{10}{s(s+1)} \tag{A-7}$$

Substituting $s = j\omega$ in Eq. (A-7), the magnitude and phase of $G(j\omega)$ at $\omega = 0$ and $\omega = \infty$ are computed as follows:

$$\lim_{\omega \to 0} |G(j\omega)| = \lim_{\omega \to 0} \frac{10}{\omega} = \infty \tag{A-8}$$

$$\lim_{\omega \to 0} \underline{/G(j\omega)} = \lim_{\omega \to 0} \underline{/\frac{10}{j\omega}} = -90° \tag{A-9}$$

$$\lim_{\omega \to \infty} |G(j\omega)| = \lim_{\omega \to \infty} \frac{10}{\omega^2} = 0 \tag{A-10}$$

$$\lim_{\omega \to \infty} \underline{/G(j\omega)} = \lim_{\omega \to \infty} \underline{/\frac{10}{(j\omega)^2}} = -180° \tag{A-11}$$

Thus the properties of the polar plot of $G(j\omega)$ at $\omega = 0$ and $\omega = \infty$ are ascertained. Next, we determine the intersections, if any, of the polar plot with the two axes of the $G(j\omega)$-plane.

If the polar plot of $G(j\omega)$ intersects the real axis, at the point of intersection, the imaginary part of $G(j\omega)$ is zero; that is,

$$\text{Im}[G(j\omega)] = 0 \tag{A-12}$$

In order to express $G(j\omega)$ as

$$G(j\omega) = \text{Re}[G(j\omega)] + j\,\text{Im}[G(j\omega)] \tag{A-13}$$

we must rationalize $G(j\omega)$ by multiplying its numerator and denominator by the complex conjugate of its denominator. Therefore, $G(j\omega)$ is written

$$G(j\omega) = \frac{10(-j\omega)(-j\omega + 1)}{j\omega(j\omega + 1)(-j\omega)(-j\omega + 1)} = \frac{-10\omega^2}{\omega^4 + \omega^2} - j\frac{10\omega}{\omega^4 + \omega^2} \tag{A-14}$$

which gives

$$\text{Im}[G(j\omega)] = \frac{-10}{\omega(\omega^2 + 1)} \tag{A-15}$$

and

$$\text{Re}[G(j\omega)] = \frac{-10}{\omega^2 + 1} \tag{A-16}$$

When we set $\text{Im}[G(j\omega)]$ to zero, we get $\omega = \infty$, meaning that the only intersect that the $G(j\omega)$ plot has with the real axis of the plane is at the origin.

Similarly, the intersection of the polar plot of $G(j\omega)$ with the imaginary axis is found by setting $\text{Re}[G(j\omega)]$ of Eq. (A-16) to zero. The only real solution for ω is also $\omega = \infty$, which corresponds to the origin of the $G(j\omega)$-plane. The conclusion is that the polar plot of $G(j\omega)$ does not intersect any one of the two axes at any finite nonzero

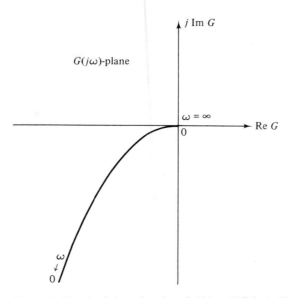

Fig. A-4. Sketch of the polar plot of $G(s) = 10/[s(s + 1)]$.

frequency. Based upon this information, as well as knowledge on the angles of $G(j\omega)$ at $\omega = 0$ and $\omega = \infty$, the polar plot of $G(j\omega)$ is easily sketched, as shown in Fig. A-4.

EXAMPLE A-2   Given the transfer function

$$G(s) = \frac{10}{s(s + 1)(s + 2)} \tag{A-17}$$

it is desired to make a rough sketch of the polar plot of $G(j\omega)$. The following calculations are made for the properties of the magnitude and phase of $G(j\omega)$ at $\omega = 0$ and $\omega = \infty$:

$$\lim_{\omega \to 0} |G(j\omega)| = \lim_{\omega \to 0} \frac{5}{\omega} = \infty \tag{A-18}$$

$$\lim_{\omega \to 0} \underline{/G(j\omega)} = \lim_{\omega \to 0} \underline{/\frac{5}{j\omega}} = -90° \tag{A-19}$$

$$\lim_{\omega \to \infty} |G(j\omega)| = \lim_{\omega \to \infty} \frac{10}{\omega^3} = 0 \tag{A-20}$$

$$\lim_{\omega \to \infty} \underline{/G(j\omega)} = \lim_{\omega \to \infty} \underline{/\frac{10}{(j\omega)^3}} = -270° \tag{A-21}$$

To find the intersections of the $G(j\omega)$ curve on the real and the imaginary axes of the $G(j\omega)$-plane, we rationalize $G(j\omega)$ to give

$$G(j\omega) = \frac{10(-j\omega)(-j\omega + 1)(-j\omega + 2)}{j\omega(j\omega + 1)(j\omega + 2)(-j\omega)(-j\omega + 1)(-j\omega + 2)} \tag{A-22}$$

After simplification, Eq. (A-22) is written

$$G(j\omega) = \frac{-30\omega^2}{9\omega^4 + \omega^2(2 - \omega^2)^2} - \frac{j10\omega(2 - \omega^2)}{9\omega^4 + \omega^2(2 - \omega^2)^2} \tag{A-23}$$

We set

$$\text{Re}[G(j\omega)] = \frac{-30}{9\omega^2 + (2 - \omega^2)^2} = 0 \qquad \text{(A-24)}$$

and

$$\text{Im}[G(j\omega)] = \frac{-10(2 - \omega^2)}{9\omega^3 + \omega(2 - \omega^2)^2} = 0 \qquad \text{(A-25)}$$

Equation (A-24) is satisfied when

$$\omega = \infty$$

which means that the $G(j\omega)$ plot intersects the imaginary axis only at the origin. Equation (A-25) is satisfied when

$$\omega^2 = 2$$

which gives the intersection on the real axis of the $G(j\omega)$-plane when $\omega = \pm\sqrt{2}$ rad/sec. Substituting $\omega = \sqrt{2}$ into Eq. (A-23) gives the point of intersection at

$$G(j\sqrt{2}) = -\tfrac{5}{3} \qquad \text{(A-26)}$$

The result of $\omega = -\sqrt{2}$ rad/sec has no physical meaning, but mathematically it simply represents a mapping point on the negative $j\omega$ axis of the s-plane. In general, if $G(s)$ is a rational function of s (a quotient of two polynomials of s), the polar plot of $G(j\omega)$ for negative values of ω is the mirror image of that for positive ω, with the mirror placed on the real axis of the $G(j\omega)$-plane.

With the information collected above, it is now possible to make a sketch of the polar plot for the transfer function in Eq. (A-17), and the sketch is shown in Fig. A-5.

Although the method of obtaining the rough sketch of the polar plot of a transfer function as described above is quite straightforward, in general, for complex transfer

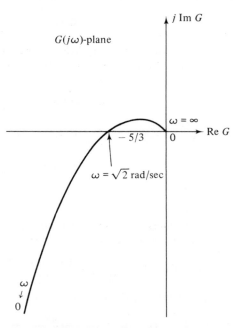

Fig. A-5. Sketch of the polar plot of $G(s) = 10/[s(s + 1)(s + 2)]$.

functions that may have multiple crossings on the real and imaginary axes in the transfer function plane, the algebraic manipulation may again be quite involved. Furthermore, the polar plot is basically a tool for analysis; it is somewhat awkward for design purposes. We shall show in the next section that approximate information on the polar plot can always be obtained from the Bode plot, which is usually sketched without any computations. Thus, for more complex transfer functions, other than using the digital computer, sketches of the polar plots are preferably obtained with the help of the Bode plots.

A.2 Bode Plot (Corner Plot) of a Transfer Function

The discussions in the last section show that the polar plot portrays a function $G(j\omega)$ in the polar coordinates in terms of its magnitude and phase as functions of ω. The Bode plot, on the other hand, contains two graphs, one with the magnitude of $G(j\omega)$ plotted in decibels versus log ω or ω, and the other with the phase of $G(j\omega)$ in degrees as a function of log ω or ω. The Bode plot is also known sometimes as the *corner plot* or the *logarithmic plot* of $G(j\omega)$. The name, corner plot, is used since the Bode plot is basically an approximation method in that the magnitude of $G(j\omega)$ in decibles as a function of ω is approximated by straight-line segments.

In simple terms, the Bode plot has the following unique characteristics:

1. Since the magnitude of $G(j\omega)$ in the Bode plot is expressed in decibels, the product and division factors in $G(j\omega)$ become additions and subtractions, respectively. The phase relations are also added and subtracted from each other in a natural way.
2. The magnitude plot of the Bode plots of most $G(j\omega)$ functions encountered in control systems may be approximated by straight-line segments. This makes the construction of the Bode plot very simple.

Since the corner plot is relatively easy to construct, and usually without point-by-point plotting, it may be used to generate data necessary for the other frequency-domain plots, such as the polar plot, or the magnitude-versus-phase plot, which is discussed later in this chapter.

In general, we may represent the open-loop transfer function of a feedback control system without pure time delay by

$$G(s) = \frac{K(s + z_1)(s + z_2) \cdots (s + z_m)}{s^i(s + p_1)(s + p_2) \cdots (s + p_n)} \qquad (A\text{-}27)$$

where K is a real constant and the zs and the ps may be real or complex numbers. As an alternative, the open-loop transfer function is written

$$G(s) = \frac{K(1 + T_1 s)(1 + T_2 s) \cdots (1 + T_m s)}{s^i(1 + T_a s)(1 + T_b s) \cdots (1 + T_n s)} \qquad (A\text{-}28)$$

where K is a real constant, and the Ts may be real or complex numbers.

In Chapter 8, Eq. (A-27) is the preferred form for root loci construction. However, for Bode plots, the transfer function should first be written in the form of Eq. (A-28). Since practically all the terms in Eq. (A-28) are of the same form, without loss of generality, we can use the following transfer function to illustrate the construction of the Bode diagram:

$$G(s) = \frac{K(1 + T_1 s)(1 + T_2 s)}{s(1 + T_a s)(1 + j2\zeta\mu - \mu^2)} \tag{A-29}$$

where K, T_1, T_2, T_a, ζ, and μ are real coefficients. It is assumed that the second-order polynomial, $1 + 2\zeta\mu - \mu^2$, $\mu = \omega/\omega_n$, has two complex-conjugate zeros.

The magnitude of $G(j\omega)$ in decibels is obtained by multiplying the logarithm to the base 10 of $|G(j\omega)|$ by 20; we have

$$
\begin{aligned}
|G(j\omega)|_{dB} = 20 \log_{10} |G(j\omega)| &= 20 \log_{10} |K| + 20 \log_{10} |1 + j\omega T_1| \\
&\quad + 20 \log_{10} |1 + j\omega T_2| - 20 \log_{10} |j\omega| \\
&\quad - 20 \log_{10} |1 + j\omega T_a| - 20 \log_{10} |1 + j2\zeta\mu - \mu^2|
\end{aligned} \tag{A-30}
$$

The phase of $G(j\omega)$ is written

$$
\begin{aligned}
\underline{/G(j\omega)} = \underline{/K} &+ \underline{/1 + j\omega T_1} + \underline{/1 + j\omega T_2} - \underline{/j\omega} \\
&- \underline{/1 + j\omega T_a} - \underline{/1 + j2\zeta\mu - \mu^2}
\end{aligned} \tag{A-31}
$$

In general, the function $G(j\omega)$ may be of higher order than that of Eq. (A-29) and have many more factored terms. However, Eqs. (A-30) and (A-31) indicate that additional terms in $G(j\omega)$ would simply produce more similar terms in the magnitude and phase expressions, so that the basic method of construction of the Bode plot would be the same. We have also indicated that, in general, $G(j\omega)$ may contain just four simple types of factors:

1. Constant factor $\qquad\qquad\qquad K$
2. Poles or zeros at the origin $\qquad (j\omega)^{\pm p}$
3. Poles or zeros not at $\omega = 0$ $\qquad (1 + j\omega T)^{\pm q}$
4. Complex poles or zeros $\qquad\quad (1 + j2\zeta\mu - \mu^2)^{\pm r}$

where p, q, and r are positive integers.

Equations (A-30) and (A-31) verify one of the unique characteristics of the Bode plot in that each of the four types of factors listed may be considered as a separate plot; the individual plots are then added or subtracted accordingly to yield the total magnitude in decibels and phase plot of $G(j\omega)$. The curves may be done on semilog graph paper or linear rectangular coordinate graph paper, depending on whether ω or $\log_{10} \omega$ is used as the abscissa.

We shall now investigate the sketching of the Bode plot of the different types of factors.

Constant Term, K

Since

$$K_{dB} = 20 \log_{10} K = \text{constant} \tag{A-32}$$

and

$$\underline{/K} = \begin{cases} 0° & K > 0 \\ 180° & K < 0 \end{cases} \tag{A-33}$$

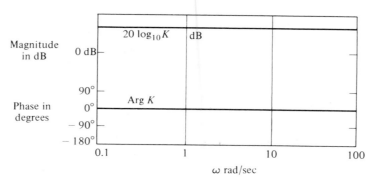

Fig. A-6. Bode plot of constant K.

the Bode plot of the constant factor K is shown in Fig. A-6 in semilog coordinates.

Poles and Zeros at the Origin, $(j\omega)^{\pm p}$

The magnitude of $(j\omega)^{\pm p}$ in decibels is given by

$$20 \log_{10} |(j\omega)^{\pm p}| = \pm 20p \log_{10} \omega \text{ dB} \qquad \text{(A-34)}$$

for $\omega \geq 0$. The last expression for a given p represents the equation of a straight line in either semilog or rectangular coordinates. The slopes of these straight lines may be determined by taking the derivative of Eq. (A-34) with respect to $\log_{10} \omega$; that is,

$$\frac{d}{d \log_{10} \omega} (\pm 20p \log_{10} \omega) = \pm 20p \qquad \text{dB} \qquad \text{(A-35)}$$

These lines all pass through the 0-dB point at $\omega = 1$. Thus a unit change in $\log_{10} \omega$ will correspond to a change of $\pm 20p$ dB in magnitude. Furthermore, a unit change in $\log_{10} \omega$ in the rectangular coordinates is equivalent to 1 decade of variation in ω, that is, from 1 to 10, 10 to 100, and so on, in the semilog coordinates. Thus the slopes of the straight lines described by Eq. (A-34) are said to be $\pm 20p$ dB/decade of frequency.

Instead of using decades, sometimes the unit *octave* is used to represent the separation of two frequencies. The frequencies ω_1 and ω_2 are separated by an octave if $\omega_2/\omega_1 = 2$. The number of decades between any two frequencies ω_1 and ω_2 is given by

$$\text{number of decades} = \frac{\log_{10} (\omega_2/\omega_1)}{\log_{10} 10} = \log_{10} \left(\frac{\omega_2}{\omega_1}\right) \qquad \text{(A-36)}$$

Similarly, the number of octaves between ω_2 and ω_1 is

$$\text{number of octaves} = \frac{\log_{10} (\omega_2/\omega_1)}{\log_{10} 2} = \frac{1}{0.301} \log_{10} \left(\frac{\omega_2}{\omega_1}\right) \qquad \text{(A-37)}$$

Thus, the relation between octaves and decades is given by

$$\text{number of octaves} = \frac{1}{0.301} \text{ decades} \qquad \text{(A-38)}$$

Substituting Eq. (A-38) into Eq. (A-35), we have

$$\pm 20p \text{ dB/decade} = \pm 20p \times 0.301 \simeq \pm 6p \text{ dB/octave} \qquad \text{(A-39)}$$

For a transfer function $G(j\omega)$ that has a simple pole at $s = 0$, the magnitude of $G(j\omega)$ is a straight line with a slope of -20 dB/decade, and passes through the 0-dB axis at $\omega = 1$.

The phase of $(j\omega)^{\pm p}$ is written

$$\underline{/(j\omega)^{\pm p}} = \pm p \times 90° \qquad \text{(A-40)}$$

The magnitude and phase curves of the function $(j\omega)^{\pm p}$ are sketched as shown in Fig. A-7 for several values of p.

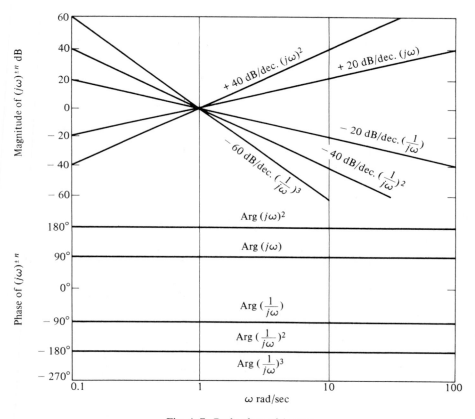

Fig. A-7. Bode plots of $(j\omega)^{\pm p}$.

Simple Zero $(1 + j\omega T)$

Let

$$G(j\omega) = 1 + j\omega T \qquad \text{(A-41)}$$

where T is a real constant. The magnitude of $G(j\omega)$ in decibels is written

$$|G(j\omega)|_{\text{dB}} = 20 \log_{10} |G(j\omega)| = 20 \log_{10} \sqrt{1 + \omega^2 T^2} \qquad \text{(A-42)}$$

To obtain asymptotic approximations to the magnitude of $G(j\omega)$, we consider both very large and very small values of ω. At very low frequencies, $\omega T \ll 1$, Eq. (A-42) is approximated by

$$|G(j\omega)|_{dB} = 20 \log_{10} |G(j\omega)| \simeq 20 \log_{10} 1 = 0 \text{ dB} \qquad \text{(A-43)}$$

since $\omega^2 T^2$ is neglected when compared with 1.

At very high frequencies, $\omega T \gg 1$, we may approximate $1 + \omega^2 T^2$ by $\omega^2 T^2$; then Eq. (A-42) becomes

$$\begin{aligned} |G(j\omega)|_{dB} = 20 \log_{10} |G(j\omega)| &\simeq 20 \log_{10} \sqrt{\omega^2 T^2} \\ &= 20 \log_{10} \omega T \end{aligned} \qquad \text{(A-44)}$$

Equation (A-44) respresents a straight line with a slope of $+20$ dB/decade of frequency. The intersect of this line with the 0-dB axis is found by equating Eq. (A-44) to zero, which gives

$$\omega = \frac{1}{T} \qquad \text{(A-45)}$$

This frequency is also the intersect of the high-frequency approximate plot and the low-frequency approximate plot which is the 0-dB line as given by Eq. (A-43). The frequency given by Eq. (A-45) is also known as the *corner frequency* of the Bode plot of the transfer function in Eq. (A-41), since the approximate magnitude plot forms the shape of a corner at that frequency, as shown in Fig. A-8. The actual magnitude curve for $G(j\omega)$ of Eq. (A-41) is a smooth curve, and deviates only slightly from the straight-line approximation. The actual values for the magnitude of the function $1 + j\omega T$ as functions of ωT are tabulated in Table A-1.

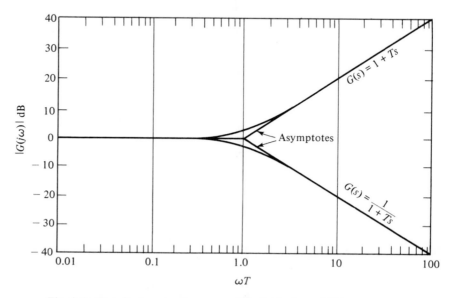

Fig. A-8. Magnitude versus frequency of the Bode plots of $G(s) = 1 + Ts$ and $G(s) = 1/(1 + Ts)$.

Table A-1

ωT	$\log_{10} \omega T$	$\|1 + j\omega T\|$	$\|1 + j\omega T\|$ (dB)	$\underline{/1 + j\omega T}$
0.01	-2	1	0	0.5°
0.1	-1	1.04	0.043	5.7°
0.5	-0.3	1.12	1	26.6°
0.76	-0.12	1.26	2	37.4°
1.0	0	1.41	3	45.0°
1.31	0.117	1.65	4.3	52.7°
1.73	0.238	2.0	6.0	60.0°
2.0	0.3	2.23	7.0	63.4°
5.0	0.7	5.1	14.2	78.7°
10.0	1.0	10.4	20.3	84.3°

Table A-2

ωT	$\|1 + j\omega T\|$ (dB)	Straight-Line Approximation of $\|1 + j\omega T\|$ (dB)	Error (dB)
0.1 (1 decade below corner frequency)	0.043	0	0.043
0.5 (1 octave below corner frequency)	1.0	0	1
0.76	2	0	2
1.0 (at the corner frequency)	3	0	3
1.31	4.3	2.3	2
2.0 (1 octave above corner frequency)	7	6	1
10 (1 decade above corner frequency)	20.043	20	0.043

Table A-2 gives a comparison of the actual values with the straight-line approximations at some significant frequencies.

The error between the actual magnitude curve and the straight-line asymptotes is symmetrical with respect to the corner frequency $1/T$. Furthermore, it is useful to remember that the error is 3 dB at the corner frequency, and 1 dB at 1 octave above $(2/T)$ and 1 octave below $(0.5/T)$ the corner frequency. At 1 decade above and below the corner frequency, the error is dropped to approximately 0.3 dB. From these facts, the procedure in obtaining the magnitude curve of the plot of the first-order factor $(1 + j\omega T)$ is outlined as follows:

1. Locate the corner frequency $\omega = 1/T$.
2. Draw the 20 dB/decade (or 6 dB/octave) line and the horizontal line at 0 dB, with the two lines interesecting at $\omega = 1/T$.
3. If necessary, the actual magnitude curve is obtained by locating the points given in Table A-1.

Usually, a smooth curve can be sketched simply by locating the 3-dB point at the corner frequency and the 1-dB points at 1 octave above and below the corner frequency.

The phase of $G(j\omega) = 1 + j\omega T$ is written as

$$\underline{/G(j\omega)} = \tan^{-1} \omega T \qquad (A\text{-}46)$$

Similar to the magnitude curve, a straight-line approximation can be made for the phase curve. Since the phase of $G(j\omega)$ varies from $0°$ to $90°$ we may draw a line from $0°$ at 1 decade below the corner frequency to $+90°$ at 1 decade above the corner frequency. As shown in Fig. A-9, the maximum deviation of the straight-line approximation from the actual curve is less than $6°$. Table A-1 gives the values of $\underline{/1 + j\omega T}$ versus ωT.

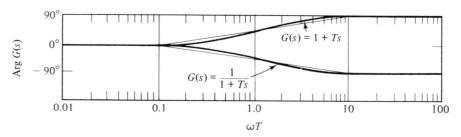

Fig. A-9. Phase versus frequency of the Bode plots of $G(s) = 1 + Ts$ and $G(s) = 1/(1 + Ts)$.

Simple Pole, $1/(1 + j\omega T)$

When

$$G(j\omega) = \frac{1}{1 + j\omega T} \qquad (A\text{-}47)$$

the magnitude, $|G(j\omega)|$ in decibels is given by the negative of the right side of Eq. (A-42), and the phase $\underline{/G(j\omega)}$ is the negative of the angle in Eq. (A-46). Therefore, it is simple to extend all the analysis for the case of the simple zero to the Bode plot of Eq. (A-47). We can write

$$\omega T \ll 1 \qquad |G(j\omega)|_{dB} \simeq 0 \text{ dB} \qquad (A\text{-}48)$$

$$\omega T \gg 1 \qquad |G(j\omega)|_{dB} = -20 \log_{10} \omega T \qquad (A\text{-}49)$$

Thus the corner frequency of the Bode plot of Eq. (A-47) is still at $\omega = 1/T$. At high frequencies, the slope of the straight-line approximation is $+20$ dB/ decade. The phase of $G(j\omega)$ is $0°$ when $\omega = 0$ and $-90°$ when ω approaches infinity. The magnitude and the phase of the Bode of Eq. (A-47) are shown in Figs. A-8 and A-9, respectively. The data in Tables A-1 and A-2 are still useful for the simple pole case, if appropriate sign changes are made to the numbers. For instance, at the corner frequency, the error between the straight-line approximation and the actual magnitude curve is -3 dB.

Quadratic Poles and Zeros

Now consider the second-order transfer function

$$G(s) = \frac{\omega_n^2}{s^2 + 2\zeta\omega_n s + \omega_n^2}$$
$$= \frac{1}{1 + (2\zeta/\omega_n)s + (1/\omega_n^2)s^2} \tag{A-50}$$

We are interested only in the cases when $\zeta \leq 1$, since otherwise, $G(s)$ would have two unequal real poles, and the Bode plot can be determined by considering $G(s)$ as the product of two transfer functions each having a simple pole.

Letting $s = j\omega$, Eq. (A-50) becomes

$$G(j\omega) = \frac{1}{[1 - (\omega/\omega_n)^2] + j2\zeta(\omega/\omega_n)} \tag{A-51}$$

The magnitude of $G(j\omega)$ in decibels is

$$20\log_{10}|G(j\omega)| = -20\log_{10}\sqrt{\left[1 - \left(\frac{\omega}{\omega_n}\right)^2\right]^2 + 4\zeta^2\left(\frac{\omega}{\omega_n}\right)^2} \tag{A-52}$$

At very low frequencies, $\omega/\omega_n \ll 1$; Eq. (A-52) may be written as

$$|G(j\omega)|_{dB} = 20\log_{10}|G(j\omega)| \cong -20\log_{10}1 = 0 \text{ dB} \tag{A-53}$$

Thus the low-frequency asymptote of the magnitude plot of Eq. (A-50) is a straight line that lies on the 0-dB axis of the Bode plot coordinates.

At very high frequencies, $\omega/\omega_n \gg 1$; the magnitude in decibels of $G(j\omega)$ in Eq. (A-50) becomes

$$|G(j\omega)|_{dB} = 20\log_{10}|G(j\omega)| \cong -20\log_{10}\sqrt{\left(\frac{\omega}{\omega_n}\right)^4}$$
$$= -40\log_{10}\left(\frac{\omega}{\omega_n}\right) \quad \text{dB} \tag{A-54}$$

This equation represents the equation of a straight line with a slope of -40 dB/decade in the Bode plot coordinates. The intersection of the two asymptotes is found by equating Eq. (A-53) with Eq. (A-54), yielding

$$-40\log_{10}\left(\frac{\omega}{\omega_n}\right) = 0 \text{ dB} \tag{A-55}$$

and from which we get

$$\omega = \omega_n \tag{A-56}$$

Thus the frequency, $\omega = \omega_n$, is considered to be the corner frequency of the second-order transfer function of Eq. (A-50), with the condition that $\zeta \leq 1$.

The actual magnitude plot of $G(j\omega)$ in this case may differ strikingly from the asymptotic lines. The reason for this is that the amplitude and phase curves of the $G(j\omega)$ of Eq. (A-50) depend not only on the corner frequency ω_n, but also on the damping ratio ζ. The actual and the asymptotic magnitude plots of $G(j\omega)$ are shown in Fig. A-10 for several values of ζ. The errors between the two sets of curves are shown in Fig. A-11 for the same set of ζs. The standard

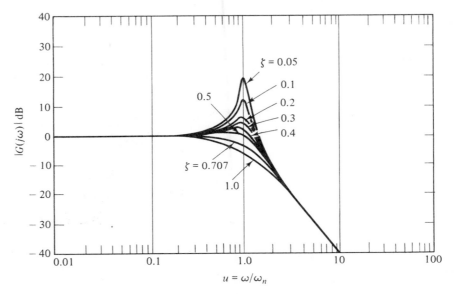

Fig. A-10. Magnitude versus frequency of Bode plot of $G(s) = 1/[1 + 2\zeta(s/\omega_n) + (s/\omega_n)^2]$.

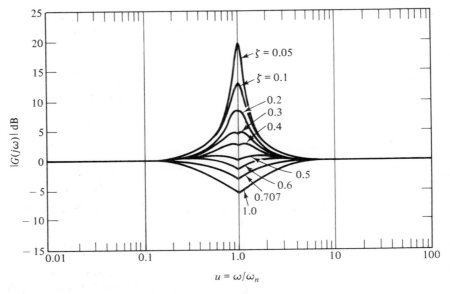

Fig. A-11. Error in magnitude versus frequency of Bode plot of $G(s) = 1/[1 + 2\zeta(s/\omega_n) + (s/\omega_n)^2]$.

procedure of constructing the magnitude portion of the Bode plot of a second-order transfer function of the form of Eq. (A-50) is to first locate the corner frequency ω_n, then sketch the asymptotic lines; the actual curve is obtained by making corrections to the asymptotes by using either the error curves of Fig. A-11 or the curves in Fig. A-10 for the corresponding ζ.

The phase of $G(j\omega)$ is given by

$$\underline{/G(j\omega)} = -\tan^{-1}\left\{\frac{2\zeta\omega}{\omega_n}\Big/\left[1 - \left(\frac{\omega}{\omega_n}\right)^2\right]\right\} \tag{A-57}$$

and is plotted as shown in Fig. A-12 for various values of ζ.

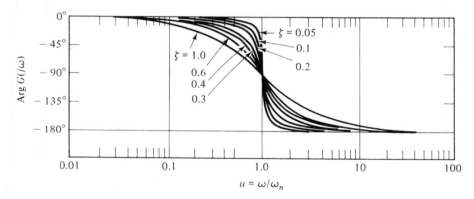

Fig. A-12. Phase versus frequency of Bode plot of $G(s) = 1/[1 + 2\zeta(s/\omega_n) + (s/\omega_n)^2]$.

The analysis of the Bode plot of the second-order transfer function of Eq. (A-50) may be applied to a second-order transfer function with two complex zeros. If

$$G(s) = 1 + \frac{2\zeta}{\omega_n}s + \frac{1}{\omega_n^2}s^2 \tag{A-58}$$

the Bode plot of $G(j\omega)$ may be obtained by inverting the curves of Figs. A-10, A-11, and A-12.

EXAMPLE A-3   As an illustrative example of the Bode plot of a transfer function, let us consider

$$G(s) = \frac{10(s + 10)}{s(s + 2)(s + 5)} \tag{A-59}$$

The first step is to express the transfer function in the form of Eq. (A-28) and set $s = j\omega$. Thus Eq. (A-59) becomes

$$G(j\omega) = \frac{10(1 + j0.1\omega)}{j\omega(1 + j0.5\omega)(1 + j0.2\omega)} \tag{A-60}$$

This equation shows that $G(j\omega)$ has corner frequencies at $\omega = 10$, 2, and 5 rad/sec. The pole at the origin gives a magnitude curve that is a straight line with a slope of -20 dB/decade and passing through the $\omega = 1$ rad/sec point on the ω axis at 0 dB. The total Bode plots of the magnitude and phase of $G(j\omega)$ are obtained by

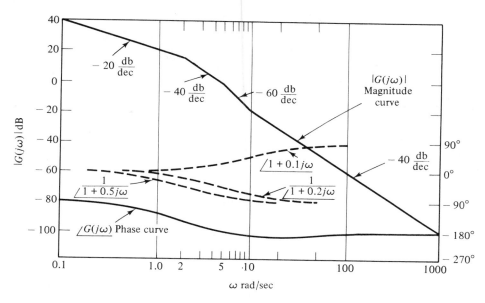

Fig. A-13. Bode plot of $G(s) = [10(s + 10)]/[s(s + 2)(s + 5)]$.

adding the component curves together, point by point, as shown in Fig. A-13. The actual magnitude curve may be obtained by considering the errors of the asymptotic curves at the significant frequencies. However, in practice, the accuracy of the asymptotic lines is deemed adequate for transfer functions with only real poles and zeros.

A.3 Magnitude-Versus-Phase Plot

The magnitude-versus-phase diagram is a plot of the magnitude of the transfer function in decibels versus its phase in degrees, with ω as a parameter on the curve. One of the most important applications of this plot is that it can be superposed on the Nichols chart (see Chapter 9) to give information on the relative stability and the frequency response of the closed-loop system. When the gain factor K of the transfer function is varied, the plot is simply raised or lowered vertically according to the value of K in decibels. However, the unique property of adding the individual plots for cascaded terms in the Bode plot does not carry over to this case. Thus the amount of work involved in obtaining the magnitude-versus-phase plot is equivalent to that of the polar plot, unless a digital computer is used to generate the data. Usually, the magnitude-versus-phase plots are obtained by first making the Bode plot, and then transferring the data to the decibel-versus-phase coordinates.

As an illustrative example, the Bode plot, the polar plot, and the magnitude-versus-phase plot of the function

$$G(s) = \frac{10(s + 10)}{s(s + 2)(s + 5)} \tag{A-61}$$

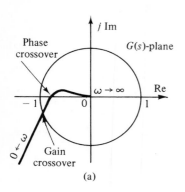

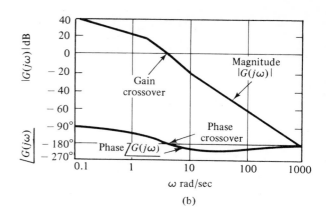

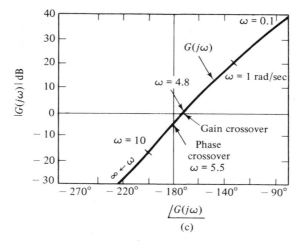

Fig. A-14. $G(s) = [10(s + 10)]/[s(s + 2)(s + 5)]$. (a) Polar plot. (b) Bode diagram. (c) Magnitude-versus-phase plot.

are sketched as shown in Fig. A-14. The Bode plot shown in Fig. A-14(a) is apparently the easiest one to sketch. The others are obtained by transferring the data from the Bode plot to the proper coordinates.

The relationships among these three plots are easily established by comparing the curves in Fig. A-14 without the need of detailed explanation. However, for the purpose of analysis and design, it is convenient to define the following terms:

Gain-crossover frequency. This is the frequency at which the magnitude of the transfer function $G(j\omega)$ is unity. In logarithmic scale, this corresponds to 0 dB. The following interpretations are made with respect to the three types of plots:

Polar Plot. The gain-crossover point (or points) is where $|G(j\omega)| = 1$ [Fig. A-14(a)].

Bode Plot. The gain-crossover point (or points) is where the magnitude curve of $G(j\omega)$ crosses the 0-dB axis [Fig. A-14(b)].

Magnitude-Versus-Phase Plot. The gain-crossover point (or points) is where the $G(j\omega)$ plot crosses the 0-dB axis [Fig. A-14(c)].

Phase-crossover frequency. This is the frequency at which the phase of $G(j\omega)$ is 180°.

Polar Plot. The phase-crossover point (or points) is where the phase of $G(j\omega)$ is 180°, or where the $G(j\omega)$ plot crosses the negative real axis [Fig. A-14(a)].

Bode Plot. The phase-crossover point (or points) is where the phase curve crosses the 180° axis [Fig. A-14(b)].

Magnitude-Versus-Phase Curve. The phase-crossover point (or points) is where the $G(j\omega)$ plot interesects the 180° axis [Fig. A-14(c)].

APPENDIX B

Laplace Transform Table

Laplace Transform $F(s)$	Time Function $f(t)$
$\dfrac{1}{s}$	$u(t)$ (unit step function)
$\dfrac{1}{s^2}$	t
$\dfrac{n!}{s^{n+1}}$	t^n (n = positive integer)
$\dfrac{1}{s+a}$	e^{-at}
$\dfrac{1}{(s+a)(s+b)}$	$\dfrac{e^{-at}-e^{-bt}}{b-a}$
$\dfrac{\omega_n{}^2}{s^2+2\zeta\omega_n s+\omega_n{}^2}$	$\dfrac{\omega_n}{\sqrt{1-\zeta^2}}\,e^{-\zeta\omega_n t}\sin\omega_n\sqrt{1-\zeta^2}\,t$
$\dfrac{1}{(1+sT)^n}$	$\dfrac{1}{T^n(n-1)!}\,t^{n-1}e^{-t/T}$
$\dfrac{\omega_n{}^2}{(1+Ts)(s^2+2\zeta\omega_n s+\omega_n{}^2)}$	$\dfrac{T\omega_n{}^2 e^{-t/T}}{1-2\zeta T\omega_n+T^2\omega_n{}^2}+\dfrac{\omega_n e^{-\zeta\omega_n t}\sin(\omega_n\sqrt{1-\zeta^2}\,t-\phi)}{\sqrt{(1-\zeta^2)(1-2\zeta T\omega_n-T^2\omega_n{}^2)}},$ where $\phi=\tan^{-1}\dfrac{T\omega_n\sqrt{1-\zeta^2}}{1-T\zeta\omega_n}$

Laplace Transform $F(s)$	Time Function $f(t)$
$\dfrac{\omega_n}{(s^2 + \omega_n{}^2)}$	$\sin \omega_n t$
$\dfrac{\omega_n}{(1 + Ts)(s^2 + \omega_n{}^2)}$	$\dfrac{T\omega_n}{1 + T^2\omega_n{}^2} e^{-t/T} + \dfrac{1}{\sqrt{1 + T^2\omega_n{}^2}} \sin (\omega_n t - \phi)$ where $\phi = \tan^{-1} \omega_n T$
$\dfrac{\omega_n{}^2}{s(s^2 + 2\zeta\omega_n s + \omega_n{}^2)}$	$1 + \dfrac{1}{\sqrt{1 - \zeta^2}} e^{-\zeta\omega_n t} \sin (\omega_n\sqrt{1 - \zeta^2}t - \phi)$ where $\phi = \tan^{-1} \dfrac{\sqrt{1 - \zeta^2}}{-\zeta}$
$\dfrac{\omega_n{}^2}{s(s^2 + \omega_n{}^2)}$	$1 - \cos \omega_n t$
$\dfrac{1}{s(1 + Ts)}$	$1 - e^{-t/T}$
$\dfrac{1}{s(1 + Ts)^2}$	$1 - \dfrac{t + T}{T} e^{-t/T}$
$\dfrac{\omega_n{}^2}{s(1 + Ts)(s^2 + 2\zeta\omega_n s + \omega_n{}^2)}$	$1 - \dfrac{T^2\omega_n{}^2}{1 - 2T\zeta\omega_n + T^2\omega_n{}^2} e^{-t/T}$ $\quad + \dfrac{e^{-\zeta\omega_n t} \sin (\omega_n\sqrt{1 - \zeta^2}t - \phi)}{\sqrt{1 - \zeta^2(1 - 2\zeta T\omega_n + T^2\omega_n{}^2)}}$ where $\phi = \tan^{-1} \dfrac{\sqrt{1 - \zeta^2}}{-\zeta} + \tan^{-1} \dfrac{T\omega_n\sqrt{1 - \zeta^2}}{1 - T\zeta\omega_n}$
$\dfrac{\omega_n{}^2}{s^2(s^2 + 2\zeta\omega_n s + \omega_n{}^2)}$	$t - \dfrac{2\zeta}{\omega_n} + \dfrac{1}{\omega_n\sqrt{1 - \zeta^2}} e^{-\zeta\omega_n t} \sin (\omega_n\sqrt{1 - \zeta^2}t - \phi)$ where $\phi = 2 \tan^{-1} \dfrac{\sqrt{1 - \zeta^2}}{-\zeta}$
$\dfrac{\omega_n{}^2}{s^2(1 + Ts)(s^2 + 2\zeta\omega_n s + \omega_n{}^2)}$	$t - T - \dfrac{2\zeta}{\omega_n} + \dfrac{T^3\omega_n{}^2}{1 - 2\zeta\omega_n T + T^2\omega_n{}^2} e^{-t/T}$ $\quad + \dfrac{e^{-\zeta\omega_n t} \sin (\omega_n\sqrt{1 - \zeta^2}t - \phi)}{\omega_n\sqrt{(1 - \zeta^2)(1 - 2\zeta\omega_n T + T^2\omega_n{}^2)}}$ where $\phi = 2 \tan^{-1} \dfrac{\sqrt{1 - \zeta^2}}{-\zeta} + \tan^{-1} \dfrac{T\omega_n\sqrt{1 - \zeta^2}}{1 - T\omega_n\zeta}$
$\dfrac{1}{s^2(1 + Ts)^2}$	$t - 2T + (t + 2T)\epsilon^{-t/T}$
$\dfrac{\omega_n{}^2(1 + as)}{s^2 + 2\zeta\omega_n s + \omega_n{}^2}$	$\omega_n\sqrt{\dfrac{1 - 2a\zeta\omega_n + a^2\omega_n{}^2}{1 - \zeta^2}} e^{-\zeta\omega_n t} \sin (\omega_n\sqrt{1 - \zeta^2}t + \phi)$ where $\phi = \tan^{-1} \dfrac{a\omega_n\sqrt{1 - \zeta^2}}{1 - a\zeta\omega_n}$
$\dfrac{\omega_n{}^2(1 + as)}{(s^2 + \omega_n{}^2)}$	$\omega_n\sqrt{1 + a^2\omega_n{}^2} \sin (\omega_n t + \phi)$ where $\phi = \tan^{-1} a\omega_n$

Laplace Transform $F(s)$	Time Function $f(t)$
$\dfrac{\omega_n^2(1 + as)}{(1 + Ts)(s^2 + 2\zeta\omega_n s + \omega_n^2)}$	$\dfrac{\omega_n}{\sqrt{1 - \zeta^2}}\sqrt{\dfrac{1 - 2a\zeta\omega_n + a^2\omega_n^2}{1 - 2T\zeta\omega_n + T^2\omega_n^2}}\, e^{-\zeta\omega_n t}$ $\times \sin(\omega_n\sqrt{1 - \zeta^2}\, t + \phi) + \dfrac{(T - a)\omega_n^2}{1 - 2T\zeta\omega_n + T^2\omega_n^2}\, e^{-t/T}$ where $\phi = \tan^{-1}\dfrac{a\omega_n\sqrt{1 - \zeta^2}}{1 - a\zeta\omega_n} - \tan^{-1}\dfrac{T\omega_n\sqrt{1 - \zeta^2}}{1 - T\zeta\omega_n}$
$\dfrac{\omega_n^2(1 + as)}{(1 + Ts)(s^2 + \omega_n^2)}$	$\dfrac{\omega_n^2(T - a)}{1 + T^2\omega_n^2}\, e^{-t/T} + \dfrac{\omega_n\sqrt{1 + a^2\omega_n^2}}{\sqrt{1 + T^2\omega_n^2}}\sin(\omega_n t + \phi)$ where $\phi = \tan^{-1} a\omega_n - \tan^{-1}\omega_n T$
$\dfrac{\omega_n^2(1 + as)}{s(s^2 + 2\zeta\omega_n s + \omega_n^2)}$	$1 + \dfrac{1}{\sqrt{1 - \zeta^2}}\sqrt{1 - 2a\zeta\omega_n + a^2\omega_n^2}\, e^{-\zeta\omega_n t}$ $\times \sin(\omega_n\sqrt{1 - \zeta^2}\, t + \phi)$ where $\phi = \tan^{-1}\dfrac{a\omega_n\sqrt{1 - \zeta^2}}{1 - a\zeta\omega_n} - \tan^{-1}\dfrac{\sqrt{1 - \zeta^2}}{-\zeta}$
$\dfrac{\omega_n^2(1 + as)}{s(1 + Ts)(s^2 + \omega_n^2)}$	$1 + \dfrac{T\omega_n^2(a - T)}{1 + T^2\omega_n^2}\, e^{-t/T} - \sqrt{\dfrac{1 + a^2\omega_n^2}{1 + T^2\omega_n^2}}\cos(\omega_n t + \phi)$ where $\phi = \tan^{-1} a\omega_n - \tan^{-1}\omega_n T$
$\dfrac{\omega_n^2(1 + as)}{s(1 + Ts)(s^2 + 2\zeta\omega_n s + \omega_n^2)}$	$1 + \sqrt{\dfrac{1 - 2\zeta a\omega_n + a^2\omega_n^2}{(1 - \zeta^2)(1 - 2T\zeta\omega_n + T^2\omega_n^2)}}\, e^{-\zeta\omega_n t}$ $\times \sin(\omega_n\sqrt{1 - \zeta^2}\, t + \phi) + \dfrac{\omega_n^2 T(a - T)}{1 - 2T\zeta\omega_n + T^2\omega_n^2}\, e^{-t/T}$ $\phi = \tan^{-1}[a\omega_n\sqrt{1 - \zeta^2}/(1 - a\zeta\omega_n)]$ $- \tan^{-1}\dfrac{T\omega_n\sqrt{1 - \zeta^2}}{1 - T\zeta\omega_n} - \tan^{-1}\dfrac{\sqrt{1 - \zeta^2}}{-\zeta}$
$\dfrac{1 + as}{s^2(1 + Ts)}$	$t + (a - T)(1 - e^{-t/T})$
$\dfrac{s\omega_n^2}{s^2 + 2\zeta\omega_n s + \omega_n^2}$	$\dfrac{\omega_n^2}{\sqrt{1 - \zeta^2}}\, e^{-\zeta\omega_n t}\sin(\omega_n\sqrt{1 - \zeta^2}\, t + \phi)$ where $\phi = \tan^{-1}\dfrac{\sqrt{1 - \zeta^2}}{-\zeta}$
$\dfrac{s}{s^2 + \omega_n^2}$	$\cos\omega_n t$
$\dfrac{s}{(s^2 + \omega_n^2)^2}$	$\dfrac{1}{2\omega_n}\, t\sin\omega_n t$
$\dfrac{s}{(s^2 + \omega_{n1}^2)(s^2 + \omega_{n2}^2)}$	$\dfrac{1}{\omega_{n2}^2 - \omega_{n1}^2}(\cos\omega_{n1} t - \cos\omega_{n2} t)$
$\dfrac{s}{(1 + Ts)(s^2 + \omega_n^2)}$	$\dfrac{-1}{(1 + T^2\omega_n^2)}\, e^{-t/T} + \dfrac{1}{\sqrt{1 + T^2\omega_n^2}}\cos(\omega_n t - \phi)$ where $\phi = \tan^{-1}\omega_n T$

Laplace Transform $F(s)$	Time Funtion $f(t)$
$\dfrac{1 + as + bs^2}{s^2(1 + T_1 s)(1 + T_2 s)}$	$t + (a - T_1 - T_2) + \dfrac{b - aT_1 + T_1{}^2}{T_1 - T_2} e^{-t/T}$ $\qquad\qquad\qquad - \dfrac{b - aT_2 + T_2{}^2}{T_1 - T_2} e^{-t/T_2}$
$\dfrac{\omega_n{}^2(1 + as + bs^2)}{s(s^2 + 2\zeta\omega_n s + \omega_n{}^2)}$	$1 + \sqrt{\dfrac{(1 - a\zeta\omega_n - b\omega_n{}^2 + 2b\zeta^2\omega_n{}^2)^2 + \omega_n{}^2(1 - \zeta^2)(a - 2b\zeta\omega_n)^2}{(1 - \zeta^2)}}$ $\qquad\qquad\qquad\qquad \times e^{-\zeta\omega_n t} \sin(\omega_n\sqrt{1 - \zeta^2}t + \phi)$ where $\qquad \phi = \tan^{-1}\dfrac{\omega_n\sqrt{1 - \zeta^2}(a - 2b\zeta\omega_n)}{b\omega_n(2\zeta^2 - 1) + 1 - a\zeta\omega_n} - \tan^{-1}\dfrac{\sqrt{1 - \zeta^2}}{-\zeta}$
$\dfrac{s^2}{(s^2 + \omega_n{}^2)^2}$	$\dfrac{1}{2\omega_n}(\sin\omega_n t + \omega_n t \cos\omega_n t)$

APPENDIX C

Lagrange's Multiplier Method

In control system design it is often desirable to maximize or minimize a function $f(x_1, x_2, \ldots, x_n)$ with the constraint $g(x_1, x_2, \ldots, x_n) = g_o$ (constant). Lagrange's method suggests that the desired value of $x_1, x_2, \ldots, x_n$ can be found by maximizing or minimizing the function

$$F = f(x_1, x_2, \ldots, x_n) + \lambda g(x_1, x_2, \ldots, x_n) \qquad \text{(C-1)}$$

where

$$\lambda = \text{Lagrange's multiplier}$$

In general, there may be m constraint equations of the form

$$g_i(\mathbf{x}) = g_{io} \qquad i = 1, 2, \ldots, m \qquad \text{(C-2)}$$

for $m < n$. The symbol $\mathbf{x}$ denotes the vector

$$\mathbf{x} = [x_1, x_2, \ldots, x_n]' \qquad \text{(C-3)}$$

With m constraint equations in Eq. (C-2) to maximize or minimize $f(x)$, F becomes

$$F = f(\mathbf{x}) + \lambda_1 g_1(\mathbf{x}) + \lambda_2 g_2(\mathbf{x}) + \ldots + \lambda_m g_m(\mathbf{x}) \qquad \text{(C-4)}$$

where $\lambda_1, \lambda_2, \ldots, \lambda_m$ are the Lagrange's multipliers.

The procedure of the Lagrange's method is outlined as follows for the single-constraint case:

1. Form the function

$$F = f(\mathbf{x}) + \lambda g(\mathbf{x}) \qquad \text{(C-5)}$$

2. Maximize or minimize F with respect to $x_1, x_2, \ldots, x_n$; that is,

simultaneously set

$$\frac{\partial F}{\partial x_i} = 0 \qquad i = 1, 2, \ldots, n \tag{C-6}$$

These n equations give the desired values of x_i, $i = 1, 2, \ldots, n$, as functions of λ. Let these functions be denoted as

$$x_i = h_i(\lambda) \qquad i = 1, 2, \ldots, n \tag{C-7}$$

3. Substitute Eq. (C-7) into the constraint equation

$$g(x_1, x_2, \ldots, x_n) = g_o \tag{C-8}$$

Equation (C-8) now gives the value of λ. Substituting this λ into Eq. (C-7) gives the desired optimal values of x_i, $i = 1, 2, \ldots, n$.
4. Substitution of the optimal values of $x_1, x_2, \ldots, x_n$ obtained from the last step into $f(\mathbf{x})$ gives the optimal value of f.

The justification of the Lagrange's method of adjoining the constraint to the function to be minimized is as follows. Let a maximum, minimum, or saddle point be described as a *stationary point*. Then the necessary condition for the existence of a stationary point of a differentiable function $f(x_1, x_2, \ldots, x_n)$ is that

$$\frac{\partial f}{\partial x_i} = 0 \tag{C-9}$$

Consider a case with just two parameters, x_1 and x_2; that is, $f(x_1, x_2)$ is to be maximized or minimized by choosing the proper values for x_1 and x_2, with the constraint that

$$g(x_1, x_2) = g_o \text{ (constant)} \tag{C-10}$$

The necessary conditions for a stationary point are

$$\frac{\partial f}{\partial x_1} = 0 \qquad \text{and} \qquad \frac{\partial f}{\partial x_2} = 0 \tag{C-11}$$

Let us assume that x_1 and x_2 are arbitrary functions of a dependent variable y; that is, f and g can be written as $f(x_1, x_2, y)$ and $g(x_1, x_2, y)$, respectively. Then, we can form the following equations:

$$\frac{df}{dy} = \frac{\partial f}{\partial x_1} \frac{dx_1}{dy} + \frac{\partial f}{\partial x_2} \frac{dx_2}{dy} \tag{C-12}$$

$$\frac{dg}{dy} = \frac{\partial g}{\partial x_1} \frac{dx_1}{dy} + \frac{\partial g}{\partial x_2} \frac{dx_2}{dy} \tag{C-13}$$

The necessary conditions for a stationary point, stated in Eq. (C-11), imply that the total derivatives in Eqs. (C-12) and (C-13) must vanish at the stationary point. Therefore,

$$df = \frac{\partial f}{\partial x_1} dx_1 + \frac{\partial f}{\partial x_2} dx_2 = 0 \tag{C-14}$$

$$dg = \frac{\partial g}{\partial x_1} dx_1 + \frac{\partial g}{\partial x_2} dx_2 = 0 \tag{C-15}$$

Let us multiply the right side of Eq. (C-15) by an undetermined parameter λ and then add the result to Eq. (C-14). We have

$$\left(\frac{\partial f}{\partial x_1} + \lambda \frac{\partial g}{\partial x_1}\right) dx_1 + \left(\frac{\partial f}{\partial x_2} + \lambda \frac{\partial g}{\partial x_2}\right) dx_2 = 0 \qquad \text{(C-16)}$$

If now λ is so chosen that

$$\frac{\partial f}{\partial x_1} + \lambda \frac{\partial g}{\partial x_1} = 0 \qquad \text{(C-17)}$$

$$\frac{\partial f}{\partial x_2} + \lambda \frac{\partial g}{\partial x_2} = 0 \qquad \text{(C-18)}$$

and

$$g(x_1, x_2) = g_o \qquad \text{(C-19)}$$

then the necessary condition of a stationary point of $f(x_1, x_2)$ is satisfied. Thus Eqs. (C-17) and (C-18) imply the maximization or minimization of $F = f + \lambda g$ with respect to x_1 and x_2.

Index